CBS PROBLEMS and SOLUTIONS SERIES

Problems and Solutions in ENGINEERING THERMODYNAMICS

Problems and Solutions in
ENGINEERING THERMODYNAMICS

- This book is suitable for examinations of all Indian universities.
- Solutions are given in most simplified manner.
- Each type of problems is included and the solutions given in detail.
- To explain the mechanism lucidly, computerised figures have been liberally added in the solutions. This volume contains more than 450 line diagrams.
- Problems from the last ten years Indian Engineering Service and Civil Service examinations are added till date in each chapter.

CBS PROBLEMS and SOLUTIONS SERIES

Problems and Solutions in ENGINEERING THERMODYNAMICS

S C SINGHAL
Engineer

CBS

CBS Publishers & Distributors Pvt. Ltd.

New Delhi • Bengaluru • Chennai • Kochi • Kolkata • Mumbai
Hyderabad • Nagpur • Patna • Pune • Vijayawada

ISBN: 81-239-1089-4

First Edition: 2004
Reprint: 2005, 2009, 2010, 2012, 2016

Published by:
Satish Kumar Jain for CBS Publishers & Distributors Pvt. Ltd.,
4819/XI Prahlad Street, 24 Ansari Road, Daryaganj, New Delhi - 110002
delhi@cbspd.com, cbspubs@airtelmail.in • www.cbspd.com
Ph.: 23289259, 23266861, 23266867 • Fax: 011-23243014

Corporate Office: 204 FIE, Industrial Area, Patparganj, Delhi - 110 092
Ph: 49344934 • Fax: 011-49344935
E-mail: publishing@cbspd.com • publicity@cbspd.com

Branches:
- ***Bengaluru:*** 2975, 17th Cross, K.R. Road, Bansankari 2nd Stage, Bengaluru - 70 • Ph: +91-80-26771678/79 • Fax: +91-80-26771680 E-mail: cbsbng@gmail.com, bangalore@cbspd.com
- ***Chennai:*** No. 7, Subbaraya Street, Shenoy Nagar, Chennai - 600030 Ph: +91-44-26681266, 26680620 • Fax: +91-44-42032115 E-mail: chennai@cbspd.com
- ***Kochi:*** Ashana House, 39/1904, A.M. Thomas Road, Valanjambalam, Ernakulum, Kochi • Ph: +91-484-4059061-65 Fax: +91-484-4059065 • E-mail: cochin@cbspd.com
- ***Kolkata:*** 6-B, Ground Floor, Rameshwar Shaw Road, Kolkata - 700014 Ph: +91-33-22891126/7/8 • E-mail: kolkata@cbspd.com
- ***Mumbai:*** 83-C, Dr. E. Moses Road, Worli, Mumbai - 400018 Ph: +91-9833017933, 022-24902340/41 • E-mail: mumbai@cbspd.com

Representatives:

- Hyderabad: 0-9885175004
- Nagpur: 0-9021734563
- Patna: 0-9334159340
- Pune: 0-9623451994
- Vijayawada: 0-9000660880

Printed at:
J.S. Offset Printers, Delhi

Preface

Engineering thermodynamics is one of the fundamental subjects for the students of mechanical engineering. The author feels great pleasure in presenting the subject of THERMODYNAMICS through typical problems and their solutions. The book covers every topic taught in thermodynamics at the undergraduate level in an engineering college. All the problems are discussed at length, and line diagrams and figures have been given liberally to help the reader in understanding the logic. This book is suitable for all university examinations. Also questions of the past ten years from various competitive examinations like Indian Engineering Service and Civil Service examinations, have been added till date.

Although every care has been taken to ensure accuracy, yet some errors might have crept in. The author will be grateful if these are brought to his notice so that these could be rectified in the subsequent printings and editions of the book. Readers are requested to send their suggestions for further improvement and revision at the following e-mail address: cbspubs@del3.vsnl.net.in.

S C Singhal

Contents

Chapter 1

Introduction

Problem 1.1 : A pump discharges a liquid into a drum at the rate of 0.032 m^3/s. The drum, 1.50 m in diameter and 4.20 m in length, can hold 3000 kg of the liquid. Find the density of the liquid and the mass flow rate of the liquid handled by the pump.

Solution :

Volume flow rate = $\dot{V}$ = 0.032 m^3/s
diameter of drum = d = 1.5 m
length of drum = L = 4.2 m
mass of liquid = m = 3000 kg

∴ Volume of drum

$$= \frac{\pi}{4} d^2 \times L = \frac{\pi}{4} (1.5)^2 \times 4.2 = 7.422 \text{ m}^3$$

∴ density of the liquid

$$= \frac{m}{\text{volume of drum}} = \frac{3000}{7.422}$$

$$= 404.2 \text{ kg/m}^3 \textbf{ Ans}$$

Mass flow rate = Volume flow rate × density of liquid
= 0.032 × 404.2 = 12.93 kg/sec. **Ans**

Problem 1.2 : The acceleration of gravity is given as a function of elevation above sea level by

$$g = 980.6 - 3.086 \times 10^{-6} H$$

where g is cm/s^2, and H is in cm. If an aeroplane weights 90,000 N at sea level, what is the gravity force upon it at 10,000 m elevation? What is the percentage difference from the sea-level weight?

Solution : Given

$$g = 980.6 - 3.086 \times 10^{-6} H \text{ cm/s}^2$$

at sea level, $H = 0$

$$\therefore \quad g = 980.6 \text{ cm/s}^2 = 9.806 \text{ m/s}^2$$

∴ mass of aeroplane

$$= \frac{\text{weight at sea level}}{g \text{ at sea level}} = \frac{90{,}000}{9.806} \text{ kg}$$

Again at $H = 10{,}000$ m = 10,00,000 cm

$$\therefore \quad g = 980.6 - 3.086 \times 10^{-6} \times 10^6$$

$$= 977.514 \text{ cm/sec}^2$$

$$= 9.77514 \text{ m/sec}^2$$

∴ gravity force at 10,000 m = mass × $g_{\text{at 10,000 m}}$

$$= \frac{90{,}000 \times 9.77514}{9.806} = 89716.77 \text{ N } \textbf{Ans}$$

$$\therefore \% \text{ difference} = \frac{90{,}000 - 89{,}716.77}{90{,}000} = 0.3157 \ \% \ \textbf{Ans}$$

Problem 1.3 : Prove that the weight of a body at an elevation H above sea-level is given by

$$W = \frac{mg}{g_0}\left(\frac{d}{d+2H}\right)^2$$

where d is the diameter of the earth.

Solution : Let m and M are the mass of body and mass of earth.

d = dia of earth

H = height above sea level

g_0 = gravity at sea level

g = gravity at elevation H

force of gravitation when m on the surface of earth = mg_0

$$= \frac{GMm}{\left(\frac{d}{2}\right)^2}$$

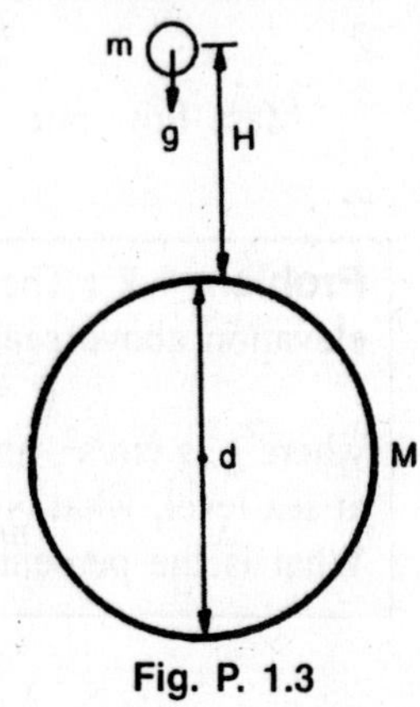

Fig. P. 1.3

Similarly, force of gravitation when 'm' is H height above the sea level.

$$mg = \frac{GMm}{\left(\frac{d}{2}+H\right)^2} = \frac{GMm}{\left(\frac{d+2H}{2}\right)^2}$$

Or
$$mg = \frac{Mg_0\left(\frac{d}{2}\right)^2}{\left(\frac{d+2H}{2}\right)^2} = Mg_0\left(\frac{d}{d+2H}\right)^2$$

a weight of body at height $H = W = mg_0\left(\frac{d}{d+2H}\right)^2$ **Ans**

> **Problem 1.4 :** The first artificial earth satellite is reported to have encircled the earth at a speed of 28,840 km/h and its maximum height above the earth's surface was stated to be 916 km. Taking the mean diameter of the earth to be 12,680 km, and assuming the orbit to be circular, evaluate the value of the gravitational acceleration at this height.
>
> The mass of the satellite is reported to have been 86 kg at sea-level. Estimate the gravitational force acting on the satellite at the operational altitude.

***Solution* :** Given :

m = mass of satellite
= 86 kg
H = 916 km
d = dia of earth
= 12,680 km
V = 28,840 km/hr
= 8011.11 m/s

Let g = be the acceleration due to gravity at height H

$$\therefore \quad mg = \frac{mv^2}{\left(H+\frac{d}{2}\right)}$$

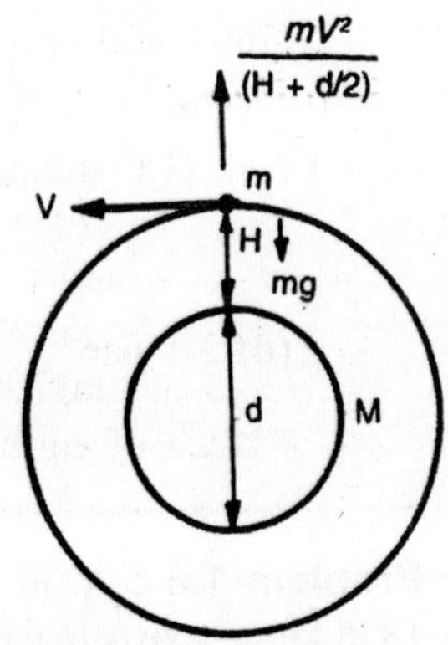

Fig. P. 1.4

$$\therefore \quad g = \frac{V^2}{\left(H + \frac{d}{2}\right)} = \frac{(8011.11)^2}{(916 + 6{,}340) \times 1{,}000}$$

$$= 8.845 \text{ m/s}^2 = 8.9 \text{ m/s}^2 \textbf{ Ans}$$

∴ gravitation force

$$= mg = 86 \times 8.906 = 765.4 \text{ N} \textbf{ Ans}$$

Problem 1.5 : Convert the following readings of pressure to kPa, assuming that the barometer reads 760 mm Hg : (a) 90 cm Hg gauge, (b) 40 cm Hg vacuum, (c) 1.2 m H_2O gauge, (d) 3.1 bar.

Solution :

Barometric pressure = 760 mmHg
= 101.325 kPa

(a) 90 cm Hg gauge
= 760 + 900
= 1,660 mm of Hg absolute

$$= \frac{1660 \times 101.325}{760}$$

= 221.32 kPa absolute **Ans**

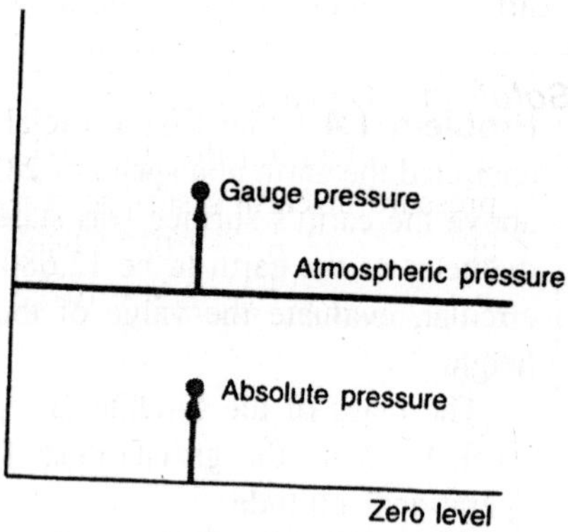

Fig. P. 1.5

(b) 40 cm Hg vacuum

$$= \frac{760 - 400}{760} \times 101.325 = 47.996 \approx 48 \text{ kPa absolute}$$

(c) 1.2 m H_2O gauge = ρgh = 9,810 × 1.2 kPa Gauge
= 11.772 kPa gauge
= 113.097 kPa (absolute)

(d) 3.1 bar = 3.1 × 1.01325 × 10^5
= 314.11 kPa (absolute)

Problem 1.6 : A 30 m high vertical column of a fluid of density 1878 kg/m^3 exists in a place where g = 9.65 m/s^2. What is the pressure at the base of the column?

Solution : Given :

$$h = \text{height of fluid column} = 30 \text{ m}$$
$$\rho = \text{density of fluid} = 1878 \text{ kg/m}^3$$
$$g = 9.65 \text{ m/s}^2$$

$\therefore$ pressure at the base of column $= \rho g h = 1878 \times 9.65 \times 30$
$= 543.681$ kPa **Ans**

Problem 1.7 : Assume that the pressure p and the specific volume v of the atmosphere are related according to the equation $pv^{1.4} = 2.3 \times 10^3$, where p is in N/m^2 abs and v is in m^3/kg. The acceleration due to gravity is constant at 9.81 m/s^2. What is the depth of atmosphere necessary to produce a pressure of 1.0132 bar at the earth's surface? Consider the atmosphere as a fluid column.

Solution : Given :

g = acceleration due to gravity = 9.81 m/s^2

pressure at earth's surface = p = 1.0132 bar

at $\quad p = 1.0132 \text{ bar} = 1.0132 \times 10^5 \; p_a.$

$$pv^{1.4} = 2.3 \times 10^{+3}$$

or
$$v = \left(\frac{2.3 \times 10^{+3}}{1.0132 \times 10^5}\right)^{\frac{1}{1.4}}$$

$$\therefore \quad \rho = \frac{1}{v} = 14.93696 \ \frac{\text{kg}}{\text{m}^3}$$

Now, $\quad p = \text{pressure at earth surface} = \rho \times g \times h$

$$\therefore \quad h = \frac{p}{\rho \times g} = \frac{1.0132 \times 10^5}{2.5763 \times 9.81}$$

Problem 1.8 : The pressure of steam flowing in a pipe line is measured with a mercury manometer, shown in Fig. P. 1.8 (a). Some steam condenses into water. Estimate the steam pressure in kPa. Take the density of mercury as 13.6×10^3 kg/m^3, density of water as 10^3 kg/m^3, the barometer reading as 76.1 cm Hg, and g as 9.806 m/s^2.

Solution **:** Given :

ρ_{Hg} = density of Hg
= 13.6×10^3 kg/m³

ρ_{H_2O} = density of H_2O
= 10^3 kg/m³

P_0 = 76.1 cm Hg
= $76.1 \times 10^{-2} \times 9.81 \times 13.6 \times 10^3$ Pa

g = 9.806 m/s²

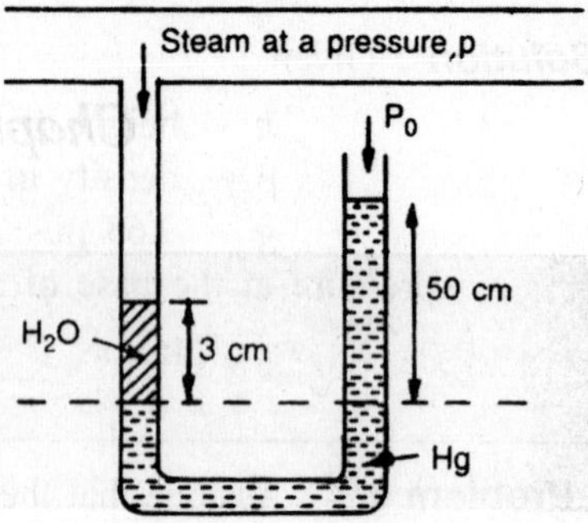

Fig. P. 1.8 (a)

Now equating pressure at x – x level of both the columns

$$P + \rho_{H_2O} \times g \times h_{H_2O} = P_0 + \rho_{Hg} \times g \times h_{Hg}$$

or ∴ $$P = P_0 + \rho_{Hg} \times g \times h_{Hg} - \rho_{H_2O} \times g \times h_{H_2O}$$

$$P = 76.1 \times 10^{-2} \times 9.8 \times 13.6 \times 10^3 + 13.6 \times 10^3 \times 9.81 \times 50 \times 10^{-2} - 3 \times 10^{-2} \times 9.81 \times 1000$$

= 167.943 kPa **Ans**

Problem 1.9 : A vacuum gauge mounted on a condenser reads 0.66 *m* Hg. What is the absolute pressure in the condenser in kPa when the atmospheric pressure is 101.3 kPa?

Solution : Given :

Reading of vacuum gauge = 0.66 *m* Hg
= $0.66 \times 13.6 \times 1000 \times 9.81$
= 88.05 kPa

atmospheric pressure = 101.3 kPa

∴ Absolute pressure = atmospheric pressure – vacuum gauge pressure
= 101.3 – 88.05 = 13.245 kPa

Chapter 2

Temperature

Problem 2.1 : The limiting value of the ratio of the pressure of gas at the steam point and at the triple point of water when the gas is kept at constant volume is found to be 1.36605. What is the ideal gas temperature of the steam point?

Solution : For a constant volume gas thermometer

$$\theta = 273.16 \frac{p}{p_t}$$

Where p_t is pressure of gas when placed in contact with water at triple point.

Given

$$\underset{P_t \to 0}{\text{Lt}} \frac{p}{p_t} = 1.36605$$

∴ ideal gas temperature of the steam point = T

$$= 273.16 \; lt \; \underset{P_t \to 0}{\text{Lt}} \frac{p}{p_t}$$

$$= 273.16 \times 1.36605 = 373.15022 \text{ K. } \textbf{Ans}$$

Problem 2.2 : The resistance of a platinum wire is found to be 11,000 ohms at the ice point, 15.247 ohms at the steam point, and 28.887 ohms at the sulphur point. Find the constants A and B in the equation

$$R = R_0 (1 + At + Bt^2)$$

and plot R against t in the range 0 to 660°C

Solution : Given

$$R = R_0 (1 + At + Bt^2),$$

where A & B are constant.

R = 11,000 ohms at ice point, i.e. t = 0°C

R = 15.247 ohms at steam point, i.e. t = 100°C

R = 28.887 ohms at sulphur point, i.e. t = 444.6°C

Now putting these values in above equation, we get,

$$11,000 = R_0 [1 + 0 + 0]$$

$$\therefore \quad R_0 = 11,000 \qquad ...(i)$$

Simply $15.247 = R_0 [1 + 100A + 10,000B]$ ...(ii)

and $28.887 = R_0 [1 + 444.6A + 197669.16B]$ ...(iii)

from (ii) $100A + 10,000B = -0.998614$

from (iii) $444.6A + 197669.16B = -0.997374$

Solving above eqn we get

$$B = 2.2469 \times 10^{-5}$$

$$A = -0.01223$$

$$\therefore \quad R = 11000 [1 - 0.01223t + 2.2469 \times 10^{-5} t^2]$$

Now graph between R and t in the range of 0°C to 660°C can be plotted.

Problem 2.3 : When the reference junction of a thermocouple is kept at the ice point and the test junction is at the Celsius temperature t, and e.m.f. ε of the thermocouple is given by the equation

$$\varepsilon = at + bt^2$$

where a = 0.20 mV/deg., and $b = -5.0 \times 10^{-4}$ mV/deg^2

(a) Compute the e.m.f. when t = –100°C, 200°C, 400°C and 500°C, and draw graph of ε against t in this range.

(b) Suppose the e.m.f. ε is taken as a thermometric property and that a temperature scale t^* is defined by the linear equation.

$$t^* = a'\varepsilon + b'$$

and that $t^* = 0$ at the ice point and $t^* = 100$ at the steam point. Find the numerical values of a' and b' and draw a graph of ε against t^*.

(c) Find the values of t^* when t = –100°C, 200°C, 400°C and 500°C and draw a graph of t^* against t.

Solution : Given

$$\varepsilon = 0.20\,t - 5.0 \times 10^{-4}\,t^2 \text{ mV}$$

(a) When t = –100°C, ε = –25 mV

When t = 200°C, ε = 20 mV

When t = 400°C, ε = 0 mV
When t = 500°C, ε = –25 mV

(b) Given $t^* = a'\varepsilon + b'$

Conditions;

t^* = 0, at ice point i.e., t = 0°C
t^* = 100, at steam point i.e., t = 100°C

$$t^* = a^* [0.2\, t - 5 \times 10^{-4}\, t^2] + b'$$

Hence $0 = a^* \times 0 + b'$

or $b' = 0$

and $100 = a^* [0.2 \times 100 - 5 \times 10^{-4} \times 100^2] + b'$

$\Rightarrow$ $100 = a^* [15] + 0$

$$\therefore \quad a^* = \frac{100}{15} = 6.67$$

$$\therefore \quad t^* = 1.334\, t - 3.335 \times 10^{-3}\, t^2$$

(c) at t = –100°C, t^* = –166.75°C

at t = 200°C, t^* = 133.36°C
at t = 400°C, t^* = –0.16°C
at t = 500°C, t^* = –166.75°C

Problem 2.4 : The temperature t on a thermometric scale is defined in terms of a property K by the relation

$$t = a \ln K + b$$

where a and b are constants.
The values of K are found to be 1.83 and 6.78 at the ice point and the steam point, the temperatures of which are assigned the numbers 0 and 100, respectively. Determine the temperature corresponding to a reading of K equal to 2.42 on the thermometer.

Solution : Given :

$$K = 1.83 \text{ at } t = 0°\text{C} \quad \text{...(i)}$$
$$K = 6.78 \text{ at } t = 100°\text{C} \quad \text{...(ii)}$$

$\therefore$ from the relation,

$$t = a \ln K + b$$

from condition (i), $0 = a \ln 1.83 + b$ ...(iii)
from condition (ii), $100 = a \ln 6.78 + b$ ...(iv)

Solving (iii) & (iv) we get,

or $100 = a \ln 6.78 - a \ln 1.83$

$\therefore$ $a = 76.36$ and $b = -46.14$

$\therefore \quad t = 76.36 \ln K - 46.14$

Now, at $K = 2.42$, $t = 21.34°C$ **Ans**

Problem 2.5 : The resistance of the windings in a certain motor is found to be 80 ohms at room temperature (25°C). When operating at full load under steady state conditions, the motor is switched off and the resistance of the windings, immediately measured again, is found to be 93 ohms. The windings are made of copper whose resistance at temperature t°C is given by

$$R_t = R_0 [1 + 0.00393 \, t]$$

where R_0 is the resistance at 0°C. Find the temperature attained by the coil during full load.

Solution : Given :

at $t = 25°C$, $R_t = 80$ ohm

$\therefore$ from relation, $R_t = R_0 [1 + 0.00393 \, t]$

or, $80 = R_0 [1 + 0.00393 \times 25]$

$R_0 = 72.843$ ohm

$\therefore \quad R_t = 72.843 [1 + 0.00393 \, t]$ ...(i)

At full load, $R_t = 93$ ohm

hence from (i), $93 = 72.843 [1 + 0.00393 \, t]$

$t = 70.41°C$ **Ans**

Chapter 3

Heat and Work Transfer

Problem 3.1 : (a) A pump forces 1 m^3/min of water horizontally from an open well to a closed tank where the pressure is 0.9 MPa. Compute the work the pump must do upon the water in an hour just to force the water into the tank against the pressure. Sketch the system upon which the work is done before and after the process.

(b) If the work done as above upon the water had been used solely to raise the same amount of water vertically against gravity without change of pressure, how many metres would the water have been elevated?

(c) If the work done in (a) upon the water had been used solely to accelerate the water from zero velocity without change of pressure or elevation, what velocity would the water have reached? If the work had been used to accelerate the water from an initial velocity of 10 m/s, what would the final velocity have been?

Solution : (a) Given :

volume flow rate, $\dot{V} = 1\ m^3/min = \frac{1}{60}\ m^3/sec$

Pressure in tank, $P_T = 0.9$ MPa

Pressure in open well, $P_a = 1$ atm $= 0.1$ MPa.

Assume the process to constant volume flow process. Work done by pump per sec.

$$= -\int_1^2 \dot{V} dp = -\dot{V}(P_T - P_a) = \dot{V}(P_a - P_T)$$

$= \frac{-1}{60} \times (0.9 - 0.1)$

$\times 10^3$ kJ/sec

$= -13.33$ kJ/sec

[–ive sign implies work is done on the system i.e., mass of water]

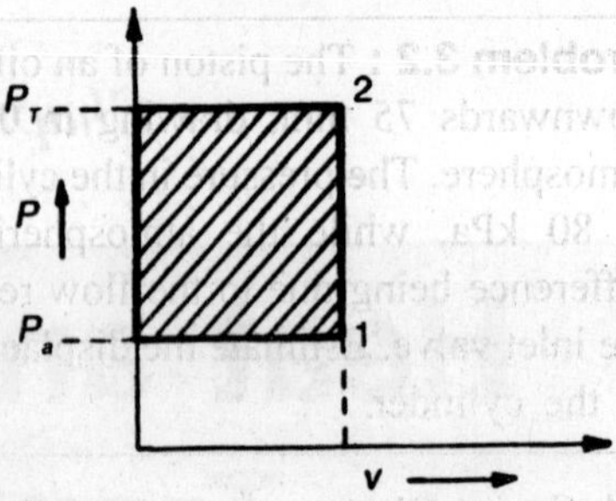

P. 3.1

(b) Say the liquid is raised upto height = h (meter)

∴ from conservation of energy,

Work done – Change in potential energy = change in K.E.

Assuming ΔK.E. = 0

∴ work done = change in P.E.

$$\Rightarrow \quad 13.33 \times 10^3 \text{ J/s} = 1000 \times \frac{1}{60} \times 9.81 \times h \, [\because (\text{mgh}) = (\rho \dot{V} \text{gh})]$$

$$\Rightarrow \quad h = \frac{13.33 \times 60}{9.81} = 81.53 \text{ m } \textbf{Ans}$$

(c) Say initial velocity of water $= V_i$

and final velocity of water $= V_f$

$$\therefore \text{ change in K.E. per sec} = \frac{1}{2} \rho \times \dot{V} \left(V_f^2 - V_i^2\right)$$

$$\therefore \text{ Work done} = \frac{1}{2} \rho \times \dot{V} \left(V_f^2 - V_i^2\right)$$

$$\Rightarrow \quad V_f = \left[\frac{2 \times W \times D}{\rho.\dot{V}} + V_i^2\right]^{1/2}$$

$$\text{if} \quad V_i = 0 \;\therefore\; V_f = \left[\frac{2 \times 13.33 \times 10^3}{1000 \times \frac{1}{60}} + 0\right]^{\frac{1}{2}} = 39.99 \text{ m/s}$$

if $\quad V_i = 10$ m/s

$$\therefore \quad V_f = \left[\frac{2 \times 13.33 \times 10^3}{1000 \times \frac{1}{60}} + 100\right]^{\frac{1}{2}} = 41.23 \text{ m/s } \textbf{Ans}$$

Problem 3.2 : The piston of an oil engine, of area 0.0045 m^2, moves downwards 75 mm, drawing in 0.00028 m^3 of fresh air from the atmosphere. The pressure in the cylinder is uniform during the process at 80 kPa, while the atmospheric pressure is 101.325 kPa, the difference being due to the flow resistance in the induction pipe and the inlet valve. Estimate the displacement work done by the air finally in the cylinder.

Solution : Given

Area of piston, A	=	0.0045 m^2
Piston displacement, L	=	75 mm = 75×10^{-3} m
Constant cyl. pressure P	=	80 kPa
Atm pressure	=	101.325 kPa
Vol of fresh air drawn	=	0.00028 m^3
Displacement work	=	$P \times A \times L$
	=	$80 \times 10^3 \times 0.0045 \times 75 \times 10^{-3}$ (const. press process.) change in vol. $\Delta V = A \times L$
	=	27 J **Ans**

Problem 3.3 : An engine cylinder has a piston of area 0.12 m^3 and contains gas at a pressure of 1.5 MPa. The gas expands according to a process which is represented by a straight line on a pressure-volume diagram. The final pressure is 0.15 MPa. Calculate the work done by the gas on the piston if the stroke is 0.30 m.

Solution : Given : Expansion process follows a st. line on P-V diagram Say 1 – 2

P_1 = 1.5 MPa,

P_2 = 0.15 MPa

Piston area

A = 0.12 m^2

Stroke length

L = 0.3 m

Here, $\Delta v = A \times L$

we know that work done

$= \int p dv$

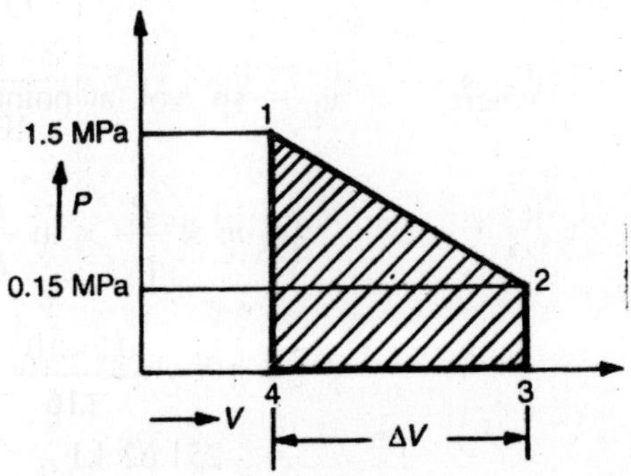

Fig. P. 3.3

= which is area on $P - V$ diagram

= Area – 1 2 3 4 1

$$= \frac{1}{2} \times (p_1 + p_2) \times A \times L$$

$$= \frac{1}{2} \times [1.5 + 0.15] \times 0.12 \times 0.3 \text{ MJ}$$

= 29.7 kJ **Ans**

Problem 3.4 : A mass of 1.5 kg of air is compressed in a quasi-static process from 0.1 MPa to 0.7 MPa for which pv = constant. The initial density of air is 1.16 kg/m³. Find the work done by the piston to compress the air.

Solution : Given :

Mass of air
$= m = 1.5$ kg

Pressure at point 1
$= P_1 = 0.1$ MPa

Pressure at point 2
$= P_2 = 0.7$ MPa

Initial air density
$= \rho_1 = 1.16$ kg/m³

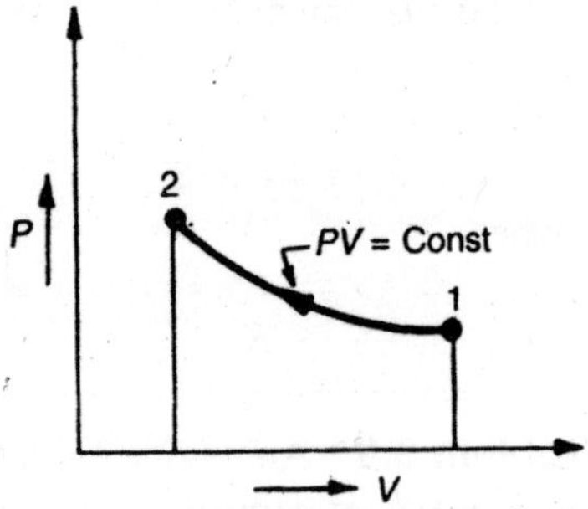

Fig. P. 3.4

pv = const.

⇒ Isothermal process

∴ Work done in isothermal process

$$= m \times P_1 v_1 \ln \frac{P_1}{P_2}$$

where v_1 = sp. vol. at point 1 $= \frac{1}{\rho_1}$

$$= m \times \frac{P_1}{\rho_1} \times \ln \frac{P_1}{P_2}$$

$$= 1.5 \times \frac{0.1 \times 10^3}{1.16} \times \ln \left(\frac{0.1}{0.7}\right)$$

$= -251.62$ kJ

[-ve sign implies compression work is negative on gas.]

Problem 3.5 : A single-cylinder, double-acting, reciprocating water pump has an indicator diagram which is a rectangle 0.075 m long and 0.05 m high. The indicator spring constant is 147 MPa per m. The pump runs at 50 rpm. The pump cylinder diameter is 0.15 m and the piston stroke is 0.20 m. Find the rate in kW at which the piston does work on the water.

Solution : Given :

K = spring constant
= 147 MPa/m
N = 50 rpm
d = cylinder dia
= 0.15 m
L = stroke length
= 0.2 m

Considering 2-stroke

∴ Power developed

$= P_m LAN \times 2$

when

$$P_m = \frac{a_d}{l_d} \times K$$

$= 0.05 \times 147$ MPa

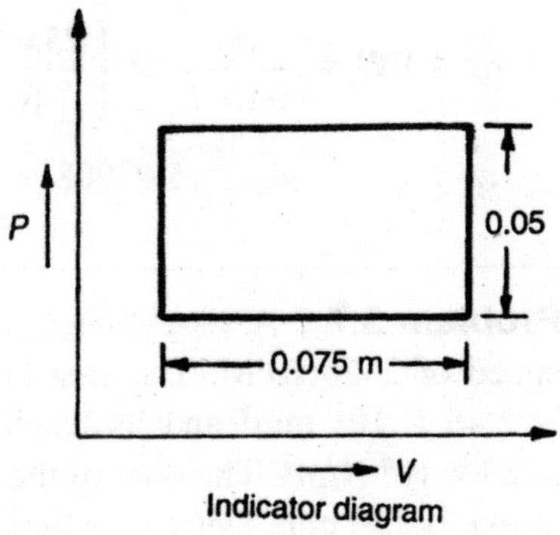

Fig. P. 3.5

∴ Power developed

$$= 0.05 \times 147 \times 10^3 \times 0.2 \times \frac{\pi}{4} \times (0.15)^2 \times \frac{50}{60}$$

= 21.65 kW **Ans**

Problem 3.6 : A single-cylinder, single-acting, 4 stroke engine of 0.15 m bore develops an indicated power of 4 kW when running at 216 rpm. Calculate the area of the indicator diagram that would be obtained with an indicator having a spring constant of 25×10^6 N/m^3. The length of the indicator diagram is 0.1 times the length of the stroke of the engine.

Solution : Given :

bore = d = 0.15 m
$I.P.$ = 4 kW

$$N = 216 \text{ rpm}$$
$$K = \text{spring constant} = 25 \times 10^6 \text{ N/m}^3$$
$$l_d = \text{length of indicator diagram} = 0.1 \times L \text{ (stroke length)}$$
$$I.P. = \frac{P_m LAN}{2} \quad \text{[since engine is 4 stroke cycle]}$$

where P_m = mean effective pressure = $\frac{a_d}{l_d} \times K$

or $$I.P. = \frac{a_d}{l_d} \times \frac{K\ LAN}{2}$$

or $$4 \text{ kW} = \frac{a_d}{0.1 \times L} \times \left(\frac{25 \times 10^6}{10^3}\right) \times L \times \frac{\pi}{4}\ (0.15)^2 \times \frac{216}{2 \times 60}$$

$$\therefore \quad a_d = 5.03008 \times 10^{-4} \text{ m}^2 = 503 \text{ mm}^2 \textbf{ Ans}$$

Problem 3.7 : A six-cylinder, 4-stroke gasoline engine is run at a speed of 2520 RPM. The area of the indicator card of one cylinder is 2.45×10^3 mm^2 and its length is 58.5 mm. The spring constant is 20×10^6 N/m^3. The bore of the cylinders is 140 mm and the piston stroke is 150 mm. Determine the indicated power, assuming that each cylinder contributes an equal power.

Solution : Given

$$n = \text{No. of cylinders} = 6$$
$$N = 2520 \text{ rpm}$$
$$a_d = 2.45 \times 10^{-3} \text{ m}^2$$
$$l_d = 58.5 \times 10^{-3} \text{ m}$$
$$K = 20 \times 10^6 \text{ N/m}^2$$
$$d = \text{dia. of cylinder} = 0.14 \text{ m}$$
$$L = \text{stroke length} = 0.15 \text{ m}$$

$$I.P. = \frac{a_d}{l_d} \times K \times LA\frac{N}{2} \times n$$

$$= \frac{2.45 \times 10^{-3}}{58.5 \times 10^{-3}} \times \left(\frac{20 \times 10^6}{10^3}\right) \times 0.15 \times \frac{\pi}{4} \times (0.14)^2 \times \frac{2520}{2 \times 60} \times 6$$

$$= 243.69 \text{ kW } \textbf{Ans}$$

Problem 3.8 : A steam turbine drives a ship's propeller through an 8 : 1 reduction gear. The average resisting torque imposed by the water on the propeller is 750×10^3 mN and the shaft power delivered by the turbine to the reduction gear is 15 MW. The turbine speed is 1450 rpm. Determine (a) the torque developed by the turbine, (b) the power delivered to the propeller shaft, and (c) the net rate of working of the reduction gear.

Solution : Given :

gear ratio = 8 : 1

T_P = average resisting torque = 750×10^3 mN

P_T = shaft power by turbine = 15 MW

N_T = turbine speed = 1450 rpm

(a) Torque developed by the turbine (T_T)

$$P_T = \frac{2\pi N_T T_T}{60}$$

$$T_T = \frac{P_T \times 60}{2\pi\, N_T} = \frac{15 \times 10^3 \times 60}{2 \times \pi \times 1450}$$

$$= 98.79 \text{ kN.m} \quad \textbf{Ans}$$

(b) $$\frac{N_P}{N_T} = \frac{1}{8},\; N_P = \frac{N_T}{8}$$

$\therefore$ P_P = power delivered to the propeller shaft

$$= \frac{2\pi N_P T_P}{60} = \frac{2\pi \times 1450}{60 \times 8} \times 750 \times 10^3$$

$$= 14.235 \text{ MW} \quad \textbf{Ans}$$

(c) $\therefore$ Net working of reduction gear

$$= P_T - P_P$$

$$= 15 - 14.235$$

$$= 0.765 \text{ MW} \quad \textbf{Ans}$$

Problem 3.9 : A fluid, contained in a horizontal cylinder fitted with a frictionless leakproof piston, is continuously agitated by means of a stirrer passing through the cylinder cover. The cylinder diameter is 0.40 m. During the stirring process lasting 10 minutes, the piston slowly moves out a distance of 0.485 m against the atmosphere. The net work done by the fluid during the process is 2 kJ. The speed of the electric motor driving the stirrer is 840 rpm. Determine the torque in the shaft and the power output of the motor.

Solution : Given :

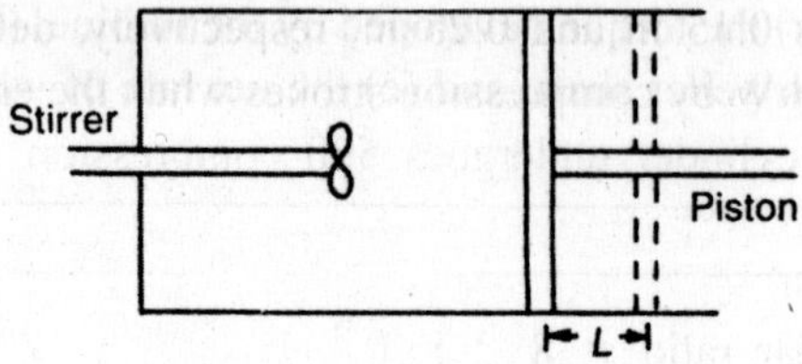

Fig. P. 3.9

cylinder dia $= d = 0.4$ m

$t = 10$ minute = time of stirring

$l = 0.485$ m

$W = 2$ kJ = Net work transfer

$N =$ speed of motor = 840 rpm

Let atmospheric pressure

$$p = 101.325 \text{ kPa}$$

$\therefore$ w_1 = work done by system against atmosphere

$$= p\left(\frac{\pi}{4}d^2\right)l$$

$$= 101.325 \times \frac{\pi}{4} \times 0.4^2 \times 0.485$$

$$= 6.175 \text{ kJ}$$

$(-)$ W_2 = work done by the stirrer in 10 minutes

$$= 2\pi NT \times 10 \text{ kJ}$$

[–ve since work is done on the system]

$\therefore$ Net work transfer

$$= w_1 - w_2 = 2 \text{ kJ}$$

$$\Rightarrow \quad w_2 = w_1 - 2 = 6.175 - 2$$

$$= 4.175 \text{ kJ}$$

or $$T = \frac{4.175}{2\pi \times 840 \times 10} = 0.08 \text{ N.m}$$ **Ans**

$\therefore$ Power input to the stirrer

$$= \frac{w_2}{10 \times 60} = \frac{4175}{600}$$

$$= 6.96 \text{ W}$$ **Ans**

Problem 3.10 : At the beginning of the compression stroke of a two-cylinder internal combustion engine the air is at a pressure of 101.325 kPa. Compression reduces the volume to 1/5 of its volume, and the law of compression is given by $pv^{1.2}$ = constant. If the bore and stroke of each cylinder is 0.15 m and 0.25 m, respectively, determine the power absorbed in kW by compression strokes when the engine speed is such that each cylinder undergoes 500 compression strokes per minute.

Solution : Given :

d = cylinder dia = 0.15 m

L = stroke length = 0.25 m

N = 500 rpm

p_1 = pressure at the beginning = 101.325 kPa

$$V_2 = \frac{V_1}{5}, \text{ compression law, } pv^{1.2} = \text{const.}$$

$$V_1 = \frac{\pi}{4} d^2 \times LN = \frac{\pi}{4} (0.15)^2 \times 0.25 \times \frac{500}{60}$$

$$= 0.03682 \text{ m}^3 = 0.037 \text{ m}^3/\text{sec}$$

$$\therefore \text{Power developed} = 2n \frac{(p_1V_1 - p_2V_2)}{n-1}$$

$$= \frac{2np_1V_1\left[1 - \frac{p_2V_2}{p_1V_1}\right]}{n-1}$$

$$= \frac{2p_1V_1n}{n-1}\left[1 - \left(\frac{V_1}{V_2}\right)^{n-1}\right]$$

$$= \frac{2 \times 1.2 \times 101.325 \times 0.037}{0.2}[1 - 5^{0.2}]$$

$$= -17.1 \text{ kW } \textbf{Ans}$$

[–ve sign shows that work is done on the system.]

Problem 3.11 : Determine the total work done by a gas system following an expansion process as shown in figure P. 3.11 a.

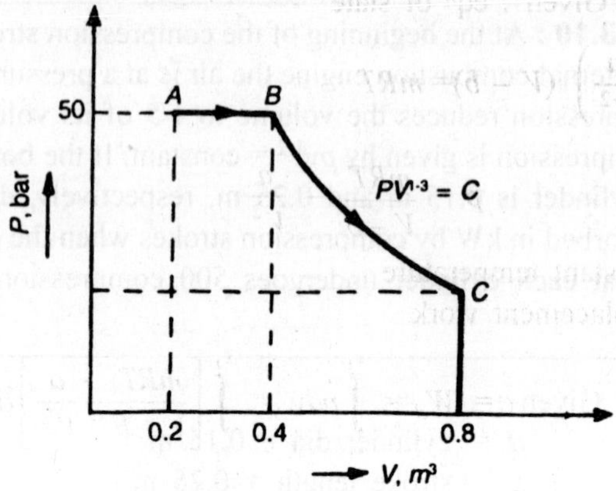

Fig. P. 3.11 (a)

Solution : Total work done by a gas system

$$= p_1 (V_B - V_A) + \frac{p_1 V_B}{n-1}\left[1 - \left(\frac{V_B}{V_C}\right)^{n-1}\right]$$

$$= 50 \times 10^5 \times (0.4 - 0.2) + \frac{50 \times 10^5 \times 0.4}{1.3 - 1}\left[1 - \left(\frac{1}{2}\right)^{0.3}\right]$$

$= 2.252$ MJ **Ans**

Problem 3.12 : A system of volume V contains a mass m of gas at pressure p and temperature T. The macroscopic properties of the system obey the following relationship :

$$\left(p + \frac{a}{V^2}\right)(V - b) = mRT$$

where a, b and R constants.

Obtain an expression for the displacement work done by the system during a constant-temperature expansion from volume V_1 to volume V_2. Calculate the work done by a system which contains 10 kg of this gas expanding from 1 m³ to 10 m³ at a temperature of 293 K. Use the values a = 15.7 × 10 Nm⁴, $b = 1.07 \times 10^{-2}$ m³, and $R = 0.278$ kJ/kg.K.

Solution : Given : eq^n of state

$$\left(p + \frac{a}{V^2}\right)(V - b) = mRT$$

or
$$p = \frac{mRT}{V-b} - \frac{a}{V^2}$$

At constant temperature

∴ displacement work

$$= W_d = \int_{v_1}^{v_2} p dv = \int_{v_1}^{v_2} \left[\frac{mRT}{V-b} - \frac{a}{V^2}\right] dv$$

$$= \left[mRT \ln(V-b) + \frac{a}{V}\right]_{V_1}^{V_2}$$

$$W_d = mRT \ln\left(\frac{V_2 - b}{V_1 - b}\right) + a\left[\frac{1}{V_2} - \frac{1}{V_1}\right] \text{ Ans}$$

Given
$$m = 10 \text{ kg}$$
$$R = 0.278 \text{ kJ/kg.K}$$
$$T = 293 \text{ K}$$
$$V_1 = 1 \text{ m}^3$$
$$V_2 = 10 \text{ m}^3$$
$$a = 157 \text{ N.m}^4$$
$$b = 0.0107 \text{ m}^3$$

$$\therefore W_d = 10 \times 0.278 \times 293 \ln\left(\frac{10 - 0.0107}{1 - 0.0107}\right) + \frac{157}{1000}\left[\frac{1}{10} - 1\right]$$

$$= 1883.3 \text{ kJ } \textbf{Ans}$$

Problem 3.13 : If a gas of volume 6000 cm^3 and at a pressure of 100 kPa is compressed quasistatically according to pV^2 = constant until the volume becomes 2000 cm^3, determine the final pressure and the work transfer.

Solution : Given :

V_1 = initial volume = 6000 cm^3

p_1 = 100 kPa = initial pressure

V_2 = final volume = 2000 cm^3

follows, pV^2 = constant

∴ final pressure

$$= p_2 = p_1 \left(\frac{V_1}{V_2}\right)^2 = 100 \times \left(\frac{6000}{2000}\right)^2$$

$$= 900 \text{ kPa } \mathbf{Ans}$$

$$\text{Work transfer} = \frac{p_1V_1}{n-1}\left[1-\left(\frac{V_1}{V_2}\right)^{n-1}\right]$$

$$= \frac{100 \times 6000 \times 10^{-6}}{1} [1 - 3]$$

$$= -1.2 \text{ kJ } \mathbf{Ans}$$

[–ve sign shows that work is done on the system.]

Problem 3.14 : The flow energy of 0.124 m^3/min of a fluid crossing a boundary to a system is 18 kW. Find the pressure at this point.

Solution : Given :

Volume flow rate

$$= v_1 = 0.124 \text{ m}^3\text{/min}$$

$$\text{flow work} = 18 \text{ kW} = pv$$

$$\therefore \quad p = \frac{18 \times 60}{0.124} = 8709.7 \text{ kPa } \mathbf{Ans}$$

Problem 3.15 : A milk chilling unit can remove heat from the milk at the rate of 41.87 MJ/h. Heat leaks into the milk from the surroundings at an average of 4.187 MJ/h. Find the time required for cooling a batch of 500 kg of milk from 45°C to 5°C. Take the c_P of milk to be 4.187 kJ/kg.K.

Solution : Given :

Heat removed from the can = Q_1 = 41.87 MJ/h

Heat leaked into the can = 4.187 MJ/h

∴ Net heat transfer = 37.683 MJ/h

Now heat transfer from the milk = $mc_P\Delta T$

= 500 × 4.187 × 50

$$= 0.104675 \text{ MJ}$$

$$\therefore \text{ time required for this transfer } = \frac{0.104675}{37.683} \text{ hr}$$

$$= 10 \text{ sec } \textbf{Ans}$$

Problem 3.16 : 680 kg of fish at 5°C are to be frozen and stored at –12°C The specific heat of fish above freezing point is 3.182, and below freezing point is 1.717 kJ/kg.K. The freezing point is –2°C, and the latent heat of fusion is 234.5 kJ/kg. How much heat must be removed to cool the fish, and what percent of this is latent heat?

Solution : Given :

m = mass of fish = 680 kg
initial temperature = 5°C
C_1 = sp. heat above freezing point = 3.182 kJ/kg.K
C_2 = sp. heat below freezing point = 1.717 kJ/kg.K
freezing point = –2°C
L = latent heat of fusion = 234.5 kJ/kg

Now heat to be removed to remain fish at –12°C

$$= m\,[C_1\,(5 - (-2)) + L + C_2\,(-2 - (-12))]$$

$$= 680\,[3.182 \times (7) + 234.5 + 1.717 \times 10]$$

$$= 186.28 \text{ MJ } \textbf{Ans}$$

$$\text{and percent of this heat to latent heat } = \frac{234.5 \times 680}{186280}$$

$$= 85.6\% \textbf{ Ans}$$

Problem 3.17 : An insulated tank of 1 m^3 volume contains air at 0.1 MPa and 300 K. The tank is connected to a high pressure line in which air at 1 MPa and 600 K flows. The tank is quickly filled with air by opening the valve between the tank and high pressure line. If the final pressure of air in the tank is 1 MPa, determine the mass of air which enters the tank and the entropy change associated with filling process. Take Universal Gas Constant $\overline{R}$ = 8.3147 kJ/kg-mole K. (IES–2002)

Solution : Given

Initial pressure = p_1 = 0.1 MPa

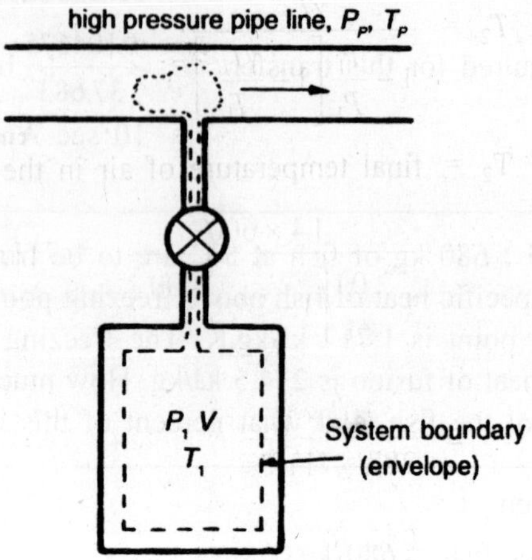

Fig. P. 3.17

volume of tank = $V = 1\ m^3$

Initial temperature T_1 = 300 K

p_p = pipeline pressure = 1 MPa

T_p = pipeline temperature = 600 K

p_2 = final pressure in tank after opening valve = 1 MPa

Let m_1 be initial mass of air and m_2 be mass of air after filling

$$\therefore \quad m_1 = \frac{p_1 V}{RT_1} \text{ and } m_2 = \frac{p_2 V}{RT_2}$$

$\therefore$ From first law

$$Q = m_2 u_2 - m_1 u_1 - (m_2 - m_1)\,[h_p]$$

Neglecting the velocity of air.

$Q = 0$ Since tank is insulated

or $\quad m_2 u_2 - m_2 h_p = m_1 u_1 - m_1 h_p$

$$\text{or} \quad \frac{p_2 V}{RT_2} \times C_v T_2 - \frac{p_2 V}{RT_2} \times C_P T_P = m_1\,[C_v T_1 - C_p T_p]$$

$$\text{or} \quad C_v - \frac{C_p T_p}{T_2} = \frac{R\,p_1 V}{p_2 V R T_1}\,[C_v T_1 - C_p T_p]$$

$$\text{or} \quad T_2 = \frac{C_p T_p}{C_v - \frac{p_1}{p_2 T_1}\left[C_v T_1 - C_p T_p\right]}$$

or $$T_2 = \frac{\gamma T_P}{1 - \frac{P_1}{P_2}\left[1 - \frac{\gamma T_P}{T_1}\right]}$$

$\therefore$ T_2 = final temperature of air in the tank

$$= \frac{1.4 \times 600}{1 - \frac{0.1}{1}\left[1 - 1.4 \times \frac{600}{300}\right]} = 711.86 \text{ K.}$$

$$\therefore\ m_2 = \frac{10^6 \times 1}{287 \times 711.86} = 4.895 \text{ kg}$$

$$m_l = \frac{0.1 \times 10^6 \times 1}{287 \times 300} = 1.161 \text{ kg}$$

$\therefore$ mass of air enters = $m_2 - m_1$ = 3.733 kg **Ans**

Change in entrophy

$$= m_1 \left[c_p \ln \frac{T_2}{T_1} - R \ln \frac{P_2}{P_1}\right] + (m_2 - m_1)\left[c_p \ln \frac{T_2}{T_p}\right]$$

$$= 1.161 \left[1.005 \ln \frac{711.86}{300} - 0.287 \ln \frac{1}{0.1}\right] + 3.733 \times 1.005 \ln \frac{711.86}{600}$$

= 0.8824 kJ/kg.K **Ans**

Chapter 4

First Law of Thermodynamics

Problem 4.1 : An engine is tested by means of a water brake at 1000 rpm. The measured torque of the engine is 10000 mN and the water consumption of the brake is 0.5 m^3/s, its inlet temperature being 20°C. Calculate the water temperature at exit, assuming that the whole of the engine power is ultimately transformed into heat which is absorbed by the cooling water.

Solution : Given :

$$N = 1000 \text{ rpm}$$

$$\text{Brake torque} = T = 10000 \text{ N.m}$$

$$\text{Volume flow rate} = \dot{V} = 0.5 \text{ m}^3/\text{s}$$

$$\text{Inlet temperature} = T_1 = 20°\text{C}$$

Let T_2 = water outlet temperature

Brake power developed by engine

$$= \frac{2\pi NT}{60} = \frac{2 \times \pi \times 1000 \times 10000}{60} = 1047.2 \text{ kW}$$

from question whole of power developed is transformed into heat

$\therefore$ Heat absorbed by the cooling water = $mCp_w\ (T_2 - 20)$

where Cp_w = sp. heat capacity of water = 4.187 kJ/kg

$\dot{m}$ = mass of cooling water circulated

$= \dot{V} \times \rho = 0.5 \times 1000$

$\therefore\ 0.5 \times 1000 \times 4.187 \times (T_2 - 20) = 1047.2$

$\therefore\ T_2 - 20 = 0.5002$

$\therefore\ T_2 = 20.5°\text{C}$ **Ans**

Problem 4.2 : In a cyclic process, heat transfers are +14.7 kJ, –25.2 kJ, –3.56 kJ and +31.5 kJ. What is the net work for this cyclic process?

Solution : From first law of thermodynamics for a cyclic process

$$\oint dQ = \oint dw$$

$$\therefore \quad \oint dw = 14.7 - 25.2 - 3.56 + 31.5 = 17.34 \text{ kJ } \textbf{Ans}$$

Problem 4.3 : A slow chemical reaction takes place in a fluid at the constant pressure of 0.1 MPa. The fluid is surrounded by a perfect heat insulator during the reaction which begins at state 1 and ends at state 2. The insulation is then removed and 105 kJ of heat flow to the surroundings as the fluid goes to state 3. The following data are observed for the fluid at states 1, 2 and 3.

State	$V\ (m^3)$	t (°C)
1	0.003	20
2	0.3	370
3	0.06	20

For the fluid system, calculate E_2 and E_3, if $E_1 = 0$

Solution : For process 1-2, constant pressure process and heat transfer = 0

∴ from 1st law

$$đQ_{1-2} = dE + đw_{1-2}$$

$$\Rightarrow \quad 0 = (E_2 - E_1) + đw_{1-2}$$

$$\Rightarrow \quad E_1 - E_2 = đw_{12} = p(V_2 - V_1) = 0.1 \times 10^3 \ (0.3 - 0.003)$$

$$\Rightarrow \quad E_2 = -29.7 \text{ kJ [since, } E_1 = 0] \ \textbf{Ans}$$

For process 2-3, heat transfer is allowed at constant pressure

∴ from first law of thermodynamics

$$dQ_{2-3} = (E_3 - E_2) + đw_{2-3}$$

$$\Rightarrow \quad đQ_{2-3} = (E_3 - E_2) + p(V_3 - V_2)$$

$$\Rightarrow \quad -105 = (E_3 + 29.7) + 0.1 \times 10^{+3} \ (0.06 - 0.3)$$

$$\Rightarrow \quad E_3 = -105 - 29.7 - 0.01 \times 10^3 \ (0.06 - 0.3)$$

$$= -110.7 \text{ kJ } \textbf{Ans}$$

Problem 4.4 : During one cycle the working fluid in an engine engages in two work interactions : 15 kJ to the fluid and 44 kJ from the fluid, and three heat interactions, two of which are known : 75 kJ to the fluid and 40 kJ from the fluid. Evaluate the magnitude and direction of the third heat transfer.

Solution : For a cyclic process, from first law of thermodynamics

$$\int dQ = \int dw$$

Let Q be the heat transfer in the 3rd process, then

$\Rightarrow \quad 75 - 40 + Q = -15 + 44$

$\therefore \quad Q = -15 + 44 - 75 + 40 = -6$ kJ **Ans**

Problem 4.5 : A domestic refrigerator is loaded with food and the door closed. During a certain period the machine consumes 1 kWh of energy and the internal energy of the system drops by 5000 kJ. Find the net heat transfer for the system.

Solution : Give data

power consumed by refrigerator = 1 kWh = 3600 kJ

change in internal energy = $(E_2 - E_1) = -5000$ kJ

$\therefore$ from first law of thermodynamics

$$dQ_{1-2} = \Delta E + dw_{1-2} = -5000 - 3600 = -8600 \text{ kJ}$$
$$= -8.6 \text{ MJ } \textbf{Ans}$$

Problem 4.6 : 1.5 kg of liquid having a constant specific heat of 2.5 kJ/kg.K is stirred in a well–insulated chamber causing the temperature to rise by 15°C. Find ΔE and W for the process.

Solution : Given : data

m = 1.5 kg

specific heat = C = 2.5 kJ/kg.K

ΔT = 15°C

For well insulated chamber, $dQ = 0$

$\therefore$ from first law of thermodynamics,

$$dQ = \Delta E + dw$$

$\Rightarrow \quad 0 = \Delta E + dw$

$\propto \quad \Delta E = -dw$

Now $\quad \Delta E = mC\,\Delta T$ [Neglecting ΔK.E. and ΔP.E.]

$= 1.5 \times 2.5 \times 15 = 56.25$ kJ **Ans**

$dw = -\Delta E = -56.25$ kJ **Ans**

Problem 4.7 : The same liquid as in problem 4.6 is stirred in a conducting chamber. During the process 1.7 kJ of heat are transferred from the liquid to the surroundings, while the temperature of the liquid is rising to 15°C. Find ΔE and W for the process.

Solution : For conducting chamber

$$đQ = -1.7 \text{ kJ}$$
$$\Delta T = 15°\text{C}$$

From previous problem, work done on the system to raise its temperature to 15°C = –56.25 kJ **Ans**

∴ From 1st law of thermodynamics

$$đQ = \Delta E + đw$$

or $$-1.7 = \Delta E - 56.25$$

∴ $$\Delta E = 56.25 - 1.7 = 54.53 \text{ kJ} \textbf{ Ans}$$

Problem 4.8 : The properties of a certain fluid are related as follows

$$u = 196 + 0.718\,t$$
$$pv = 0.287\,(t + 273)$$

where u is the specific internal energy (kJ/kg), t is in °C, p is pressure (kN/m^2), and v is specific volume (m^3/kg).
For this fluid, find C_v and C_p.

Solution : Given :

$$u = 196 + 0.718\,t$$
$$pv = 0.287\,(t + 273)$$

We know that

$$C_v = \left(\frac{\partial u}{\partial t}\right)_V = \left[\frac{\partial}{\partial t}(196 + 0.718t)\right]_v$$
$$= 0.718 \text{ kJ/kg.K} \textbf{ Ans}$$

Again $$C_p = \left(\frac{\partial h}{\partial t}\right)_p$$

Now $$h = u + pv$$
$$= 196 + 0.718\,t + 0.287\,(t + 273)$$

$$\left(\frac{\partial h}{\partial t}\right)_p = 0 + 0.718 + 0.287 \times (1 + 0)$$

$$C_p = 0.718 + 0.287 = 1.005 \text{ kJ/kg.K} \textbf{ Ans}$$

Problem 4.9 : A system composed of 2 kg of the above fluid expands in a frictionless piston and cylinder machine from an initial state of 1 MPa, 100°C to a final temperature of 30°C. If there is no heat transfer, find the net work for the process.

Solution : Given :

mass of fluid = m = 2 kg

initial pressure = p_1 = 1 MPa

initial temperature = T_1 = 100°C

final temperature = T_2 = 30°C

heat transfer = $dQ = 0$

∴ From 1st law of thermodynamics

$$dQ = du + dw$$

$$\Rightarrow \quad 0 = m(u_2 - u_1) + dw$$

$$\therefore \quad dQ = m(u_1 - u_2)$$

$$= m[196 + 0.718 \times T_1] - m[196 + 0.718 \times T_2]$$

$$= 0.718 [T_1 - T_2] \times 2$$

$$= 0.718 [100 - 30] \times 2$$

$$= 2 \times 50.26 \text{ kJ} = 100.52 \text{ kJ } \textbf{Ans}$$

Problem 4.10 : If all the work in the expansion of problem 4.9 is done on the moving piston, show that the equation representing the path of the expansion in the *pv*–plane is given by $pv^{1.4}$ = constant.

Solution : Let process 1–2 show expansion process following pv^n = constant

$$\therefore \quad W_{1-2} = \int_{v_1}^{v} pdv$$

$$= \frac{p_1V_1 - p_2V_2}{n-1}$$

or, W_{1-2}

$$= \frac{mR(T_1 - T_2)}{n-1}$$

Given : W_{1-2}

$= 50.26$ kJ

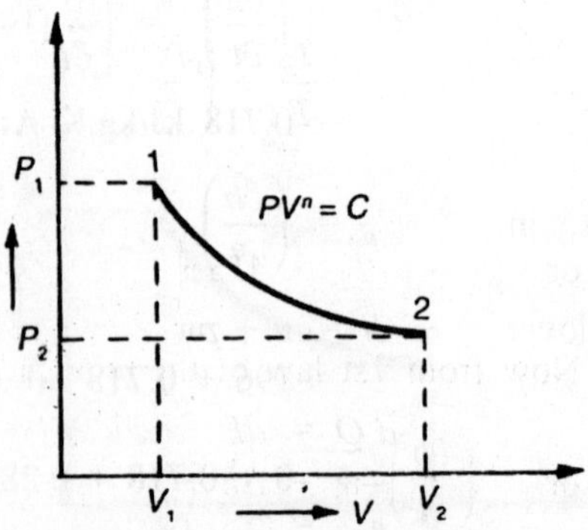

Fig. P. 4.10

$$m = 2 \text{ kg}$$
$$T_1 = 100°C$$
$$T_2 = 30°C$$
$$R = 0.287 \text{ kJ/kg.K}$$

$$\therefore \quad W_{1-2} = \frac{mR(T_1 - T_2)}{n-1}$$

$$100.52 = \frac{2 \times 0.287(100-30)}{n-1}$$

$$\therefore \quad n = 1.399 = 1.4$$

Hence law of expansion follows $pv^{1.4}$ = constant **Ans**

> **Problem 4.11** : A stationary system consisting of 2 kg of the fluid of problem 4.8 expands in an adiabatic process according to $pv^{1.2}$ = constant. The initial conditions are 1 MPa and 200°C, and the final pressure is 0.1 MPa. Find W and ΔE for the process. Why is the work transfer not equal to $\int pdV$.

Solution : Given :

mass of fluid = m = 2 kg

p_1 = initial pressure = 1 MPa

T_1 = initial temperature = 200°C = 473 K

p_2 = final pressure = 0.1 MPa

Law of expansion,

$$pv^{1.2} = \text{Constant}$$

$$\therefore \quad \frac{T_2}{T_1} = \left(\frac{P_2}{P_1}\right)^{\frac{n-1}{n}}$$

$$\text{or,} \quad T_2 = 473\left(\frac{0.1}{1}\right)^{\frac{0.2}{1.2}} = 322.25 \text{ K}$$

Now from 1st law of thermodynamics

$$đQ = dE + đw \quad [đQ = 0]$$

$$\text{or} \quad -w = \Delta E = m\, C_v\, (T_2 - T_1)$$

$$= 2 \times 0.718\, (322.25 - 473) = -216.48 \text{ kJ}$$

$\therefore \quad w = +216.48 \text{ kJ}, \Delta E = -216.48 \text{ kJ}$ **Ans**

$$\int_1^2 pdv = \frac{p_1V_1 - p_2V_2}{n-1} = \frac{mR(T_1 - T_2)}{0.2}$$

$$= \frac{2 \times 0.287 \times (473 - 322.25)}{.02}$$

$= 432.65$ kJ **Ans**

$\int pdv$ work is calculated when the process is quasi-static or reversible. But here reversibility of process is not shown, therefore $\int pdv$ will not be used in this problem.

Problem 4.12 : A mixture of gases expands at constant pressure from 1 MPa, 0.03 m^3 to 0.06 m^3 with 84 kJ positive heat transfer. There is no work other than that done on a piston. Find ΔE for the gaseous mixture.
The same mixture expands through the same state path while a stirring device does 21 kJ of work on the system. Find ΔE, W and Q for the process.

Solution : Given :

pressure $= p = 1$ MPa
initial volume $= V_1 = 0.03$ m^3
final volume $= V_2 = 0.06$ m^3
heat added $= Q = +84$ kJ
w = work done in the process
$= p(V_2 - V_1)$
$= 1 \times 10^3 (0.06 - 0.03) = 30$ kJ

$\therefore$ from 1st law of thermodynamics,

$Q = \Delta E + w$

$\therefore \quad \Delta E = Q - w = 84 - 30 = 54$ kJ **Ans**

Now work done on system by stirring

$w = -21$ kJ

For the same state path,

$\Delta E = 54$ kJ

$\therefore$ From 1st law of thermodynamics

$Q = \Delta E + w = 54 - 21 = 33$ kJ **Ans**

Problem 4.13 : A mass of 8 kg gas expands within a flexible container so that the *p–v* relationship is of the form $pv^{1.2}$ = constant. The initial pressure is 1000 kPa and the initial volume is 1 m³. The final pressure is 5 kPa. If specific internal energy of the gas decreases by 40 kJ/kg, find the heat transfer in magnitude and direction.

Solution : Given :

$$\text{mass of gas} = 8 \text{ kg}$$
$$\text{initial pressure} = p_1 = 1000 \text{ kPa}$$
$$\text{initial volume} = V_1 = 1 \text{ m}^3$$
$$\text{final pressure} = p_2 = 5 \text{ kPa}$$
$$\Delta U = (U_2 - U_1) \times m = -8 \times 40 \text{ kJ/kg}$$
$$= -40 \times 8 = -320 \text{ kJ}$$

law of expansion

$$pv^{1.2} = \text{constant}$$

$$\therefore \quad W_{1-2} = \frac{p_1V_1 - p_2V_2}{n-1} = \frac{p_1V_1\left[1 - \left(\frac{p_2}{p_1}\right)^{\frac{n-1}{n}}\right]}{n-1}$$

$$= \frac{1000 \times 1}{0.2}\left[1 - \left(\frac{5}{1000}\right)^{\frac{0.2}{1.2}}\right] = 2932.41 \text{ kJ}$$

From 1st law of thermodynamics.

$$đQ = dE + đw_{1-2}$$
$$= -320 + 2932.41 = +2612.41 \text{ kJ } \textbf{Ans}$$

i.e., Heat will be transferred to the system

Problem 4.14 : A gas of mass 1.5 kg undergoes a quasi–static expansion which follows a relationship $p = a + bV$, where a and b are constants. The initial and final pressures are 1000 kPa and 200 kPa respectively and the corresponding volumes are 0.20 m³ and 1.20 m³. The specific internal energy of the gas is given by the relation.

$$u = 1.50\, pv - 85 \text{ kJ/kg}$$

where p is in kPa and v is in m³/kg. Calculate the net heat transfer and the maximum internal energy of the gas attained during expansion.

Solution : Given :

initial pressure = p_1 = 1000 kPa
final pressure = p_2 = 200 kPa
initial volume = V_1 = 0.20 m³
final volume = V_2 = 1.20 m³
mass of gas = m = 8 kg

Expansion law,

$$p = a + bV, \text{ } a \text{ and } b \text{ are constants}$$

Now $1000 = a + b \times 0.2$ ...(a)

and $200 = a + b \times 1.2$ ...(b)

Solving equation (a) and (b)

$$a = 1160$$

$$b = -800$$

$$\therefore \quad P = 1160 - 800\ V$$

Hence $w_{1-2} = \int_{0.2}^{1.2} P\,dv$

$$= \int_{0.2}^{1.2} [1160 - 800V]\,dV \quad \left[1160\,V - \frac{800}{2}V^2\right]_{0.2}^{1.2}$$

$$= 1160\ (1.2 - 0.2) - 400\ (1.2^2 - .02^2)$$

$$= 600 \text{ kJ}$$

$$m\,(U_2 - U_1) = \text{change in internal energy}$$

$$= 1.5 \times [1.5\ p_2V_2 - 85] - [1.5\ p_1V_1 - 85] \times 1.5$$

$$= 1.5\ [p_2V_2 - p_1V_1]$$

$$= 1.5\ [200 \times 1.2 - 1000 \times 0.2]$$

$$= 60 \text{ kJ/kg}$$

$\therefore$ From 1st law of thermodynamics,

$$dQ = \Delta U + w_{1-2}$$

$$= 60 + 600 = 660 \text{ kJ } \textbf{Ans}$$

(b) $U = 1.5\ pv - 85\ \dfrac{\text{kJ}}{\text{kg}}$ or, $U = mu$

$$U = 1.5 \times 1.5\ pv - 85 \times 1.5 \text{ kJ}$$

$$= 1.5\ pV - 8.5 \times 1.5$$

$U = 1.5\ (a + bV)\ V - 8.5 \times 1.5$ kJ $[\because\ p = a + bV]$ ...(c)

For maximum internal energy

$$\frac{dU}{dV} = 0 = 1.5\,[a + 2bV]$$

$$\therefore \quad V = \frac{-a}{2b} = \frac{-1160}{(-)2 \times 800}$$

$$= \frac{1160}{1600} = 0.725\ m^3$$

$\therefore$ from (c), we get

$$U = 1.5\,(a + bV)V - 85 \times 1.5$$
$$= 1.5\,(1160 - 800V)V - 85 \times 1.5$$
$$= 1.5\,[1160 - 800 \times 0.725] \times 0.725 - 85 \times 1.5$$
$$= 503.25 \text{ KJ } \textbf{Ans}$$

Problem 4.15 : The heat capacity at constant pressure of a certain system is a function of temperature only and may be expressed as

$$C_p = \frac{41.87}{t + 100} \text{ J/°C}$$

where t is the temperature of the system in °C. The system is heated while it is maintained at a pressure of 1 atmosphere until its volume increases from 2000 cm^3 to 2400 cm^3 and its temperature increases from 0°C to 100°C. (a) Find the magnitude of the heat interaction. (b) How much does the internal energy of the system increase?

Solution : Given :

$$T_1 = 0°C,$$
$$T_2 = 100°C$$
$$V_1 = 2000\ cm^3 = 0.002\ m^3$$
$$V_2 = 2400\ cm^3 = 0.0024\ m^3$$
$$p = 1 \text{ atm} = 1.013 \times 10^5\ N/m^2$$

(a) From first law of thermodynamics

$$dQ_{1-2} = dH - Vdp \text{ kJ}$$

or

$$Q_{1-2} = \int_{t_1}^{t_2} C_p dt \quad [\text{at constant pressure, } dp = 0]$$

$$= \int_{0°C}^{100°C} \left(2.093 + \frac{41.87}{t + 100}\right) dt$$

$$= [2.093\ t + 41.87 \ln (t + 100)]_{0°C}^{100°C}$$

$$= 2.093\ (100 - 0) + 41.87 \ln \frac{200}{100}$$

$$= 238.32 \text{ J } \textbf{Ans}$$

(b) We know

$$H = U + pv$$

$$\therefore \quad U = H - pv$$

$$\text{or} \quad dU = dH - pdV$$

$$\text{or} \quad (U_2 - U_1) = \int C_p dt - \int pdV$$

$$= \int_{0°C}^{100°C} \left(2.093 + \frac{41.87}{t+100}\right) dt - p \int_{0.002}^{0.0024} dv$$

$$= [2.093\ (100 - 0) + 41.87 \ln \frac{100+100}{100+0}] - 1.013 \times 10^5\ [0.0024 - 0.002]$$

$$= 197.79 \text{ J } \textbf{Ans}$$

Problem 4.16 : An imaginary engine receives heat and does work on a slowly moving piston at such rates that the cycle of operation of 1 kg of working fluid can be represented as a circle 10 cm in diameter on a p–v diagram on which 1 cm = 300 kPa and 1 cm = 0.1 m^3/kg. (a) How much work is done by each kg of working fluid for each cycle of operation and heat input in a cycle. If the heat rejected by the engine in a cycle is 1000 kJ per kg of working fluid, what would be its thermal efficiency?

Solution : Given :

dia of circle = 10 cm

mass of fluid = 1 kg

1 cm = 300 kPa

1 cm = 0.1 m^3/kg

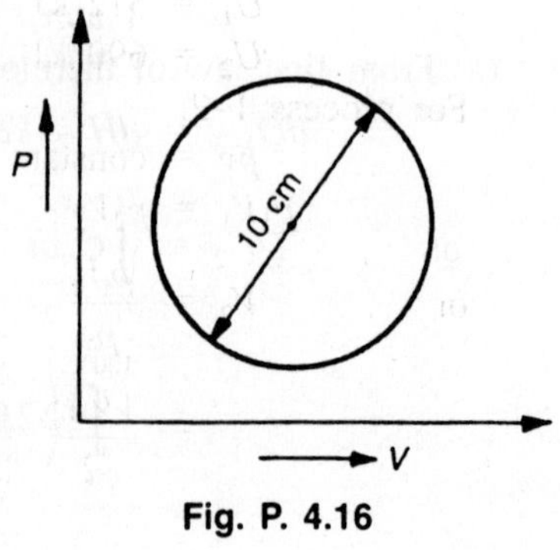

Fig. P. 4.16

(a) Work done by working fluid

= area under the curve

$$= \frac{\pi}{4} d^2 = \frac{\pi}{4} \times 100 \text{ cm}^2$$

$$= \frac{\pi}{4} \times 100 \times (300 \times 0.1) \text{ kJ/kg}$$

$$= 2356.19 \text{ kJ/kg}$$

(b) Q = Heat rejected = 1000 kJ

$$\because \text{ thermal efficiency} = \frac{\text{Work done}}{\text{heat supplied}}$$

Now for a cyclic engine

heat supplied = work done + heat rejected

$\therefore$ thermal efficiency

$$= \frac{\text{Work done}}{\text{work done + heat supplied}}$$

$$= \frac{2356.19}{2356.19 + 1000} = 0.702 \textbf{ Ans}$$

> **Problem 4.17 :** A gas undergoes a thermodynamic cycle consisting of three processes beginning at an initial state where p_1 = 1 bar, V_1 = 1.5 m³ and U_1 = 512 kJ. The processes are as follows :
> (i) Process 1-2 : Compression with pV = constant to p_2 = 2 bar, U_2 = 690 kJ
> (ii) Process 2-3 : W_{23} = 0, Q_{23} = –150 kJ, and
> (iii) Process 3-1 : W_{31} = +50 kJ. Neglecting KE and PE changes, determine the heat interactions Q_{12} and Q_{31}.

Solution : Given :

$$p_1 = 1 \text{ bar},$$
$$p_2 = 2 \text{ bar}$$
$$V_1 = 1.5 \text{ m}^3$$
$$U_1 = 512 \text{ kJ}$$
$$U_2 = 690 \text{ kJ}$$

For process 1–2,

$$pv = \text{constant}$$
$$p_1V_1 = p_2V_2$$

or $$V_2 = \frac{p_1V_1}{p_2}$$

$$= \frac{1 \times 1.5}{2} = 0.75 \text{ bar}$$

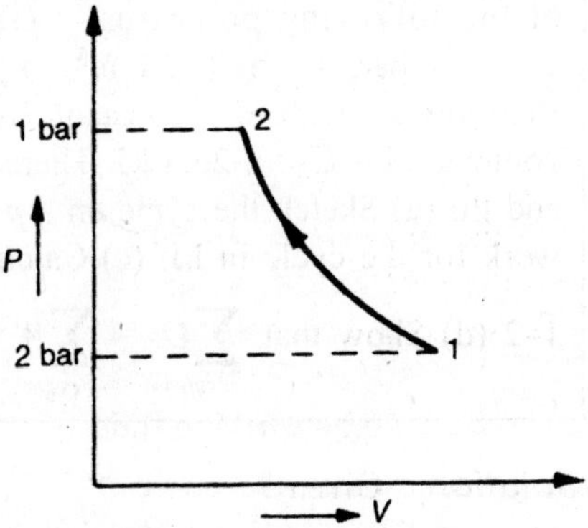

Fig. P. 4.17

∴ Applying 1st law of thermodynamics for process 1-2

$$đQ = du + đw$$

or $$Q_{1-2} = (U_2 - U_1) + W_{1-2}$$

$$= (U_2 - U_1) + p_1V_1 \ln \frac{V_2}{V_1}$$

$$= (690 - 512) + 1 \times 10^2 \times 1.5 \times \ln \frac{0.75}{1.5}$$

$$= 74.03 \text{ kJ } \textbf{Ans}$$

For process 2-3

$$W_{2-3} = 0$$

$$Q_{2-3} = -150 \text{ kJ}$$

∴ Applying 1st law of thermodynamics,

$$đQ = du + đw$$

or $$Q_{2-3} = (U_3 - U_2) + W_{2-3}$$

or $$-150 = U_3 - 690 + 0$$

∴ $$U_3 = 690 - 150 = 540 \text{ kJ}$$

For process 3-1

$$W_{3-1} = 50 \text{ kJ}$$

Applying 1st law of thermodynamics

$$đQ = du + đw$$

or, $$Q_{3-1} = (U_1 - U_3) + W_{3-1}$$

$$= (512 - 540) + 50 = 22 \text{ kJ } \textbf{Ans}$$

Problem 4.18 : A gas undergoes a thermodynamic cycle consisting of the following processes : (i) Process 1-2 : Constant pressure $p = 1.4$ bar, $V_1 = 0.028$ m^3, $W_{12} = 10.5$ kJ, (ii) Process 2-3 : Compression with pV = constant, $U_3 = U_2$, (iii) Process 3-1 : Constant volume, $U_1 - U_3 = -26.4$ kJ. There are no significant changes in KE and PE (a) Sketch the cycle on a p–V diagram. (b) Calculate the net work for the cycle in kJ. (c) Calculate the heat transfer for process 1–2 (d) Show that $\sum_{cycle} Q = \sum_{cycle} W$

Solution : Given :

$$p_1 = p_2 = 1.4 \text{ bar}$$

$$V_1 = 0.028 \text{ m}^3$$

$$W_{1-2} = 10.5 \text{ kJ}$$

$$U_3 = U_2$$
$$U_1 - U_3 = -26.4 \text{ kJ}$$

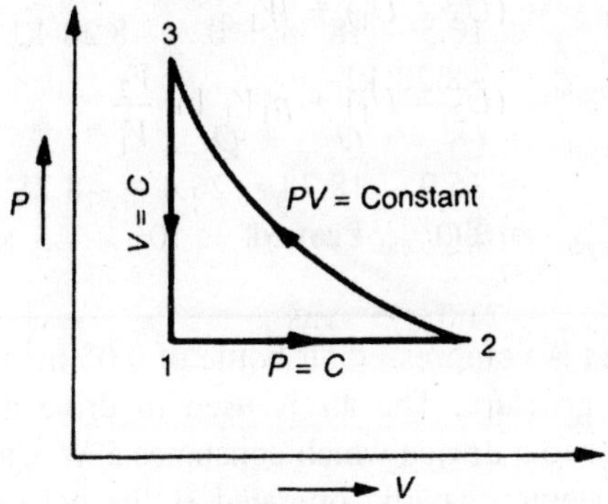

Fig. P. 4.18

For process 1-2

$$W_{1-2} = p_1(V_2 - V_1)$$

or, $10.5 = 1.4 \times 10^2 \, (V_2 - 0.028)$

$\therefore$ $V_2 = 0.103 \text{ m}^3$

Now $U_1 - U_3 = -26.4$ kJ

or, $U_1 - U_2 = -26.4 \text{ kJ} \; [\because U_3 = U_2]$

or, $U_2 - U_1 = 26.4$ kJ

Applying 1st law of thermodynamics for process 1-2

(c) $Q_{1-2} = (U_2 - U_1) + W_{1-2}$

$= 26.4 + 10.5 = 36.9$ kJ **Ans**

For process 2-3, pv = constant

$$W_{2-3} = p_2V_2 \ln \frac{V_3}{V_2} = p_2V_2 \ln \frac{V_1}{V_2}$$

or, $W_{2-3} = 1.4 \times 10^2 \times 0.103 \times \ln \dfrac{0.028}{0.103}$

$= -18.78$ kJ

From 1st law of thermodynamics,

$$Q_{2-3} = (U_3 - U_2) + W_{2-3}$$
$$= 0 - 18.78 = -18.78 \text{ kJ}$$

For process 3-1

$$V = \text{constant}$$
$$W_{3-1} = 0$$

$\therefore$ Applying 1st law of thermodynamics

$$Q_{3-1} = (U_1 - U_3) + W_{3-1}$$

$= -26.4 + 0 = -26.4$ kJ

(b) Net work for the cycle

$= W_{1-2} + W_{2-3} + W_3 - 1$

$= 10.5 - 18.78 + 0 = -8.28$ kJ **Ans**

(d) $\Sigma W_{cycle} = -8.28$ kJ

and $\Sigma Q_{cycle} = Q_{1-2} + Q_{2-3} + Q_{3-1}$

$= 36.9 - 18.78 - 26.4 = -8.28$ kJ

Hence $\Sigma W_{cycle} = \Sigma Q_{cycle}$ **Proved**

Problem 4.19 : A compressed air bottle of 0.05 m^3 volume contains air at 3.5 atm pressure. The air is used to drive a turbogenerator supplying power to a device which consumes 5 W. Calculate the time for which the device can be operated if the actual output of the turbogenerator is 60% of the maximum theoretical output. The ambient pressure is 1 atm.
For air $(C_p/C_v) = 1.4$ (CES – 93)

Solution :

$V = 0.05\ m^3$;

$P = 3.5 \times 10^5$ Pa

Output = 5 W;

$n_{generator} = 60\%$

$P_{ambient} = 1 \times 10^5$ Pa

To find : The time for which the device can be operated.

Energy available in the compressed air bottle

$= 3.5 \times 10^5 \times 0.05$

Energy at dead state $= 1 \times 10^5 \times 0.05$ J

Net available energy $= (3.5 - 1) \times 10^5 \times 0.05 = 12500$ J

$$\text{Energy used/sec} = \frac{5}{0.6} \text{ Joules}$$

Time for which the turbogenerator can be operated with 5 W output

$$= \frac{12500}{5/0.6} = 1500 \text{ sec} = 25 \text{ minutes}$$

Problem 4.20 : The diameter and stroke of a gas engine cylinder are 18 cm and 30 cm respectively. The ratio of expansion is 6. The pressure and temperature of the mixture at the beginning of compression are 1.04 kg.f/cm^2 and 100°C respectively. Find the index of the compression process and the weight of the mixture in the cylinder, if the pressure at the end of compression is 7 kg.f/cm^2. Also calculate the work done and heat transferred during the process, indicating the direction of flow. Assume R = 29.3 kg.m/kg.K and ratio of specific heats equal to 1.4 for the mixture. (CES – 1987)

Solution : Let 1 and 2 represent the initial and final condition of gas.

Volume of the cylinder = V_1

$$= \frac{\pi}{4} \times 18^2 \times 30 \text{ cm}^3 = 7634 \text{ cm}^3$$

$$P_1 = 1.04 \text{ kg/cm}^2$$

$$T_1 = 100 + 273 = 373 \text{ K}$$

$$R = 29.3 \text{ kg.m/kg.K}$$

$$m = \frac{P_1 V_1}{RT_1} = \frac{1.04 \times 7634}{29.3 \times 373 \times 100} \text{ kg}$$

$$= 7.26 \times 10^{-3} \text{ kg.}$$

Mass of gas $= 7.26 \times 10^{-3}$ kg

$$\frac{V_1}{V_2} = 6.$$

$$\frac{P_2}{P_1} = \frac{7}{1.04}$$

$$P_1 V_1^n = P_2 V_2^n$$

$$\left(\frac{P_2}{P_1}\right) = \left(\frac{V_1}{V_2}\right)^n$$

$$\log\left(\frac{P_2}{P_1}\right) = n \log\left(\frac{V_1}{V_2}\right)$$

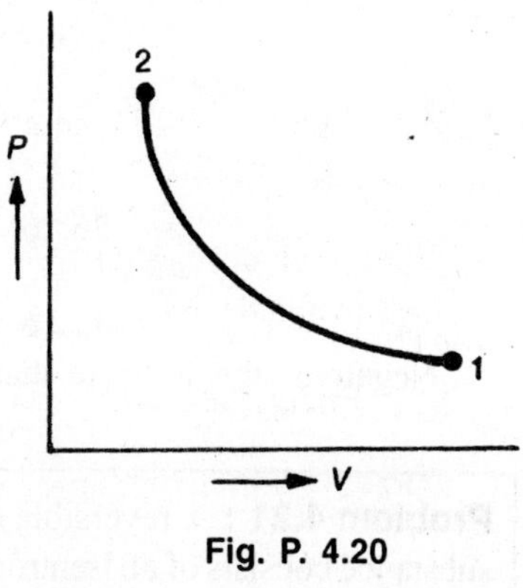

Fig. P. 4.20

Index of compression

$$n = \frac{\log\left(\frac{P_2}{P_1}\right)}{\log\left(\frac{V_1}{V_2}\right)} = \frac{\log\left(\frac{7}{1.04}\right)}{\log(6)}$$

$$n = 1.18$$

$$\text{Work done} = \frac{P_1V_1 - P_2V_2}{(n-1)}$$

$$= \frac{1.04 \times 7634 - 7 \times \dfrac{7634}{5}}{(1.18-1) \times 100} \text{ kg.m}$$

$$= -152.68 \text{ kg.m/cycle}$$

Negative sign indicate work is done on the system.

$$\left(\frac{T_1}{T_2}\right) = \left(\frac{P_1}{P_2}\right)^{\frac{n-1}{n}}$$

$$T_2 = \frac{373}{\left(\dfrac{1.04}{7}\right)^{\frac{0.18}{1.18}}} = 498.9 \text{ K}$$

$$\text{Heat added} = C_v\left(\frac{\gamma - n}{1-n}\right)(T_2 - T_1)$$

$$= C_v \times \frac{(1.4-1.18)}{(1-1.18)} (498.9 - 373)$$

$$= -153.87 \times 0.17 \text{ kcal/kg.K}$$

$$= -26.16 \frac{\text{kcal}}{\text{kg}}$$

$$= -12.16 \times 7.26 \times 10^{-3} \text{ kcal/cycle}$$

Negative sign indicate that the heat is rejected by the system.

Problem 4.21 : A reversible cycle using an ideal gas as the working substance consists of an isentropic compression from initial temperature to 555 K, a constant volume process from 555 K to 835 K, a reversible adiabatic expansion to 555 K, a constant pressure expansion from 555 K to 835 K followed by a constant volume process to the initial temperature. Draw the cycle on *P–V* and *T–S* diagrams and determine the initial temperature (γ = 1.4). Also compute the work done.

(CES – 1989)

Solution :

$$\frac{V_3}{V_5} = \frac{V_2}{V_1} \quad ...(1)$$

$$\frac{V_2}{V_1} = \left(\frac{T_1}{T_2}\right)^{\frac{1}{\gamma-1}}$$

$$= \left(\frac{T_1}{255}\right)^{2.5} \quad ...(2)$$

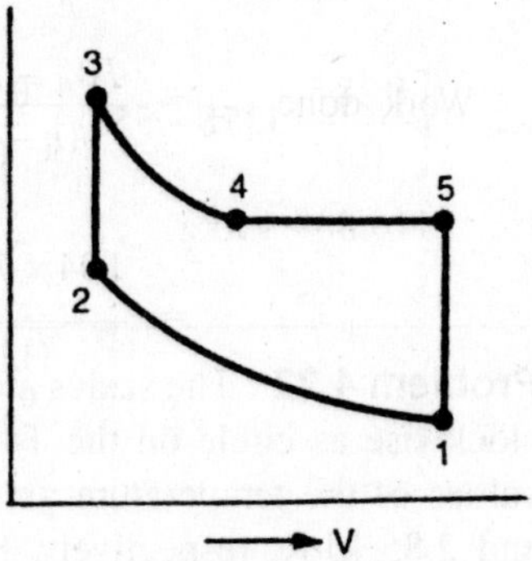

Fig. P. 4.21 (a)

Process 4-5

$$\frac{P_4V_4}{T_4} = \frac{P_5V_5}{T_5};$$

$$\frac{V_4}{V_5} = \frac{T_4}{T_5} = \frac{555}{835} \quad ...(3)$$

Process 3-4

$$\left(\frac{V_3}{V_4}\right) = \left(\frac{T_3}{T_4}\right)^{2.5}$$

$$= \left(\frac{835}{555}\right)^{2.5} \quad ...(4)$$

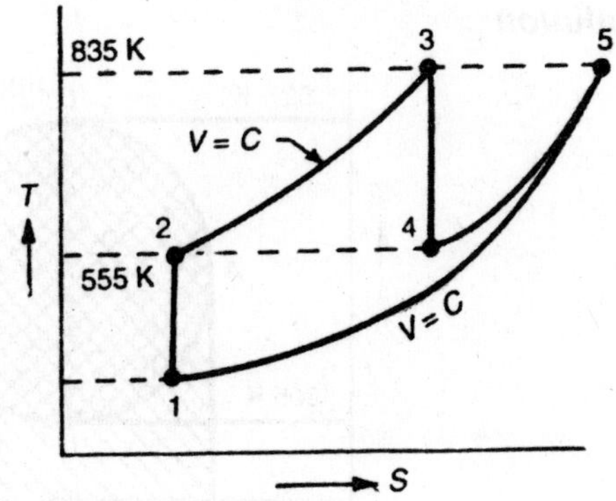

Fig. P. 4.21 (b)

equation (3) × (4)

$$= \frac{V_3}{V_5} = \left(\frac{555}{835}\right) \times \left(\frac{835}{555}\right)^{2.5}$$

$$= \left(\frac{835}{555}\right)^{1.5} = 1.845$$

Substituting this in (1)

$$\left(\frac{T_1}{255}\right)^{2.5} = 1.845;$$

$$T_1 = 325.8 \text{ K}$$

$$\text{Work done} = \left[\frac{R}{\gamma-1}(T_1 - T_2) + 0 + \frac{R}{\gamma-1}(T_3 - T_4) + R(T_5 - T_4)\right]$$

$$= \left[\frac{0.287}{0.4}(325.8-555)+\frac{0.287}{0.4}(835-555)+0.287(835-55)\right]$$

$= 116.809$ kJ/kg

Problem 4.22 : The scales are so chosen that a reversible cycle plots clockwise as circle on the T–S plane. The minimum and maximum values of the temperature are 305 and 627 K and the entropy 1.23 and 2.85 kJ/K, respectively. Find the cycle work and efficiency. (CES – 1990)

Solution :

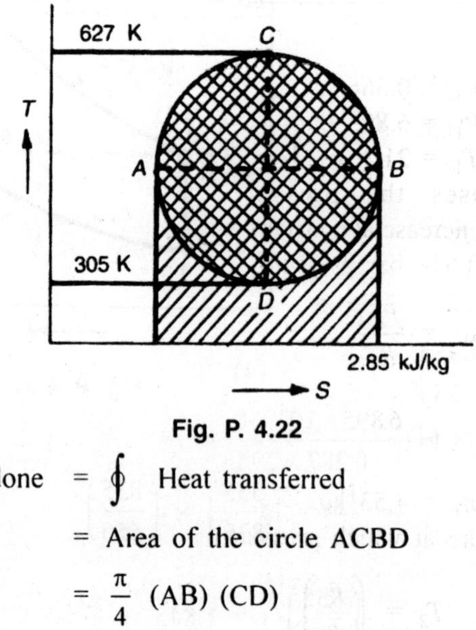

Fig. P. 4.22

$$\oint \text{Work done} = \oint \text{Heat transferred}$$

$= $ Area of the circle ACBD

$$= \frac{\pi}{4}\,(AB)\,(CD)$$

$$= \frac{\pi}{4}\,(2.85-1.23)\,\frac{kJ}{K}\,(627-305)\,(K)$$

$= 409.86$ kJ

$$\eta \text{ of the cycle} = \frac{\text{Work done}}{\text{Heat supplied}}$$

Work done/cycle = Area of the circle ACBD = 409.86 kJ

$$\text{Heat supplied} = \frac{409.86}{2} + (2.85-1.23)\times\left(\frac{627+305}{2}\right)$$

$$= 204.93 + 754.92 = 959.85 \text{ kJ}$$

$$\eta = \frac{409.86}{959.85} = 42.70\%$$

Problem 4.23 : A rigid tank of 0.566 m³ volume contains air at 6.895 bar and 21.1°C. The tank is equipped with a relief valve that opens at a pressure of 8.618 bar and remains open until the pressure drops to 8.274 bar. If a fire causes the valve to operate once as described, determine the air temperature just before the valve opens and the mass of air lost due to the fire. Assume that the temperature of the air remains constant during discharge and air in the tank behaves as an ideal gas. (CES – 1990)

Solution :

$$V_1 = 0.566 \text{ m}^3$$
$$P_1 = 6.895 \text{ bar}$$
$$T_1 = 21.1 + 273 = 294.1 \text{ K}$$

Fire causes the pressure and temperature to increase at constant volume till it reaches 8.618 bar pressure.

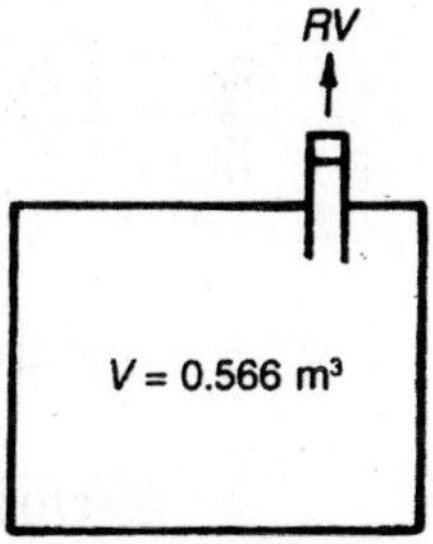

Fig. P. 4.23

$$\text{Mass of air} = \frac{P_1V_1}{RT_1}$$

$$= \frac{6.895 \times 10^4 \times 0.566}{0.287 \times 294.1} \times \frac{4.18}{427}$$

$$m_1 = 4.53 \text{ kg.}$$

Temperature at which pressure is 8.618 bar

$$T_2 = \left(\frac{P_2}{P_1}\right) T_1 = \left(\frac{8.628}{6.895}\right) 294.1 = 367.4 \text{ K}$$

Air temperature just before the valve open = 367.6 K

Mass of air at the time of opening valve = m_1 = 4.53 kg

Mass of air at the time of closing of valve at pressure 8.274 bar and temperature 367.6 K = m_2

$$= \frac{8.274 \times 10^4 \times 0.566}{0.287 \times 367.6} \times \frac{4.18}{427} = 4.345 \text{ kg}$$

Mass of air escaped = 4.53 – 4.345 = 0.185 kg.

Problem 4.24 : 0.3 kg of air at 1.5 bar pressure and 87°C temperature at condition 1 is compressed polytropically to condition 2 at pressure 7.5 bar, index of compression being 1.2. It is then cooled at constant pressure to condition 3 and then finally heated at constant temperature to its original condition 1. Find the net work done and heat transferred.

$R = 0.287$ kJ/kg.K
$C_p = 1.005$ kJ/kg.K
$\gamma = 1.4$ (CES – 1992)

Solution :

$P_1 = 1.5$ bar
$P_2 = 7.5$ bar
$T_1 = 273 + 87$
$= 360$ K

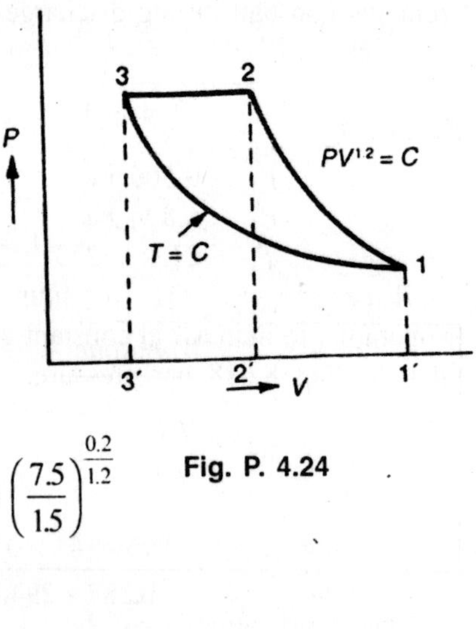

Fig. P. 4.24

Process 1-2
$PV^{1.2} = C$

$$\frac{T_2}{T_1} = \left(\frac{P_2}{P_1}\right)^{\frac{n-1}{n}}$$

$$T_2 = T_1 \left(\frac{P_2}{P_1}\right)^{\frac{n-1}{n}} = 360 \left(\frac{7.5}{1.5}\right)^{\frac{0.2}{1.2}}$$

$= 470.8$ K

Process 2-3

$T_3 = T_1$
$= 273 + 87 = 360$ K

$P_2 = P_3 = 7.5$ bar

Process 1-2

Work done $= W_{1-2}$

$$= \frac{R}{n-1}(T_1 - T_2)$$

$$= \frac{0.287}{(1.2-1)}(360 - 470.8) = -159 \text{ kJ/kg}$$

Heat transfer

$$Q_{1-2} = C_v = \left(\frac{\gamma - n}{1-n}\right)(T_2 - T_1)$$

$$= 0.717 \frac{1.4 - 1.2}{1 - 1.2} (470.8 - 360) = -79.44 \text{ kJ/kg}$$

Process 2-3

Workdone W_{2-3}

$$= R(T_3 - T_2) = 0.287 (360 - 470.8)$$
$$= -31.8 \text{ kJ/kg}$$

Heat transferred

$$Q_{2-3} = C_p (T_3 - T_2) = 1.005 (360 - 470.8)$$
$$= -111.35 \text{ kJ/kg}$$

Process 3-1

$$\text{Workdone } W_{3-1} = RT_3 \ln \left(\frac{P_2}{P_1}\right) = 0.287 \times 360 \ln \left(\frac{7.5}{1.5}\right)$$
$$= 166.3 \text{ kJ/kg}$$

Heat transferred

$$Q_{3-1} = W_{3-1} \text{ as } \Delta U = 0 \text{ (Isothermal process)}$$
$$Q_{3-1} = 166.3 \text{ kJ/kg}$$

Table P4.24

Process	*Workdone kJ/kg*	*Heat transferred kJ/kg*
1–2	–159	–79.44
2–3	–31.8	–111.35
3–1	+166.3	+166.3
Total	–24.5	–24.49

$$\text{Net work done} = \text{Net heat transferred} = \int dQ = \int dW$$
$$= -24.5 \times 0.3 \qquad (\because 0.3 \text{ kg} = \text{mass of air})$$
$$= -7.35 \text{ kJ}$$

Problem 4.25 : A gas is enclosed in a cylinder by a frictionless piston at a pressure of 300 kPa. A spring is connected to the piston that resists its outward movement from the cylinder. The atmospheric air on the spring side of the piston is at 1 bar. If the gas expands from 0.1 to 0.5 m^3, what is work done by the gas?

Solution : The pressure exerted by the spring,

$$P_s = \text{Gas pressure} - \text{Atmospheric pressure}$$
$$= 300 - 100 = 200 \text{ kPa}$$

Thus spring force is given by

$F_s = P_s \times A$,

where A = piston area in m^2.

This force gives spring displacement (compression) of

$$x_1 = \frac{\text{Volume}}{\text{Area}} = \frac{0.1}{A}$$

And thus spring constant is given by

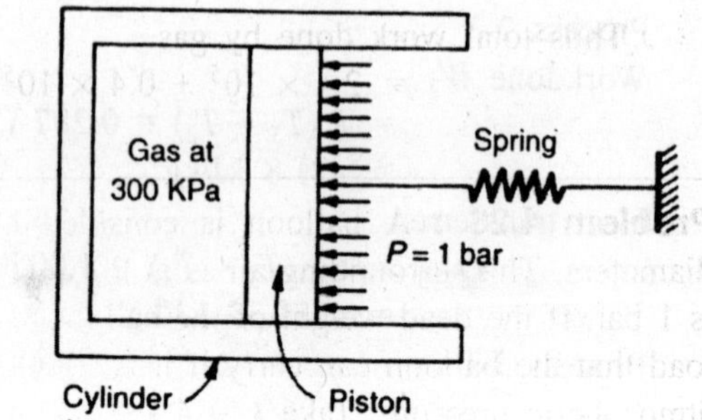

Fig. P. 4.25

$$K_s = \frac{200 \times 10^3 \times A}{0.1 / A} = 2 \times 10^6 \times A^2 \text{ N/m.}$$

The gas, when it expands has to do work against the atmosphere and also against the spring.

The workdone against atmospheric pressure

$$W_{(\text{against atmosphere})} = 1 \times 10^5 \times (0.5 - 0.1)$$
$$= 0.4 \times 10^5 \text{ N-m.}$$

The work done against spring

$$W_{(\text{against spring})} = \int_{x_1}^{x_2} 2 \times 10^6 \times A^2 \times x \, dx$$

$$= 2 \times 10^6 \times A^2 \left[\frac{x^2}{2}\right]_{x_1}^{x_2}$$

where $x_1 = \dfrac{0.1}{A}$

and $x_2 = \dfrac{0.5}{A}$

Thus $W_{(\text{against spring})} = 2 \times 10^6 \times A^2 \times \dfrac{1}{2}\left[\dfrac{(0.5)^2}{A^2} - \dfrac{(0.1)^2}{A^2}\right]$

$$= \frac{2 \times 10^6 \times 0.24}{2} = 2 \times 10^5 \text{ N-m}$$

∴ Thus total work done by gas

$$= 2.4 \times 10^5 + 0.4 \times 10^5 = 280 \text{ N-m}$$

Problem 4.26 : A balloon is considered a sphere of 15 m in diameters. The surrounding air is at 20°C and atmospheric pressure is 1 bar. If the dead weight of the balloon is 4870 N, determine the load that the balloon can carry if it is filled with H_2 at 20°C and at atmospheric pressure. Take k = 4.157 kJ/kg.K for H_2.

Solution :

$$\text{Volume of balloon} = 1767.15 \text{ m}^3$$

$$\text{Density of air} = \frac{1.02 \times 10^5}{293 \times 287} \qquad \left(\because \rho = \frac{p}{RT}\right)$$

$$= 1.213 \text{ kg/m}^3$$

$$\text{Density of } H_2 = \frac{1.02 \times 10^5}{293 \times 4157} = 0.0837 \text{ kg/m}^3$$

$$\therefore \text{ Thrust on balloon} = 1767.15 \times (1.213 - 0.0837) \times 9.807$$

$$= 19570 \text{ N}$$

Less weight of balloon

$$= 4870 \text{ N}$$

∴ Net weight that can be carried

= 14700 N **Ans**

Chapter 5

First Law Applied to Flow Processes

Problem 5. 1 : A blower handle 1 kg/s of air at 20°C and consumes a power of 15 kW. The inlet and outlet welocities of air are 100 m/s and 150 m/s respectively. Find the exit air temperature, assuming adiabatic conditions. Take c_p of air is 1.005 kJ/kg-K.

Solution : Given data :

$\dot{m} = \dot{m}_1$

$= \dot{m}_2 = 1$ kg/s

$T_1 = 20°C$

$= 293$ K

$W = 15$ kW

$V_1 = 100$ m/s

$V_2 = 150$ m/s

$T_2 = ?$

$Q = 0$

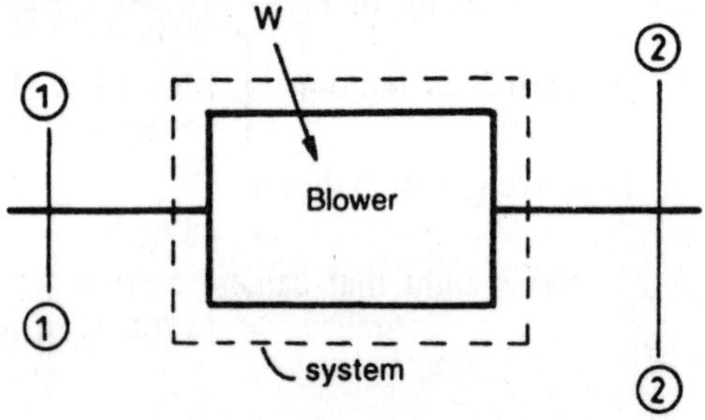

Fig. P. 5.1

Applying S.F.E.E. between 1 and 2 neglecting ΔP

$$\frac{dQ}{d\tau} + \dot{m}\left[h_1 + \frac{V_1^2}{2}\right] = \frac{dw}{d\tau} + \dot{m}\left[h_2 + \frac{V_2^2}{2}\right]$$

$$\dot{m}\ [(h_1 - h_2) + \frac{V_1^2 - V_2^2}{2}\ ] = \frac{dw}{d\tau}$$

$$1\left[C_P(T_1 - T_2) + \frac{V_1^2 - V_2^2}{2}\right] = \frac{dw}{d\tau}$$

$$C_P\,(20 - T_2) + \frac{100^2 - 150^2}{2000} = -15$$

$$\Rightarrow \qquad T_2 = 28.7°C \textbf{ Ans}$$

Problem 5.2 : A turbine operates under steady flow conditions, receiving steam at the following state: Pressure 1.2 MPa, temperature 188°C, enthalpy 2785 kJ/kg, velocity 33.3 m/s and elevation 3 m. The steam leaves the turbine at the follwing state; Pressure 20 kPa, enthalpy 2512 kJ/kg, velocity 100 m/s, and elevation 0 m. Heat is lost to the surroundings at the rate of 0.29 kJ/s. If the rate of steam flow through the turbine is 0.42 kg/s, what is the power output of the turbine in kW?

***Solution* :** Given data :

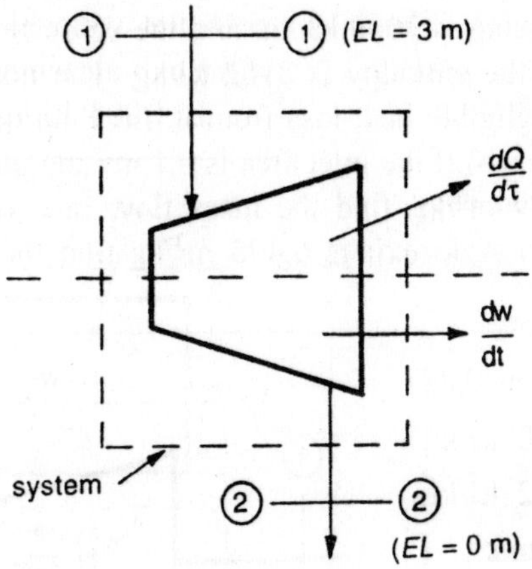

Fig. P. 5.2

$p_1 = 1.2$ MPa

$T_1 = 188°C = 461$ K

$h_1 = 2785$ kJ/kg

$V_1 = 33.3$ m/s

$Z_1 = 3$ m

$p_2 = 20$ kPa

$h_2 = 2512$ kJ/kg

$V_2 = 100$ m/s

$Z_2 = 0$

$$\frac{dQ}{d\tau} = 0.29 \text{ kJ/s}$$

$$\dot{m}_1 = \dot{m}_2 = \dot{m} = 0.42 \text{ kg/s}$$

$$\frac{dw}{d\tau} = ?$$

Applying S.F.E.E. between (1) and (2)

$$\frac{dQ}{dt} + \dot{m}\left[h_1 + \frac{V_1^2}{2} + Z_1 g\right] = \dot{m}\left[h_2 + \frac{V_2^2}{2} + Z_2 g\right] + \frac{dw}{dt}$$

$$\therefore \frac{dw}{dt} = \frac{dQ}{dt} + \dot{m}\left[(h_1 - h_2) + \frac{V_1^2 - V_2^2}{2} + (Z_1 - Z_2)g\right]$$

$$= -0.29 + 0.42\left[(2785 - 2512) + \frac{33.3^2 - 100^2}{2000} + \frac{3 \times 9.8}{1000}\right]$$

$$= 112.51 \text{ kW}$$

Problem 5.3 : A nozzle is a device for increasing the velocity of a steadily flowing stream. At the inlet to a certain nozzle, the enthalpy of the fluid passing is 3000 kJ/kg. and the velocity is 60 m/s. At the discharge end, the enthalpy is 2762 kJ/kg. The nozzle is horizontal and there is negligible heat loss from it. (a) Find the velocity at exit from the nozzle. (b) If the inlet area is 0.1 m² and the specific volume at inlet is 0.187 m³/kg find the mass flow rate. (c) If the specific volume at the nozzle exit is 0.498 m³/kg find the exit area of the nozzle.

Solution : Given data

$h_1 = 3000$ kJ/kg

$h_2 = 2762$ kJ/kg

$V_1 = 60$ m/s

$\frac{dQ}{d\tau} = 0$ and

$\frac{dw}{d\tau} = 0, \; Z_1 = Z_2$

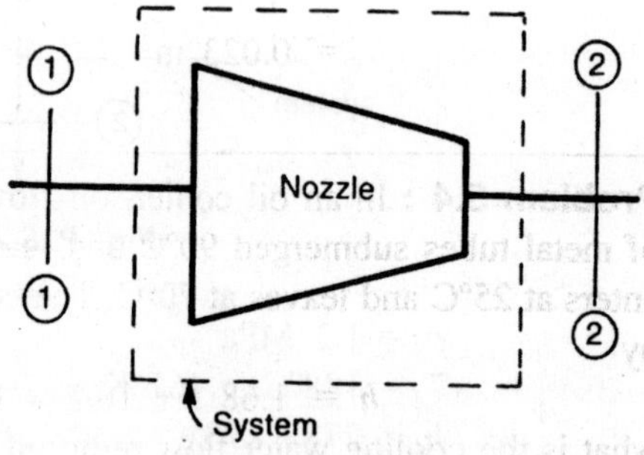

Fig. P. 5.3

Hence applying S.F.E.E. between (i) and (ii)

$$\dot{m}_1\left[h_1 + \frac{V_1^2}{2} + Z_1 g\right] + \frac{dQ}{d\tau} = \dot{m}_2\left[h_2 + \frac{V_2^2}{2} + Z_2 g\right] + \frac{dw}{d\tau}$$

$$\Rightarrow \qquad h_1 + \frac{V_1^2}{2} = h_2 + \frac{V_2^2}{2} \qquad [\because \dot{m}_1 = \dot{m}_2]$$

(a)
$$V_2 = \sqrt{\left[(h_1 - h_2) + \frac{V_1^2}{2}\right] \times 2}$$

$$= \sqrt{\left[(3000 - 2762) + \frac{60^2}{2000}\right] \times 2 \times 1000}$$

$$= 692.5 \text{ m/s}$$

(b) a_1 = inlet area = 0.1 m^2

Spectific volume $v_1 = 0.187$ m^3/kg

$\therefore$ mass flow rate

$$= \rho a_1 V_1 = \frac{a_1 V_1}{v_1} = \frac{0.1 \times 60}{0.187} = 32.08 \text{ kg/s}$$

(c) $v_2 = 0.498$ m^3/kg

$$\dot{m} = \frac{a_2 V_2}{v_2}$$

$$\Rightarrow \qquad a_2 = \frac{\dot{m} v_2}{V_2} = \frac{32.08 \times 0.498}{692.5}$$

$$= 0.023 \text{ m}^2$$

Problem 5.4 : In an oil cooler, oil flows steadily through a bundle of metal tubes submerged 90°C and leaves at 30°C, while the water enters at 25°C and leaves at 70°C. The enthalpy of oil at t°C is given by

$$h = 1.68\, t + 10.5 \times 10^{-4}\, t^2 \text{ kJ/kg}$$

what is the cooling water flow required for cooling 2.78 kg/s of oil?

Solution :

Given data :

T_1 = 90°C

T_2 = 30°C

T_3 = 25°C

$$T_4 = 70°C$$
$$h_{oil} = 1.68\ t + 10.5 \times 10^{-4}\ t^2 \text{ kJ/kg}$$
$$\therefore \quad h_1 = 1.68 \times 90 + 10.5 \times 10^{-4} \times 90^2$$
$$= 159.\ 71 \text{ kJ/kg}$$
$$h_2 = 1.68 \times 30 + 10.5 \times 10^{-4} \times 30^2$$
$$= 51.35 \text{ kJ/kg}$$
$$h_3 = C_{pw}\ T_3$$
$$h_4 = C_{pw}\ T_4$$
where $C_{pw} = 4.187$ kJ/kg.K

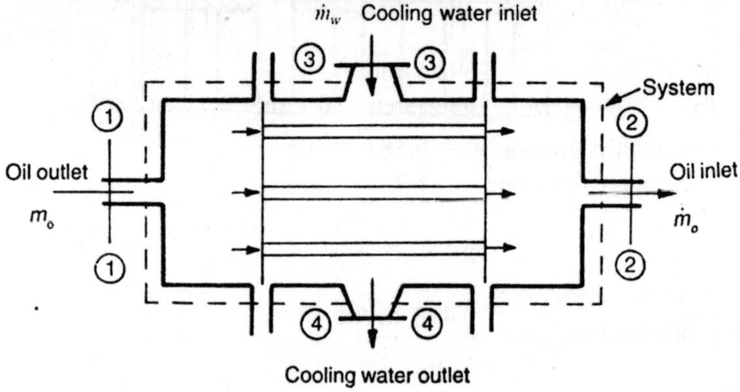

Fig. P. 5.4

Now applying S.F.E.E. between (1), (2), (3) & (4) neglecting, ΔK.E, ΔP.E

and $\frac{dQ}{d\tau} = 0$ & $\frac{dw}{d\tau} = 0$

$$\dot{m}_0\ (h_1) + \dot{m}_w (h_3) = \dot{m}_0\ (h_2) + \dot{m}_w (h_4)$$

or, $\dot{m}_0\ (h_1 - h_2) = \dot{m}_w (h_4 - h_3)$

or, $$\dot{m}_w = \frac{\dot{m}_0(h_1 - h_2)}{C_p w(T_4 - T_3)} = \frac{2.78(159.71 - 51.35)}{4.187 \times (70 - 25)}$$

$$= 1.598 \text{ kg/s} = 1.6 \text{ kg/s}$$ **Ans**

Problem 5.5 : A thermoelectric generator consists of a series of semiconductor elements (Fig.p.5.5), heated on one side and cooled on the other. Electric current flow is produced as a result of energy transfer as heat. In a particular experiment t.. e current was measured to be 0.5 amp and the electrostatic potential at (1) was 0.8 volt above that at (2). Energy transfer as heat to the hot side of the generator was taking place at a rate of 5.5 watts. Determine the rate of energy transfer as heat from the cold side and the energy conversion efficiency.

Solution : Given data :

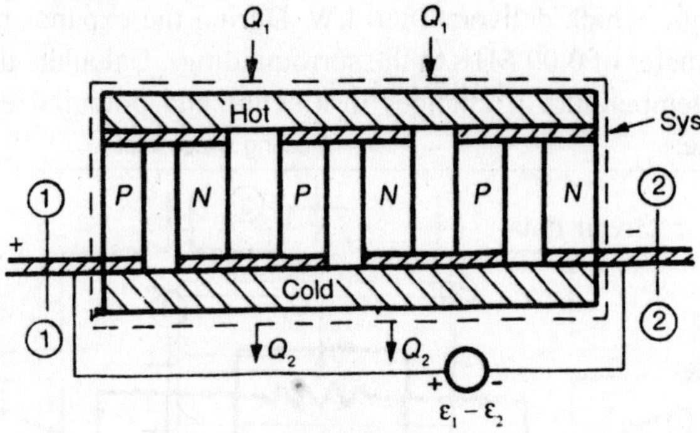

Fig. P. 5.5

$$\text{current} = I = 0.5 \text{ amp}$$

electrostatic potenial difference

$$= V_1 - V_2 = 0.8 \text{ V}$$

$$Q_1 = 5.5 \text{ W}$$

$$Q_2 = ?$$

$\dfrac{dw}{d\tau}$ = rate of work transfer due to flow of electric current

$$= (V_1 - V_2)\,\text{I} = 0.5 \times 0.8 = 0.4 \text{ W}$$

here $\dot{m}_1 = \dot{m}_2 = 0$

∴ Applying S.F.E.E. for the system

$$\frac{dQ}{d\tau} = \frac{dw}{dt}$$

$$\Rightarrow \quad \frac{dQ_1}{d\tau} - \frac{dQ_2}{d\tau} = \frac{dw}{dt}$$

$$\Rightarrow \quad 5.5 - \frac{dQ_2}{d\tau} = 0.4$$

$$\therefore \quad \frac{dQ_2}{d\tau} = 5.5 - 0.4 = 5.1 \text{ W } \textbf{Ans}$$

$$\text{Energy Conversion efficiency} = \frac{\text{work transfer rate}}{\text{energy transfer rate}}$$

$$= \frac{0.4}{5.5} = 0.073 \textbf{ Ans}$$

Problem 5.6 : A turbocompressor delivers 2.33 m³/s at 0.276 MPa, 43°C which is heated at this pressure is 430°C and finally expanded in a turbine which delivers 1860 kW. During the expansion, there is a heat transfer of 0.09 MJ/s to the surrouundings. Calculate the turbine exhaust temperature if changes in kinetic and potential energy are negligible.

Solution : Given data :

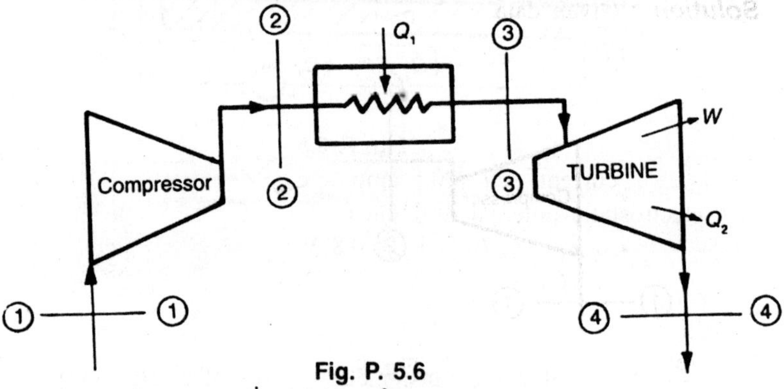

Fig. P. 5.6

Volume flow rate $\dot{V} = 2.33 \text{ m}^3/\text{sec.}$

$$P_2 = 0.276 \text{ MPa} = p_3$$
$$T_2 = 43°\text{C} = 316 \text{ K}$$
$$T_3 = 430°\text{C} = 703 \text{ K}$$
$$W = 1860 \text{ kW}$$
$$Q_2 = 0.09 \text{ MW}$$

Applying ideal gas equation at (2)

$$p_2 \dot{V}_2 = \dot{m} RT_2$$

$$\therefore \quad \dot{m} = \frac{p_2 \dot{V}_2}{RT_2} = \frac{0.276 \times 10^3 \times 2.33}{0.287 \times 316}$$

$$= 7.09 \text{ kg/s}$$

Now Applying S.F.E.E fatubine between (3) & (4)

$$\Rightarrow \quad \dot{m}(h_1) + \frac{dQ}{d\tau} = \dot{m}(h_2) + \frac{dw}{d\tau}$$

$$\Rightarrow \qquad \dot{m}\, C_p\, (T_2 - T_1) = \frac{dQ}{dt} - \frac{dw}{d\tau} = -90 - 1860 = -1950$$

$$\therefore \qquad T_2 = T_1 - \frac{1950}{7.09 \times 1.005} = 156.33°\text{C} \ \textbf{Ans}$$

Problem 5.7 : A reciproacating air compressor takes in 2 m³/min at 0.11 MPa, 20°C which it delivers at 1.5 MPa, 111°C to an after cooler where the air is cooled at constant pressure to 25°C. The power absorbed by the compressor is 4.15 kW. Determine the heat transfer in (a) the compressor, and (b) the cooler, state your assumptions.

Solution : Given data

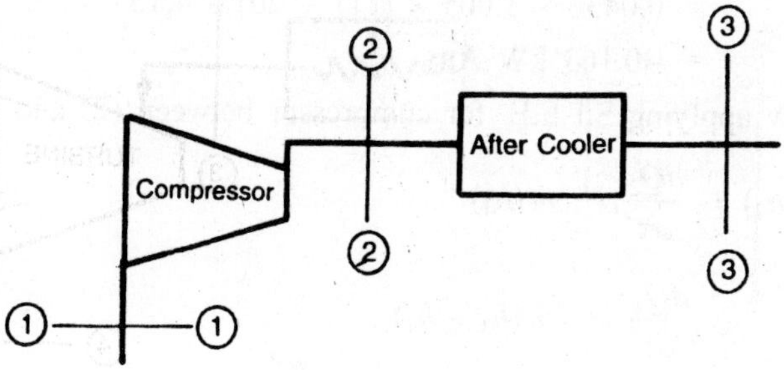

Fig. P. 5.7

$\dot{V}$ = Volume flow rate = 2 m³/min
p_1 = 0.11 MPa
T_1 = 20°C = 293 K
P_2 = 1.5 MPa
T_2 = 111°C = 384 K
T_3 = 25°C = 298 K
W = 4.15 kW

Assumption :

(i) Neglecting ΔK.E. and ΔP.E

(ii) Assume flow is steady.

(iii) There is no work transfer in after cooler.

Applying ideal gas equation at (1)

$$P_1 \dot{V} = \dot{m} R T_1$$

$$\therefore \quad \dot{m} = \frac{P_1 \dot{V}}{RT_1} = \frac{0.11 \times 10^3 \times 2}{0.287 \times 293 \times 60}$$

$$= 0.0436 \text{ kg/s}$$

Now applying S.F.E.E. for compressor between (1) and (2)

$$\dot{m}(h_1) + \frac{dQ_1}{d\tau} = \dot{m}(h_2) + \frac{dw}{d\tau}$$

$$\therefore \quad \frac{dQ_1}{d\tau} = \dot{m}(h_1 - h_2) + \frac{dw}{d\tau}$$

$$= m C_{P_a} (T_2 - T_1) + \frac{dw}{d\tau}$$

$$= 0.0436 \times 1.005 \times (111 - 20) - 4.15$$

$= -0.163$ kW **Ans**

Now applying S.F.E.E. for compressor between (2) and (3)

$$\dot{m}(h_2) + \frac{dQ_2}{d\tau} = \dot{m}(h_3)$$

or
$$\frac{dQ_2}{d\tau} = \dot{m}(h_3 - h_4)$$

$$= \dot{m} C_{P_a} (T_3 - T_4)$$

$$= 0.0436 \times 1\ 1.005\ (25 - 111)$$

$= -3.76$ kW **Ans**

Problem 5.8 : In a water cooling tower air enters at a height of 1 m above the ground level and leaves at a height of 7 m. The inlet and outlet vellocities are 20 m/s and 30 m/s respectively. Water enters at a height of 8 m and leaves at a height of 8 m and leavea at a height of 0.8 m. The velocity of water at entry and exit are 3 m/s and 1 m/s respectively. Air temperatures are 80°C and 30°C at the entry ad exit respectively. Air temperatures are 30°C and 70°C at the entry and exit respectively. The cooling tower is well insulated and a fan of 2.25 kW drives the air through the cooler. Find the amount of air per second required for 1 kg/s of water flow. The values C_p of air and water are 1.005 and 4.187 kJ/kg.K respectively.

Solution : Given data :

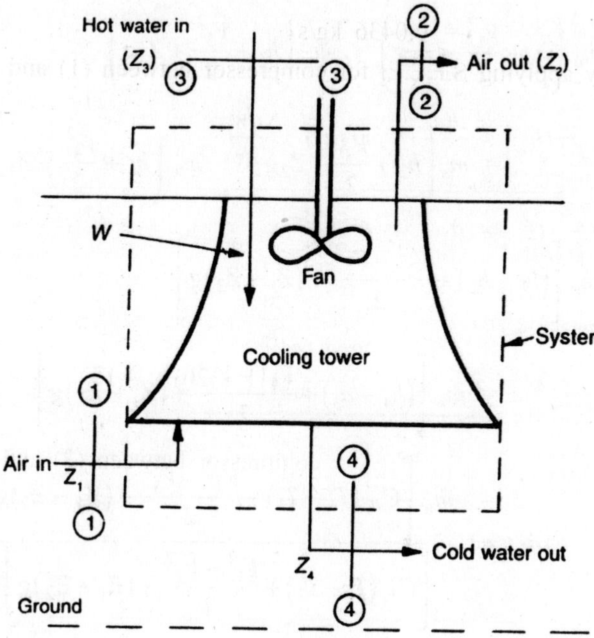

Fig. P. 5.8

$Z_1 = 1$ m
$Z_2 = 7$ m
$V_1 = 20$ m/s
$V_2 = 30$ m/s
$Z_3 = 8$ m
$Z_4 = 0.8$ m
$V_3 = 3$ m/s
$V_4 = 1$ m/s
$T_3 = 80$°C
$T_4 = 50$°C
$T_1 = 30$°C
$T_2 = 70$°C
w = 2.2 5 kW

let $\dot{m}_a$ and $\dot{m}_w$ are mass flow of air and water

$$\dot{m}_w = 1 \text{ kg/s}$$

Applying S.F.E.E for system neglecting $\frac{dQ}{d\tau} = 0$

$$\Rightarrow \dot{m}_a \left(h_1 + \frac{V_1^2}{2} + Z_1 g \right) + \dot{m}_w \left(h_3 + \frac{V_3^2}{2} + Z_3 g \right)$$

$$= \dot{m}_a \left(h_2 + \frac{V_2^2}{2} + Z_2 g \right) + \dot{m}_w \left(h_4 + \frac{V_4^2}{2} + Z_4 g \right)$$

$$\Rightarrow \dot{m}_a \left[(h_1 - h_2) + \frac{V_1^2 - V_2^2}{2} + (Z_1 - Z_2) g \right]$$

$$= \dot{m}_w \left[(h_4 - h_3) + \frac{V_4^2 - V_3^2}{2} + (Z_4 - Z_3) g \right]$$

or

$$\dot{m}_a = \frac{\dot{m}_w \left[C_{P_w} (T_4 - T_3) + \frac{V_4^2 - V_3^2}{2} + (Z_4 - Z_3) g \right]}{\left[C_{P_a} (T_1 - T_2) + \frac{V_1^2 - V_2^2}{2} + (Z_1 - Z_2) g \right]}$$

$$= \frac{1 \left[4.187(50 - 80) + \frac{1^2 - 4^2}{2000} + \frac{(0.8 - 8) \times 9.81}{1000} \right]}{\left[1.005(30 - 70) + \frac{20^2 - 30^2}{2000} + \frac{(1 - 7) \times 9.81}{1000} \right]}$$

$= 3.10$ kg/s **Ans**

Problem 5.9 : Air at 101.325 kPa, 20° C is taken into a gas turbine power plant at a velocity of 140 m/s through opening of 0.15 m^2 cross-sectional area. The air is compressed heated, expanded through a turbine, and exhausted area at 0.18 MPa , 150°C through an opening of 0.10 m^2 cross-sectional. The power output is 375 kW. Calculate the net amount of heat added to the air in kJ/kg. Assume that air obeys the law $pv = 0.287$ (t + 273), where p is the pressure in KPa, v is the specific volume in m^3/kg, and t is the temperature in °C. Take C_p = 1.005 kJ/kg.K.

Solution : Given data :

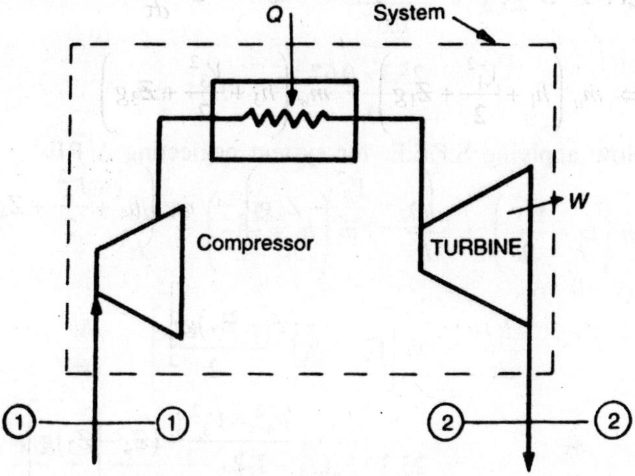

Fig. P. 5.9

$$P_1 = 101.325 \text{ kPa}$$
$$T_1 = 20°\text{C} = 293 \text{ K}$$
$$V_1 = 140 \text{ m/s}$$
$$a_1 = \text{opening cross section area at inlet} = 0.15 \text{ m}^2$$
$$P_2 = 0.18 \text{ MPa}$$
$$T_2 = 150°\text{C} = 423 \text{ K}$$
$$a_2 = \text{opening c/s at outlet} = 0.1 \text{ m}^2$$
$$W = 375 \text{ kW.}$$

At inlet condition

$$P_1 v_1 = 0.287\ (T_1 + 273)$$

$$\therefore \quad v_1 = \frac{0.287 \times 293}{101.325} = 0.83 \text{ m}^3/\text{kg}$$

$$\text{Now mass flow rate} = \dot{m} = \frac{a_1 V_1}{v_1} = \frac{0.15 \times 140}{.083} = 25.3$$

For outlet condition

$$v_2 = \frac{0.287\,(150 + 273)}{0.18 \times 10^3} = 0.67 \text{ m}^3/\text{kg}$$

$$\therefore \quad \text{mass flow rate} = \frac{a_2 V_2}{v_2}$$

or $$25.3 = \frac{0.1 \times V_2}{0.67}$$

$$\therefore \quad V_2 = \frac{25.3 \times 0.67}{0.1} = 170.64 \text{ m/s}$$

Now applying S.F.E.E. for system neglecting Δ P.E

$$\dot{m}\left(h_1 + \frac{V_1^2}{2}\right) + \frac{dQ}{d\tau} = \dot{m}\left(h_2 + \frac{V_2^2}{2}\right) + \frac{dw}{d\tau}$$

$$\frac{dQ}{d\tau} = \dot{m}\left[(h_2 - h_1) + \frac{v_2^2 - v_1^2}{2}\right] + \frac{dw}{d\tau}$$

$$= 25.3\left[c_p(T_2 - T_1) + \frac{v_2^2 - v_1^2}{2}\right] + \frac{dw}{d\tau}$$

$$= 25.3\left[1.005(150 - 20) + \frac{170.64^2 - 140^2}{2 \times 1000}\right] + 375$$

$= 3801.02$ kW **Ans**

Problem 5.10 : A gas flow steadily through a rotary compressor. The gas enters the compressor at a temperagture of 16°C, a pressure of 100 kPa, and an enthalpy of 391.2 kJ/kg. The gas leaves the compressor at a temperature of 245° C, a pressure of 0.6 MPa, and an enthalpy of 534 kJ/kg. There is no heat transfer to or from the gas as it flows through the compressor. (a) Evaluate the external work done per unit mass of gas assuming the gas velocities at entry and exit to be negligible. (b) Evaluate the external work done per unit mass of gas when the gas velocity at entry is 80 m/s and that at exit is 160 m/s.

Solution : Given

$T_1 = 16°C$
$p_1 = 100$ kPa
$h_1 = 391.2$ kJ/kg
$T_2 = 245°C$
$p_2 = 0.6$ MPa

$$h_2 = 534.5 \text{ kJ/kg}$$

$$\frac{dQ}{d\tau} = 0$$

(a) Neglecting ΔP.E. and ΔK.E.

Applying S.F.E.E. for system

$$\dot{m}(h_1) = \dot{m}(h_2) + \frac{dw}{d\tau}$$

$$\Rightarrow \quad \frac{dw}{d\tau} = \dot{m}(h_1 - h_2)$$

$$\text{or,} \quad \frac{dw}{dm} = h_1 - h_2$$

$$= 391.2 - 534.5$$

$$= -143.3 \text{ kJ/kg } \textbf{Ans}$$

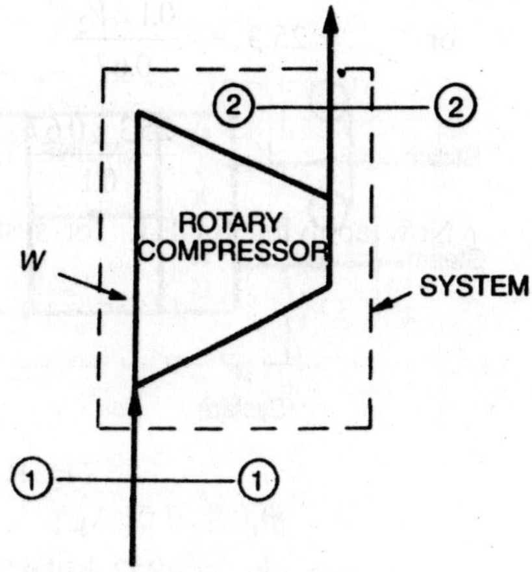

Fig. P. 5.10

(b) Neglecting only ΔP.E., $V_1 = 80$ m/s, $V_2 = 160$ m/s

Applying S.F.E.E. for system

$$\dot{m}\left(h_1 + \frac{V_1^2}{2}\right) = \dot{m}\left(h_2 + \frac{V_2^2}{2}\right) + \frac{dw}{d\tau}$$

$$\therefore \quad \frac{dw}{dm} = (h_1 - h_2) + \frac{V_1^2 - V_2^2}{2}$$

$$= (391.2 - 534.5) + \frac{80^2 - 160^2}{2 \times 1000}$$

$$= -152.9 \text{ kJ/kg } \textbf{Ans}$$

Problem 5.11 : The steam supply to an engine comprises two streams which mix before entering the engine. One stream is supplied at the rate of 0.01 kg/s with an enthalpy of 2952 kJ/kg and a velocity of 20 m/s. The other stream is supplied at the rate of 0.1 kg/s with an enthalpy of 2569 kJ/kg and a velocity of 120 m/s. At the exit from the engine the fluid leaves as two streams, one of water at the rate of 0.001 kg/s with an enthalpy of 420 kJ/kg and the other of steam; the fluid velocities at the exit are negligible. The engine develops a shaft power of 25 kW. The heat transfer is negligible. Evaluate the enthalpy of the second exit stream.

Solution : Given data :

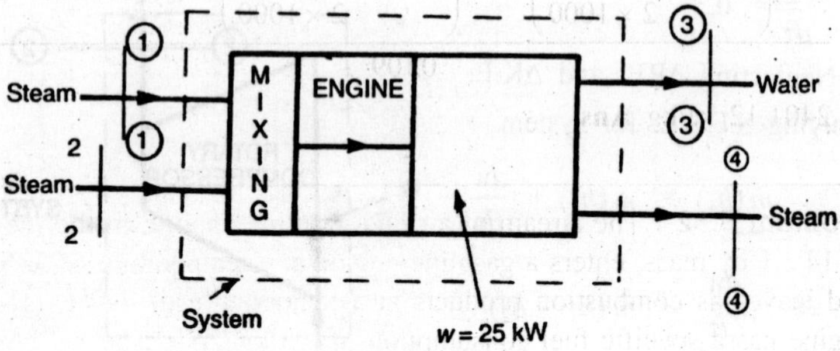

Fig. P. 5.11

$$\dot{m}_1 = 0.01 \text{ kg/s}$$
$$h_1 = 2952 \text{ kJ/kg}$$
$$V_1 = 20 \text{ m/s}$$
$$\dot{m}_2 = 0.1 \text{ kg/s}$$
$$h_2 = 2569 \text{ kJ/kg}$$
$$V_2 = 120 \text{ m/s}$$
$$\dot{m}_3 = 0.001 \text{ kg/s}$$
$$h_3 = 420 \text{ kJ/kg}$$
$$\frac{dQ}{d\tau} = 0$$

By mass balance

$$\dot{m}_1 + \dot{m}_2 = \dot{m}_3 + \dot{m}_4$$

$$\therefore \quad \dot{m}_4 = 0.01 + 0.1 - 0.001$$
$$= 0.11 - 0.001 = 0.109 \text{ kg/s}$$

Now applying S.F.E.E. for the system considered, neglecting change in potential energy

$$\dot{m}\left[h_1 + \frac{V_1^2}{2}\right] + \dot{m}_2\left[h_2 + \frac{V_2^2}{2}\right]$$
$$= m_3 h_3 + m_4 h_4 + \frac{dw}{d\tau}$$

$$\therefore h_4 = \frac{\dot{m}_1\left(h_1 + \frac{V_1^2}{2}\right) + \dot{m}_2\left(\frac{V^2}{2} + h_2\right) - m_3 h_3 - \frac{dw}{d\tau}}{m_4}$$

$$= \frac{0.01\left(2952 + \frac{20^2}{2 \times 1000}\right) + 0.1\left(2569 + \frac{120^2}{2 \times 1000}\right) - 0.001 \times 400 - 25}{0.109}$$

$= 2401.12$ kJ/kg **Ans**

> **Problem 5.12 :** The stream of air and gasoline vapour, in the ratio of 14 : 1 by mass, enters a gasoline engine at a temperature of 30°C and leaves as combustion products at a temperature of 790°C. The engine has a specific fuel consumption of .03 kg/kWh. The net heat transfer rate from the fuel–air stream to the jacket cooling water and to the surroundings is 35 kW. THe shaft power delivered by the engine is 26 kW. COmpute the increase in the specific enthalpy of the fuel–air stream, assuming the changes in kinetic energy and in elevation to be negligible.

Solution : Given data

$$\frac{\dot{m}_a}{\dot{m}_g} = 14$$

$$T_1 = T_2 = 30°C = 303 \text{ K}$$

$$T_3 = 790°C = 1063 \text{ K}$$

Sp. fuel consumption = 0.3 kg/kWh

$$Q = \text{Net heat transfer} = 35 \text{ kW}$$

$$W = \text{shaft power developed} = 26 \text{ kW}$$

Change in Kinetic energy and in elevation are neglected

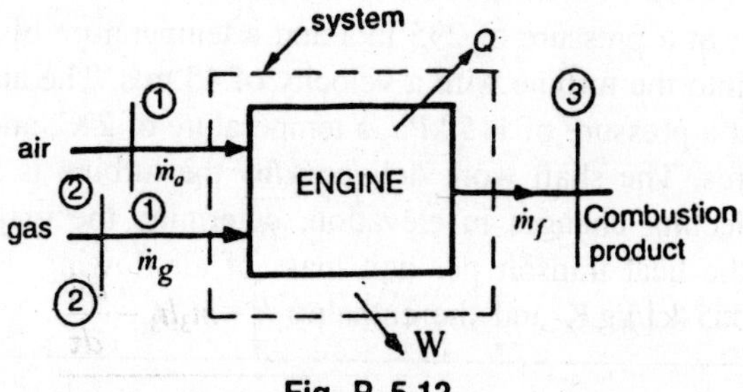

Fig. P. 5.12

Now $\dot{m}_g = 0.3 \text{ kg/kWh} \times \dfrac{26 \text{ kW}}{3600 \text{ sec}} \times h$

$= 2.1667 \times 10^{-3} \text{ kg/s}$

$\therefore \quad \dot{m}_a = 0.0303 \text{ kg/s}$

$\therefore \quad \dot{m}_f = \dot{m}_a + \dot{m}_g$

$= 2.1667 \times 10^{-3} + 0.0303$

$= 0.0325 \text{ kg/s}$

Now applying S.F.E.E. for the system considered

$$\dot{m}_a (h_1) + \dot{m}_g (h_2) + \frac{dQ}{d\tau} = \dot{m}_f (h_3) + \frac{dw}{d\tau}$$

$$\therefore \dot{m}_f h_3 - (\dot{m}_a h_1 + \dot{m}_g h_2)$$

$$= \frac{dQ}{d\tau} - \frac{dw}{d\tau} = -35 - 26 = -61 \text{ kW}$$

$$\text{or,} \left[h_3 - \left(\frac{\dot{m}_a}{\dot{m}_f} h_1 + \frac{\dot{m}_g}{\dot{m}_f} h_2 \right) \right]$$

$= \text{increase in sp. enthalpy}$

$= -\dfrac{61}{\dot{m}_f}$ kJ/kg of mixture

$= -\dfrac{61}{0.0325} = -1876.93$

$= -1877$ kJ/kg of mixture **Ans**

Problem 5.13 : An air turbine forms part of an aircraft refrigerating plant. Air at a pressure of 295 kPa and a temperature of 58°C flows steadily into the turbine with a velocity of 45 m/s. The air leaves the turbine at a pressure of 115 kPa, a temperature of 2°C, and a velocity of 150 m/s. The shaft work delivered by the turbine is 54 kJ/kg of air. Neglecting changes in elevation, determine the magnitude and sign of the heat transfer per unit mass of air flowing. For air, take $C_p = 1.005$ kJ/kg.K and the enthalpy $h = C_p t$.

Solution : Given data :

$p_1 = 295$ kPa

$T_1 = 58°C$

$V_1 = 54$ m/s

$p_2 = 115$ kPa

$T_2 = 2°C$

$V_2 = 150$ m/s

$W = 54$ kJ/kg of air

$\dot{m}_a$ = mass flow rate of air

Now applying S.F.E.E. for system considered neglecting the change in elevation

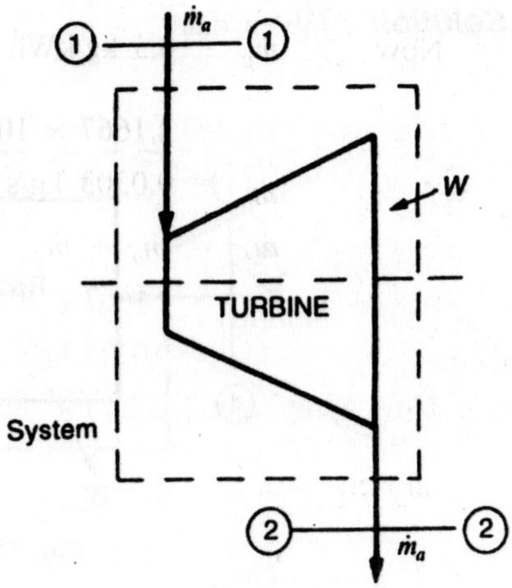

Fig. P. 5.13

$$\dot{m}_a\left(h_1+\frac{V_1^2}{2}\right)+\frac{dQ}{d\tau}=\dot{m}_a\left(h_2+\frac{V_2^2}{2}\right)+\frac{dw}{d\tau}$$

$$\frac{dQ}{d\tau}=\dot{m}_a\left[(h_2-h_1)+\frac{V_2^2-V_1^2}{2}\right]+\frac{dw}{d\tau}$$

or

$$\frac{dQ}{dm_a}=\left[(h_2-h_1)+\frac{V_2^2-V_1^2}{2}\right]+\frac{dw}{dm_a}$$

$$=\left[C_P(T_2-T_1)+\frac{V_2^2-V_1^2}{2}\right]+\frac{dw}{dm_a}$$

$$=\left[1.005(2-58)+\frac{150^2-45^2}{2000}\right]+54$$

$$=+\ 9.96 \text{ kJ/kg of air}$$

+ve sign indicates that heat flow to the system. **Ans**

Problem 5.14 : A room for four persons has two fans, each consuming 0.18 kW power, and three 100 W lamps. Ventilation air at the rate of 80 kg/h enters with an enthalpy of 84 kJ/kg and leaves with an enthalpy of 59 kJ/kg. If each person puts out heat at the rate of 630 kJ/h, determine the rate at which heat is to be removed by a room cooler, so that a steady state is maintained in the room.

Solution : Given data :

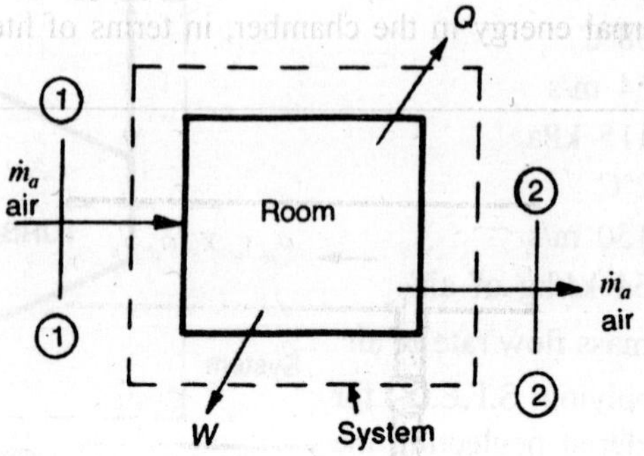

Fig. P. 5.14

$$\text{Power consumed by fan} = 2 \times 0.18 = 0.36 \text{ kW}$$

$$\text{Power consumed by lamps} = 3 \times 100 = 300 \text{ W} = 0.3 \text{ kW}$$

$$\dot{m}_a = \text{mass flow rate of air} = 80 \text{ kg/h}$$

$$h_1 = 84 \text{ kJ/kg}$$

$$h_2 = 59 \text{ kJ/kg}$$

$$\frac{dQ_1}{d\tau} = \text{Net heat puts out by man}$$

$$= 4 \times 630 \text{ kJ/h} = 0.7 \text{ kW}$$

Now total power consumed = w = 0.36 + 0.3 = 0.66 kW

Let Q be the net heat transfer by cooler then applying S.F.E.E. for system considered neglecting ΔK.E. and ΔP.E.

$$\dot{m}_a (h_1) + \frac{dQ}{d\tau} + \frac{dQ_1}{d\tau} = \dot{m}_a (h_2) + \frac{dw}{d\tau}$$

or
$$\frac{dQ}{d\tau} = \dot{m}_a (h_2 - h_1) + \frac{dw}{d\tau} - \frac{dQ_1}{d\tau}$$

$$= \frac{80}{3600} (59 - 84) - 0.66 - 0.7$$

$$= -1.92 \text{ kW } \textbf{Ans}$$

Problem 5.15 : Steam flowing in a pipeline is at a steady state represented by P_p, t_p, u_p, v_p, h_p and V_p. A small amount of the total flow is led through a small tube to an evacuated chamber which is allowed to fill slowly until the pressure is equal to the pipeline pressure. If there is no heat transfer, derive an expression for the final specific internal energy in the chamber, in terms of hte properties in the pipeline.

Solution :

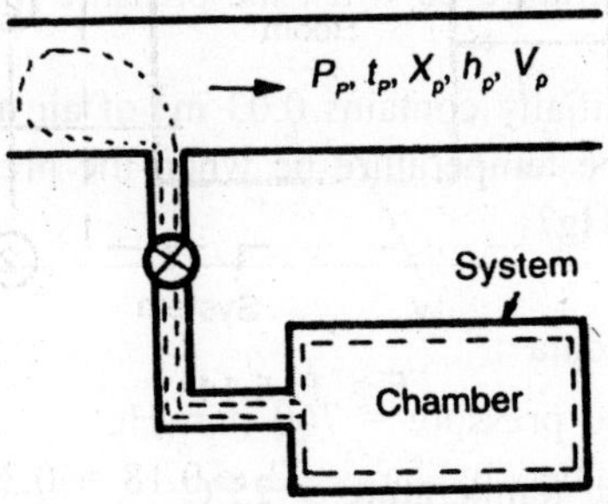

Fig. P. 5.15

We know that energy equation for the process is

$$Q = m_2u_2 - m_1u_1 - (m_2 - m_1)\left(\frac{Vp^2}{2} + h_p\right)$$

where u_2 = internal energy of chamber after filling
u_1 = internal energy of chamber before filling
m_1 = mass of gas in chamber before filling
m_2 = mass of gas in chamber after filling

Here $m_1 = 0$
$Q = 0$

Hence equation reduces to

$$0 = m_2u_2 - 0 - m_2\left(\frac{Vp^2}{2} + h_p\right)$$

or $$u_2 = \left(\frac{Vp^2}{2} + h_p\right)$$

Problem 5.16 : The internal energy of air is given, at ordinary temperatures, by

$$u = u_0 + 0.718\ t$$

where u is in kJ/kg, u_0 is any arbitrary value of u at 0°C, kJ/kg, and t is temperature in °C.

Also for air, $pv = 0.287\ (t + 273)$
where p is in kPa and v is in m^3/kg.
(a) An evacuated bottle is fitted with a valve through which air from the atmosphere, at 760 mm Hg and 25°C, is allowed to flow slowly to fill the bottle. If no heat is transferred to or from the air in the bottle, what will its temperature be when the pressure in the bottle reaches 760 mm Hg?
(b) If the bottle initially contains 0.03 m^3 of air at 400 mm Hg and 25°C, what will the temperature be when the pressure in the bottle reaches 760 mm Hg?

Solution : Given data :

(a) p_p = atmospheric pressure = 760 mm Hg
T_p = atmospheric temperature = 25°C
V_p = velocity of atmospheric air = 0
$Q = 0$
h_p = enthalpy of atmospheric air

Before filling
m_1 = mass of air = 0
u_1 = internal energy = 0

After filling
m_2 = mass of air
u_2 = internal energy

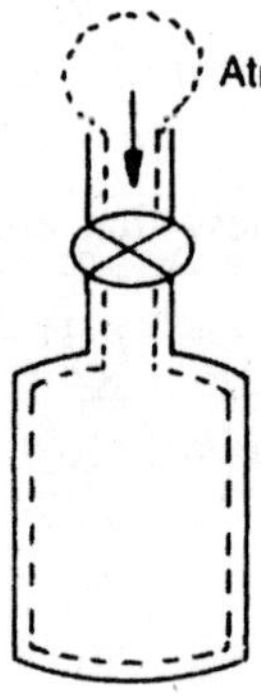

Fig. P. 5.16

∴ Energy equation for the process in a considered system

$$Q = m_2u_2 - m_1u_1 - (m_2 - m_1)\left(h_p + \frac{V_p^2}{2}\right)$$

$$\Rightarrow \quad 0 = m_2u_2 - 0 - (m_2)\left(h_p + \frac{V_p^2}{2}\right) \qquad [\because V_p = 0]$$

$$\Rightarrow \quad u_2 = h_p = v_p + (pv)_p$$

$$= u_0 + 0.718 \times 25 + 0.287 \times (25 + 273)$$

$$u_2 = u_0 + 103.476 \text{ kJ/kg.K}$$

$$u_0 + 0.718\,t = u_0 + 103.476$$

$$\therefore \quad t = \frac{103.476}{0.718} = 144.16°\text{C } \textbf{Ans}$$

(b) If $m_1 \neq 0$

Before filling,

$$V_1 = 0.03 \text{ m}^3$$
$$p_1 = 400 \text{ mm Hg}$$
$$T_1 = 25°\text{C}$$
$$pV_1 = m \times 0.287\ (t + 273)$$

$$\therefore \quad m_1 = \frac{p_1 V_1}{0.287(t+273)}$$

$$= \frac{0.400 \times 13.6 \times 9810 \times 0.03}{0.287 \times 298}$$

$$= 18.72 \text{ kg}$$

$$u_1 = u_0 + 0.718 \times 25 = u_0 + 17.95 \text{ kJ/kg}$$

After filling

$$V = 0.03 \text{ m}^3$$

when $p_2 = 760$ mm Hg. $= 0.76 \times 9810 \times 13.6$ N/m^2

$$T_2 = \text{unknown}$$

$$\therefore \quad p_2 V = m_2 \times 0.287\ (t_2 + 273)$$

$$m_2 = \frac{p_2 V}{0.287(t_2+273)}$$

$$= \frac{0.760 \times 13.6 \times 9810 \times 0.03}{0.287(t_2+273)}$$

$$= \frac{10598.90}{(t_2+273)} \text{ kg}$$

$$u_2 = u_0 + 0.718 \times t_2 = u_0 + 0.718\ t_2 \text{ kJ/kg}$$

Hence from energy equation for changing process.

$$Q = m_2 u_2 - m_1 u_1 - (m_2 - m_1)\left(h_p + \frac{V_p^2}{2}\right)$$

$$\therefore \quad Q = 0$$
$$V_p = 0$$

$$\therefore \quad m_2 u_2 - m_1 u_1 = (m_2 - m_1)\ h_p$$

$\therefore$ Where $h_p = u_p + (pv)_p$

$$= u_0 + 0.718 \times 25 + 0.287 \times 298$$
$$= (u_0 + 103.476) \text{ kJ/kg}$$

or $\quad m_2u_2 - m_1u_1 = m_2h_p - m_1h_p$

$\Rightarrow \quad m_2(u_2 - h_p) = m_1(u_1 - h_p)$

$$\Rightarrow \frac{10598.90}{t_2 + 273}[u_0 + 0.718 \times t_2 - u_0 - 103.476]$$

$$= 18.72\,[u_0 + 17.95 - u_0 - 103.476]$$

$$\Rightarrow \frac{[0.718t_2 - 103.476]}{t_2 + 273} = -0.1511$$

$\Rightarrow 0.718t_2 - 103.476$

$$= -0.1511t_2 - 41.239$$

$\Rightarrow \quad 0.8691t_2 = 62.237$

$\therefore \quad t_2 = 71.61°C$ **Ans**

Problem 5.17 : A pressure cylinder of volume V contains air at pressure p_0 and temperature T_0. It is to be filled from a compressed air line maintained of constant pressure p_1 and temperature T_1. Show that the temperature of the air in the cylinder after it has been charged to the pressure of the line is given by

$$T = \frac{\gamma T_1}{1 + \frac{p_0}{p_1}\left(\gamma \frac{T_1}{T_0} - 1\right)}$$

Solution : Given data :

For pipeline

p_1 = pressure

T_1 = temperature

Let h_1 = enthalpy of air in pipeline = C_pT_1

Before filling the tank

p_0 = initial pressure

T_0 = initial temperature

V = volume of tank

From ideal gas equation

$$m_1 = \frac{p_0V}{RT_0}$$

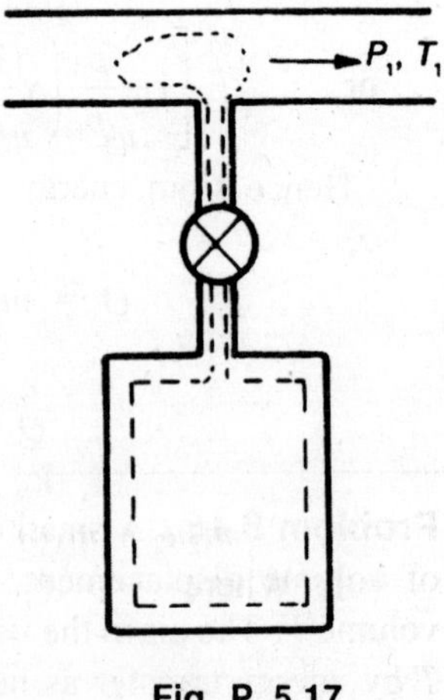

Fig. P. 5.17

$$u_1 = C_V T_0$$

After filling the tank

Final pressure = p_1 and

Final temperature = T

∴ from ideal gas equation,

$$m_2 = \frac{p_1 V}{RT}$$

$$u_2 = C_v T$$

Now energy equation for the system considered

$$\Rightarrow \quad Q = m_2 u_2 - m_1 u_1 - (m_2 - u_1)\left(h_p + \frac{V_p^2}{2}\right)$$

here

$$Q = 0$$

$$V_p = 0$$

$$\therefore \quad m_2 u_2 - m_1 u_1 = m_2 h_p - m_1 h_p$$

$$\Rightarrow \quad m_2 (u_2 - h_p) = m_1 (u_1 - h_p)$$

$$\Rightarrow \quad \frac{p_1 V}{RT}[C_v T - C_p T_1] = \frac{p_0 V}{RT_0}[C_v T_0 - C_p T_1]$$

$$\Rightarrow \quad (T - \gamma T_1) = \frac{p_0}{p_1} \times \frac{T}{T_0}[T_0 - \gamma T_1]$$

$$T + \frac{p_0}{p_1}\frac{T}{T_0}[\gamma T_1 - T_0] = \gamma T_1$$

$$\text{or,} \quad T\left[1 + \frac{p_0}{p_1}\left(\gamma \frac{T_1}{T_0} - 1\right)\right] = \gamma T_1$$

$$\therefore \quad T = \frac{\gamma T_1}{\left[1 + \frac{p_0}{p_1}\left(\gamma \frac{T_1}{T_0} - 1\right)\right]} \quad \textbf{Ans}$$

Problem 5.18 : A small reciprocating vacuum pump having the rate of volume displacement V_d is used to evacuate a large vessel of volume V. The air in the vessel is maintained at a constant temperature T by energy transfer as heat. If the initial and final pressures are p_1

and p_2 respectively, find the time taken for the pressure drop and the necessary energy transfer as heat during evacuation. Assume that for air, $pV = mRT$, where m is the mass and R is the constant, and u is a function of T only.

Solution : Given :

Initial pressure of air in tank = P_1
final pressure of air in tank = P_2
volume of vessel = V
temperature of vessel = T
volume displacement of pump = V_d

Now for air

$$pV = mRT$$

or

$$m = \frac{pV}{RT}$$

In a small period of time 'dt', dm be the flow of air extracted from tank.

$$\therefore \quad dm = +\frac{Vdp}{RT} \quad \text{...(i)}$$

From pump analysis

$$\text{mass of air removed} = dm = +\frac{-P(V_d dt)}{RT} \quad \text{...(ii)}$$

∴ Equating (i) and (ii)

$$\frac{dP}{P} = -\left(\frac{V_d}{V}\right) dt$$

Integrating both sides between 0 to t, we get

$$\ln \frac{P_2}{P_1} = -\frac{V_d}{V} t$$

or

$$t = \frac{V}{V_d} \ln \frac{p_1}{p_2} \quad \textbf{Ans}$$

During evacuation process

From first law of thermodynamics

$$\therefore \quad dQ = dh - Vdp$$

$$dQ = -Vdp \quad [\because dh = 0 = \text{at constant temperature}]$$

$$= V(p_1 - p_2) \quad \textbf{Ans}$$

Chapter 6

Second Law of Thermodynamics

Problem 6.1 : An inventor claims to have developed an engine that takes in 105 MJ at a temperature of 400 K, rejects 42 MJ at a temperature of 200 K, and delivers 15 kWh of mechanical work. Would you advise investing money to put this engine in the market?

Solution : Given :

Q_1 = heat received from reservoir at 400 K
= 105 MJ

Q_2 = heat rejected from reservoir at 200 J
= 42 MJ

W = work output = 15 kWh
= 15 kJ/s × 3600 = 54 MJ

∴ n_{thermal} of engine

$$= \frac{W}{Q_1} = \frac{54}{105} = 51.43\ \%$$

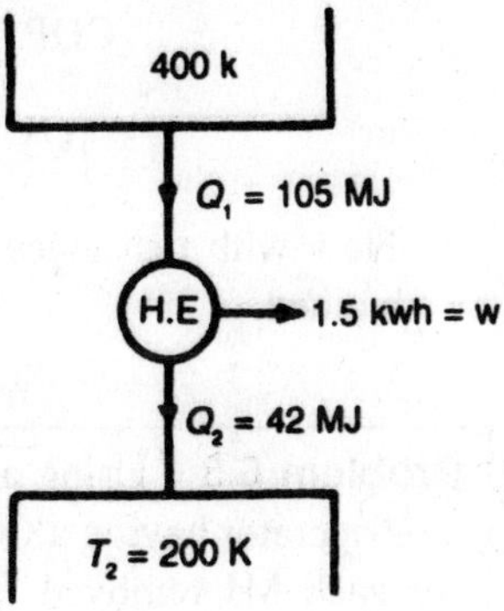

Fig. P. 6.1

Now $$n_{\text{carnot}} = 1 - \frac{T_2}{T_1} = 1 - \frac{200}{400} = 50\ \%$$

here $n_{\text{carnot}} < n_{\text{engine}}$ which is violation of carnot theorem.
∴ Engine is not feasible.

Problem 6.2 : If a refrigerator is used for heating purposes in winter so that the atmosphere becomes the cold body and the room to be heated becomes the hot body, how much heat would be available for heating for each kW input to the driving motor? The COP of the refrigerator is 5, and the electromechanical efficiency of the motor is 90%. How does this compare with resistance heating?

Solution : Let Q_1 be the heat drawn from the cold atmosphere and Q_2 be the heat rejected to the hot space and W be the work input to the refrigerator.

Given :

$$w = 1 \text{ kW}$$
$$\text{efficiency of motor} = 90\%$$
$$\text{COP} = 5$$
$$\therefore \quad w_{\text{actual}} = 0.9 \text{ kW}$$
$$\therefore \quad \text{COP} = 5 = \frac{Q_1}{W}$$

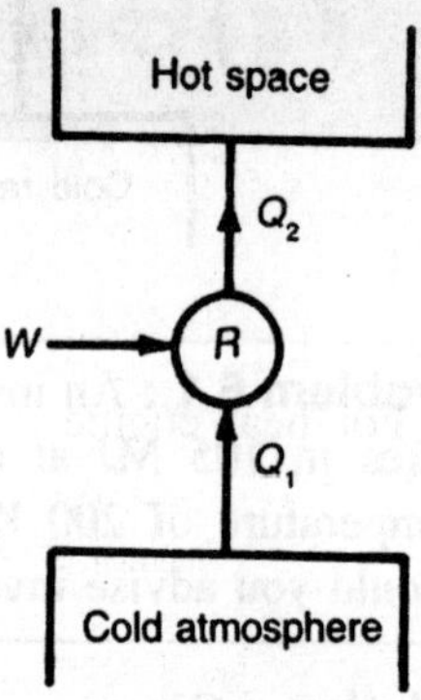

Fig. P. 6.2

From energy balance

$$w + Q_1 = Q_2$$
$$\therefore \quad Q_1 = Q_2 - w$$
$$\therefore \quad \text{COP} = 5 = \frac{Q_2 - w}{w}$$
$$\text{or} \quad Q_2 = 5w + w = 6w = 6 \times 0.9$$
$$= 5.4 \text{ kW} \quad \textbf{Ans}$$

Now with resistance heating only 1 kW of equivalent heat would be obtained.

Problem 6.3 : Using an engine of 30% thermal efficiency to drive a refrigerator having a COP of 5, what is the heat input into the engine for each MJ removed from the cold body by the refrigerator?
If this system is used as a heat pump, how many MJ of heat would be available for heating for each MJ of heat input to the engine?

Solution : For refrigerator

$$\text{COP} = \frac{\text{heat removed}}{\text{work input}}$$
$$\Rightarrow \quad 5 = \frac{1 \text{ MJ}}{w}$$
$$\therefore \quad w = 0.2 \text{ MJ}$$

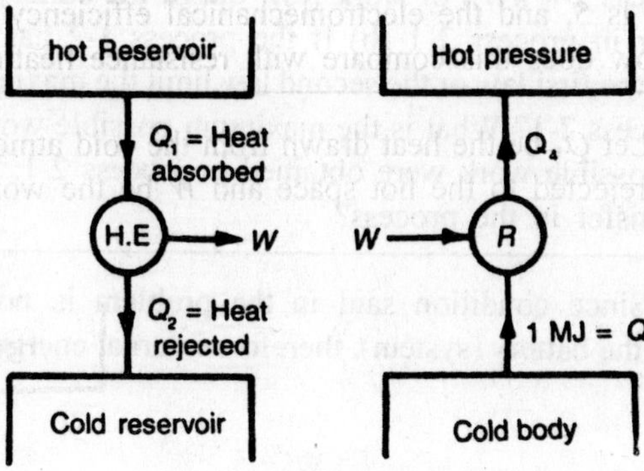

Fig. P. 6.3

For heat engine

$$n_{\text{thermal}} = 0.30 = \frac{w}{Q_1}$$

or, $$Q_1 = \frac{w}{n_{\text{thermal}}} = \frac{0.2}{0.3} = 666.67 \text{ kJ } \textbf{Ans}$$

For heat pump system

Given $Q_1 = 1$ MJ

For engine

$$\therefore \quad n_{\text{thermal}} = 0.30 = \frac{w}{Q_1} = \frac{w}{1MJ}$$

$$\therefore \quad w = 0.3 \text{ MJ}$$

For heatpump

We know that

$$(\text{COP})_{\text{H.P.}} = (\text{COP})_{\text{Ref}} + 1 = 5 + 1 = 6$$

$$\therefore \quad (\text{COP})_{\text{H.P.}} = 6 = \frac{Q_4}{w} = \frac{Q_4}{0.3}$$

$$\therefore \quad Q_4 = 1.8 \text{ MJ } \textbf{Ans}$$

Problem 6.4 : An electric storage battery which can exchange heat only with a constant temperature atmosphere goes through a complete cycle of two processes. In process 1-2, 2.8 kWh of electrical work flow into the battery while 732 kJ of heat flow out to the atmosphere.

During process 2-4 kWh of work flow out of the battery (a) find the heat transfer in process 2-1 (b) If the process 1-2 has occurred as above, does the first law or the second law limit the maximum possible work of process 2-1? What is the maximum possible work? (c) If the maximum possible work were obtained in process 2-1, what will be the heat transfer in the process?

Solution : Since condition said in the problem is not steady state condition for the battery (system), therefore internal energy of the system will increase.

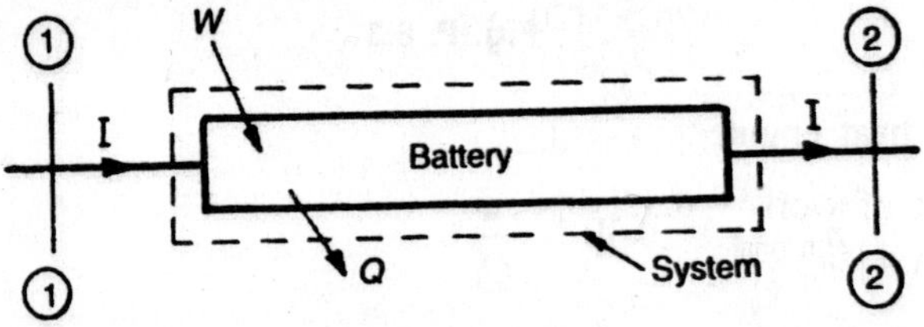

Fig. P. 6.4

Here both Q_{1-2} & w_{1-2} are negative then from first law

$$U_2 - U_1 = w_{1-2} - Q_{1-2}$$

given $w_{1-2} = 2.8 \text{ kWh} = 10.080 \text{ kJ}$

$Q_{1-2} = 732 \text{ kJ}$

$\therefore \quad U_2 - U_1 = 10080 - 732 = 9348 \text{ kJ}$

For process 2-1

$w_{2-1} = 2.4 \text{ kWh} = 8640 \text{ kJ}$

and $U_1 - U_2 = -9348 \text{ kJ}$

From first law

(a) $Q_{2-1} = (U_1 - U_2) + w_{2-1} = -9348 + 8640$

$= -708$ kJ **Ans**

(b) From second law, complete conversion of heat into work is not possible.

For process 2–1

$Q_{2-1} = -708 \text{ kJ}$

i.e., heat rejected to the atmosphere.

If $Q_{2-1} = 0$,

Then for PMM_2, w_{2-1} will be maximum

the $w_{2-1} = (U_2 - U_1) = 9348$ kJ **Ans**

(c) $Q_{2-1} = 0$ **Ans**

Problem 6.5 : A household refrigerator is maintained at a temperature of 2°C. Every time the door is opened, warm material is placed inside, introducing an average of 420 kJ, but making only a small change in the temperature of the refrigerator. The door is opened 20 times a day, and the refrigerator operates at 15% of the ideal COP. The cost of work is 32 paise per kWh. What is the monthly bill for this refrigerator? The atmosphere is at 30°C.

Solution : Given data :

In 24 hour

Total heat absorbed from cold space

$$= 20 \times 420 = 8400 \text{ kJ}$$

Given refrigerator COP

$$= 15\% \text{ (Ideal COP}_{\text{Ref.}})$$

Cost of work = 0.32 rupee per Kwh.

Now ideal

$$\text{COP}_{\text{Ref}} = \frac{T_2}{T_1 - T_2}$$

$$= \frac{(2+273)}{30-2}$$

$$= \frac{275}{28} = 9.82$$

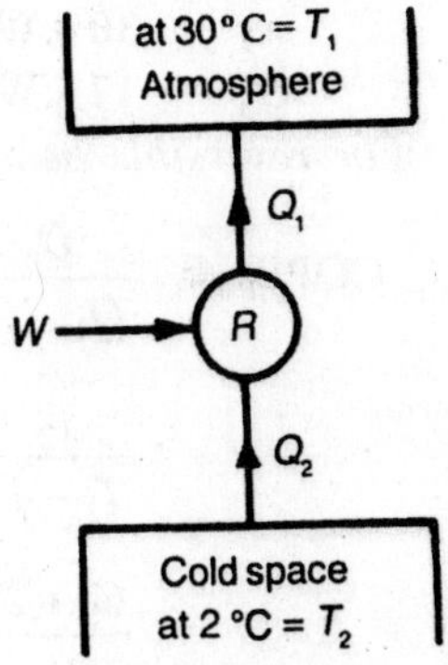

Fig. P. 6.5

∴ Given refrigerator

$$\text{COP} = 9.82 \times 0.15 = 1.47$$

Now $\quad \text{COP} = \dfrac{Q_2}{w}$

$$\therefore \quad w = \frac{Q_2}{\text{COP}} = \frac{8400}{1.47}$$

$$= 5714.29 \text{ kJ}$$

∴ Cost of work on monthly basis

$$= \frac{5714.29 \times 0.32 \times 30}{3600}$$

$$= \text{Rs. } 15.24 \textbf{ Ans}$$

Problem 6.6 : A heat pump working on the carnot cycle takes in heat from a reservoir at 5°C and delivers heat to a reservoir at 60°C. The heat pump is driven by a reversible heat engine which takes in heat from a reservoir at 840°C and rejects heat to a reservoir at 60°C. The reversible heat engine also drives a machine that absorbs 30 kW. If the heat pump extracts 17 kJ/s from the 5°C reservoir, determine (a) the rate of heat supply from the 840°C source, and (b) the rate of heat rejection to the 60°C sink.

Solution : Given :

$$w_1 = 30 \text{ kW}$$
$$Q_3 = 17 \text{ kW}$$

For reversible heat pump

$$\text{COP}_{\text{HP}} = \frac{Q_4}{Q_4 - Q_3}$$
$$= \frac{T_2}{T_2 - T_3}$$
$$= \frac{60+273}{60-5}$$
$$= 6.05$$

$$\text{or, COP}_{\text{HP}} = \frac{Q_4}{w_2}$$
$$= \frac{w_2 + Q_3}{w_2}$$
$$= 6.05$$

$$\therefore \quad w_2 = \frac{Q_3}{5.05} = \frac{17}{5.05} = 3.37 \text{ kW}$$

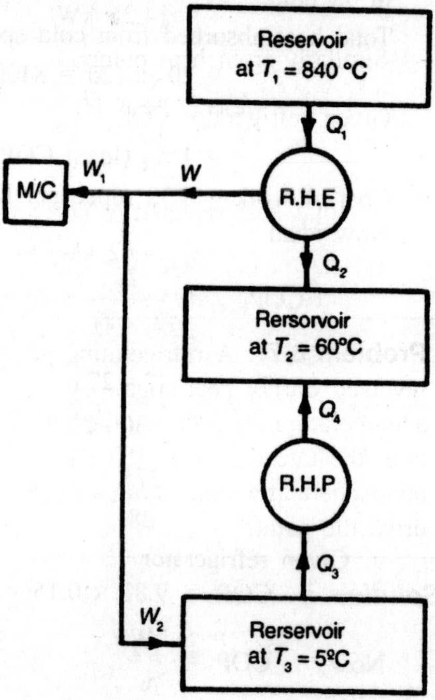

Fig. P. 6.6

For reversible heat engine

$$w = w_1 + w_2 = 30 + 3.37$$
$$= 33.37 \text{ kW}$$

$$\therefore \quad n_{\text{reversible}} = 1 - \frac{T_2}{T_1}$$
$$= 1 - \frac{60+273}{840+273} = 0.7008$$

or, $\frac{w}{Q_1} = 0.7008$

$\therefore \quad Q_1 = \frac{w}{0.7008} = \frac{33.37}{0.7008}$

$= 47.62$ kW **Ans**

Applying 1st law of thermodynamics in heat engine

$\therefore \quad Q_2 = Q_1 - w$

$= 47.62 - 33.37$

$= 14.25$ kW

Similarly from heat pump

$Q_4 = w_2 + Q_3$

$= 3.37 + 17$

$= 20.37$ kW

$\therefore$ Total heat rejected to 60°C sink

$= 34.62$ kW **Ans**

Problem 6.7 : A refrigerating plant for a food store operates as a reversed Carnot heat engine cycle. The store is to be maintained at a temperature of –5°C, and the heat transfer from the store to the cycle is at the rate of 5 kW. If heat is transferred from the cycle to the atmosphere at a temperature of 25°C, calculate the power required to drive the plant.

Solution : Given :

$Q_2 = 5$ kW

$(COP)_{R.Carnot} = \frac{T_2}{T_1 - T_2}$

$= \frac{-5+273}{30}$

$= 8.93$

$\therefore \quad w = \frac{Q_2}{8.93}$

$= 0.56$ kW **Ans**

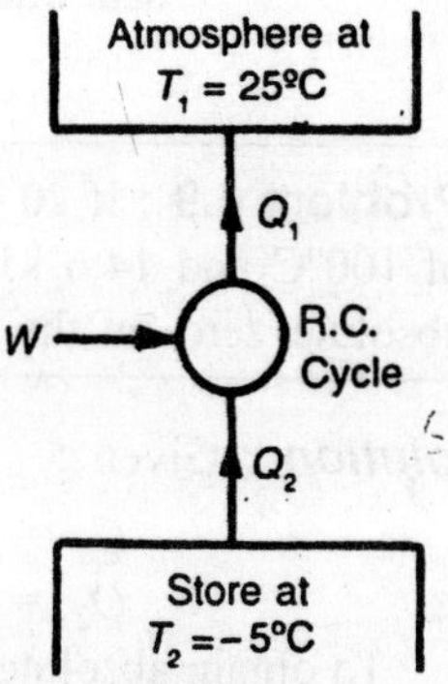

Fig. P. 6.7

Problem 6.8 : A heat engine is used to drive a heat pump. The heat transfers from the heat engine and from the heat pump are used to heat the water circulating through the radiators of a building. The efficiency of the heat engine is 27% and the COP of the heat pump is 4. Evaluate the ratio of the heat transfer to the circulating water to the heat transfer to the heat engine.

Solution : Given data :

$$n_{H.E.} = 27\%$$

or $$\frac{w}{Q_1} = 0.27$$

or $$\frac{w}{w+Q_2} = 0.27$$

$$\therefore \quad Q_2 = \frac{0.73w}{0.27}$$

$$= 2.7 \text{ kW}$$

and $$Q_1 = 3.7 \text{ kW}$$

again $$\text{COP} = 4$$

or $$\frac{Q_4}{w} = 4$$

$$\therefore \quad Q_4 = 4\,w$$

$\therefore$ Total heat rejected to building $= Q = Q_2 + Q_4$

$$= 4 + 2.7 = 6.7 \text{ kW}$$

$$\therefore \quad \frac{\text{heat tranfer to building}}{\text{heat transfer to heat engine}} = \frac{6.7}{3.7} = 1.81 \textbf{ Ans}$$

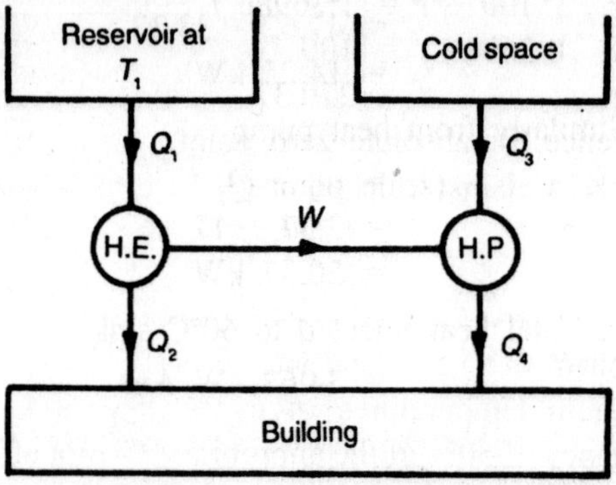

Fig. P. 6.8

Problem 6.9 : If 20 kJ are added to a Carnot cycle at a temperature of 100°C and 14.6 kJ are rejected at 0°C, determine the location of absolute zero on the Celsius scale.

Solution : Given :

$$Q_1 = 20 \text{ kJ}$$

$$Q_2 = 14.6 \text{ kJ}$$

To obtain absolute zero point, i.e., 0°K, let t be the arbitrary choose value at kelvin scale then from Carnot cycle

$$\frac{T_1 + t}{T_2 + t} = \frac{Q_1}{Q_2} = \frac{20}{14.6}$$

$$= 1.369863014$$

$$\approx 1.36986$$

or, $\frac{100 + t}{0 + t} = 1.36986$

or, $100 + t = 1.36986\ t$

or, $0.36986\ t = 100$

or, $t = 270.37$

Hence for absolute zero point,

0°K, Celsius scale point

$= -270.37°C$ **Ans**

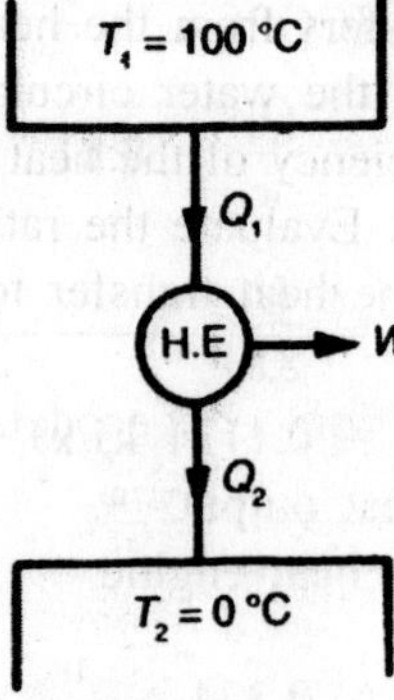

Fig. P. 6.9

Problem 6.10 : A heat engine operates between the maximum and minimum temperatures of 671°C and 60°C respectively, with an efficiency of 50% of the appropriate Carnot efficiency. It drives a heat pump which uses river water at 4.4°C to heat a block of flats in which the temperature is to be maintained at 21.1°C. Assuming that a temperature difference of 11.1°C exists between the working fluid and the river water, on the one hand, and the required room temperature on the other, and assuming the heat pump to operate on the reversed Carnot cycle, but with a COP of 50% of the ideal COP, find the heat input to the engine per unit heat output from the heat pump. Why is direct heating thermodynamically more wasteful?

Solution : Given data

$$n_{\text{H.E.}} = n_{\text{carnot}} \times 0.5$$

$$= \left[1 - \left(\frac{60+273}{671+273}\right)\right] \times 0.5 = 0.324$$

$$\text{COP}_{\text{HP}} = (\text{COP})_{\text{R.C.C.}} \times 0.5$$

$$= 0.5 \times \left(\frac{T_4}{T_4 - T_3}\right) = \left(\frac{21.1+273}{(21.1-4.4)}\right) \times 0.5$$

$$= 8.81$$

for $Q_4 = 1 \text{ unit} = 1 \text{ kJ}$

$Q_1 = ?$

Now $(COP)_{HP}$

$$= \frac{Q_4}{w} = 8.81$$

$$\therefore w = \frac{Q_4}{8.81}$$

$$= 0.1134 \text{ kJ/kJ}$$

of heat output

From heat engine

$$\eta_{HE} = 0.324 = \frac{w}{Q_1}$$

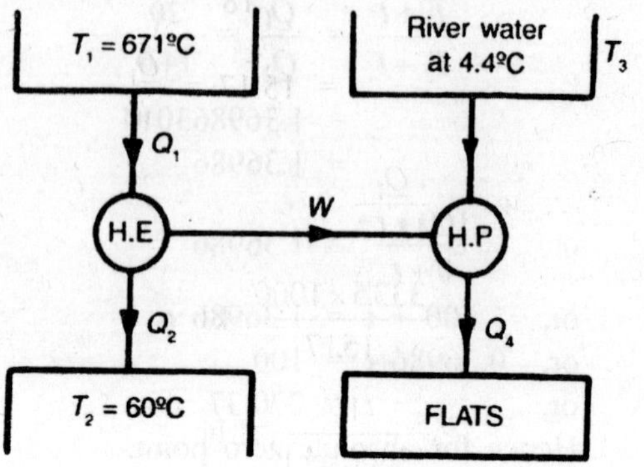

Fig. P. 6.10

$$Q_1 = \frac{w}{0.324}$$

$$= \frac{0.1134}{0.324}$$

$$= 0.351 \text{ kJ/kJ of output}$$

Problem 6.11 : An ice-making plant produces ice at atmospheric pressure and at 0°C from water. The mean temperature of the cooling water circulating through the condenser of the refrigerating machine is 18°C. Evaluate the minimum electrical work in kWh required to produce 1 tonne of ice. (The enthalpy of fusion of ice at atmospheric pressure is 333.5 kJ/kg).

Solution : Given :

mass of ice = 1000 kg

L = Latent heat of fusion of ice

= 333.5 kJ/kg

$\therefore$ heat required to convert ice

$= Q_1$

$= m\ L$

$= 1000 \times 333.5 \text{ kJ} = 333.5 \text{ MJ}$

For minimum work input, COP should be equal to reversed carnot cycle COP for refrigeration.

$$\therefore \quad COP = \frac{T_1}{T_2 - T_1}$$

$$= \frac{0+273}{18}$$

$$= 15.17 = \frac{Q_1}{w}$$

$$\therefore \; w = \frac{Q_1}{15.17}$$

$$= \frac{333.5 \times 1000}{15.17} \text{ kJ} \times \frac{1}{(3600\,\text{sec})} \times \text{hr}$$

$$= 6.11 \text{ kWh } \textbf{Ans}$$

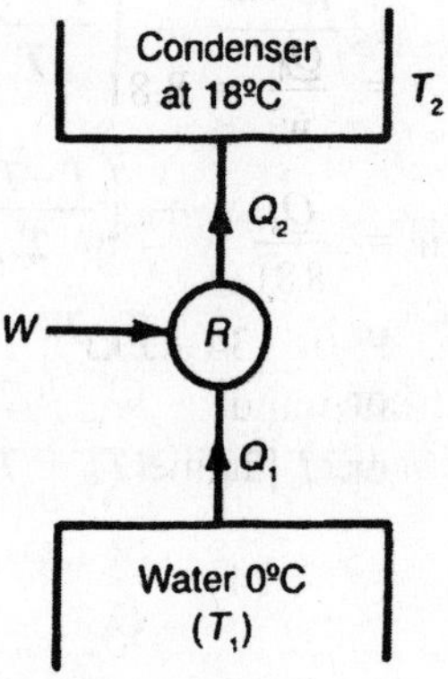

Fig. P. 6.11

> **Problem 6.12 :** A reversible engine operates between temperatures T_1 and T ($T_1 > T$). The energy rejected from this engine is received by a second reversible engine at the same temperature T. The second engine rejects energy at temperature T_2 ($T_2 < T$). Show that (a) temperature T is the arithmetic mean of temperatures T_1 and T_2 if the engines produce the same amount of work output, and (b) temperature T is the geometric mean of temperatures T_1 and T_2 if the engines have the same cycle efficiencies.

Solution : (a) For same work output of both the engine, i.e., $w_1 = w_2$

For 1st H.E.

$$n_{carnot} = \frac{T_1 - T}{T_1} = \frac{w_1}{Q_1}$$

or,
$$w_1 = \left(\frac{T_1 - T}{T_1}\right) Q_1$$

For 2nd H.E.

$$n_{carnot} = \frac{T - T_2}{T} = \frac{w_2}{Q_2}$$

$$\therefore \qquad w_2 = \left(\frac{T - T_2}{T}\right) Q_2$$

$$\because \quad w_1 = w_2$$

$$\therefore \quad \frac{T_1 - T}{T_1} = \left(\frac{T - T_2}{T}\right)\frac{Q_2}{Q_1}$$

$$= \left(\frac{T - T_2}{T}\right)\frac{T}{T_1}$$

or $\quad T_1\, T - T^2 = T^2 - T_2 T$

or, $\quad 2T^2 = (T_1 + T_2)\, T$

or, $T\,[2T - (T_1 + T_2)] = 0$

$$\therefore \quad T = \frac{T_1 + T_2}{2} \quad \textbf{Ans}$$

(b) For same efficiency

$$n_1 = n_2$$

$$\Rightarrow \quad \frac{T_1 - T}{T_1} = \frac{T - T_2}{T}$$

or, $\quad T_1 \times T - T^2 = T \times T_1 - T_1 \times T_2$

or, $\quad T^2 = T_1 \times T_2$

$$\therefore \quad T = \sqrt{T_1 \times T_2} \quad \textbf{Ans}$$

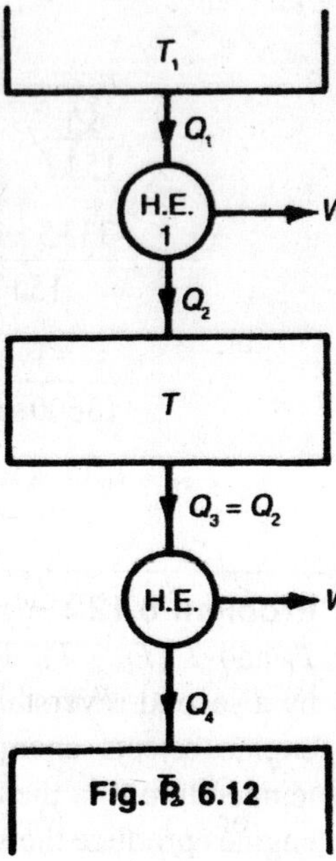

Fig. P. 6.12

Problem 6.13 : Two Carnot engines A and B are connected in series between two thermal reserviors maintained at 1000 K and 100 K respectively. Engine A receives 1680 kJ of heat from the high–temperature reservoir and rejects heat to the Carnot engine B. Engine B takes in heat rejected by engine A and rejects heat to the low–temperature reservoir. If engines A and B have equal thermal efficiencies, determine (a) the heat rejected by engine B, (b) the temperature at which heat is rejected by engine A, and (c) the work done during the process by engines A and B respectively. If engines A and B deliver equal work, determine (d) the amount of heat taken in by engine B, and (e) the efficiencies of engines A and B.

Solution **:** *For same efficiency*

(b) $\quad T = \sqrt{T_1\, T_2}$

$$= \sqrt{1000 \times 100}$$
$$= 316.23 \text{ K}$$

(a) For first engine

$$n_{carnot} = 1 - \frac{T}{T_1}$$

$$= 1 - \frac{Q_2}{Q_1}$$

or, $$Q_2 = \frac{T}{T_1} \times Q_1$$

$$= \frac{316.23}{1000} \times 1680$$
$$= 531.26 \text{ kJ}$$

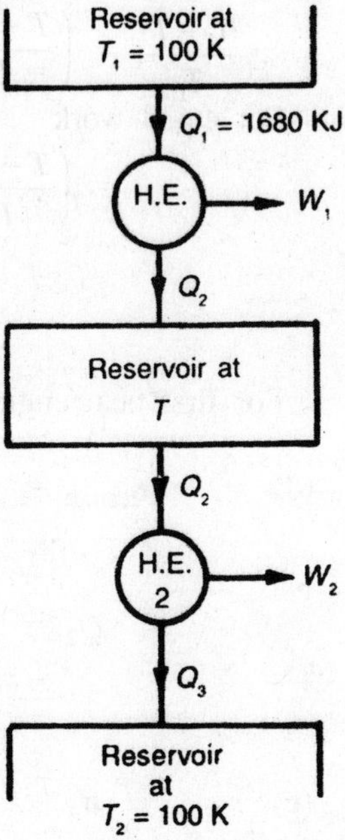

Fig. P. 6.13

For 2nd engine

$$n_{carnot} = 1 - \frac{T_2}{T} = 1 - \frac{Q_3}{Q_2}$$

or $$Q_3 = \frac{T_2}{T} Q_2$$

$$= \frac{100}{316.23} \times 531.26 = 167.998 = 168 \text{ kJ } \textbf{Ans}$$

(c) For 1st engine,

$$n_{carnot} = \frac{T_1 - T}{T_1} = \frac{w_1}{Q_1}$$

$$\therefore \quad w_1 = \left(\frac{T_1 - T}{T_1}\right) Q_1$$

$$= \left(\frac{1000 - 316.23}{1000}\right) \times 1680 = 1148.73 \text{ kJ } \textbf{Ans}$$

For 2nd engine

$$n_{carnot} = \frac{T_1 - T_2}{T} = \frac{w_2}{Q_2}$$

$$w_2 = \left(\frac{T - T_2}{T}\right) Q_2$$

$$= \left(\frac{316.23 - 100}{316.23}\right) \times 531.26$$

$$= 363.26 \text{ kJ } \textbf{Ans}$$

For equal work

$$T = \frac{T_1 + T_2}{2} = \frac{1000 + 100}{2}$$

$$= \frac{1100}{2} = 550 \text{ K}$$

For first heat engine

(d) $$n_{carnot} = 1 - \frac{T}{T_1} = 1 - \frac{Q_2}{Q_1}$$

$$\therefore \quad Q_2 = \frac{T}{T_1} Q_1 = \frac{550}{1000} \times 1680$$

$$= 924 \text{ kJ } \textbf{Ans}$$

(e) $$n_A = 1 - \frac{T}{T_1} = 1 - \frac{550}{1000}$$

$$= 45\% \textbf{ Ans}$$

$$n_B = 1 - \frac{T_2}{T} = 1 - \frac{100}{550} = 81.8\% \textbf{ Ans}$$

Problem 6.14 : A heat pump is to be used to heat a house in winter and then reversed to cool the house in summer. The interior temperature is to be maintained at 20°C. Heat transfer through the walls and roof is estimated to be 0.525 kJ/s per degree temperature difference between the inside and outside. (a) If the outside temperature in winter is 5°C, what is the minimum power required to drive the heat pump? (b) If the power output is the same as in part (a), what is the maximum outer temperature for which the inside can be maintained at 20°C?

Solution : Given data for winter

(a) Outside temp. = 5°C

Inside temp. = 20°C

$\therefore \quad Q_2$ = heat transfer to house

$$= (20 - 5) \times 0.525 = 7.875 \text{ kJ/s}$$

$$(\text{COP})_{\text{H.P.}} = \frac{Q_2}{w} = \frac{T_2}{T_2 - T_1}$$

$$\Rightarrow \quad w = Q_2 \frac{(T_2 - T_1)}{T_2}$$

$$= 7.875 \frac{(20-5)}{20+273} = 403.16 \text{ W } \textbf{Ans}$$

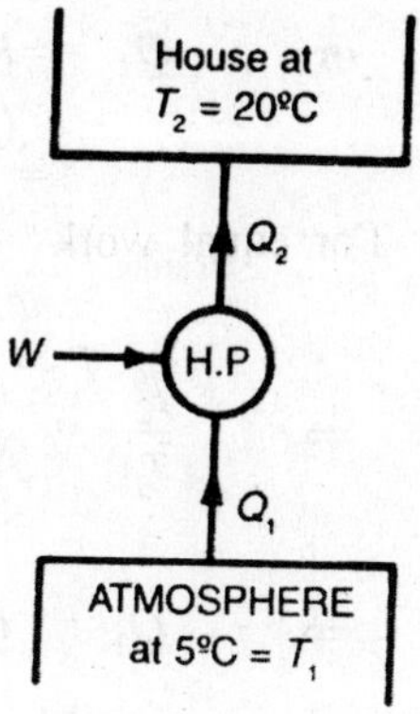

Fig. P. 6.14 (a)

(b) Suppose T be maximum temperature if refrigerator will be reversible.

$$\therefore \quad Q_1 = (T - 20) \times 0.525 \text{ kW}$$

$$w = 0.40316 \text{ kW}$$

$$(\text{COP})_{\text{H.P.}} = \frac{20+273}{20-5} = 19.53$$

$$\therefore \quad (\text{COP})_{\text{Ref}} = 18.53$$

$$\text{Now}(\text{COP})_{\text{Ref.}} = \frac{Q_1}{w}$$

or $\quad Q_1 = (T - 20) \times 0.525$

$$= w \times (\text{COP})_{\text{ref.}}$$

or $\quad (T - 20) = \dfrac{0.40316 \times 18.53}{0.525}$

$$T = 34.2°\text{C } \textbf{Ans}$$

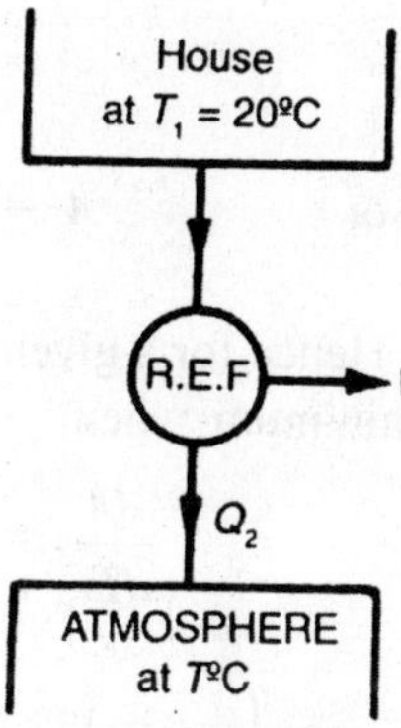

Fig. P. 6.14 (b)

Problem 6.15 : Consider an engine in outer space which operates on the Carnot cycle. The only way in which heat can be transferred from the engine is by radiation. The rate at which heat is radiated is proportional to the fourth power of the absolute temperature and to the area of the radiating surface. Show that for a given power output and a given T_1, the area of the radiator will be a minimum when

$$\frac{T_2}{T_1} = \frac{3}{4}$$

Solution : Given

$$Q_2 \propto A\,T_2^4 \quad \text{Where, } A = \text{Area of radiating surface}$$

or $Q_2 = KA\,T_2^4$, K

$= $ Constant of proportion

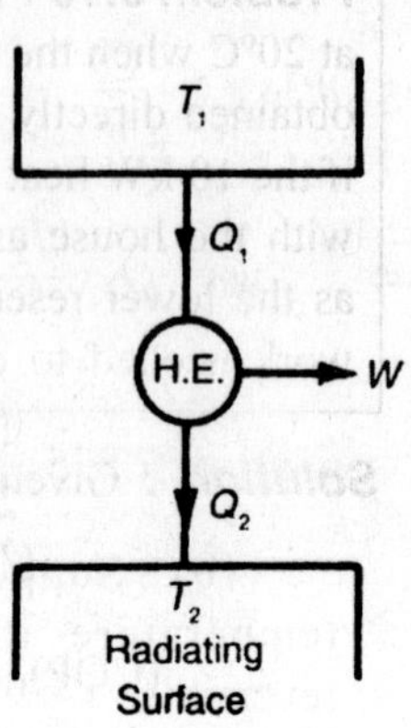

Fig. P. 6.15

$$n_{carnot} = \frac{T_1 - T_2}{T_1} = \frac{Q_1 - Q_2}{Q_1}$$

$$\Rightarrow \quad \frac{T_2}{T_1} = \frac{Q_2}{Q_1}$$

$$\Rightarrow \quad Q_1 = Q_2 \times \frac{T_1}{T_2} = \frac{KA\,T_2^4 \times T_1}{T_2}$$

$$\Rightarrow \quad Q_1 = KA\,T_2^3\,T_1$$

$\therefore$ work output $= w$

$$= Q_1 - Q_2 = KA\,T_2^3(T_1 - T_2)$$

or
$$A = \frac{w}{KT_2^3(T_1 - T_2)}$$

Hence for a given work output and T_1, area of radiating surface will be minimum when

$$\frac{dA}{dT_2} = 0$$

or
$$\frac{d\left(\dfrac{w}{K\,T_2^3(T_1 - T_2)}\right)}{dT_2} = 0$$

or,
$$\frac{d\left[T_2^{-3}(T_1 - T_2)^{-1}\right]}{dT_2} = 0$$

$$\text{or } (-)\,(T_1 - T_2)^{-2} \times (-)\ T_2^{-3} + (T_1 - T_2)^{-1} \times -3\ T_2^{-4} = 0$$

$$\Rightarrow (T_1 - T_2)^{-2}\ T_2^{-3} = (T_1 - T_2)^{-1}\ 3\ T_2^{-4}$$

$$\Rightarrow (T_1 - T_2)^{-1} = 3\,T_2^{-1}$$

$$\Rightarrow \quad T_2 = 3(T_1 - T_2) = 3\ T_1 - 3T_2$$

$$\Rightarrow \quad 4T_2 = 3\ T_1 \Rightarrow \frac{T_2}{T_1} = \frac{3}{4}$$ **Ans**

Problem 6.16 : It takes 10 kW to keep the interior of a certain house at 20°C when the outside temperature is 0°C. This heat flow is usually obtained directly by burning gas or oil. Calculate the power required if the 10 kW heat flow were supplied by operating a reversible engine with the house as the upper reservoir and the outside surroundings as the lower reservoir, so that the power were used only to perform work needed to operate the engine.

Solution : Given

$$Q_1 = 10 \text{ kW}$$

$$\therefore \quad (\text{COP})_{HP} = \frac{Q_1}{w}$$

$$= \frac{T_1}{T_1 - T_2}$$

or
$$w = \frac{Q_1(T_1 - T_2)}{T_1}$$

$$= \frac{10(20-0)}{(20+273)}$$

$$= \frac{200}{293} = 682.6 \text{ W} \textbf{ Ans}$$

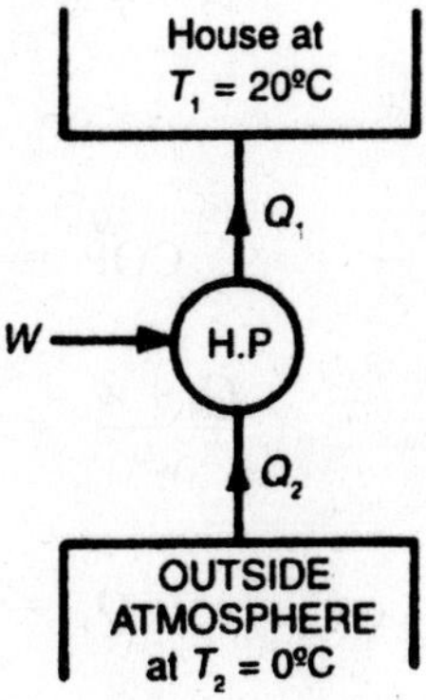

Fig. P. 6.16

Problem 6.17 : A reversible power cycle is used to drive a reversible heat pump cycle. The power cycle takes in Q_1 heat units at T_1 and rejects Q_2 at T_2. The heat pump abstracts Q_4 from the sink at T_4 and discharges Q_3 at T_3. Develop an expression for the ratio Q_4/Q_1 in terms of the four temperatures.

Solution : *For H.E.*

$$n_{\text{carnot}} = \frac{T_1 - T_2}{T_1} = \frac{w}{Q_1}$$

or,
$$w = \frac{(T_1 - T_2)}{T_1} \times Q_1 \qquad \ldots\text{(i)}$$

For H.P.

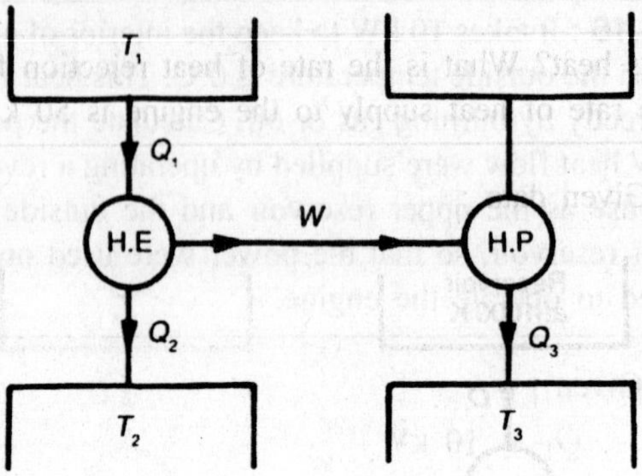

Fig. P. 6.17

$$\text{COP} = \frac{Q_3}{w} = \frac{Q_4 + w}{w} \qquad [\because w = Q_3 - Q_4]$$

$$\frac{Q_4 + w}{w} = \frac{T_3}{T_3 - T_4}$$

or,
$$Q_4 = \frac{T_3}{T_3 - T_4} w - w = \frac{T_4}{T_3 - T_4} w$$

$$\therefore \quad w = \left(\frac{T_3 - T_4}{T_4}\right) Q_4 \qquad \ldots(ii)$$

Now equating both the work

$$\therefore \left(\frac{T_1 - T_2}{T_1}\right) Q_1 = \left(\frac{T_3 - T_4}{T_4}\right) Q_4$$

or
$$\frac{Q_4}{Q_1} = \frac{T_4}{T_1} \times \left(\frac{T_1 - T_2}{T_3 - T_4}\right) \quad \textbf{Ans}$$

Problem 6.18 : A heat engine operating between two reserviors at 1000 K and 300 K is used to drive a heat pump which extracts heat from the reservoir at 300 K at a rate twice that at which the engine rejects heat to it. If the efficiency of the engine is 40% of the maximum possible and the COP of the heat pump is 50% of the maximum possible, what is the temperature of the reservoir to which the heat pump rejects heat? What is the rate of heat rejection from the heat pump if the rate of heat supply to the engine is 50 kW?

Solution : Given data :

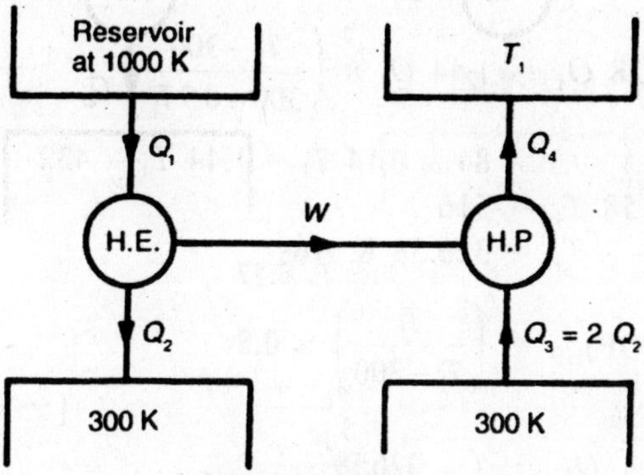

Fig. P. 6.18

$$n_{HE} = (n_{carnot}) \times 0.4$$

$$COP_{HP} = (n_{R.Carnot}) \times 0.5$$

Now $$n_{HE} = \left(\frac{1000 - 300}{1000}\right) \times 0.4$$

or, $$\frac{w}{Q_1} = 0.28$$

$\therefore$ $$w = (0.28)\,Q_1 \quad \text{...(i)}$$

Similarly,

$$COP_{HP} = \left(\frac{T_1}{T_1 - 300}\right) \times 0.5 = \frac{0.5\,T_1}{T_1 - 300}$$

$\Rightarrow$ $$\frac{Q_4}{w} = \frac{w + Q_3}{w} = 1 + \frac{Q_3}{w} = \frac{0.5\,T_1}{T_1 - 300}$$

or, $$\frac{Q_3}{w} = \frac{300 - 0.5\,T_1}{T_1 - 300}$$

$\therefore$ $$w = \frac{Q_3(T_1 - 300)}{(300 - 0.5\,T_1)} \quad \text{...(ii)}$$

$$Q_3 = 2Q_2$$

and from heat engine energy balance

$$Q_1 = w + Q_2$$

$$\therefore \quad Q_2 = Q_1 - w = Q_1 - 0.28\,Q_1$$

$$= 0.72\,Q_1 \qquad [\because w = 0.28\,Q_1]$$

$$\therefore \quad Q_3 = 2Q_2 = 1.44\,Q_1$$

∴ from (ii)

$$0.28\,Q_1 = 1.44\,Q_1 \times \left(\frac{T_1 - 300}{300 - 0.5\,T_1}\right)$$

or $$84 - 0.14\,T_1 = 1.44\,T_1 - 432$$

or $$1.58\,T_1 = 516$$

$$\therefore \quad T_1 = 326.58 \text{ K } \textbf{Ans}$$

(b) $$(\text{COP})_{HP} = \left(\frac{T_1}{T_1 - 300}\right) \times 0.5$$

$$\Rightarrow \quad \frac{Q_4}{w} = \left(\frac{326.58}{326.58 - 300}\right) \times 0.5$$

$$\therefore \quad Q_4 = \left(\frac{326.58}{326.58 - 300}\right) 0.5 \times 0.28 \times Q_1$$

$$= \left(\frac{326.58}{326.58 - 300}\right) 0.5 \times 0.28 \times 50$$

$$= 86 \text{ kW } \textbf{Ans}$$

Problem 6.20 : A solar-powered heat pump receives heat from a solar collector at T_h rejects heat to the atmosphere at T_a and pumps heat from a cold space at T_C. The three heat transfer rates are Q_h Q_a and Q_C respectively. Derive an expression for the minimum ratio Q_h/Q_c in terms of the thee temperatures.
If T_h = 400 K, T_a = 300 K, T_C = 200 k, Q_C = 12 kW, what is the minimum Q_h? If the collector captures 0.2 kW/m², what is the minimum collector area required?

Solution : Given :

$$T_h = 400 \text{ K}$$

$$T_a = 300 \text{ K}$$

$$T_C = 200 \text{ K}$$

$$Q_C = 12 \text{ kW}$$

$$\therefore \frac{Q_h}{12} = \left(\frac{300-200}{400-300}\right) \times \frac{400}{200}$$

$\therefore Q_h$ = 24 kW **Ans**

$\therefore$ Area of solar collector

$$= \frac{Q_h}{0.2} = \frac{24}{0.2}$$

$= 120 \text{ m}^2$ **Ans**

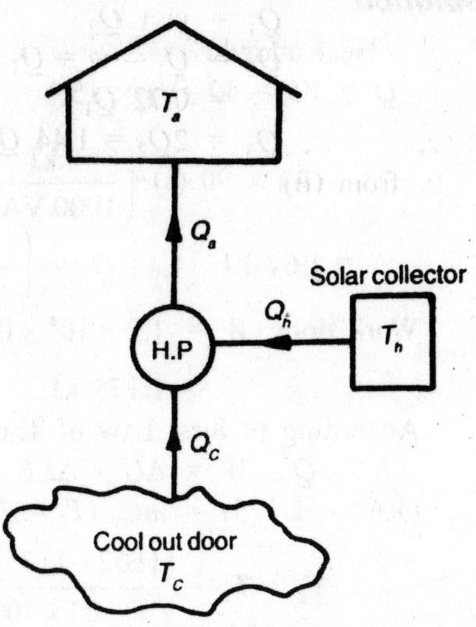

Fig. P. 6.20

From energy balance

$$Q_a = Q_h + Q_C \quad \ldots\text{(i)}$$

We know that $(\Delta S)_{\text{universe}} \geq 0$

$$\text{or} \quad \frac{Q_a}{T_a} - \frac{Q_h}{T_h} - \frac{Q_c}{T_c} \geq 0$$

$$\text{or} \quad \frac{Q_a}{T_a} \geq \frac{Q_h}{T_h} + \frac{Q_c}{T_c}$$

$$\text{or} \quad \frac{Q_h + Q_c}{T_a} \geq \frac{Q_h}{T_h} + \frac{Q_c}{T_c} \qquad [\because Q_a = Q_h + Q_c]$$

$$\text{or} \quad Q_h\left[\frac{T_h - T_a}{T_h \times T_a}\right] \geq Q_C\left[\frac{T_a - T_c}{T_c \times T_a}\right]$$

For minimum ratio $\frac{Q_h}{Q_C}$

$$\frac{Q_h}{Q_C} = \left(\frac{T_a - T_C}{T_h - T_a}\right)\frac{T_h}{T_C} \quad \textbf{Ans}$$

Problem 6.21 : A vertical cylinder with a freely floating piston contains 0.1 kg air at 1.2 bar and a small electrical resistor. The resistor is wired to an external 12 volt battery. When a current of 1.5 A is passed through the resistor for 90 seconds, the piston sweeps a volume of 0.01 m^3. Assume (i) piston and the cylinder are insulated and (ii) air behaves as an ideal gas with C_v = 700 J/kg.K. Find the rise in temperature of air.

Solution :

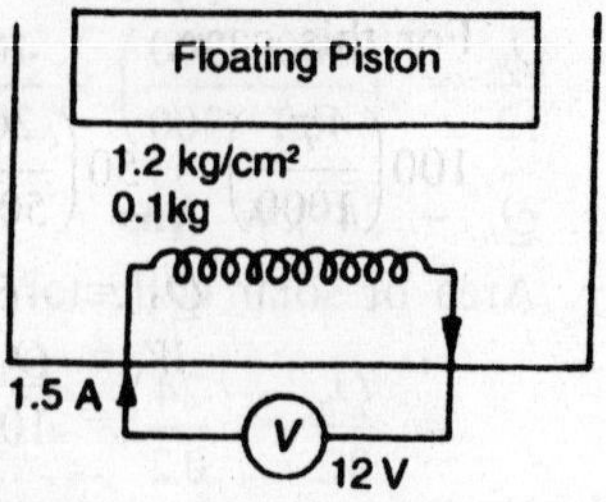

Fig. P. 6.21

Heat added

$$Q = VIt = 12\ (\text{V}) \times 15\ (\text{A}) \times 90\ (\text{s}) \left(\frac{1\ \text{kJ}}{1000\ \text{VAs}}\right)$$

$$= 1.62\ \text{kJ}$$

Work done $W = 1.2 \times 10^4 \times 0.01 \times \dfrac{4.18}{427}$

$$= 1.175\ \text{kJ}$$

According to First Law of Thermodynamics for a closed system

$$Q - W = \Delta U + \Delta KE + \Delta PE = \Delta U$$

$$(1.62 - 1.175) = mC_v\,(T_2 - T_1)$$

$$T_2 - T_1 = \frac{(1.62 - 1.175)1000}{0.1 \times 700} = 6.357°\text{C}$$

Problem 6.22 : P 6.22 shows a reversible heat engine E_R having heat interactions with three constant temperature systems. Calculate the thermal efficiency of the heat engine.

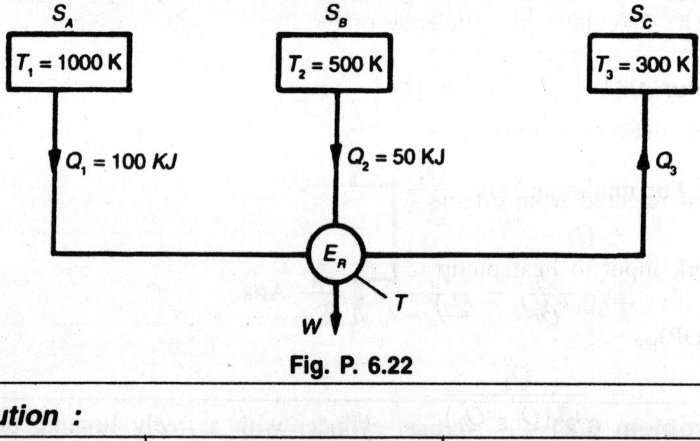

Fig. P. 6.22

Solution :

Source	*Work done*	*Heat rejected*
S_A	$100\left(\dfrac{1000 - T}{1000}\right)$	$100\left(\dfrac{T}{1000}\right)$
S_B	$50\left(\dfrac{500 - T}{500}\right)$	$50\left(\dfrac{T}{500}\right)$
S_C	$Q_3\left(\dfrac{300 - T}{300}\right)$	$Q_3\left(\dfrac{T}{300}\right)$

For this case,

$$100\left(\frac{T}{1000}\right) + 50\left(\frac{T}{500}\right) + Q_3\left(\frac{T}{300}\right) = 0$$

$$Q_3 = -60 \text{ kJ}$$

$$W = Q_1 + Q_2 + Q_3$$

$$= 100 + 50 - 60 = 90 \text{ kJ}$$

$$n_{\text{thermal}} = \frac{\text{Heat added} - \text{Heat rejected}}{\text{Heat added}}$$

$$= \frac{100 + 50 - 60}{100 + 50} = 60\%$$

Problem 6.23 : A heat engine drives a heat pump at COP equal to 5, by supplying 70 per cent of the work output. The heat rejected by the heat engine and the heat pump are used to heat water in a certain chemical plant. The total heat received by the chemical plant is twice the amount of heat supplied by the heat engine.

(a) Give a schematic diagram of the complete system.

(b) Calculate the efficiency of the heat engine.

Solution :

Heat input to engine

$= Q_1$

Heat rejected from engine

$= Q_2$

Work input to heat pump

$= 0.7\ (Q_1 - Q_2)$

$(\text{COP})_{HP} = 5$

$$= \frac{Q_3}{0.7\,(Q_1 - Q_2)}$$

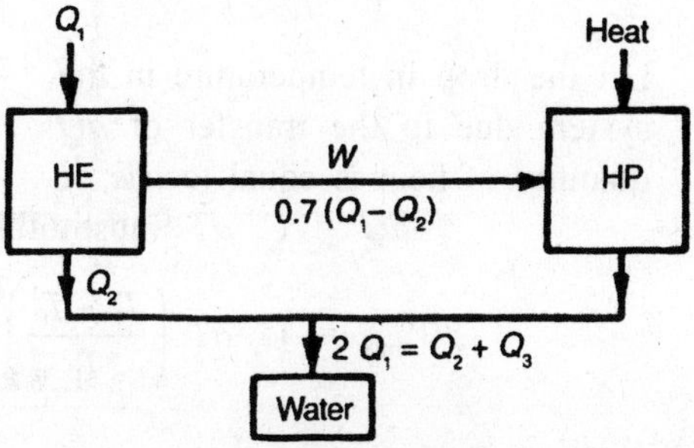

Fig. P. 6.23

$\therefore \quad Q_3 = 3.5\ (Q_1 - Q_2)$

$$\text{Heat given to water} = 3.5\ (Q_1 - Q_2) + Q_2$$

$$3.5\ (Q_1 - Q_2) + Q_2 = 2Q_1$$

$$3.5\ Q_1 - 2.5Q_2 = 2Q_1$$

$$2.5\ Q_2 = 1.5Q_1$$

$$\frac{Q_2}{Q_1} = \frac{1.5}{2.5} = 0.6$$

$$n_{HE} = \frac{Q_1 - Q_2}{Q_1} = 1 - \frac{Q_2}{Q_1}$$

$$= 1 - 0.6 = .04$$

Problem 6.24 : A system maintained at constant volume is initially at temperature T_1. If a heat reservoir at temperature T_0 which is less than T_1 is available, determine the maximum work obtainable as the system is cooled to T_0 in terms of T_0, T_1 and C_v. (CES – 1986)

Solution : A reversible heat engine operating between the system and receiver will give the maximum work obtainable from the system.

Let dQ be the heat taken out of the system and dW_{max} is the corresponding work output

$$dW_{max} = dQ \times n_{engine}$$

$$n_{engine} = \frac{T_1 - T_0}{T_1}$$

= efficiency of reversible engine operating between (1) and (2)

$$dW_{max} = dQ \times \left(\frac{T_1 - T_0}{T_1}\right) \quad ...(1)$$

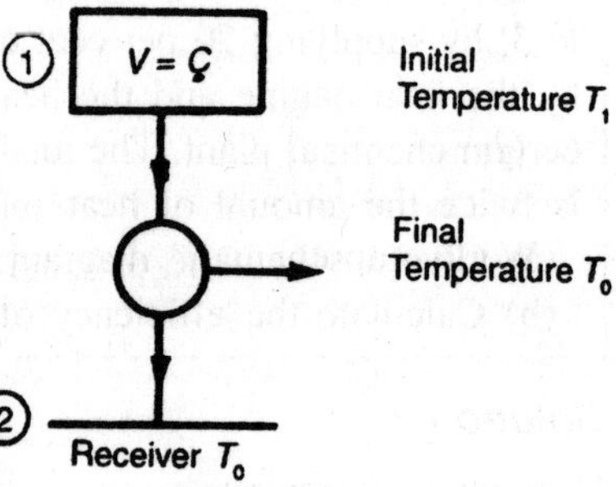

Fig. P. 6.24

Let the drop in temperature in the system due to the transfer of dQ quantity of heat is equal to dT.

$dQ = C_v\ dT$ Substituting this in (1)

$$dW_{max} = C_v\ dT \left(\frac{T_1 - T_0}{T_1}\right) = C_v \left(1 - \frac{T_0}{T_1}\right) dT$$

$$W_{max} = \int_{T_1}^{T_0} dW = \int_{T_1}^{T_0} C_v \left(1 - \frac{T_0}{T_1}\right) dT$$

$$= C_v \left[T - T_0 \log T\right]_{T_1}^{T_0}$$

$$W_{max} = C_v \left[(T_0 - T_1) - T_0 \log\left(\frac{T_0}{T_1}\right)\right]$$

Problem 6.25 : An engine in outer space operates on the carnot cycle. The only way in which heat is rejected by the engine to the surroundings is by radiation which is proportional to the product of the fourth power of the absolute temperature of the radiating surface and its area. For a given power output of the engine, hot reservoir temperature T_1, determine the ratio $\left(\frac{T_1}{T_2}\right)$ for which the area of the radiating surface is minimum. (CES – 1986)

Solution : Let 'A' be area of the radiating surface

$$Q_2 = KAT_2^4$$

$$\frac{Q_1}{T_1} = \frac{Q_2}{T_2},$$

$$Q_1 = \left(\frac{T_1}{T_2}\right) Q_2$$

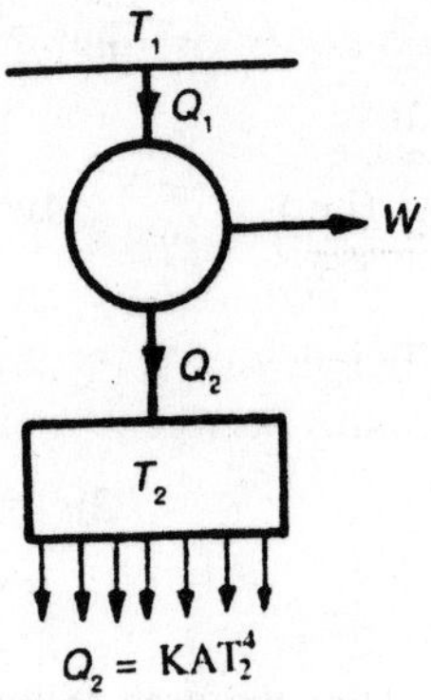

Fig. P. 6.25 (a)

Work output

$$= Q_1 - Q_2 = \left(\frac{T_1}{T_2}\right) Q_2 - Q_2$$

$$= Q_2 \left[\frac{T_1}{T_2} - 1\right] = KAT_2^4 \left[\frac{T_1}{T_2} - 1\right]$$

Let $$\frac{T_1}{T_2} = y,$$

$$T_2 = \frac{T_1}{y}$$

$$\text{Work output} = KA\left(\frac{T_1}{y}\right)^4 (y - 1)$$

For a particular output, work done is constant.

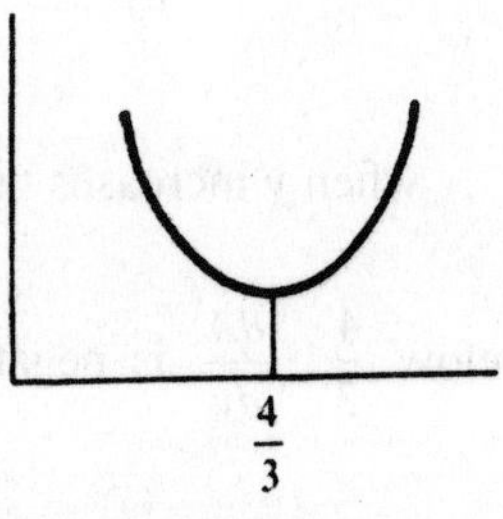

Fig. P. 6.25 (b)

$$\frac{KAT_1^4(y-1)}{y^4} = C$$

$$A = \frac{Cy^4}{KT_1^4(y-1)}$$

For the minimum area of radiating surface, the ratio of $\frac{T_1}{T_2}$ can be found out by the equation

$$\frac{dA}{d\left(\frac{T_1}{T_2}\right)} = 0; \quad \frac{dA}{dy} = 0$$

$$\frac{dA}{dy} = \frac{C}{KT_1^4}\left[\frac{(y-1)\,4 \times y^3 - y^4\,(1)}{(y-1)^2}\right] = 0$$

(i.e.)

$$4y^3\,(y-1) - y^4 = 0$$

$$4y^3\,(y-1) = y^4$$

$$4\,(y-1) = y$$

$$3y = 4; \; y = \frac{4}{3}$$

For the area to be minimum $\frac{d^2A}{dy^2}$ should be positive or the slope on the right side of the curve is positive and left side is negative.

$$\frac{dA}{dy} = \frac{C}{KT_1^4}\frac{\left(3y^4 - 4y^3\right)}{(y-1)^2} = \frac{C\left(y^3\right)(3y-4)}{KT_1^4(y-1)^2}$$

when y increases beyond $\frac{4}{3}$, $\frac{dA}{dy}$ is positive, and when y decreases below $\frac{4}{3}$, $\frac{dA}{dy}$ is negative. Hence $y = \frac{4}{3}$ is the minimum point.

Hence area is minimum when the ratio of source to sink $\frac{T_1}{T_2}$ is $\frac{4}{3}$.

Problem 6.26 : A spherical balloon has a diameter of 0.3 m and contains air at a pressure of 1.5 kg/cm². The diameter of the balloon increases to 0.4 m due to heating and during this process the pressure proportional to diameter. Calculate the work done during this process.

Solution :

Let d_1 and d_2 be the initial and final diameters of the balloon.

$P_1 = kd_1$ (As pressure at anytime is proportional to the diameter)

$$P_2 = kd_2$$
$$P_1 = 1.5 \text{ kg/cm}^2$$
$$d_1 = 0.3 \text{ m}$$
$$d_2 = 0.4 \text{ m}$$

$$k = \frac{1.5 \times 10^4}{0.3} \frac{\text{kg}}{\text{m}^3}$$

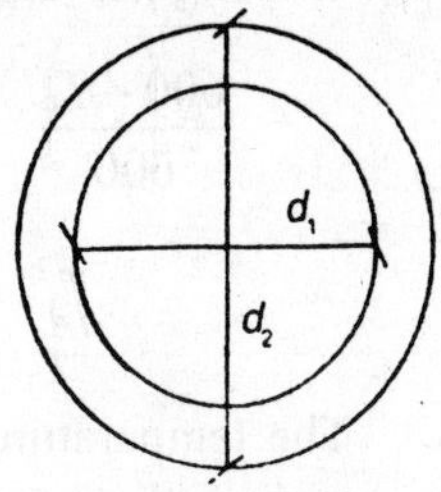

Fig. P. 6.26

$$\text{Work done} = \int P\, dV;\ P = kd$$

$$V = \frac{4}{3}\pi r^3 = \frac{4}{3}\pi\left(\frac{d}{2}\right)^3$$

$$dV = \frac{4}{3}\pi \times 3 \times \left(\frac{d}{2}\right)^2 \frac{1}{2} \times dd$$

$$\text{Work done} = \int_{0.3}^{0.4} kd \times \frac{4}{8}\pi d^2 \times dd = \frac{k\pi}{2}\left[\frac{d^4}{4}\right]_{0.3}^{0.4}$$

$$= \frac{1.5 \times 10^4}{0.3} \times \frac{\pi}{2}\left[\frac{0.4^4}{4} - \frac{0.3^4}{4}\right]$$

$$= 343.6 \text{ kg.m}$$

Problem 6.27 : Two reversible heat engines E_1 and E_2 are kept in series between a hot reservoir at a temperature T_1 of 600 K and cold reservoir at a temperature T_2 of 300 K. Engine E_1 receives 500 kJ of heat from the reservoir T_1. The thermal efficiency of both the engines is the same. Determine the

(i) temperature at which heat is rejected by engine E_1 and received by engine E_2

(ii) work done by engine E_1
(iii) work done by engine E_2
(iv) heat rejected by engine E_2 to the cold reservoir and
(v) efficiency of the engines. (CES – 1985)

Solution : Let T_3 be the temperature at which heat is rejected by engine E_1 to engine E_2 and Q_2 be the heat rejected by engine E_2 to the cold reservoir at temperature T_2 = 300 K. Since the thermal efficiencies of both the engines are same

$$\frac{600 - T_3}{600} = \frac{T_3 - 300}{T_3}$$

$$T_3^2 = 180000;$$

$$T_3 = 424.2 \text{ K}$$

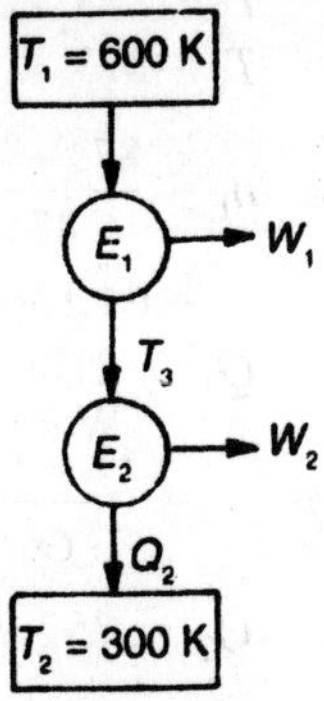

Fig. P. 6.27

The temperature at which heat is rejected by engine E_1 to engine E_2 is 424.2 K.

Work done by engine E_1

$$\frac{W.D}{500} = \eta_{E_1} = \frac{600 - 424.2}{600}$$

$$W.D_{E_1} = \frac{(600 - 424.2)}{600} \times 500$$

$$= 146.5 \text{ kJ}$$

Work done by engine E_2

Heat received by engine E_2 = 500 – 146.5 = 353.5 kJ

Work done by engine E_2

$$\frac{W.D}{353.5} = \frac{424.2 - 300}{424.2},$$

$$\text{W. } D_{E_2} = 103.5 \text{ kJ}$$

Heat rejected to cold source by engine E_2 is :

353.5 – 103.5 = 250 kJ.

Efficiency of engine E_1 = efficiency of engine E_2

$$= \frac{600 - 424.2}{600} = 29.3\%$$

Problem 6.28 : A reversible heat engine operates between 600°C and 40°C and drives a reversible refrigerator operating between 40°C and –18°C. Still there is net output of work equal to 370 kJ, while the heat received by the engine is 2100 kJ. Determine the cooling effect.

Solution :

$T_1 = 600 + 273 = 873$ K

$T_2 = 273 + 40 = 313$ K

$T_3 = 273 - 18 = 255$ K

$T_4 = 273 + 40 = 313$ K

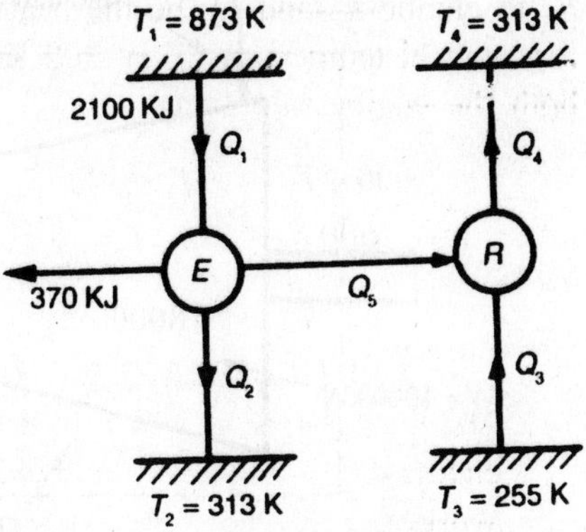

Fig. P. 6.28

$$n_E = \frac{873 - 313}{873}$$

$$= 64.15\%$$

$Q_1 = 2100$ kJ

$$= \frac{Q_1 - Q_2}{Q_1} = \frac{2100 - Q_2}{2100}$$

$Q_2 = 752.9$ kJ

$$W = Q_1 - Q_2$$
$$= 2100 - 752.9$$
$$= 1347.1 \text{ kJ}$$
$$= Q_5 + 370 \text{ kJ}$$

$\therefore$ $Q_5 = 977.1$ kJ

For the refrigerator

$$W_{in} = Q_4 - Q_3$$

$$COP_R = \frac{Q_3}{W_{in}} = \frac{Q_3}{Q_4 - Q_3} = \frac{1}{\frac{Q_4}{Q_3} - 1}$$

$$= \frac{1}{\frac{T_4}{T_3} - 1} = \frac{1}{\left(\frac{313}{255} - 1\right)} = \frac{1}{0.2275} = 4.4$$

$$Q_3 = 4.4 \times W_{in}$$
$$= 4.4 \times Q_5 = 4.4 \times 977.1$$
$$= 4299 \text{ kJ}$$

Cooling effect $= Q_3 = 4299$ kJ

Problem 6.29 : A compressor takes in 500 kg/minute of air at 0.98 bar and 68°C. The diameters of inlet and delivery pipes are 450 mm and 200 mm. The power input is 1000 kW. Determine the rate and direction of heat flow. (C_P = 1.005 kJ/kg°C) (CES – 1995)

Solution :

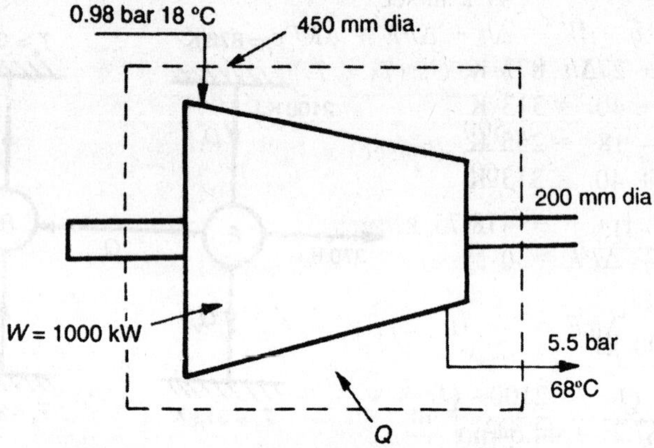

Fig. P. 6.29

$$\text{Mass flow rate} = \frac{500}{60}\ \text{kg/sec}$$

$$= \frac{25}{3}\ \text{kg/sec}$$

$$P_1V_1 = mRT_1$$

$$T_1 = 273 + 18 = 291\ \text{K}$$

$$V_1 = \frac{mRT_1}{P_1}$$

$$= \frac{\frac{25}{3}(\text{kg / sec}) \times 0.287\left(\frac{\text{kPa} \times \text{m}^3}{\text{kg} \times \text{K}}\right) \times 291\,(\text{K})}{98\ (\text{kPa})} = 7.1\ \text{m}^3/\text{sec.}$$

$$\text{Velocity at 1} = \frac{7.1}{\left(\frac{\pi}{4} \times 0.45^2\right)} = 44.64\ \text{m/sec.}$$

$$T_2 = 68 + 273 = 341\ \text{K}$$

$$\text{Velocity at 2} = \frac{\frac{mRT_2}{P_2}}{\frac{\pi}{4}\times(0.2)^2} = \frac{\frac{25}{3}\times\frac{0.287\times 341}{550}}{\frac{\pi}{4}\times(0.2)^2}$$

$$= 47.2 \text{ m/sec.}$$

$$q - W = \Delta h + \Delta PE + \Delta KE$$

$$\Delta h = m \times C_P\,(T_2 - T_1)$$

$$= \frac{500}{60}\ (\text{kg/sec}) \times 1.005 \left(\frac{\text{kJ}}{\text{kg}^\circ\text{C}}\right) (68 - 18)\ (^\circ\text{C})$$

$$= 418.75 \text{ kJ/sec.}$$

$$\Delta PE = 0$$

$$\Delta KE = \frac{1}{2}\left(V_2^2 - V_1^2\right)$$

$$= \frac{1}{2}\ (47.2^2 - 44.64^2)\left(\frac{\text{m}^2}{\text{sec}^2}\right) \times \left(\frac{1\text{ kJ/kg}}{1000\text{m}^2/\text{s}^2}\right) = 0.1176 \text{ kJ/kg}$$

$$= 0.1176 \times \frac{500}{60} \text{ kJ/sec} = 0.9796 \text{ kJ/sec}$$

W = 1000 kW = 1000 kJ/sec = –1000 kJ/sec.

(as work is done on the system)

$$q - W = \Delta h + \Delta PE + \Delta KE$$

$$q + 1000 = 418.75 + 0 + 0.9796$$

$$q = -580.3 \text{ kJ/sec.}$$

(negative sign indicates that the heat has flown out of the system)

Problem 6.30 : A reversible engine receives heat from a mixture of water vapour and liquid water at 1 atm and rejects 3775 kJ/hr of heat at 100 K below temperature of a mixture of ice and liquid water at 1 atm. It delivers 0.386 kW power. Find the number of degrees separating absolute zero and ice point on Kelvin scale. (CES–1994)

Solution :

Let T_1 and T_2 be the source and sink temperatures $(T_1 - T_2)$ = 100 K

$$n = \frac{T_1 - T_2}{T_1} = \frac{100}{T_1}$$

$$= \frac{Q_1 - Q_2}{Q_1} = \frac{W}{(Q_2 + W)}$$

$$= \frac{1389.6}{(3775 + 1389.6)}$$

$$\frac{100}{T_1} = \frac{1389.6}{(3775 + 1389.6)}$$

$$T_1 = 371.66$$
$$T_2 = 371.66 - 100$$
$$= 271.66 \text{ K}$$

Degrees separating absolute zero and ice point on Kelvin

$$= 27316 - 271.66$$
$$= 1.50 \text{ degrees.}$$

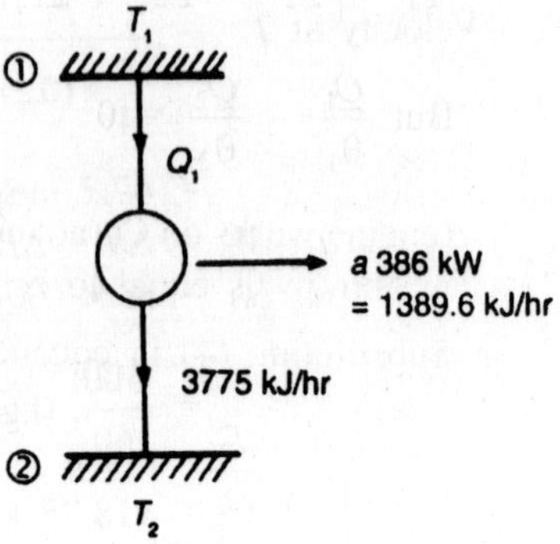

Fig. P. 6.30

Problem 6.31 : A working fluid undergoes through a Carnot cycle of operations, the upper absolute temperature of the fluid being θ_1 and the lower absolute temperature being θ_2. The amount of heat taken in and rejected by the working fluid are Q_1 and Q_2 respectively (on account of losses of heat due to conduction etc., the heat source temperature T_1 is higher than θ_1 and the sink temperature T_2 is lower than θ_2. If $T_1 = (\theta_2 + KQ_1)$ and $T_2 = (\theta_2 - KQ_2)$, where K is the same constant for both the equations, show that the efficiency of the plant

$$n = 1 - \frac{T_2}{(T_1 - 2KQ_1)}$$ (CES – 1997)

Solution : Given :

$$T_1 = \theta_1 + KQ_1 \quad \ldots(1)$$
$$T_2 = \theta_2 - KQ_2 \quad \ldots(2)$$

From (1) and (2)

$$K = \frac{T_1 - \theta_1}{Q_1} = \frac{\theta_2 - T_2}{Q_2}$$

$$\frac{T_1}{Q_1} - \frac{\theta_1}{Q_1} = \frac{\theta_2}{Q_2} - \frac{T_2}{Q_2} = K \quad \ldots(3)$$

Adding the first two terms

$$\frac{T_1}{Q_1} - \frac{T_2}{Q_2} + \frac{\theta_2}{Q_2} - \frac{\theta_1}{Q_1} = 2K$$

But $\frac{Q_1}{\theta_1} - \frac{Q_2}{\theta_2} = 0$ ...(4)

(engine works on Carnot cycle, change in entropy is equal to zero)

Substituting (4) in equation (3)

$$\frac{T_1}{Q_1} - \frac{T_2}{Q_2} = 2K \quad ...(5)$$

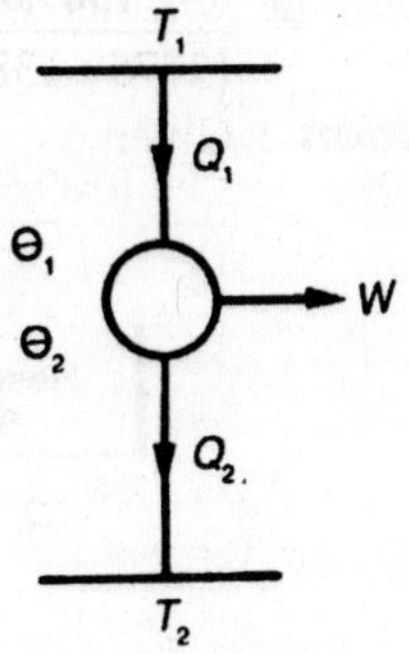

Fig. P. 6.31

Efficiency $n = \frac{Q_1 - Q_2}{Q_1} = 1 - \frac{Q_2}{Q_1}$

$$n = 1 - \frac{Q_2}{Q_1}$$

To prove that $n = 1 - \frac{T_2}{T_1 - 2KQ_1} = 1 - \frac{Q_2}{Q_1}$

As per equation (5) $2K = \frac{T_1}{Q_1} - \frac{T_2}{Q_2}$

Substituting this in the above equation

$$R.H.S. = n = 1 - \frac{T_2}{T_1 - \left[\frac{T_1}{Q_1} - \frac{T_2}{Q_2}\right]Q_1} = 1$$

$$= 1 - \frac{T_2}{T_1 - \left[T_1 - T_2\frac{Q_1}{Q_2}\right]}$$

$$= 1 - \frac{T_2}{T_1 - T_1 + T_2\frac{Q_1}{Q_2}} = 1 - \frac{Q_2}{Q_1} = \text{L.H.S.}$$

Problem 6.32 : An inventor claims that he has developed a heat engine which absorbs 1200 kJ and 800 kJ from reservior at 800 K and 600 K respectively and rejects 600 kJ and 200 kJ as heat to reservior at 400 K and 300 K respectively. It delivers 1200 kJ work. Determine whether the heat engine is theoretically possible.

Solution : Given :

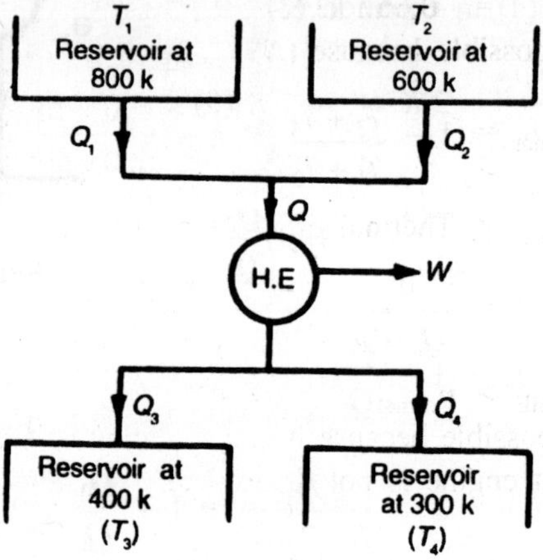

Fig. P. 6.32

$$Q_1 = 1200 \text{ kJ}$$
$$Q_2 = 800 \text{ kJ}$$
$$Q_3 = 600 \text{ kJ}$$
$$Q_4 = 200 \text{ kJ}$$
$$W = 1200 \text{ kJ}$$

(a) From entropy principle

Change in entropy of Reservoir at 500 K

$$= (\Delta S)_1 = -\frac{1200}{800} \text{ kJ/K}$$

Change in entropy of reservoir at 60C

$$= (\Delta S)_2 = -\frac{800}{600} \text{ kJ/K}$$

Change in entropy of reservoir at 400 K

$$= (\Delta S)_3 = \frac{600}{400} \text{ kJ/K}$$

Change in entropy of reservoir at 300 K

$$= (\Delta S)_4 = \frac{200}{300} \text{ kJ/K}$$

Change in entropy of H.E. = $(\Delta S)_5 = 0$

$$\therefore \quad (\Delta S)_{\text{universe}} = (\Delta S)_1 + (\Delta S)_2 + (\Delta S)_3 + (\Delta S)_4 + (\Delta S)_5$$

$$= -\frac{1200}{800} - \frac{800}{600} + \frac{600}{400} + \frac{200}{300}$$

$$= -0.667 \text{ kJ/K} < 0$$

which is impossible because $(\Delta S)_{\text{universe}} \geq 0$

(b) $$n_{\text{carnot}} = 1 - \frac{T_3 + T_4}{T_1 + T_2} = 1 - \frac{400 + 300}{600 + 800} = 0.5$$

Now $$n_{\text{HE}} = \text{Thermal efficiency of engine}$$

$$= \frac{W}{Q_{added}} = \frac{W}{Q_1 + Q_2} = \frac{1200}{2000} = 0.6$$

Here $n_{\text{HE}} > n_{\text{carnot}}$

which is impossible because it is a violation of Carnot's theorem. Hence, given heat engine is not theoretically possible.

Chapter 7

Entropy

Problem 7.1 : Two kg of water at 80°C are mixed adiabatically with 3 kg of water at 30°C in a constant pressure process of 1 atmosphere. Find the increase in the entropy of the total mass of water due to the mixing process (Cp of water = 4.187 kJ/kg.K)

Solution : Given data :

Water A; m_A = 2 kg, T_A = 80°C = 80 + 273 = 353 K

Water B; m_B = 3 kg, T_B = 30°C = 30 + 273 = 303 K

Atmospheric pressure = P = 1 atm

C_P of water = 4.187 kJ/kg.K

Let at equilibrium. t_f be the final temperature and

$$T_B < T_f < T_A$$

from heat balance,

heat loss = heat gain

$$\Rightarrow \quad m_A C_P (T_A - T_f) = m_B C_P (T_f - T_B)$$

$$\Rightarrow \quad T_f = \frac{m_A T_A + m_B T_B}{m_A + m_B} = \frac{2 \times 353 + 3 \times 303}{2 + 3}$$

$$= 323 \text{ K}$$

Now from property relation

$$Tds = dH - VdP$$

$$ds = \frac{dH}{T} - R\frac{dP}{P}$$

For constant pressure process, we get,

$$S_f - S_i = \Delta S = \int_i^f \frac{mC_p dT}{T} = mC_p \ln \frac{T_f}{T_i}$$

Hence $(\Delta S)_A$ = change in entropy of water A = $m_A C_P \ln \dfrac{T_f}{T_A}$

$$\& \ (\Delta S)_B = \text{change in entropy of water } B = m_B C_P \ln \frac{T_f}{T_B}$$

$$\therefore (\Delta S)_{\text{Total mass}} = \Delta S_A + \Delta S_B$$

$$= m_A C_P \ln \frac{T_f}{T_A} + m_B C_p \ln \frac{T_f}{T_B}$$

$$= 2 \times 4.187 \ln \frac{323}{353} + 3 \times 4.187 \ln \frac{323}{303}$$

$$= 0.05915 \text{ kJ/K}$$

= +*ve*, hence there is an increase in entropy. **Ans**

Problem 7.2 : In a Carnot cycle, heat is supplied at 350°C and rejected at 27°C. The working fluid is water which, while receiving heat, evaporates from liquid at 350°C to steam at 350°C. The associated entropy change is 1.44 kJ/kg.K. (a) If the cycle operates on a stationary mass of 1 kg of water, how much is the work done per cycle, and how much is the heat supplied? (b) If the cycle operates in steady flow with a power output of 20 kW, what is the steam flow rate?

Solution : Given data :

T_1 = 350°C = 350 + 273 = 623 K

T_2 = 27°C = 27 + 273 = 300 K

$S_1 - S_4$ = 1.44 kJ/kg.K

(a) For stationary mass system.

Q_1 = heat supplied at reversible isothermal process 1-4

$= T_1 (S_1 - S_4)$

$$= 623 \times 1.44 \text{ K} \times \frac{\text{kJ}}{\text{kg.K}}$$

= 897.12 kJ/kg. **Ans**

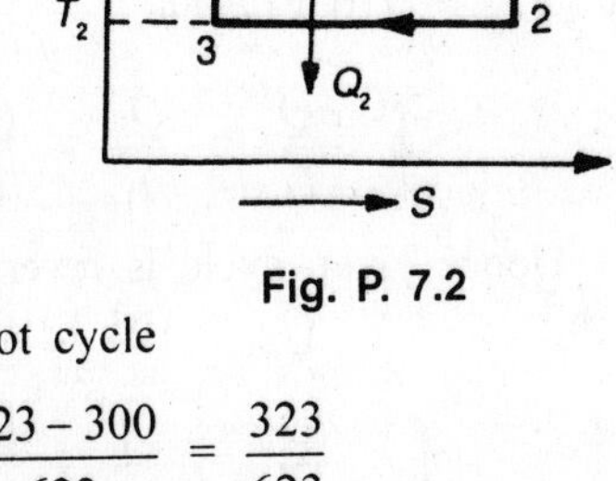

Fig. P. 7.2

Now thermal efficiency of carnot cycle

$$n_{th} = \frac{T_1 - T_2}{T_1} = \frac{623 - 300}{623} = \frac{323}{623}$$

Again $\eta_{th} = \dfrac{\text{work done by the system}}{\text{heat supplied to the system}} = \dfrac{w}{Q_1}$

$$\Rightarrow \quad w = \frac{323}{623} \times 987.12 = 465.12 \text{ kJ/kg} \textbf{ Ans}$$

(b) For steady state flow system

Say, $\dot{m}$ = steam rate flow

$$= \frac{\text{power output}}{(\text{work done by system / kg})} = \frac{20 \text{ kW}}{465.12 \text{ kJ / kg}} = 0.043 \text{ kg/s} \textbf{ Ans}$$

Problem 7.3 : A heat engine receives reversibly 420 kJ/cycle of heat from a source at 327°C, and rejects heat reversibly to a sink at 27°C. There are no other heat transfers. For each of the three hypothetical amounts of heat rejected, in (a), (b), and (c) below, compute the cyclic integral of dQ/T. From these results show which case is irreversible, which reversible, and which impossible. (a) 210 kJ/cycle rejected, (b) 105 kJ/cycle rejected, (c) 315 kJ/cycle rejected.

Solution : Given data :

Q_1 = heat received
= 420 kJ/cycle

T_1 = source temperature
= 327°C = 327 + 273
= 600 K

T_2 = Sink temperature
= 27°C = 27 + 273
= 300 K

Q_2 = heat rejected to sink

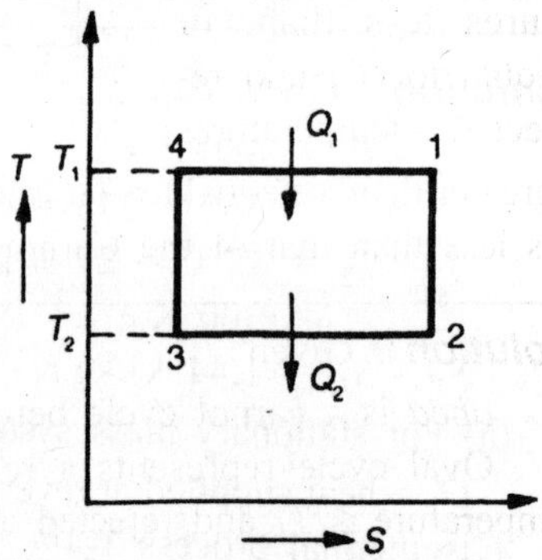

Fig. P. 7.3

(a) Q_2 = 210 kJ/cycle.

$$\oint \frac{dQ}{T} = \frac{Q_1}{T_1} - \frac{Q_2}{T_2} = \frac{420}{600} - \frac{210}{300} = 0$$

Hence given cycle is reversible.

(b) Q_2 = 105 kJ/cycle

$$\oint \frac{dQ}{T} = \frac{Q_1}{T_1} - \frac{Q_2}{T_2} = \frac{420}{600} - \frac{105}{300} = 0.35$$

Here $\oint \frac{dQ}{T} \geq 0$, hence cycle is impossible.

(c) $Q_2 = 315$ kJ/cycle

$$\oint \frac{dQ}{T} = \frac{Q_1}{T_1} - \frac{Q_2}{T_2} = \frac{420}{600} - \frac{315}{300} = -0.35$$

Here $\oint \frac{dQ}{T} \leq 0$, hence cycle is irreversible.

Problem 7.4 : abcd represents a Carnot cycle bounded by two reversible adiabatics and two reversible isotherms at temperatures T_1 and T_2 ($T_1 > T_2$). The oval figure is a reversible cycle, where heat is absorbed at temperatures less than, or equal to, T_1, and rejected at temperatures greater than, or equal to, T_2. Prove that the efficiency of the oval cycle is less than that of the Carnot cycle.

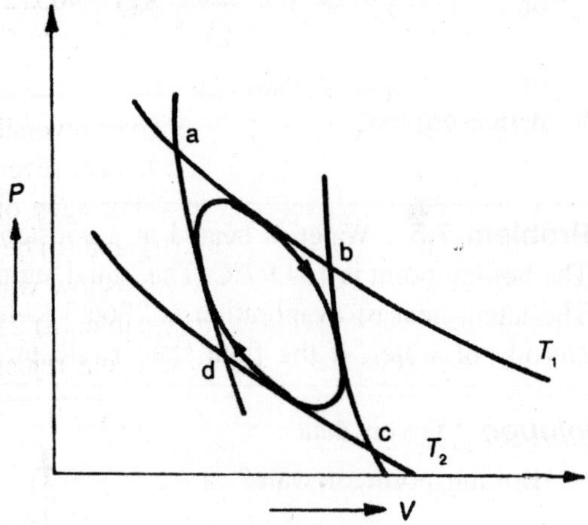

Fig. P. 7.4

Solution : Given :

abcd is a carnot cycle between T_1 and T_2 where $T_1 > T_2$.

Oval cycle represents a reversible cycle where heat received at temperature $\leq T_1$ and rejected at $\geq T_2$.

Now, Let Q_1 = heat received by engine.

& Q_2 = heat rejected by engine for the oval cycle.

$\therefore$ $(\Delta S)_I$ = change in entropy of reservior at T_1 temperature $= -\frac{Q_1}{T_1}$

& $(\Delta S)_{II}$ = change in entroy of reservior at T_2 temperature $= -\frac{Q_2}{T_2}$

$\therefore$ Now, $(\Delta S)_{universe} \geq 0$

$$\Rightarrow \quad -\frac{Q_1}{T_1} + \frac{Q_2}{T_2} \geq 0$$

or, $$\frac{Q_2}{T_2} \geq \frac{Q_1}{T_1}$$

or, $$\frac{Q_1}{T_1} \leq \frac{Q_2}{T_2}$$

or $$\frac{T_2}{T_1} \leq \frac{Q_2}{Q_1}$$

or $$1 - \frac{T_2}{T_1} \geq 1 - \frac{Q_2}{Q_1}$$

or $$n_{\text{carnot}} \geq n_{\text{oval cycle}}$$

hence proved

> **Problem 7.5 :** Water is heated at a constant pressure of 0.7 MPa. The boiling point is 164.97°C. The initial temperature of water is 0°C. The latent heat of evaporation is 2066.3 kJ/kg. FInd the increase of entropy of water, if the final state is steam.

Solution : Given data :

Boiling point of water $= T_2 = 164.97°C$

$= 164.97 + 273 = 437.97$ K

initial temperature of water $= T_1 = 0°C = 273$ K

latent heat of vaporisation $= L = 2063.3$ kJ/kg

From property relation, $$ds = \frac{dH}{T} - \frac{V}{T}\,dp$$

For constant pressure process, $dp = 0$, we get,

$$ds = \frac{dH}{T}$$

$$\Delta S = S_f - S_i = \int_i^f mC_p \frac{dT}{T}$$

entropy change of water as heated from 0°C to 164.97°C

$$\Delta S_1 = mC_P \ln \frac{T_2}{T_1}$$

$$= 1 \times 4.187 \times \ln \frac{437.97}{273}$$

$$= 1.9791 \text{ kJ/kg.K}$$

entropy change of water at 164.97°C to steam at 164.97°C

$$\Delta S_{II} = \frac{L}{T_2} = \frac{2063.3}{437.97} = 4.7111 \text{ kJ/kg.K}$$

∴ total entropy change of water

$$= \Delta S_I + \Delta S_{II} = 6.6902 \text{ kJ/Kg.K } \textbf{Ans}$$

Problem 7.6 : Ten grammes of water at 20°C is converted into ice at –10°C at constant atmospheric pressure. Assuming the specific heat of liquid water to remain constant at 4.2 J/g.K and that of ice to be half of this value, and taking the latent heat of fusion of ice at 0°C to be 335 J/g, calculate the total entropy change of the system.

***Solution* :** Given data :

mass of water = m = 10 gm

latent heat of fusion of ice at 0°C = 335 J/g

sp. heat of water, C_1 = 4.2 J/g.K

sp. heat of water, C_2 = 2.1 J/g.K

From property relation, $ds = \frac{dH}{T} - \frac{V}{T} dp$

For constant pressure process, $dp = 0$

$$ds = \frac{dH}{T}$$

$$\Rightarrow \quad \Delta S = S_f - S_i = \int_i^f mC_p \frac{dT}{T}$$

∴ entropy change of system as water at 20°C converts into water at 0°C,

$$\Delta S_I = mC_P \ln \frac{T_f}{T_i} = 10 \times 4.2 \ln \frac{273}{293}$$

$$= -2.9694 \text{ J/K}$$

entropy change of system as water at 0°C converts into ice at 0°C

$$\Delta S_{II} = -\frac{335}{273} \times 10 = -12.2711 \text{ J/K}$$

entropy change of system as ice at 0°C converts into ice at –10°C

$$\Delta S_{III} = \int_{273}^{263} mC_p \frac{dT}{T}$$

$$= 10 \times 2.1 \ln \frac{263}{273} = -0.7837 \text{ J/K}$$

∴ total entropy change of system water converts into ice

$$(\Delta S)_{total} = \Delta S_I + \Delta S_{II} + \Delta S_{III}$$

$$= 16.02416 \text{ J/K } \textbf{Ans}$$

Problem 7.7 : Calculate the entropy change of the universe as a result of the following processes :

(a) A copper block of 600 g mass and with C_P of 150 J/K at 100°C is placed in a lake at 8°C

(b) The same block, at 8°C, is dropped from a height of 100 m into the lake.

(c) Two such blocks, at 100 and 0°C, are joined together.

***Solution* :** Given data :

mass of copper block = m = 600 gm

C_P = 150 J/K

(a) Temperature of copper block = T_2

= 100°C = 100 + 273 = 373 K

Lake temperature = T_1 = 8°C = 8 + 273 = 281 K

Considering reversible process,

$$\Delta s = \int \frac{dQ}{T} = \int \frac{mC_p dT}{T}$$

entropy change of copper block, as it is cooled from 100°C to 8°C

$$\Delta S_I = \int_{373\,K}^{281\,K} C_P \frac{dT}{T} = C_P \ln \frac{281}{373}$$

$$= 150 \ln \frac{281}{373} = -42.48 \text{ J/K}$$

Heat lost by copper block = heat transferred to the lake

$$Q = C_P \, \Delta T$$

$$= 150 \times (100 - 8) = 13800 \text{ J}$$

∴ entropy change of lake

$$\Delta S_{II} = \frac{Q}{T_1} = \frac{13800}{281} = 49.11 \text{ J/K}$$

∴ Entropy change of universe,

$$(\Delta S)_{univ.} = \Delta S_I + \Delta S_{II} = 6.63 \text{ J/K } \textbf{Ans}$$

(b) Copper block temperature = 8°C = 273 + 8 = 281 K

Lake temperature = 8°C = 273 + 8 = 281 K

Considering isothermal dissipation of work, entropy change of copper block

$$(\Delta S)_I = 0$$

Now this isothermal work transfer dissipates into heat transfer to the surrounding

$$w = Q = mgh = 0.6 \times 9.81 \times 100 = 588.6 \text{ J}$$

∴ entropy change of lake

$$= (\Delta S)_{II}$$

$$= \frac{Q}{T} = \frac{588.6}{281} = 2.095 \text{ J/K. } \textbf{Ans}$$

(c) T_1 = first copper block temperature

= 100°C = 100 + 273 = 373 K

T_2 = 2nd copper block temperature

= 0°C = 273 K

Let T_f be the final temperature then by heat balance,

$$mC_P (T_1 - T_f) = mC_P (T_f - T_2)$$

$$\therefore \quad T_f = \frac{T_1 + T_2}{2} = \frac{373 + 273}{2} = 323 \text{ K}$$

Considering reversible process,

$$\Delta S = \int \frac{dQ}{T} = \int mCp \frac{dT}{T}$$

Change in entropy of 1st copper block,

$$\Delta S_I = C_P \int_{373}^{323} \frac{dT}{T} = 150 \times \ln \frac{323}{373}$$

Similarly change in entropy of 2nd copper block

$$\Delta S_{II} = C_P \int_{273}^{323} \frac{dT}{T} = 150 \times \ln \frac{323}{273}$$

$\therefore$ total change in entropy of universe

$$(\Delta S)_{univ.} = \Delta S_I + \Delta S_{II} = 150 \left[\ln \frac{323}{373} + \ln \frac{323}{273} \right]$$

$$= 3.64 \text{ J/K } \textbf{Ans}$$

Problem 7.8 : A system maintained at constant volume is initially at temperature T_1, and a heat reservior at the lower temperature T_0 is available. Show that the maximum work recoverable as the system is cooled to T_0 is $w = C_V \left[(T_1 - T_0) - T_0 \ln \frac{T_1}{T_0} \right]$

Solution : Given data :

initial temperature of system = T_1

Heat reservoir temperature = T_0

w = work recovered from the system as system at T_1 cooled to T_0 = kJ/kg

From energy balance

$$Q_1 = w + Q_2$$

where Q_1 = heat transferred from the system, kJ/kg

$$\Rightarrow \qquad Q_2 = Q_1 - w \qquad \text{...(i)}$$

Q_2 = heat rejected to the reservior, kJ/kg.

Considering reversible process,

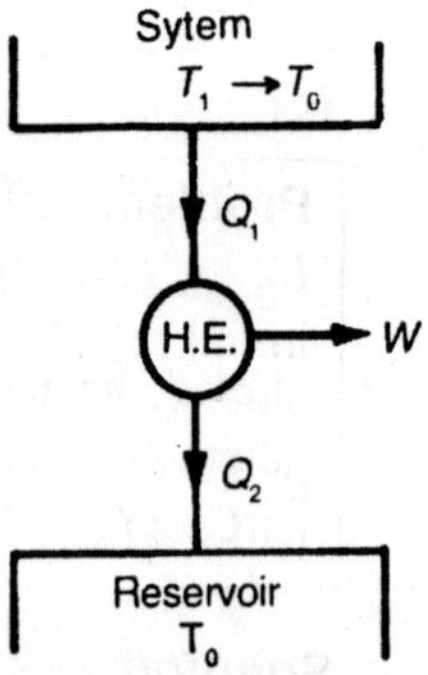

Fig. P. 7.8

$$\Delta S = \text{change in entropy} = \int_{T_{initial}}^{T_{final}} \frac{dQ}{T}$$

$\therefore$ change in entropy of system as it cools from T_1 to T_0

$$\Delta S_I = C_v \int_{T_1}^{T_0} \frac{dT}{T} = C_v \ln \frac{T_0}{T_1} \text{ kJ/kg.K}$$

Now, $\qquad Q_1 = C_v (T_1 - T_0)$

$\therefore$ from (i) $Q_2 = C_v (T_1 - T_0) - w$ kJ/kg

Now, entropy change of the reservior

$$(\Delta S)_{II} = \frac{Q_2}{T_0} = \frac{C_v(T_1 - T_0) - w}{T_0} \text{ kJ/kg.K}$$

Now, by entropy principles,
entropy change of universe

$$= (\Delta S)_{univ} \geq 0$$

$$\Rightarrow \Delta S_I + \Delta S_{II} \geq 0$$

$$\Rightarrow \quad C_V \ln \frac{T_0}{T_1} + \frac{C_v(T_1 - T_0)}{T_0} - \frac{w}{T_0} \geq 0$$

$$\Rightarrow \quad w \leq T_0 C_v \ln \frac{T_0}{T_1} + C_v (T_1 - T_0)$$

$\therefore$ For maximum work recoverable

$$w_{max} = C_v \left[T_0 \ln \frac{T_0}{T_1} + (T_1 - T_0)\right]$$

$$w_{max} = C_v \left[(T_1 - T_0) - T_0 \ln \frac{T_1}{T_0}\right] \text{ kJ/kg } \textbf{Ans}$$

Problem 7.9 : Each of three identical bodies satisfies the equation $U = CT$, where C is the heat capacity of each of the bodies. Their initial temperatures are 200 K, 250 K, and 540 K. If $C = 8.4$ kJ/K, what is the maximum amount of work that can be extracted in a process in which these bodies are brought to a final common temperature?

Solution : Given data :

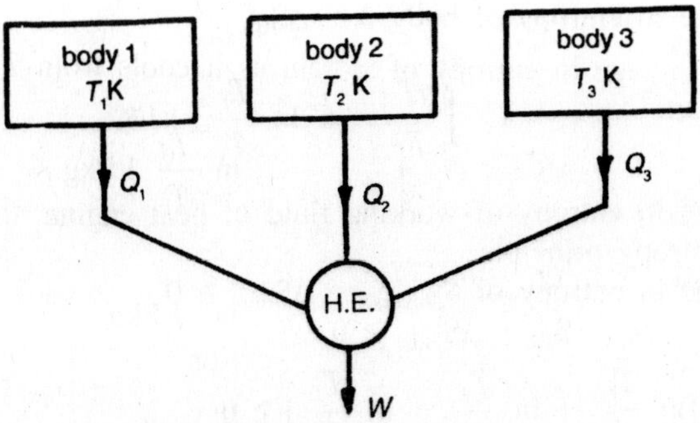

Fig. P. 7.9

$U = CT,$

$C =$ heat capacity kJ/K

initial temperature of body 1 = T_1 K

initial temperature of body 2 = T_2 K

Initial temperature of body 3 = T_3 K

Say, final temperature of the bodies in T_f

then considering the reversible process,

$$\Delta S = \int_{T_{initial}}^{T_{final}} \frac{dQ}{T} \quad ...(i)$$

from 1st law of thermodynamics, for a reversible process

$$dQ = du + pdv$$

Considering no pdv work i.e., constant volume process of a system as a body

$$dQ = du$$

$\therefore$ from (i)

$$\Delta S = \int_{T_{initial}}^{T_{final}} \frac{du}{T} = C \int_{T_{initial}}^{T_{final}} \frac{\Delta T}{T}$$

Now change in entropy of body 1 = ΔS_I.

$$= C \int_{T_1}^{T_f} \frac{dT}{T} = C \ln \frac{T_f}{T_1}, \text{ kJ/K}$$

Change in entropy of body 2 = ΔS_{II}

$$= C \int_{T_2}^{T_f} \frac{dT}{T} = C \ln \frac{T_f}{T_2}, \text{ kJ/K}$$

change in entropy of body 3 = ΔS_{III}

$$= C \int_{T_3}^{T_f} \frac{dT}{T} = C \ln \frac{T_f}{T_3}, \text{ kJ/K}$$

change in entropy of working fluid of heat engine = $\Delta S_{H.E.} = 0$

By entropy principle

change in entropy of $S_{universe} = \Delta S_{univ} \geq 0$

$\Rightarrow S_I + S_{II} + S_{III} + \Delta S_{H.E.} \geq 0$

$$\Rightarrow C [\ln \frac{T_f}{T_1} + \ln \frac{T_f}{T_2} + \ln \frac{T_f}{T_3}] \geq 0$$

$$\Rightarrow \ln \frac{T_f^3}{T_1 T_2 T_3} \geq 0 \qquad \text{...(ii)}$$

Now for a cyclic process, $\oint dw = \oint dQ$

$\Rightarrow$ w = work extracted in a process

$= Q_1 + Q_2 + Q_3$

where Q_1 = heat in or out from body 1 kJ

Q_2 = heat in or out from body 2 kJ

Q_3 = heat in or out from body 3 kJ

$\therefore \quad w = C[(T_1 - T_f) + (T_2 - T_f) + (T_3 - T_f)]$

$= C[T_1 + T_2 + T_3 - 3T_f]$

Hence for maximu, w_{max}, T_f should be minimum

$$\therefore \ln \frac{T_f^3}{T_1 T_2 T_3} = 0 \Rightarrow T_f = \sqrt[3]{T_1 T_2 T_3}$$

$$\therefore \quad w_{max} = C[T_1 + T_2 + T_3 - \sqrt[3]{T_1 T_2 T_3} \times 3]$$

Now putting

$T_1 = 200$ K

$T_2 = 250$ K

$T_3 = 540$ K

$$T_f = \sqrt[3]{200 \times 250 \times 540} = 299.99 \text{ K} = 300 \text{ K}$$

$\therefore w_{max} = 8.4$ kJ/K $[200 + 250 + 540 - 300 \times 3]$ K = 756 kJ **Ans**

$\therefore \quad Q_1$ = heat transferred to the body 1

Q_2 = heat transferred to the body 2

Q_3 = heat transferred from the body 3.

> **Problem 7.10 :** A reversible engine, as shown in figure 7.10, during a cycle of operation draws 5 MJ from the 400 K reservior and does 840 kJ of work. Find the amount and direction of heat interaction with other reserviors. (IES – 1999)

Solution : Now considering reversible process

$$\Rightarrow \quad \Delta S = \frac{dQ}{T} \text{ kJ/K}$$

$\therefore$ Change in entropy of reservoir at 400 K

$$\Delta S_1 = \frac{-Q_1}{400}$$

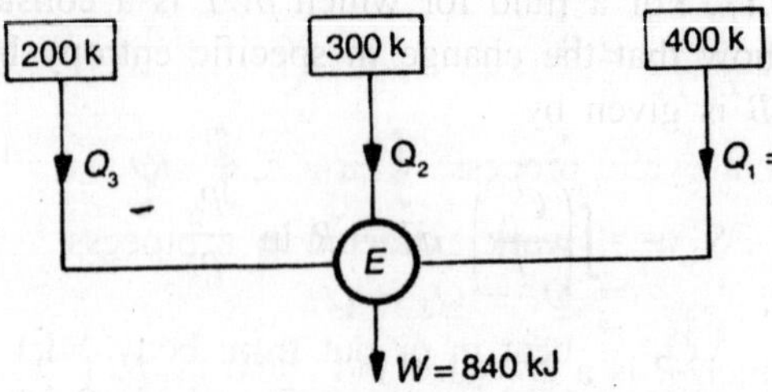

Fig. P. 7.10

$$= -\frac{5}{400} \text{ MJ/K}$$

Suppose Q_2 and Q_3 are heat transferred to the reservior then change in entropy of reservior at 300 K

$$\Delta S_2 = \frac{Q_2}{300} \text{ MJ/K}$$

and change in entropy of reservior at 200 K

$$\Delta S_3 = \frac{Q_3}{200} \text{ MJ/K}$$

∴ By entropy principle, for a reversible engine total change of entropy of universe $\Delta S_{univ} = 0$

$$\Rightarrow \frac{-5}{400} + \frac{Q_2}{300} + \frac{Q_3}{200} = 0$$

$$\Rightarrow \frac{Q_2}{300} + \frac{Q_3}{200} = \frac{5}{400} \quad \text{...(i)}$$

From 1st law of thermodynamics, for a cyclic process $\oint dw = \oint dQ$

$$\Rightarrow \quad w = Q_1 - Q_2 - Q_3 \text{ MJ} \quad \text{...(ii)}$$

from (i)

$$2Q_2 + 3Q_3 = \frac{5}{4} \times 6 = 7.5 \text{ MJ} \quad \text{...(a)}$$

from (ii)

$$Q_2 + Q_3 = 5 \text{ MJ} - 0.84 \text{ MJ} = 4.16 \text{ MJ} \quad \text{...(b)}$$

Now (a) – 2 × (b), we get

$$Q_3 = 7.5 - 8.32 = -0.82 \text{ MJ}$$

$$Q_2 = 4.16 - Q_3 = +4.98 \text{ MJ}$$

Hence, (Q_3) heat is transferred from the reservoir at 200 K and (Q_2) heat transferred to the reservior at 300 K **Ans**

Problem 7.11 : For a fluid for which pv/T is a constant quantity equal to R, show that the change in specific entropy between two states A and B is given by

$$S_B - S_A = \int_{T_A}^{T_B} \left(\frac{C_p}{T}\right) dT - R \ln \frac{P_B}{P_A}$$

A fluid for which R is a constant and equal to 0.287 kJ/kg.K, flows steadily through an adiabatic machine, entering and leaving through two adiabatic pipes. In one of these pipes the pressure and temperature are 5 bar and 450 K and in the other pipe the pressure and temperature are 1 bar and 300 K respectively. Determine which pressure and temperature refer to the inlet pipe.

For the given temperature range, C_P is given by

$C_P = a \ln T + b$

Where T is the numerical value of the absolute temperature and a = 0.026 kJ/kg.K, b = 0.86 kJ/kg.K.

Solution : Given :

$$\frac{pv}{T} = \text{const.} = R$$

Now,

from property relation,

$$Tds = dh - Vdp$$

$$ds = \frac{dh}{T} - \frac{V}{T} dP$$

$$= \frac{dh}{T} - R\frac{dp}{P}$$

on integrating, we get

$$\int_A^B ds = S_B - S_A$$

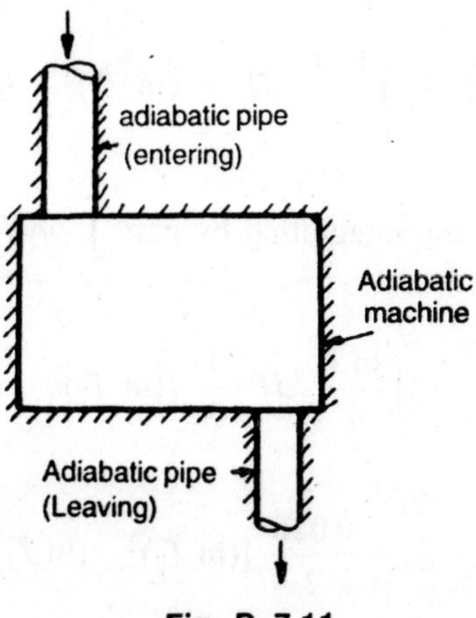

Fig. P. 7.11

$$= \int_{T_1}^{T_B} c_p \frac{dT}{T} - R \int_{P_1}^{P_B} \frac{dP}{P} = \int_{T_A}^{T_B} \left(\frac{C_P}{T}\right) dT - R \ln \frac{P_B}{P_A}$$ kJ/kg.K **Proved**

2nd part,

Given data, $R = 0.287$ kJ/kg.K

For first pipe,

$$p_1 = 5 \text{ bar}, T_1 = 450 \text{ K}$$

For second pipe,

$$p_2 = 1 \text{ bar}, T_2 = 300 \text{ K}$$

$$C_P = 0.026 \ln T + 0.86 \quad \text{kJ/kg.K}$$

Let first pipe is the inlet, and 2nd pipe is the outlet

$$\therefore \quad S_2 - S_1 = \int_{T_1}^{T_2}\left(\frac{C_P}{T}dT\right) - 0.287 \ln \frac{P_2}{P_1}$$

$$= \int_{T_1}^{T_2}\left(0.26\frac{\ln T}{T}dT\right) + \int_{T_1}^{T_2}\frac{0.86}{T}dT - 0.287 \ln \frac{P_2}{P_1}$$

$$\text{Now} \quad \int_{T_1}^{T_2}\frac{\ln T}{T}dT = [\ln T \times \ln T]_{T_1}^{T_2} - \int_{T_1}^{T_2}\frac{1}{T} \times \ln T\, dT$$

$$\Rightarrow 2\int_{T_1}^{T_2}\frac{\ln T}{T}\, dT = (\ln T_2)^2 - (\ln T_1)^2$$

[by integrating by part $\int uvdx = u\int vdx - \int\left(\frac{du}{dx}\right)\int Vdx\, dx$]

$$\therefore \quad \int_{T_1}^{T_2}\frac{\ln T}{T}\, dT = \frac{1}{2}\,[(\ln T_2)^2 - (\ln T_1)^2]$$

$$\therefore S_2 - S_1 = \frac{0.026}{2}\,[(\ln T_2)^2 - (\ln T_1)^2] + 0.84 \ln \frac{T_2}{T_1} - R \ln\frac{P_2}{P_1}, \text{ kJ/kg.K}$$

$$= \frac{0.026}{2}\,[(\ln 300)^2 - (\ln 450)^2] + 0.86 \ln \frac{300}{450} - 0.287 \ln \frac{1}{5} \text{ kJ/kg.K}$$

$= 0.050942$ kJ/kg.K $= +ve$ value

Hence our assumption of fluid intensity through pipe 1 is correct.

Ans

Problem 7.12 : Two vessels, A and B, each of volume 3 m^3 may be connected by a tube of negligible volume. Vessel A contains air at 0.7 MPa, 95°C, while vessel B contains air at 0.35 MPa, 205°C. Find the change of entropy when A is connected to B by working from the first principles and assuming the mixing to be complete and adiabatic. For air take the relations $PV = 0.287\ T$.

Solution : Given data :

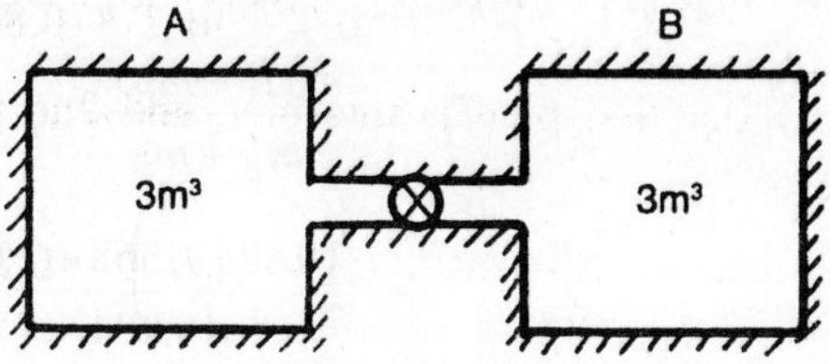

Fig. P. 7.12

Vessel A,

Initial Pressure

$$P_1 = 0.7 \text{ MPa}$$
$$T_1 = 95°\text{C} = 95 + 273 = 368 \text{ K} = \text{initial temperature}$$

Vessel B,

$$P_2 = \text{initial pressure} = 0.35 \text{ MPa}$$
$$T_2 = \text{initial temperature} = 205 + 273 = 478 \text{ K}$$
$$PV = 0.287\ T,$$
$$P = \textit{in} \text{ kPa}$$
$$v = \text{m}^3/\text{kg}$$
$$T = \text{K}$$
$$C_P = 1.005 \text{ kJ/kg.K}$$
$$C_v = 0.718 \text{ kJ/kg.K}$$

Now for vessel A,

$$P_1V = m_ART_1$$

where V in $m^3 = vm$

and m = mass of air

$$\therefore \quad m_A = \frac{p_1V}{RT_1} = \frac{0.7 \times 10^3 \times 3}{0.287 \times 368} = 19.884 \text{ kg}$$

Similarly for vessel B,

$$m_B = \text{mass of air in vessel } B = \frac{p_2V}{RT_2}$$

$$= \frac{0.35 \times 10^3 \times 3}{0.287 \times 478} = 7.654 \text{ kg}$$

Let T_f and P_f are the final temperature and pressure after mixing the air of two vessel.

From 1st law of thermodynamics

$$đQ = du + đw$$

here $đQ = 0,\ đw = 0$

$$\therefore \quad du = 0 \Rightarrow v_{\text{initial}} = v_{\text{final}}$$

$$m_A C_v T_1 + m_B C_v T_2 = (m_A + m_B)\, C_v T_f$$

$$\therefore \quad T_f = \frac{m_A T_1 + m_B T_2}{m_A + m_B}$$

$$= \frac{19.884 \times 368 + 7.654 \times 478}{19.884 + 7.654}$$

$$= 398.574 \text{ K}$$

Now applying

$$p_f V = mRT_f$$

$$V = \text{total volume} = 3 + 3 = 6 \text{ m}^3$$

$$m = \text{total mass} = m_A + m_B = 27.538 \text{ Kg}$$

$$\Rightarrow \quad p_f \times 6 = 27.538 \times .287 \times 398.574$$

$$\Rightarrow \quad p_f = 0.525 \text{ MPa}$$

From property relation

$$ds = \frac{dH}{T} - \frac{mR}{P}\, dP \quad \text{kJ/K},$$

$$\Delta S = mC_P \int_1^2 \frac{dT}{T} - mR \int_1^2 \frac{dP}{P}$$

Change in entropy of air of vessal A

$$(S_f - S_1) = \Delta S_I$$

$$= m_A C_P \ln \frac{T_f}{T_1} - m_A\, 0.287 \times \ln \frac{P_f}{P_1}$$

$$= 19.884 \left[1.005 \ln \frac{398.574}{368} - 0.287 \ln \frac{0.52502}{0.7} \right]$$

$$= 3.2364 \text{ kJ/K}$$

Change in entropy of air of vessel B

$$(S_f - S_2) = \Delta S_{II} = m_B C_P \ln \frac{T_f}{T_2} - m_B \times 0.287 \times \ln \frac{p_f}{P_2}$$

$$= 7.654\left[1.005 \ln \frac{398.574}{478} - 0.287 \ln \frac{0.52507}{.35}\right]$$

$= -2.2886$ kJ/K

total change in entropy

$= \Delta S_I + \Delta S_{II}$

$= 3.2364 - 2.2886 = 0.9478$ kJ/K **Ans**

Problem 7.13 : A resistor of 30 ohms is maintained at a constant temperature of 27°C while a current of 10 amperes is allowed to flow for 1 sec. Determine the entropy change of the resistor and the universe. If the resistor initially at 27°C is now insulated and same current is passed for same time, determine the entropy change of the resistor and the universe. The specific heat of the resistor is 0.9 kJ/kg.K and the mass of the resistor is 10 gm.

Solution : Given data :

R = Resistance = 30 Ω

T = temp. of resistor

= 27°C

= 27 + 273 = 300 K

I = Current = 10 A

t = time of flow of current

= 1 sec.

Q = heat transfer to surrounding

$= I^2Rt = 10^2 \times 30 \times t$

= 3000 J

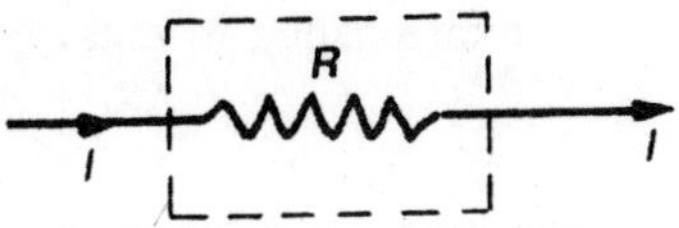

Fig. P. 7.13

∴ change in intropy of surrounding

$$\Delta S_I = \frac{Q}{T} = 10 \text{ J/K}$$

and change in entropy of resistor = 0 **Ans**

∴ total change in entropy

$= (\Delta S)_{univ}$

$= \Delta S_I + \Delta S_{II}$

$= 0 + 10 = 10$ J/K **Ans**

When resistor is insulated

Heat produced by resistor = Q

= 3000 J = 3 kJ

$= m \times C \times (T_2 - T_1)$

where m = mass of resistor
= 10 gm
C = sp. heat of coil
= 0.9 kJ/kg.K
T_2 = final temp. of resistor

$\Rightarrow 0.01 \times 0.9 \times (T_2 - 300) = 3$

$\Rightarrow T_2 = 633.33$ K

$\therefore$ For a reversible process,

$$-\Delta S = \int \frac{dQ_{Rw}}{T} = mC\int_{T_1}^{T_2} \frac{dT}{T}$$

$\therefore$ change in entropy of resistor $= \Delta S_I = m\ C \ln \frac{T_2}{T_1}$

$$= 0.01 \times 0.9 \ln \frac{633.333}{300}$$

$= 0.006725$ kJ/K

& change in entropy of atmosphere

$= \Delta S_{II} = 0$

$\therefore$ total change in entropy

$= \Delta S_I + \Delta S_{II} = 6.72$ J/K **Ans**

Problem 7.14 : A copper rod is of length 1 m and diameter 0.01 m. One end of the rod is at 100°C and the other at 0°C. The rod is perfectly insulated along its length and the thermal conductivity of copper is 380 W/m.K. Calculate the rate of heat transfer along the rod and the rate of entropy production due to irreversibility of this heat transfer.

Solution : Given data :

L = length copper rod
= 1 m

d = dia of copper rod
= 0.01 m

T_1 = one end temperature
= 100°C = 100 + 273
= 373 K

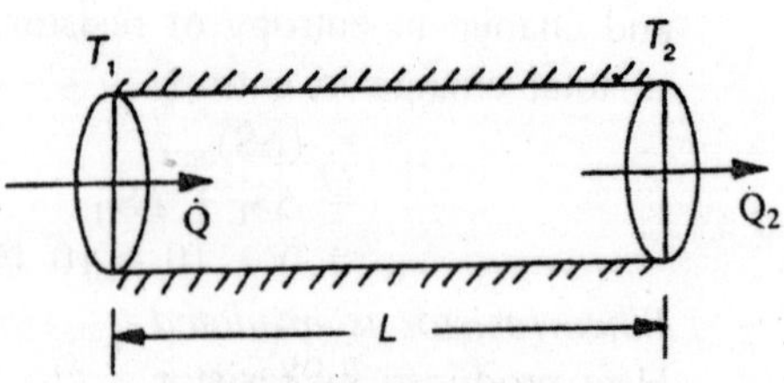

Fig. P. 7.14

T_2 = other end temperature
= 0°C = 0 + 273 = 273 K
K = Thermal conductivity of copper
= 380 W/m.K

Condensing the steady state rate of heat flow through the system from Fourier law

$$Q = \text{rate of heat transfer} = KA\frac{\Delta T}{L}$$

[∵ A = area of cross section ⊥r to the flow direction = $\frac{\pi}{4}d^2$]

$$= 380 \times \frac{\pi}{4}(0.01)^2 \times \frac{100}{1}$$

$= 2.985\ W$ **Ans**

Now rate of entropy production = change in entropy – entropy transfer of system

$$= (S_2 - S_1) - \int_1^2 \frac{dQ}{T} \quad \text{kW/K}$$

Now $S_2 - S_1$ = change in entropy of system
= 0 = due to steady state heat flow rate.

$$\text{Entropy transfer} = \int_1^2 \frac{dQ}{T} = \frac{\dot{Q}}{T_1} - \frac{\dot{Q}}{T_2}$$

= since it is the boundarhy phenomena

∴ rate of entropy produce

$$= 0 - \left[\frac{\dot{Q}}{T_1} - \frac{\dot{Q}}{T_2}\right]$$

$$= \dot{Q}\left[\frac{T_1 - T_2}{T_1 \times T_2}\right]$$

$$= 2.985 \times \frac{(373 - 273)}{373 \times 273}$$

= 0.002931 W/K

Problem 7.15 : A lump of ice with a mass of 1.5 kg at an initial temperature of 260 K melts at the pressure of 1 bar as a result of heat transfer from the environment. After some time has elapsed the resulting water attains the temperature of the environment, 293 K. Calculate the entropy production associated with this process. The latent heat of fusion of ice is 333.4 kJ/kg, the specific heat of ice and water are 2.07 and 4.2 kJ/kg.K respectively, and ice melts at 273.15 K.

Solution : Given data :

mass of ice = 1.5 kg

T_1 = initial temperature of ice
= 260 K

pressure = 1 bar

T_2 = final temp. of water
= 293 K

L = latent heat of fusion of ice
= 333.4 kJ/kg.K

sp. heat of ice = C_1
= 2.07 kJ/kg.K

sp. heat of water = C_2
= 4.2 kJ/kg.K

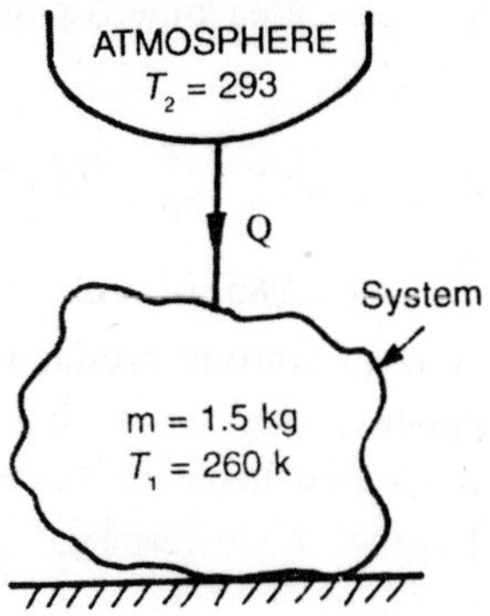

Fig. P. 7.15

Melting point of ice = 273.15 K.

Now, entropy production during process
= change in entropy of system – entropy transfer

$$= (S_2 - S_1) - \int_1^2 \frac{dQ}{T}$$

Now Q = heat transfer from atmosphere to ice lump
= heat required to convert ice at 260 K to water at 293 K
$= mC_1\,(T_{273.11} - T_{260}) + mL + mC_2\,(T_{293} - T_{273.15})$
$= 1.5 \times 2.07\,(273.15 - 260) + 1.5 \times 333.4 + 1.5 \times 4.2\,(293 - 273.15)$
= 665.9858 kJ

$$\therefore \text{ entropy transfer } = \frac{Q}{T_2} = \frac{665.9858}{293} = 2.2730 \text{ kJ/K}$$

Entropy change of system = entropy change of system as ice at 260 K converts into ice at 273.15 K + entropy of system as ice at 273 K melts into water at 0°C + entropy system as water at 273.15 K converts into water at 293 K.

$$= mC_1 \ln \frac{273.15}{260} + m\frac{L}{273.15} + mC_2 \ln \frac{293}{273}$$

$$= 1.5\left[2.07 \ln \frac{273.15}{260} + \frac{333.9}{273.15} + 402 \ln \frac{293}{273.15}\right]$$

$$= 2.4260 \text{ kJ/K}$$

Entropy production during the process

$$= (S_2 - S_1) - \int_1^2 \frac{dQ}{T} = 0.1530 \text{ kJ/K} \quad \textbf{Ans}$$

> **Problem 7.16** : An ideal gas is compressed reversibly and adiabatically from state a to state b. It is then heated reversibly at constant volume to state c. After expanding reversibly and adiabatically to state d such that $T_b = T_d$, the gas is again reversibly heated at constant pressure to state e such that $T_e = T_c$. Heat is then rejected reversibly from the gas at constant volume till it returns to state a. Express T_a in terms of T_b and T_c. If T_B = 555 K and T_c = 835 K, estimate T_a. Take γ = 1.4

Solution : Given data :

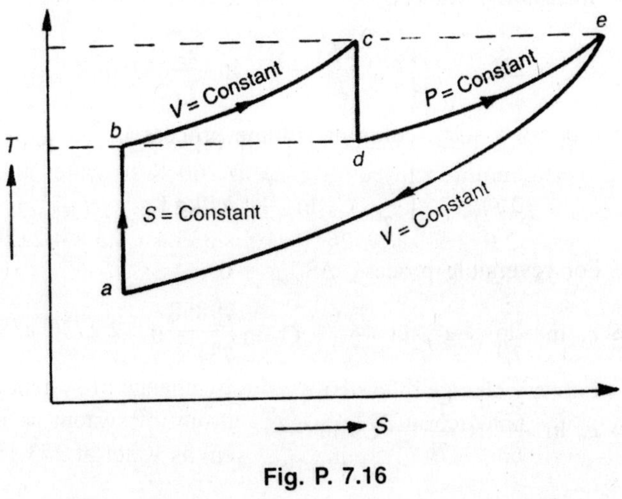

Fig. P. 7.16

$$\therefore \quad T_b = T_d$$

and $\quad T_c = T_e$

In process $a - b$, reversibly and adiabatically, i.e., isentropic process

$$\Delta S_I = 0$$

In process $b - c$; constant volume process from 1st law of thermodynamics

$$đQ = du + PdV$$

at constant volume $dv = 0$

$$\therefore \quad đQ = du$$

$$\therefore \quad (\Delta S)_{II} = \int_b^c \frac{dQ}{T} = \int_b^c \frac{du}{T}$$

$$= C_v \ln \frac{T_c}{T_b} \text{ kJ/kg.K.}$$

In process $c - d$: reversibly and adiabatically, i.e., isentropic process. $(\Delta S)_{III} = 0$

In process $d - e$: constant pressure process from property relation

$$ds = \frac{dh}{T} - \frac{V}{T} dp$$

at constant pressure,

$$dp = 0$$

$$\therefore \quad ds = \frac{dh}{T}$$

⇒ Integrating we get,

$$\Delta S_{IV} = \int_d^e C_p \frac{dT}{T} = C_P \ln \frac{T_e}{T_d} \text{ kJ/kg.K}$$

In process $e - a$: constant volume process

$$(\Delta S)_v = \int_e^a \frac{du}{T} \; C_v \ln \frac{T_a}{T_e} \text{ kJ/kg.K}$$

∴ For reversible process, $\Delta S_{total} = 0$

$$\Rightarrow c_v \ln \frac{T_C}{T_b} + C_P \ln \frac{T_e}{T_d} + C_v \ln \frac{T_a}{T_e} = 0$$

$$\Rightarrow c_v \ln \frac{T_c}{T_b} \times \frac{T_a}{T_e} = C_P \ln \frac{T_d}{T_e}$$

$$\Rightarrow \ln \frac{T_C \times T_a}{T_b \times T_C} = \frac{C_P}{C_v} \ln \frac{T_b}{T_C} = \ln \left(\frac{T_b}{T_c}\right)^\gamma$$

$$\Rightarrow \qquad \frac{T_a}{T_b} = \frac{T_b^\gamma}{T_c^\gamma}$$

$$\Rightarrow \qquad T_\alpha = \frac{T_b^{\gamma+1}}{T_c^\gamma}$$

If $\quad T_b = 555$ K,

$T_c = 835$ K, $\gamma = 1.4$

$$T_a = \frac{555^{2.4}}{835^{1.4}} = 313.29 \text{ K}$$

Problem 7.17 : Two blocks of metal, each having a mass of 10 kg and a specific heat of 0.4 kJ/kg.K, are at a temperature of 40°C. A reversible refrigerator receives heat from one block and rejects heat to the other. Calculate the work required to cause a temperature difference of 100°C between the two blocks.

Solution : Given data :

mass of each block

$= m = 10$ kg

sp. heat $= C$

$= .04$ kJ/kg.K

initial temp. of block $= T_1$

$= 40°C$

$= 40 + 273$

$= 313$ K

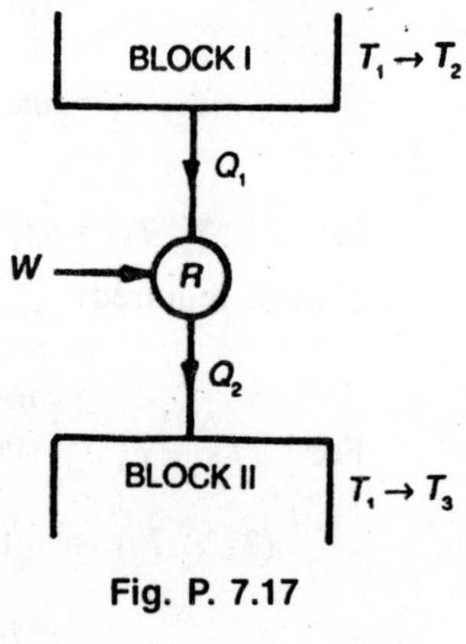

Fig. P. 7.17

Let after finite no. of cycle temp. of first block becomes T_2 and that of 2nd block is T_3 where $T_3 > T_2$ and $T_3 - T_2 = 100°C$

Q_1 = heat absorbed from block I, kJ

$= mC\,(T_1 - T_2)$

Q_2 = heat rejects to block II, kJ

$$= mC\ (T_3 - T_1)$$

w = work input to the refrigerator, KJ

then by energy balance,

$$Q_1 + W = Q_2$$

$$\therefore \quad W = Q_2 - Q_1 \quad \text{...(i)}$$

Now for reversible process

$$\Delta S = \int_{T_{initial}}^{T_{final}} \frac{dQ}{T}$$

$\therefore$ ΔS = change of entropy of block I

$$= mc \ln \frac{T_2}{T_1}$$

ΔS = change of entropy of block II

$$= mc \ln \frac{T_3}{T_1}$$

change of entropy of working fluid of refrigerator

$$= \oint \Delta s = 0$$

$$\therefore \quad \Delta S_{total} = \Delta S_1 + \Delta S_2 = mc \left[\ln\frac{T_2}{T_1} + \ln\frac{T_3}{T_1} \right] = 0$$

= for reversible system.

$$= mc \ln \frac{T_2\, T_3}{T_1^2} = 0$$

$$\Rightarrow \quad T_2\, T_3 = T_1^2 \quad \text{...(ii)}$$

$\therefore$ work required= $Q_2 - Q_1$ = from (i)

$$= mc\ [T_3 - T_1 - T_1 + T_2]$$

$$= mc\ [T_3 + T_2 - 2T_1]$$

For $T_3 - T_2 = 100$ K

$$(T_3 \times T_2) = \sqrt{(T_3 - T_2)^2 + 4\, T_2\, T_3}$$

$$= \sqrt{100^2 + 4 \times 313^2} = 633.94 \text{ K}$$

$\therefore$ work required

$$= 10 \times 0.4 \times (633.94 - 2 \times 313)$$

$$= 31.76 \text{ kJ}$$ **Ans**

Problem 7.18 : The value of C_p for a certain substance can be represented by $C_p = a + bT$. (a) Determine the heat absorbed and the increase in entropy of a mass m of the substance when its temperature is increased at constant pressure from T_1 to T_2. (b) Find the increase in the molal specific entropy of copper, when the temperature is increased at constant pressure from 500 to 1200 K. Given for copper: when $T = 500$ K, $C_p = 25.2 \times 10^3$ and when $T = 1200$ K, $C_p = 30.1 \times 10^3$ J/kg mol K.

Solution : Given :

$$C_p = a + bT$$

(a) Condition : heating at constant pressure from T_1 to T_2 from property relation

$$đQ = dh - Vdp$$

or $$đQ = dh = mCpdT$$

or $$Q = \int_{T_1}^{T_2} mC_p dT = m\int_{T_1}^{T_2}(a+bT)\,dT$$

$$= m\left[aT + \frac{bT^2}{2}\right]_{T_1}^{T_2}$$

$$= ma\,(T_2 - T_1) + mb\left(\frac{T_2^2 - T_1^2}{2}\right) \quad \textbf{Ans}$$

Similarly at constant pressure from property relation

$$ds = mC_P\frac{dT}{T}$$

or $$S_2 - S_1 = \int ds = m\int_{T_1}^{T_2} C_P\frac{dT}{T}$$

$$= m\int_{T_1}^{T_2}(a + b\,T)\frac{dT}{T}$$

$$= m\,[a \ln T + bT]_{T_1}^{T_2}$$

$$= m\left[a \ln\frac{T_2}{T_1} + b(T_2 - T_1)\right] \quad \textbf{Ans}$$

(b) Given

$T_1 = 500$ K

$T_2 = 1200$ K

Now $C_P = a + bT$

Conditions are,

at $T = 500$ K,

$C_P = 25.2 \times 10^3$ J/K mole.K

at $T = 1200$ K,

$C_P = 30.1 \times 10^3$ J/K mole.K

or $25.2 = a + b \times 500$

and $30.1 = a + b \times 1200$

solving for (a) and (b) we get,

$$b = \frac{4.9}{700} = 7 \times 10^{-3} \text{ kJ/kg mole.K}$$

$$\therefore \quad a = 21.7 \text{ kJ/kg mole.K}$$

$$\therefore \quad \Delta S = a \ln \frac{T_2}{T_1} + b\,(T_2 - T_1)$$

$$= 21.7 \ln \frac{1200}{500} + 7 \times 10^{-3}\,(1200 - 500)$$

$$= 23.9 \text{ kJ/kg mole.K}$$ **Ans**

Problem 7.19 : A thermally insulated 50–ohm resistor carries a current of 1 A for 1 s. The initial temperature of the resistor is 10°C. Its mass is 5 g and its specific heat is 0.85 J/g.K. (a) What is the change in entropy of the resistor? (b) What is the change in entropy of the universe?

Solution : Given :

m = mass of resistor
= 5 g

C = specific heat
= 0.85 J/g.K

T_1 = initial temperature = 10°C

R = Resistance = 50 ohm

d = Current = 1A

t = time of flow of current = 1 sec.

for a steady flow

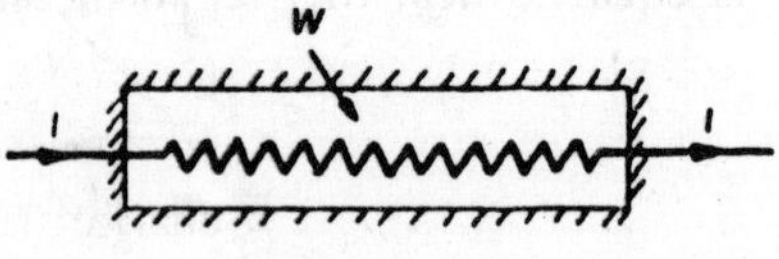

Fig. P. 7.17

(w) work supplied to the resistor by electric current = heat produced (Q)

$$\therefore \quad Q = I^2Rt = 1^2 \times 50 \times 1 = 50 \text{ J}$$

$$\text{Now} \quad Q = mC\,\Delta T$$

$$\text{or} \quad 50 = 5 \times 0.85\,(T_2 - T_1)$$

$$\therefore \quad T_2 = 21.76°\text{C}$$

(a) Change in entropy of resistor $= mC \ln \dfrac{T_2}{T_1}$

$$= 5 \times 0.85 \ln \frac{2176 + 273}{10 + 273} = 0.173 \text{ J/K}$$

(b) Since resistor is insulated

$\therefore$ change in entropy of the atmosphere = 0

$\therefore$ change in entropy of resistor = 0.173 J/K **Ans**

> **Problem 7.20 :** A 50 ohm resistor carrying a constant current of 1 A is kept at a constant temperature of 27°C by a stream of cooling water. In a time interval of 1 s, (a) What is the change in entropy of the resistor? (b) What is the change in entropy of the universe?

Solution : Given :

R = resistance = 50 ohm

I = current = 1A

T = 27°C = 300 K

t = time interval = 1s

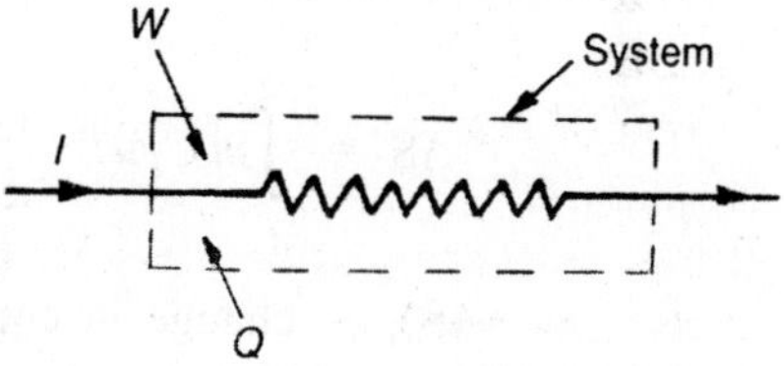

Fig. P. 7.20

Considering the isothermal dissipation of work through system. Hence at steady state work transfer (due to flow of current) is equal to heat transfer to the surrounding.

$$\therefore \quad w = Q = I^2Rt$$

$$\therefore \quad 1^2 \times 50 \times 1 = 50\text{J}$$

$$\therefore \quad (\Delta S)_{surr.} = \text{change in entropy of surrounding}$$

$$= \frac{Q}{T} = \frac{50}{300} = 0.167 \text{ J/K}$$

$$\& \quad (\Delta S)_{resistor} = 0 \text{ at steady state } \textbf{Ans}$$

$$\therefore \quad (\Delta S)_{univ} = \text{change in entropy of universe}$$

= change in entropy of surrounding + change in entropy of resistor (system)

= 0.167 + 0

= 0.167 J/K **Ans**

Problem 7.21 : 36 g of water at 30°C are converted into steam at 250°C at constant atmospheric pressure. The specific heat of water is assumed constant at 4.2 J/g K and the latent heat of vaporizatin at 100°C is 2260 J/g. For water vapour, assume $PV = mRT$ where R = 0.4619 kJ/kg.K, and

$$\frac{C_P}{R} = a + bT + cT^2,$$

where a = 3.634, b = 1.195 × 10^{-3}. K^{-1} and c = 0.135 × 10^{-6} K^{-2} Calculate the entropy change of the system.

Solution : Given :

m = mass of water = 36 g

T_1 = temperature of water = 30°C

T_2 = temperature of steam = 250°C

L = latent heat of vaporisation = 2260 J/g

Cp_v = specific heat of vapour

= $R\,(a + bT + CT^2)$

At constant pressure

$$\Delta S = \int_{T_i}^{T_f} m\,C_P dT$$

∴ $(\Delta S)_I$ = change in entropy of system as water at 30°C heated to 100°C water

$$= \int_{30°C}^{100°C} mC_{P_w} dT = 36 \times 4.2 \ln \frac{100+273}{30+273}$$

= 31.4263 J/K

$(\Delta S)_{II}$ = change in entropy of system as water at 100°C converted into steam at 100°C

$$= \frac{m \times L}{T} = \frac{36 \times 2260}{373} = 218.1233 \text{ J/K}$$

$(\Delta S)_{III}$ = change in entropy of system as steam at 100°C heated to 250°C

$$= \int_{100^\circ C}^{250^\circ C} mC_{P_v} \frac{dT}{T} = mR \int_{100^\circ C}^{250^\circ C} \frac{\left(a + bT + cT^2\right)}{T} dT$$

$$= mR \left[a \ln T + bT + \frac{C}{2} T^2 \right]_{100^\circ C}^{250^\circ C}$$

$$= mR \left[a \ln \frac{250 + 273}{100 + 273} + b(250 - 100) + \frac{C}{2}\left\{(250 + 273)^2 - (100 + 273)^2\right\}\right]$$

$= 36 \times 0.4619$

$$\left[3.634 \ln \frac{523}{373} + 150 \times 1.195 \times 10^{-3} + \frac{0.135 \times 10^{-6}}{2} \times \left\{(523)^2 - (373)^2\right\}\right]$$

$= 23.5562$ J/K

$\therefore \quad \Delta S$ = total entropy change of the system

$= (\Delta S)_I + (\Delta S)_{II} + (\Delta S)_{III}$

$= 31.5263 + 218.1233 + 23.5562$

$= 273.11$ J/K **Ans**

Problem 7.22 : A body of finite mass is originally at a temperature T_1, which is higher than that of a heat reservior at a temperature T_2. An engine operates in infinitesimal cycles between the body and the reservior until it lowers the temperature of the body from T_1 to T_2. In this process there is a heat flow Q out of the body. Prove that the maximum work obtainable from the engine is $Q + T_2 (S_1 - S_2)$, where $S_1 - S_2$ is the decrease in entropy of the body.

Solution : Given :

$T_1 > T_2$

Q = heat extractes from body

$S_1 - S_2$ = decrease in entropy of body.

Let Q_1 be heat rejected to the reservior and w is the work obtainable.

$\therefore$ from first law of thermodynamics

$$Q = Q_1 + w$$

or, $\quad Q_1 = Q - w$

$\therefore \quad (\Delta S)_{body} = S_1 - S_2$

$$(\Delta S)_{reservior} = \frac{Q_1}{T_2} = \frac{Q-w}{T_2}$$

$$(\Delta S)_{engine} = 0$$

$\therefore \quad (\Delta S)_{universe} > 0$

or $(\Delta S)_{body} + (\Delta S)_{reservoir} + (\Delta S)_{engine} \geq 0$

or $(S_1 - S_2) + \dfrac{Q-w}{T_2} \geq 0$

or $[T_2 (S_1 - S_2) + Q] \geq w$

or $\quad w \leq [T_2 (S_1 - S_2) + Q]$

$\therefore$ For maximum work obtainable

$$w = [T_2 (S_1 - S_2) + Q]$$ **Ans**

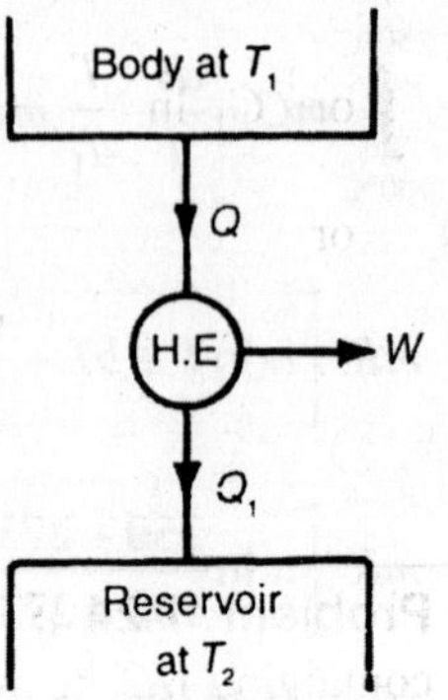

Fig. P. 7.22

Problem 7.23 : Show that if two bodies of thermal capacities C_1 and C_2 at temperatures T_1 and T_2 are brought to the same temperature T by means of a reversible heat engine, then

$$\ln T = \frac{C_1 \ln T_1 + C_2 \ln T_2}{C_1 + C_2}$$

Solution : Given :

C_1 = thermal capacity of 1st body

C_2 = thermal capacity of 2nd body

T = final temperature of both the body.

Let $T_1 > T_2$

$\therefore (\Delta S)_I$ = change in entropy of 1st body

$$= C_1 \ln \frac{T}{T_1}$$

$\therefore (\Delta S)_{II}$ = change in entropy of 2nd body

$$= C_2 \ln \frac{T}{T_2}$$

Now a reversible heat engine

$(\Delta S)_{universe} = 0$

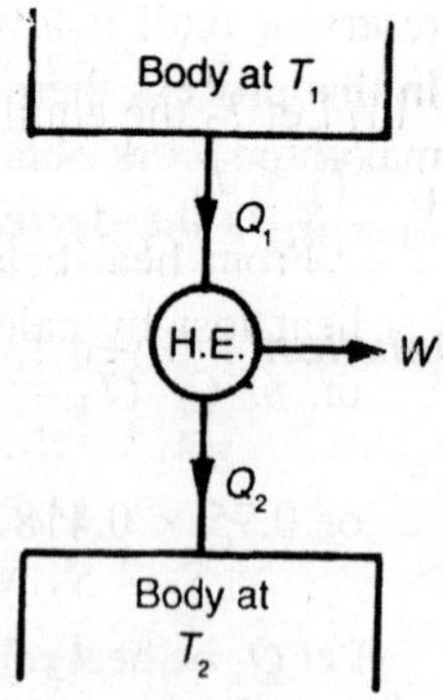

Fig. P. 7.23

or $(\Delta S)_I + (\Delta S)_{II} = 0$

or $C_1 \ln \frac{T}{T_1} + C_2 \ln \frac{T}{T_2} = 0$

or $\ln T (C_1 + C_2) - [C_1 \ln + C_2 \ln T_2]$

$\therefore \qquad \ln T = \frac{C_1 \ln T_1 + C_2 \ln T_2}{C_1 + C_2}$ **Ans**

Problem 7.24 : An insulated 0.75 kg copper calorimeter can containing 0.2 kg water is in equilibrium at a temperature of 20°C. An experimenter now place 0.05 kg of ice at 0°C in the calorimeter and encloses the latter with a heat insulating shield. (a) When all the ice has melted and equilibrium has been reached, what will be the temperature of water and the can? The specific heat of copper is 0.418 kJ/kg.K and the latent heat of fusion of ice is 333 kJ/kg: (b) Compute the entropy increase of the universe resulting from the process. (c) What will be the minimum work needed by a stirrer to bring back the temperature of water to 20°C?

Solution : Given :

m_c = mass of copper coloremeter = 0.75 kg

m_w = mass of water in calorimeter = 0.2 kg

T_1 = initial temperature of caloriemeter and water at equilibrium = 20°C = 293 K

m_2 = mass of ice = 0.05 kg

T_2 = temperature of ice = 0°C = 273 K

(a) Let T_2 the equillibrium temperature after ice has melted down and $T_i < T_2 < T_1$

$\therefore$ From heat balance,

heat lost by calorimeter and water = heat gained by ice

or, $m_C C_C (T_1 - T_2) + m_w C_w (T_1 - T_2)$

$= M_i L_{\text{fusion}} + m_2 C_w (T_2 - T_1)$

or $0.75 \times 0.418 \times (20 - T_2) + 0.2 \times 4.187 \times (20 - T_2)$

$= 0.05 \times 333 + 0.05 \times 4.187 \times (T_2 - 0)$

$\therefore \qquad T_2 = 4.68°\text{C}$ **Ans**

(b) $(\Delta S)_{\text{colorimeter}}$ = change in entropy of calorimeter

$$= m_C C_C \ln \frac{T_2}{T_1} = 0.75 \times 0.418 \ln \frac{(4.68 + 273)}{293}$$

$= -0.016836$ kJ/K

$(\Delta S)_{water}$ = change in entropy of water

$$= m_w C_w \ln \frac{T_2}{T_1}$$

$$= 0.2 \times 4.187 \times \ln \frac{(4.68 + 273)}{293}$$

$= -0.044971$ kJ/K

$(\Delta S)_I$ = change in entropy of ice as it melted at 0°C

$$= \frac{m_i L}{T_i} = \frac{0.05 \times 333}{273} = 0.06099 \text{ kJ/K}$$

$(\Delta S)_{II}$ = change in entropy of ice as water at 0°C heated to 4.68°C

$$= m_i C_w \ln \frac{T_2}{T_i}$$

$$= 0.05 \times 4.187 \ln \frac{(4.68 + 273)}{273}$$

$= 0.0035584$ kJ/K

& $(\Delta S)_{surrounding} = 0$

$\therefore \quad (\Delta S)_{universe} = (\Delta S)_{calorimeter} + (\Delta S)_{water} + (\Delta S)_I + (\Delta S)_{II} + (\Delta S)_{surrounding}$

$= -0.016836 - 0.044971 + 0.06099 + 0.0035584$

$= 0.0027414$ kJ/K **Ans**

(c) From property relation

$$đQ = du + dw$$

since $\quad đQ = 0$

$\therefore \quad -đw = du$

$\therefore$ minimum work required to stirr to bring back the temperature of water to 20°C

$= (\Delta u)_{calorimeter} + (\Delta u)_{water} + (\Delta u)_{ice}$

$= m_C C_C (20 - T_2) + m_w C_w (20 - T_2) + m_i C_C (20 - T_2)$

$= 0.75 \times .0418 \times (20 - 4.68) + 0.2 \times 4.187 (20 - 4.68) + 0.05 \times 4.187 (20 - 4.68)$

$= 20.84$ kJ **Ans**

Problem 7.25 : A body of constant heat capacity C_P and at a temperature T_i is put in contact with a reservior at a higher temperature T_f. The pressure remains constant while the body comes to equilibrium with the reservior. Show that the entropy change of the universe is equal to

$$C_P\left[\frac{T_i - T_f}{T_f} - \ln\left(1 + \frac{T_i - T_f}{T_f}\right)\right]$$

Prove that this entropy change is positive.

Given : $\ln(1 + x) = x - \frac{x^2}{2} + \frac{x^3}{3} - \frac{x^4}{4} + \ldots\ldots$

where $x < 1$.

Solution : Given :

T_i = initial temperature of body
T_f = Reservior temperature
C_p = heat capacity of body
Let Q = heat transferred from the reservior
$= C_P (T_f - T_i)$

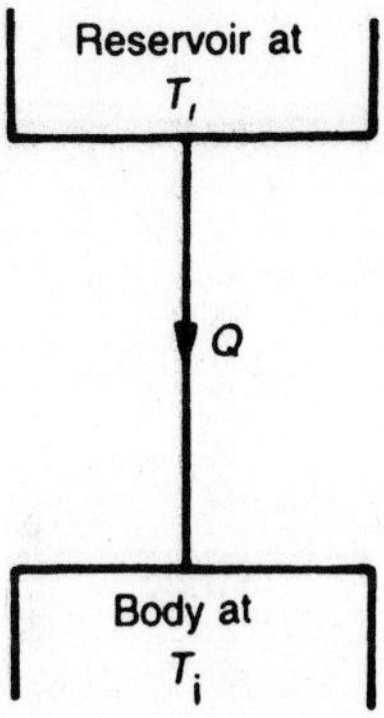

Fig. P. 7.25

From property relation, at constant pressure

$$\Delta S = \int_{T_i}^{T_f} mC_P \frac{dT}{T} = \int_{T_i}^{T_f} C_P \frac{dT}{T} \text{ kJ/kg.K}$$

$\therefore$ $(\Delta S)_{body}$ = change in entropy of body

$$= C_P \ln \frac{T_f}{T_i}$$

$$\& \quad (\Delta S)_{reservoir} = -\frac{Q}{T} = -\frac{C_P(T_f - T_i)}{T_f}$$

$$\therefore \quad (\Delta S)_{universe} = (\Delta S)_{body} + (\Delta S)_{reservior} = C_p \ln \frac{T_f}{T_i} - \frac{C_P(T_f - T_i)}{T_f}$$

$$= C_P\left[\frac{T_i - T_f}{T_f} - \ln \frac{T_i}{T_f}\right]$$

$$= C_P\left[\frac{T_i - T_f}{T_f} - \ln\left(1 + \frac{T_i - T_f}{T_f}\right)\right]$$

Now $\ln \dfrac{T_i}{T_f} = \ln\left(1 + \dfrac{T_i - T_f}{T_f}\right)$

Here $\dfrac{T_i - T_f}{T_f} < 1$

∴ from logarithmic expansion

$$\ln\left[1 + \left(\frac{T_i - T_f}{T_f}\right)\right]$$

$$= \frac{T_i - T_f}{T_f} - \frac{(T_i - T_f)^2}{2T_f^{\,2}} + \frac{(T_i - T_f)^3}{3T_f^{\,3}} \ldots$$

$$= \frac{T_i - T_f}{T_f}\left[1 - \frac{(T_i - T_f)}{2T_f} + \frac{(T_i - T_f)^2}{3T_f^{\,2}} \bar{0} \ldots\ldots\right]$$

Since $T_f > T_i$

∴ $\dfrac{T_i - T_f}{T_f} = -Ve$

∴ $\ln\left[1 + \dfrac{(T_i - T_f)}{T_f}\right]$

$= (-V_e)$ quantity × [1 + $(+V_e)$ quantity]

∴ $\dfrac{T_i - T_f}{T_f} - \ln\left[1 + \left(\dfrac{T_i - T_f}{T_f}\right)\right] > 0$

∴ $(\Delta S)_{universe} > 0$ **Ans**

Problem 7.26 : Two bodies of equal heat capacities C and temperatures T_1 and T_2 form an adiabatically closed system. What will the final temperature be if one lets this system come to equilibrium (a) freely? (b) reversibly? (c) What is the maximum work which can be obtained from this system?

Solution : Given :

for 1_{st} body

Initial temperature $= T_1$

Heat capacity $= C$

For 2_{nd} body,

Initial temperature $= T_2$

Heat capacity $= C$

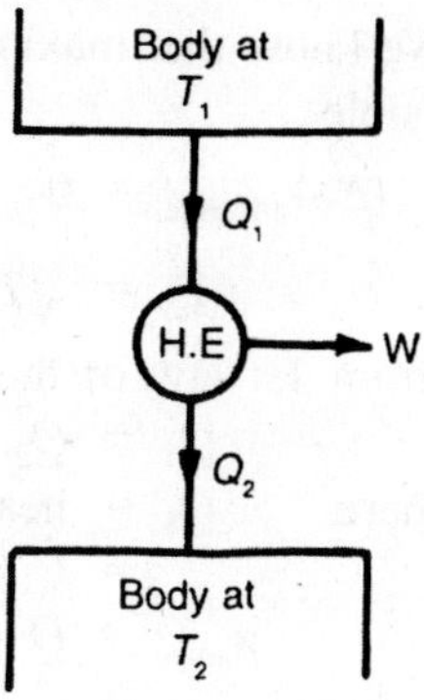

Fig. P. 7.26

(a) Let T_3 be the final temperature when the two body come to equillibrium as $T_1 > T_3 > T_2$

For free equillibrium system heat lost by 1st body = heat gained by 2nd body

$$\Rightarrow \quad C\,(T_1 - T_3) = C\,(T_3 - T_2)$$

or $T_3 = \dfrac{T_1 + T_2}{2}$ **Ans**

(b) For reversible equillibrium system

$$(\Delta S)_{system} = 0$$

$\therefore$ $(\Delta S)_I$ = change in entropy of 1st body

$$= C \ln \frac{T_3}{T_1}$$

and $(\Delta S)_{II}$ = change in entropy of 2nd body

$$= C \ln \frac{T_3}{T_2}$$

$$\therefore \quad (\Delta S)_{system} = (\Delta S)_I + (\Delta S)_{II} = 0$$

$$= C\left[\ln\frac{T_3}{T_1} + \ln\frac{T_3}{T_2}\right] = 0$$

or $\ln \dfrac{T_3^2}{T_1 T_2} = 0$

or $\dfrac{T_3^2}{T_1 T_2} = e^0 = 1$

$\therefore \quad T_3 = \sqrt{T_1 T_2}$ **Ans**

(c) We know that maximum work obtained is possible when engine is reversible

i.e. $(\Delta S)_{universe} = 0$

i.e. $T_3 = \sqrt{T_1 T_2}$

$\therefore$ from 1st law of thermodynamic

$$Q_1 = Q_2 + w$$

Where Q_1 = heat transferred by body at temperature T_1

$$= C\,(T_1 - T_3)$$

$$\therefore \quad w_{max} = Q - Q_2$$

$$= C\,[(T_1 - T_3) - (T_3 - T_2)]$$

$$= C\,[T_1 - \sqrt{T_1 T_2} - \sqrt{T_1 T_2} + T_2]$$

$$= C\left[\left(\sqrt{T_1}\right)^2 - 2\sqrt{T_1}\,\sqrt{T_2} + \left(\sqrt{T_2}\right)^2\right]$$

$$= C\left[\sqrt{T_1} - \sqrt{T_2}\right]^2$$ **Ans**

Problem 7.27 : In the temperature range between 0°C and 100°C a particular system maintained at constant volume has a heat capacity.

$$C_V = A + 2\,BT$$

with $A = 0.014$ J/K and $B = 4.2 \times 10^{-4}$ J/K^2

A heat reservior at 0°C and a resersible work source are available, what is the maximum amount that can be transferred to the reversible work source as the system is cooled from 100°C to the temperature of the reservoir.

Solution **:** Given :

Heat transfer is maintained at constant volume

$$C_v = A + 2BT, = 0.014 + 8.4 \times 10^{-4}\,T \text{ J/K}$$

Final temperature of source = T_2
= 0°C

Q_1 = heat transfer from the source and

Q_2 = heat received by the reservior at T_2°C

Now for source at constant volume

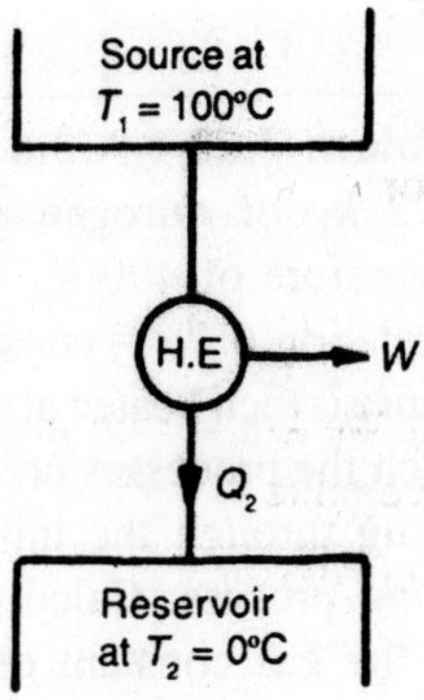

Fig. P. 7.27

$$Q_1 = \int_{T_1}^{T_2} C_v dT$$

$$= \int_{T_1}^{T_2} \left[0.014 + 8.4 \times 10^{-4} T\right] dT$$

$$= [0.014\ (T_2 - T_1) + \frac{8.4 \times 10^{-4}}{2}\ (T_2^2 - T_1^2)]$$

$$= [0.014\ (273 - 373) + 4.2 \times 10^{-4}\ (273^2 - 373^2)] = -28.53 \text{ J}$$

–ve sign indicate that heat is lost from source.

Now $(\Delta S)_{source}$ = change in entropy of source

$$= \int_{T_1}^{T_2} C_v \frac{dT}{T} = \int_{T_1}^{T_2} (A + 2BT) \frac{dT}{T}$$

$$= [A \ln T + 2\ BT]_{T_1}^{T_2} = A \ln \frac{T_2}{T_1} + 2\ B\ (T_2 - T_1)$$

$$= 0.014 \ln \frac{273}{573} + 2 \times 4.2 \times 10^{-4}\ (273 - 373) = -0.08837 \text{ J/K}$$

$$(\Delta S)_{reservior} = \frac{Q_2}{T_2} = \frac{Q_1 - w}{T_2} \quad [\because Q_1 = Q_2 + w]$$

$\because$ $(\Delta S)_{univserse} \geq 0$

$\therefore$ $(\Delta S)_{source} + (\Delta S)_{reservior} \geq 0$

or $$-0.08837 + \frac{Q_1 - w}{T_2} \geq 0$$

or $$-T_2 \times 0.08837 + Q_1 \geq w$$

∴ for maximum work obtainable

$$W_{max} = -273 \times 0.8837 + 28.53 = 4.405 \text{ J } \textbf{Ans}$$

Problem 7.28 : An ideal gas mixture consisting of 1 kg of oxygen and 2 kg of nitrogen is initially at a pressure of 700 kPa and a temperature of 400 K. The mixture expands reversibly according to the relation $pV^{1.2}$ = constant, until the final temperature is 300 K. The mixture is then heated at constant pressure back to its initial temperature. Sketch the processes on a p–V diagram clearly indicating the isotherm passing through the initial and final states of the constant pressure heating process. Calculate:

(a) the gas constant of the mixture,

(b) the entropy change per unit mass of the mixture for the combined process.

Molecular weights : O_2 = 32; N_2 = 28.

Solution :

$$m_{O_2} = 1 \text{ kg}; \; m_{N_2} = 2 \text{ kg}$$

$$R_{mix} = \frac{1\left(\frac{8.314}{32}\right) + 2\left(\frac{8.314}{28}\right)}{3}$$

$$= .0285 \text{ kJ/kg.K}$$

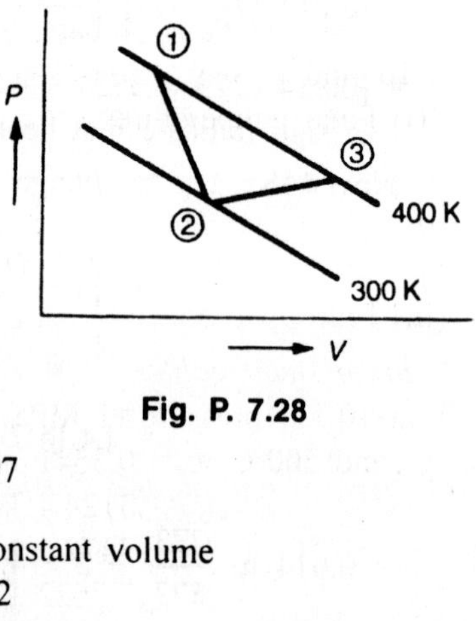

Fig. P. 7.28

Specifric heat of the mixture at constant pressure

$$= \frac{1 \times 0.91 + 2 \times 1.04}{3} = 0.997$$

Specific heat of the mixtur eat constant volume

$$= C_P - R = 0.712$$

$$\Delta S_{1-2} = C_v \left(\frac{\gamma - n}{1 - n}\right) \ln \frac{T_2}{T_1}$$

$$= 0.712 \times \left(\frac{1.4 - 1.2}{1 - 1.2}\right) \ln \frac{300}{400} = 0.205$$

$$\Delta S_{2-3} = 0.997 \times \ln \frac{T_3}{T_2} = 0.997 \ln \frac{400}{300}$$

$$= 0.287$$

$$\Delta S_{1-3} = 0.492 \text{ kJ/kg K}$$

Problem 7.29 : An insulated vessel is divided into two compartments connected by a valve. Initially, one compartment contains steam at 10 bar, 500°C and the other is evacuated. The valve is opened and the steam is allowed to fill the entire volume, achieving a final pressure of 1 bar. Determine the final temperature, in °C, the percentage of the vessel volume initially occupied by steam, and the amount of entropy produced in kJ/kg.K. (CES – 1993)

Solution : Given :

Initial

$$P_A = 10 \text{ bar}$$
$$P_B = 0$$
$$T_A = 500 \text{ °C}$$

Final

$$P_A = 1 \text{ bar}$$
$$P_B = 1 \text{ bar}$$

To find

(i) Final temperature

(ii) $\dfrac{V_A}{(V_A + V_B)}$ (i.e.) $\dfrac{V_A}{V_{AB}}$

(iii) Entropy produced

From steam tables

at 10 bar pressure (1 MPa) $T_{sat} = 179.91$°C

and 500°C $v = 0.3541 \text{ m}^3\text{/kg}$

$$u = 3124.4 \text{ kJ/kg}$$
$$h = 3478.5 \text{ kJ/kg}$$
$$S = 7.7622 \text{ kJ/kg.K}$$

Final pressure : 1 bar = 0.1 MPa

As there is no heat transfer and work transfer $\Delta Q = \Delta W = 0$

But $\Delta Q = \Delta U + \Delta W$

$\therefore \quad \Delta U = 0$

(i) Since there is no change in ΔU, final temp = initial temp = 500°C

(ii) From steam tables

for 0.1 MPa pressure and $U = 3124.4$ kJ/kg

u $\left(\frac{kJ}{kg}\right)$	v $\left(\frac{m^3}{kg}\right)$	h $\left(\frac{kJ}{kg}\right)$	S $\left(\frac{kJ}{kg.K}\right)$
2967.9 3131.6	3.103 3.565	3278.2 3488.1	8.5435 8.8342

$$v = 3.545 \frac{m^3}{kg}$$

$$h = 3478.9 \frac{kJ}{kg}$$

$$S = 8.8214 \frac{kJ}{kg.K}$$

Percentage of vessel volume initially occupied by steam :

$$\frac{V_A}{V_{AB}} = \frac{0.3541}{3.545} = \frac{1}{10} = 10\%$$

(iii) Entropy produced

$$= 8.8214 - 7.7622$$
$$= 1.0592 \text{ kJ/kg.K}$$

Problem 7.30 : 0.14 m^3 of steam at 20 bar and 250°C is expanded reversibly and polytropically to 2 bar. Find the final temperature, work done, heat transferred and change of entropy if the index of expansion is 1.25. (CES – 1996)

Solution :

$$\frac{T_1}{T_2} = \left(\frac{P_1}{P_2}\right)^{\frac{n-1}{n}} = (10)^{0.25/1.25} = 1.585$$

$$T_2 = \frac{T_1}{1.585} = \frac{(273+250)}{1.585} = 330 \text{ K}$$

$$\text{Work done} = \frac{R(T_1 - T_2)}{n-1} = \frac{0.287(523-330)}{0.25}$$

$$= \frac{0.287 \times 193}{0.25}$$

$$= 223 \text{ kJ/kg}$$

Heat transferred,

$$Q = \left(\frac{\gamma - n}{\gamma - 1}\right) w$$

$$= \frac{(1.4 - 1.25)}{1.4 - 1} \times 223$$

$$= \frac{0.15}{0.4} \times 223$$

$$= 83.5 \text{ kJ/kg}$$

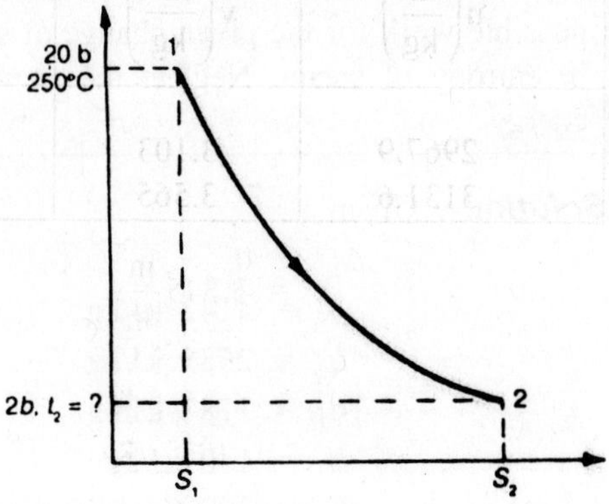

Fig. P. 7.30

Change in entropy

$$= S_2 - S_1 = R \log_e \frac{v_2}{v_1}$$

We know that

$$p_1 v_1^n = p_2 v_2^n$$

$$\left(\frac{v_2}{v_1}\right)^n = \frac{p_1}{p_2} = 10$$

$$\left(\frac{v_2}{v_1}\right)^{1.25} = 10$$

$$\left(\frac{v_2}{v_1}\right) = 6.31$$

$$\therefore \quad S_2 - S_1$$

$$= 0.287 \log_e 6.31 = 0.287 \times 1.8421$$

$$= 0.53 \text{ kJ/kg.K}$$

Problem 7.31 : Steam flows through an adiabatic steady flow turbine. The enthalpy at entrance is 4142 kJ/kg and at exit is 2585 kJ/kg. The values of flow availability of steam at entrance and exit are 1787 kJ/kg and 140 kJ/kg respectively. If the dead state temperature

T_0 is 300 K, determine per kg of steam, the actual work, the maximum possible work for the given change of state of steam and the change in entropy of steam. Neglect the changes in kinetic and potential energy. (CES – 93)

Solution : Given :

$$\Delta Q = 0$$
$$h_1 = 4142 \text{ kJ/kg}$$
$$h_2 = 2585 \text{ kJ/kg}$$
$$\phi_1 = 1787 \text{ kJ/kg}$$
$$\phi_2 = 140 \text{ kJ/kg}$$
$$T_0 = 300 \text{ K}, \Delta KE = 0; \Delta PE = 0$$

To find out :

(i) Actual work/kg of steam

(ii) Maximum possible work/kg of steam

(iii) Change in entropy

(i) Actual work/kg of steam

$$Q - W = m\,(\Delta h + \Delta PE + \Delta KE)$$

Actual work/kg of steam

$$= W_u = -\Delta h = -(2585 - 4142)$$
$$= 1557 \text{ kJ}$$

(ii) Maximum possible work/kg of steam :

$$W_{reversible} = (\phi_1 - \phi_2) = 1787 - 140$$
$$= 1647 \text{ kJ}$$

(iii) $T_0 S_{generated} = W_{rev} - W_u$

Change in Entropy

$$= S_{generated} = \frac{1647 - 1557}{300} = 0.3 \text{ kJ/kg.K}$$

Problem 7.32 : During some integral number of complete cycles a reversible heat engine absorbs 2800 kJ from a heat reservior at 1000 K and performs 800 kJ of mechanical work. The engine exchanges heat with two other heat reserviors one of which is 5400 K and the other at 600 K. Determine the heat exchanged (whether absorbed or rejected) with these two reserviors, the change in entropy of each of the three reserviors and the change in the entropy of the universe. Draw a neat sketch of the system. (CES – 1986)

Solution :

$2800 + Q_1 = Q_2 + 800$ ...(1)

$\frac{2800}{1000} + \frac{Q_1}{5400} - \frac{Q_2}{600}$

$= 0$ (for reversible process)

$(280 \times 54) + Q_1 - 9Q_2$

$= 0$ ...(2)

Solving (1) and (2)

$Q_1 = -360$ kJ

$Q_2 = 1640$ kJ

Heat is rejected both at Q_1 and Q_2

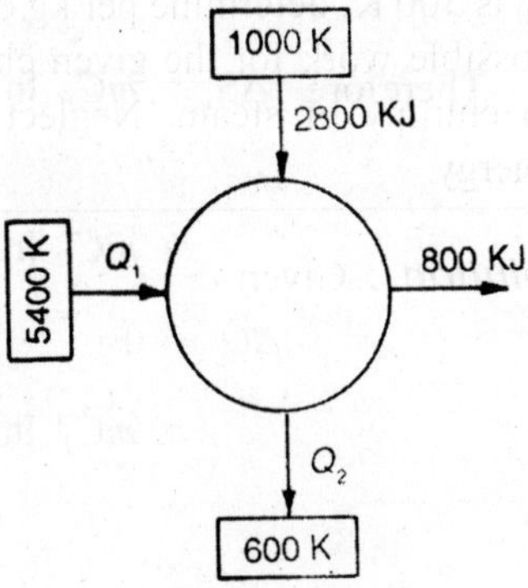

Fig. P. 7.32

Change in entropy of 1000 K reservior $= -\frac{2800}{1000} = +2.80$ kJ/K

Change in entropy of 5400 K reservior $= -\frac{360}{5400} = -0.0667$ kJ/K

Change in entropy of 600 K reservior $= \frac{1640}{600} = -2.7333$ kJ/K

Change in entropy of the universe = +2.80 − 0.0667 − 2.777 = 0

(Since all the processes are reversible the total change in entropy is zero.)

Problem 7.33 : A mass m of water at T_1 is isobarically and adiabatically mixed with an equal mass of water at T_2. Show that

$$\Delta S_{total} = 2m\, C_P \ln \frac{\left(\frac{T_1 + T_2}{2}\right)}{\sqrt{T_1 T_2}} \qquad \text{(CES – 1988)}$$

Solution : Since both the masses of water are same, the equilibrium temperature $= \left(\frac{T_1 + T_2}{2}\right)$

Change in entropy is given by

$$\Delta S = mC_P \int_{T_1}^{\frac{T_1+T_2}{2}} \frac{dT}{T} - mC_P \int_{\frac{T_1+T_2}{2}}^{T_2} \frac{dT}{T}$$

$$\text{Therefore } \Delta S = mC_P \ln \frac{(T_1 + T_2)}{2T_1} - mC_P \ln \frac{2T_2}{(T_1 + T_2)}$$

$$= mC_P \ln \frac{T_1 + T_2}{2T_1} + mC_P \ln \frac{T_1 + T_2}{2T_2}$$

$$= mC_P \ln \frac{(T_2 + T_2)^2}{4T_1 T_2}$$

$$= mC_P \ln \left[\frac{T_1 + T_2}{2\sqrt{T_1 T_2}} \right]^2$$

$$= 2mC_P \ln \left[\frac{T_1 + T_2}{2\sqrt{T_1 T_2}} \right]$$

Since the arithmetic mean $\frac{T_1 + T_2}{2}$ > the geometric mean; $\sqrt{T_1 \, T_2}$

change in entropy = 2 mC_P ln [a quantity > 1]

= *+ve.*

Entropy of the universe increases.

Problem 7.34 : Two identical bodies of constant heat capacity are at the same initial temperature T_1. A refrigerator operates between these two bodies until one of the body is cooled to the temperature T_2. If the bodies remain at constant pressure and undergo no change of phase, obtain an expression for the minimum amount of work required to achieve this. (CES – 1987)

Solution : Let C_P be the specific heat of the body at constant pressure.

For minimum work, the refrigerator should operate on a reversible cycle. If it is so, then the entropy change for the total cycle is equal to zero.

$$\oint \frac{dQ}{T} = 0$$

or

$$\frac{C_P dT_1}{T_1} + \frac{C_P dT_2}{T_2} = 0$$

If T_f is the final temperature of the high level reservior, then the above equation can be written as

$$C_P \int_{T_1}^{T_f} \frac{dT_1}{T_1} + C_P \int_{T_1}^{T_f} \frac{dT_2}{T_2} = 0$$

$$\therefore \qquad \log_e\left(\frac{T_f}{T_1}\right) = \log_e\left(\frac{T_1}{T_2}\right)$$

$$T_f = \frac{T_1^2}{T_2}$$

Cyclic workdone = Cyclic heat added

$$\oint dW = \oint dQ$$

$\therefore$ work done = heat rejected at the higher temperature – heat picked up at the lower temperature.

$$= C_P \int_{T_1}^{T_f} dT - C_P \int_{T_2}^{T_1} dT$$

$$= C_P\,(T_f + T_2 - 2T_1)$$

Substituting $T_f = \dfrac{T_1^2}{T_2}$ = in the above equaiton, we get

$$W_{min} = C_P\left(\frac{T_1^2}{T_2} + T_2 - 2T_1\right)$$

$$\therefore \text{ Minimum work done} = C_P\left(\frac{T_1^2}{T_2} + T_2 - 2T_1\right)$$

Problem 7.35 : A system operating in a reversible cycle receives 25 kJ of heat at 225°C (1–2). This is followed by an adiabatic expansion (2–3) to 115°C at which temperature 15 kJ of heat is received (3–4). Then a further adiabatic expansion (4–5) to 32°C occurs after which 25 kJ of heat is rejected at constant temperature (5–6) followed by an adiabatic compression (6–7). Find the change in entropy during the last process (7–1) which occurs after adiabatic compressions (6–7) to 115°C. (CES – 1995)

Solution :

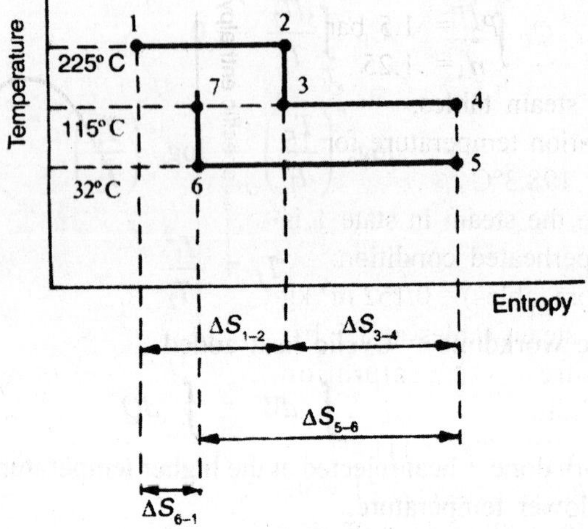

Fig. P. 7.35

$$\Delta S_{1-2} = \frac{dQ}{T} = \frac{25}{(225+273)} = 0.0502 \text{ kJ/K}$$

$$\Delta S_{3-4} = \frac{15}{(115+273)} = 0.0387 \text{ kJ/K}$$

$$\Delta S_{5-6} = \frac{15}{(32+273)} = 0.0820 \text{ kJ/K}$$

Change in entropy in the process 7-1 is

$= 0.0502 + 0.0387 - 0.0820 = 0.0069$ kJ/K

Since there is a decrease in entropy the change during process 7–1 is **–0.0069 kJ/K**

Problem 7.36 : The steam, initially at a pressure of 15 bar and a temperature of 250°C, expand reversibly and polytropically to 1.5 bar. Find the final temperature, workdone and change of entropy, if the index of expansion is 1.25. State the assumptions made (1 bar = 1 kgf/cm²) (IES – 1991)

Solution : Let 1 and 2 represent the initial and final condition of steam.

$$P_1 = 15 \text{ bar}$$
$$T_1 = 250°\text{C}$$
$$P_2 = 1.5 \text{ bar}$$
$$n = 1.25$$

From steam tables,
Saturation temperature for 15 bar = 198.3°C
Hence the steam in state 1 is in superheated condition.
v_1 (from tables) = 0.152 m³/kg
From steam tables at 1.5 bar pressure saturation temperature

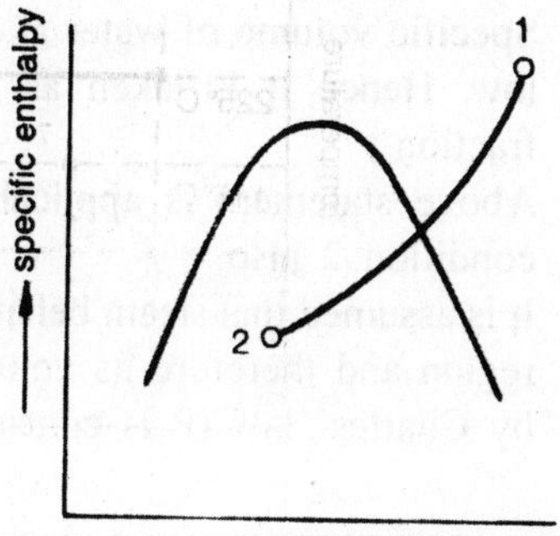

Fig. P. 7.36

$$T_2 = 111.4°\text{C}$$

Specific volume (V_f) of water = 0.001 m³kg
Specific volume (V_g) of steam = 1.1590 m³/kg
Specific volume of steam at 2

$$= v_2 = v_1 \left(\frac{P_1}{P_2}\right)^{\frac{1}{n}}$$

$$= 0.152 \; (10)^{\frac{1}{1.25}} = 0.959 \text{ m}^3/\text{kg}$$

Hence dryness fraction

$$= x = \frac{0.959}{1.159} = 82.7\%$$

$$\text{Work done} = \frac{P_2 v_2 - P_1 v_1}{427(n-1)} \text{ kcal/kg}$$

$$= \frac{(1.5 \times 0.959 - 15 \times 0.152)}{427 \times (1 - 1.25)} \text{ kcal/kg}$$

$$= -7.88 \text{ kcal/kg}$$

Change in entropy.
Initial entropy S_1 = 6.710 kJ/kg.K
Final entropy S_2 = 1.434 + (0.83 × 5.790)
= 6.2397 kJ/kg.K

Change in entropy

$$S_2 - S_1 = 6.2397 - 6.710$$
$$= -0.4703 \text{ kJ/kg.K}$$

Assumptions made

1 Specific volume of water at condition 2 (1.5 bar pressure) is very low. Hence it is taken as zero. While calculating the dryness fraction.

2 Above statement is applicable for the calculation of entropy at condition 2 also.

3 It is assumed that steam behaves like an ideal gas in the superheated region and therefore its volume can be approximately determined by Charles' law (P is contstant).

Problem 7.37 : A mass m of fluid at temperature T_1 is mixed with an equal mass of same fluid at temperature T_2. Prove that the resultant change entropy of the universe is

$$2mc \frac{(T_1 + T_2)/2}{\sqrt{T_1 T_2}}$$

IES – 2001

Solution : Let T_f be the final temperature after mixing.

Considering adiabatic process of mixing.

internal energy before mixing
= internal energy after mixing

or, $\quad mcT_1 + mcT_2 = 2mcT_f$

$$\therefore \quad T_f = \frac{T_1 + T_2}{2}$$

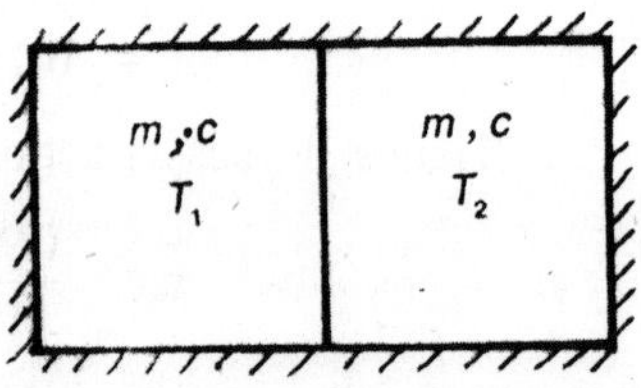

Fig. P. 7.37

∴ Change in entropy of

∴ Change in entropy for the fluid in subsystem 1

$$= (\Delta S)_1 = mc \ln \frac{T_f}{T_1}$$

change in entropy for the fluid in subsystem 2

$$= (\Delta S)_2 = mc \ln \frac{T_f}{T_2}$$

∴ Change in entropy of universe

$$= (\Delta S_1) + (\Delta S)_2$$

$$= mc \ln \frac{T_f}{T_1} + mc \ln \frac{T_f}{T_2}$$

$$= mc \ln \frac{T_f^{\,2}}{T_1 T_2} = 2mc \ln \frac{T_f}{\sqrt{T_1 T_2}}$$

$$= 2mc \, \frac{(T_1 + T_2)/2}{\sqrt{T_1 T_2}} \quad \textbf{Ans}$$

Chapter 8

Availability and Irreversibility

Problem 8.1 : In a steam generator, water is evaporated at 260°C, while the combustion gas C_p = 1.08 kJ/kg.K) is cooled from 1300°C to 320°C. The surroundings are at 30°C. Determine the loss in available energy due to the above heat transfer per kg of water evaporated. (Latent heat of vaporization of water at 260°C = 1662.5 kJ/kg.)

Solution : Given data :

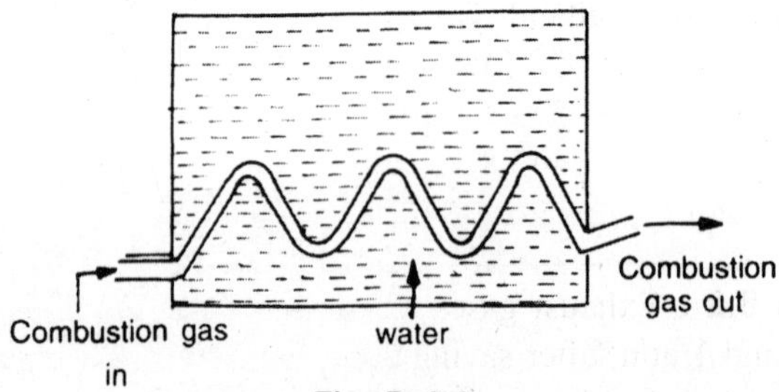

Fig. P. 8.1

Evaporation temp. of water $= T_1$

$= 260 + 273 = 533$ K

Combustion gas inlet temp $= T_i = 1300°C$

$= 1300 + 273 = 1573$ K

Combustion gas outlet temp. $= T_f = 320°C$

$= 320 + 273 = 593$ K

T_0 = Surrounding temperature = 30°C

= 30 + 273 = 303 K

L = latent heat of vaporization at 260°C = 1662.5 kJ/kg

$$C_P = 1.08 \text{ kJ/kg.K}$$

Now considering the reversible heat transfer

heat released by gas = heat used to evaporate the water at 260°

$$\Rightarrow \quad m_g C_{P_g} (1300 - 320) = m_a L$$

$$\Rightarrow \quad m_g = \frac{1 \times 1662.5}{1.08 \times (1300 - 320)} = 1.57 \text{ kg}$$

Now $\Delta S = \int_1^2 \frac{dQ}{T}$ = for reversible process

Change in entropy of gas as it cools from 1300°C to 320°C

$$(\Delta S)_1 = mC_p \ln \frac{T_2}{T_1} = m_g C_{p_g} \ln \frac{320 + 273}{1300 + 273}$$

$$= -1.6541 \text{ kJ/K}$$

Change in entropy of water as it evaporated at 260°C

$$(\Delta S)_2 = \frac{dQ}{T_1} = \frac{m_a \times L}{T_1}$$

$$= \frac{1 \times 1662.5}{533} = 3.1191 \text{ kJ/K}$$

$\therefore$ Loss of availability = T_0 ($\Delta S_{\text{universe}}$)

$= T_0 (\Delta S_l + \Delta S_2)$

$= 533 (-1.6541 + 3.1191)$

$= 443.8$ kJ **Ans**

Problem 8.2 : Exhaust gases leave an internal combustion engine at 800°C and 1 atm, after saving done 1050 kJ of work per kg of gas in the engine (C_P of gas = 1.1 kJ/kg.K). The temperature of the surroundings is 30°C. (a) How much available energy per kg of gas is lost by throwing away the exhaust gases? (b) What is the ratio of the lost available energy to the engine work?

Solution : Given data :

T_2 = exhaust gas temperature

= 800°C = 800 + 273 = 1073 K

P_2 = Exhaust gas pressure = 1 atm

W = done delivered by engine = 1050 kJ/kg of gas

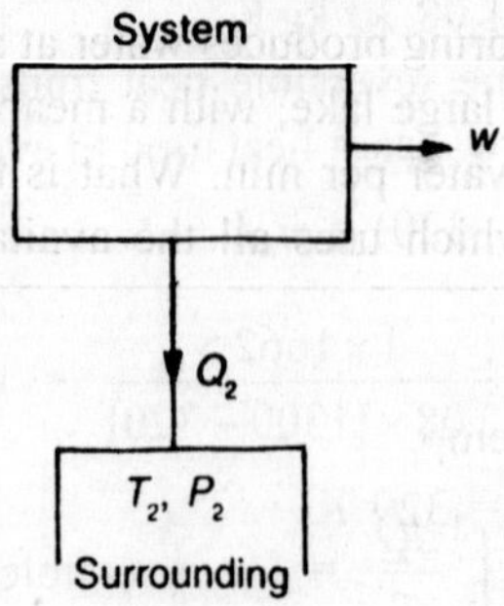

Fig. P. 8.2

$$C_P \text{ of gas} = 1.1 \text{ kJ/kg.K}$$
$$T_0 = \text{temperature of surrounding} = 30 + 273 = 303 \text{ K}$$

Now heat rejected to atmosphere by throwing the exhausted gas

$$Q = mC_P\Delta T$$
$$= 1 \times 1.1 \times (1073 - 303) = 847 \text{ kJ/kg}$$

∴ Change of entropy of atmosphere

$$= \Delta S_I = \frac{Q}{T_0} = \frac{847}{303} = 2.7954 \text{ kJ/kg.K}$$

Change of entropy system (engine) in throwing away the gas

$$\Delta S_2 = \int_{T_2}^{T_0} \frac{dQ_{rev}}{T} = mC_p \ln \frac{T_0}{T_2}$$

$$= 1 \times 1.1 \times \ln \frac{303}{1073}$$
$$= -1.3909 \text{ kJ/kg.K}$$

(a) ∴ Loss of available energy $= T_0 (\Delta S_{\text{total change of entropy}})$

$$= T_0 (\Delta S_I + \Delta S_2)$$
$$= 303 (2.7954 - 1.3909) = 425.55 \text{ kJ/kg of gas}$$

(b) $$\text{Ratio} = \frac{\text{lost available enregy}}{\text{engine work}}$$

$$= \frac{425.55}{1050} = 0.405$$ **Ans**

> **Problem 8.3 :** A hot spring produces water at a temperature of 56°C. The water flows into a large lake, with a mean temperature of 14°C, at a rate of 0.1 m^3 of water per min. What is the rate of working of an ideal heat engine which uses all the available energy?

Solution : Given data :

T_1 = hot spring water temp.

= 56°C = 56 + 273 = 329 K

T_2 = Lake water tempr.

= 14°C = 14 + 273 = 287 K

$\dot{Q}$ = rate of flowing of hotspring water = 0.1 m^3/min

$\dot{m}$ = mass of flow = $\dot{Q} \times \rho$

$= \dfrac{0.1 \times 1000}{60}$ kg/s

ρ = density of water

= 1000 kg/m^3

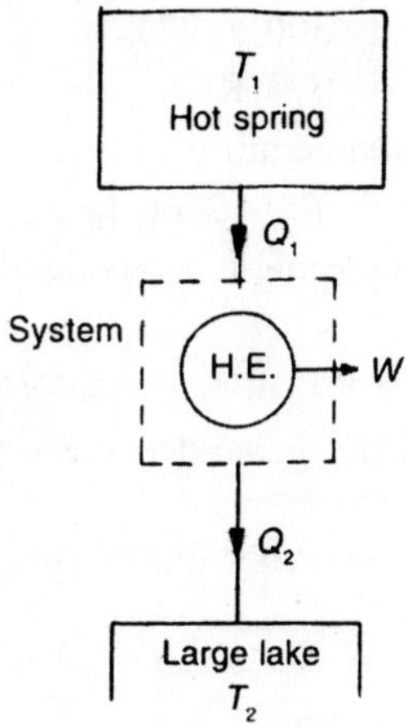

Fig. P. 8.3

From general expression of maximum of an open system

$$dw = d\dot{m}_1 \ [h_1 - T_0S_1 + \frac{V_1^2}{2} + Z_1 g] - d\dot{m},$$

$$[h_2 - T_0S_2 + \frac{V_2^2}{2} + Z_2 g] - d\ [u - T_0S + p_0V + \frac{V_2^2}{2} + Zg]$$

Neglecting Δ.K.E & P.E. and there is no change in system boundary (control volume)

and $d\dot{m}_1 = d\dot{m}_2 = \dot{m}$

$$\Rightarrow \quad dw = \dot{m}\,[(h_1 - h_2) + T_0\,(S_2 - S_1)]$$

$$= \dot{m}\,[C(T_1 - T_2) + T_0\ C\ \ln \frac{T_2}{T_1}]$$

$$= \frac{0.1 \times 1000 \times 9.2}{60}\ [(56 - 14) + 287\ \ln \frac{287}{329}]$$

= 19.62 kW **Ans**

Problem 8.4 : 0.2 kg of air at 300°C is heated reversibly at constant pressure to 2066 K. Find the available and unavailable energies of the heat added. Take T_0 = 30°C and C_p = 1.0047 kJ/kg.K.

Solution : Given data :

mass of air = 0.2 kg

initial temperature of air

$= T_1 = 300°C$

$= 300 + 273$

$= 573 \text{ K}$

final temperature of air

$= T_2 = 2066 \text{ K}$

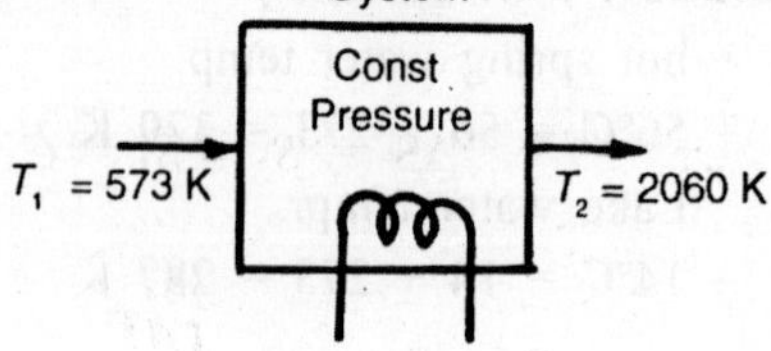

Fig. P. 8.4

atmospheric temperature $= T_0 = 30°C$

$= 30 + 273 = 303 \text{ K}$

$C_p = 1.0047 \text{ kJ/kg.K}$

from general expression of maximum work of an open system

$$dw = d\dot{m}_1\left[h_1 - T_0S_1 + \frac{V_1^2}{2} + Z_1g\right] - d\dot{m}_2$$

$$\left[h_2 - T_0S_2 + \frac{V_2^2}{2} + Z_2g\right] - d\,[u - T_0S + p_0V +$$

$$\frac{V^2}{2} + Zg\,]$$

for an open system,

$d\dot{m}_1 = d\dot{m}_2 = \dot{m}$, neglecting K.E. and P.E. and

$$d\left[u - T_0S + p_0V + \frac{V^2}{2} + Zg\right] = 0$$

$$\therefore \quad dw = \dot{m}\,[h_1 - T_0S_1 - h_2 + T_0S_2]$$

∴ available energy

$$dw = \dot{m}\,[(h_1 - h_2) + T_0(S_2 - S_1)$$

$$= \dot{m}\,[C_P\,(T_1 - T_2) + T_0\,(S_2 - S_1)]$$

Now $(S_2 - S_1)$ = increase in entropy of air in the system from property relation,

$$ds = \frac{dH}{T} - R\frac{dP}{P}$$

at constant pressure process, $dp = 0$

$$\therefore \quad ds = \frac{dh}{T}$$

$$\therefore \quad \Delta S = S_2 - S_1 = C_P \int_1^2 \frac{dT}{T} \text{ kJ/kg.K}$$

$$S_2 - S_1 = C_p \int_1^2 \frac{dT}{T} = C_p \ln \frac{T_2}{T_1}$$

$$= 1.0047 \times \ln \frac{2066}{573} \text{ kJ/kg.K} \quad \text{...(i)}$$

$$dw = \dot{m}\left[C_P(T_1 - T_2) + T_0\left(C_P \ln\frac{T_2}{T_1}\right)\right]$$

∴ available energy

$= dw = 0.2$

$$\left[1.0047(573 - 2066) + 303 \times 1.0047 \ln\frac{2066}{573}\right] = -221.9 \text{ kJ } \textbf{Ans}$$

–ve sign indicates the increase in available energy.

Unavailable energy

$$= T_0\,(\Delta S_{universe}) = 303\,[(S_2 - S_1)]$$

$$= 303 \times 0.2 \times 1.0047 \times \ln\frac{2066}{573} \quad \text{[from (i)]}$$

$$= 78.1 \text{ kJ } \textbf{Ans}$$

Problem 8.5 : Eighty kg of water at 100°C are mixed with 50 kg of water at 60°C, while the temperature of the surroundings is 15°C. Determine the decrease in available energy due to mixing.

Solution : Given data :

for first fluid m_1 = mass of water = 80 kg

T_1 = tempr of water (initial) = 100°C = 100 + 273 = 373 K for 2nd fluid

m_2 = mass of water
= 50 kg

T_2 = tempr of water (initial)
= 60°C = 60 + 273
= 333 K

temperature of surrounding
= 15°C = 15 + 273
= 288 K

Let T_f be the final tempr after mixing at equilibrium, then from heat balance

heat lost by 1st fluid
= heat gained by 2nd fluid

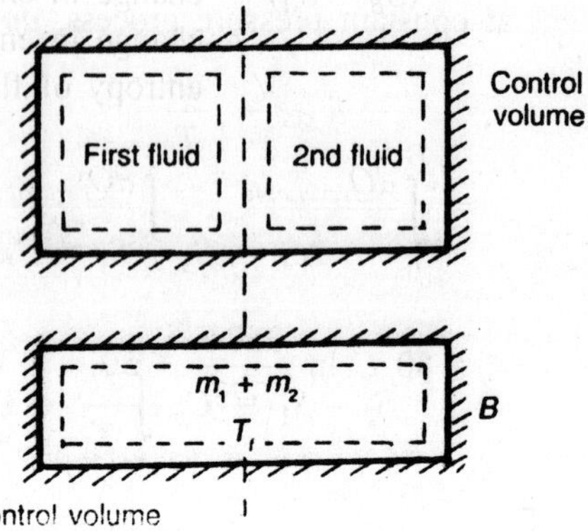

Fig. P. 8.5

$$\Rightarrow \quad m_1 C (T_1 - T_f) = m_2 C (T_f - T_2)$$

$$\therefore \quad T_f = \frac{m_1 T_1 + m_2 T_2}{m_1 + m_2}$$

$$= \frac{80 \times 373 + 50 \times 333}{80 + 50}$$

$$= 357.62 \text{ K}$$

Considering above mixing as closed system, then from general equation of maximum work

$$dw = d\dot{m}_1 \left[h_1 - T_0 S_1 + \frac{V_1^2}{2} + Z_1 g \right] - d\dot{m}_2$$

$$\left[h_2 - T_0 S_2 + \frac{V_2^2}{2} + Z_2 g \right] - d\left[u - T_0 S + p_0 V + \frac{V^2}{2} + Zg \right]$$

for an closed system

$$d\dot{m}_1 = d\dot{m}_2 = 0$$

$$\therefore \quad dw = \text{change in available energy}$$

$$= -d\,[u - T_0 S + p_0 V + V^2/2 + Zg]$$

Neglecting ΔK.E. and ΔP.E. & any change in system boundary

$$dw = \text{change in available energy} = -d\,[u - T_0 S]$$

$$= -[u_B - T_0 S_B - u_A + T_o S_A]$$

$$= (u_A - u_B) + T_0 (S_B - S_A)$$

where u_A = internal energy of fluid before mixing

u_B = internal energy of fluid after mixing

$(S_B - S_A)$ = change in entropy of the system

= change in entropy of fluid 1 + change in entropy of fluid 2

$$= \int_{T_1}^{T_f} \frac{dQ_{reversible}}{T} + \int_{T_2}^{T_f} \frac{dQ_{reversible}}{T_1} \quad m_1 C \ln \frac{T_f}{T_1} + m_2 C \ln \frac{T_f}{T_2}$$

$$= 80\ C \ln \frac{357.6}{373} + 50 \times C \ln \frac{357.6}{333}$$

$$\therefore \quad dw = (u_A - u_B) + T_0 (S_B - S_A)$$

$$= \left[(m_1 T_1 + m_2 T_2) + T_0\left(80 \times \ln\frac{357.6}{373} + 50 \times \ln\frac{357.6}{333} - (m_1 + m_2)T_f\right)\right] \times C$$

= 238.9 kJ **Ans**

Problem 8.6 : A lead storage battery used in an automobile is able to deliver 5.2 MJ of electrical energy. This energy is available for starting the car.

Let compressed air be considered for doing an equivalent amount of work in starting the car. The compressed air is to be stored at 7 MPa, 25°C. What is the volume of the tank that would be required to let the compressed air have an availability of 5.2 MJ? For air, $pv = 0.287\ T$, where T is in K, p in kPa, and v in m^3/kg.

Solution : Given data :

initial pressure of tank air

= p_1 = 7 MPa

initial temperature of tank air

= T_1 = 25°C = 25 +273

= 298 K

Availability

= 5.2 MJ

For air pv

= 0.287 T, T in K

P in kPa

v in m^3/kg

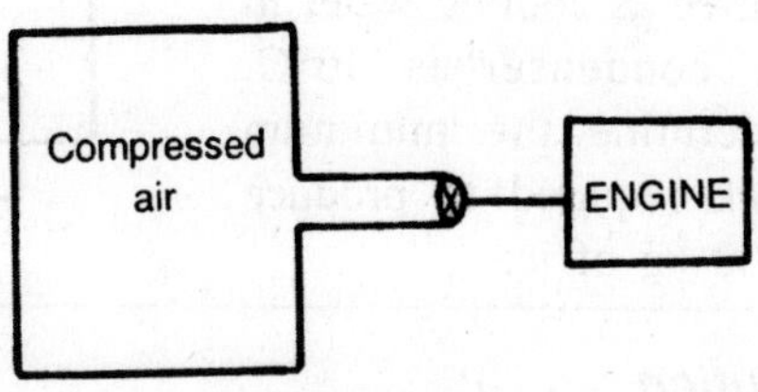

Fig. P. 8.6

Let the tank volume is V m³ and surrounding tempr. be 25°C and pressure 1 bar.

$$\therefore \text{ mass of air} = m = \frac{V}{v} = \text{kg} = \frac{p_1 V}{0.287\, T_1} \quad [\because v = \frac{0.287 \times T_1}{p_1}]$$

$$= \frac{0.287 \times 298}{7000} = .012218 \text{ m}^3/\text{kg}$$

$$V_0 = \frac{0.287 \times T_{atmosphere}}{p_{atmosphere}} = \frac{0.2887 \times 298}{100}$$

$$= 0.85526 \text{ m}^3/\text{kg}$$

$$\therefore \text{ Availability} = (u - u_0) - T_0 (S - S_0) + p_0 (V - V_0)$$

$$= 0 - 298 \left[0.287 \ln \frac{298}{298} - 0.287 \ln \frac{7000}{100} \right] + 100$$

$$(0.012218 - .085526)$$

$$= 279 \text{ kJ/kg}$$

From question, availability = 5.2 MJ

$$\therefore \quad m = \frac{5200}{279} = 18.635 \text{ kg}$$

$$\therefore \text{tank volume} = V = m \times v$$

$$= 18.635 \times 0.012218 = 0.2277 \text{ m}^3 \textbf{ Ans}$$

Problem 8.7 : Ice is to be made from water supplied at 15°C by the process shown in figure P8.7. The final temperature of the ice is −10°C, and the final temperature of the water that is used as cooling water in the condenser is 30°C. Determine the minimum work required to produce 1000 kg of ice.

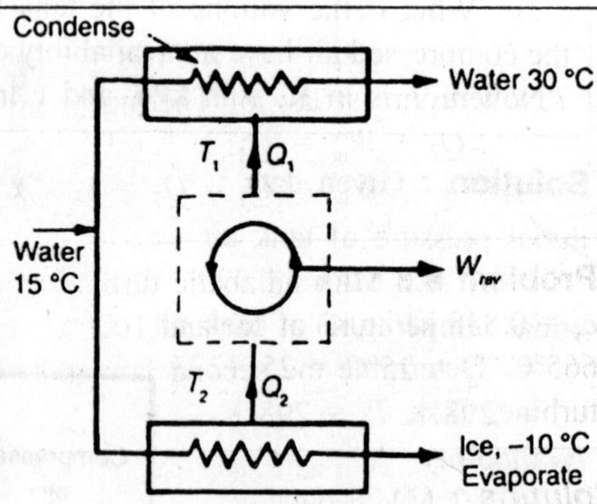

Fig. P. 8.7

Solution :

Fusion tempr of ice is 0°C and latent heat of fusion is 334 kJ/kg.

Q_2 = heat transfer from the evaporation
= heat required to convert 1000 Kg of water at 15° to 1000 kg of ice at –10°C
= 1000 × 4.2 × (15 – 0) +1000 × 334 + 1000 × 2.1 (0 + 10)
= 418000 kJ

and

Q_1 = heat rejected in condenser
= heat required to convert m kg of water at 15°C to water at 30°C
$= \dot{m} \times C_P \times (30 - 15)$
$= \dot{m} \times 4.2 \times 15$
$= 63\ \dot{m}$ kJ ...(i)

Considering the reversibility in the system for minimum work input

$$|\Delta S_{evaporate}| = |\Delta S_{condenses}|$$

= i.e. change in entropy of evaporate = change in entropy of condenser

$$\Rightarrow m_{\text{ice}} \times 4.2 \ln \frac{288}{273} + \frac{m_{ice} \times 334}{273} + m_{\text{ice}} \times 2.1 \ln \frac{273}{263}$$

$$= m_{\text{water}} \times 4.2 \times \ln \frac{303}{288}$$

$$m_{\text{water}} = 7158.3 \text{ kg}$$

∴ from (i)

∴ $Q_1 = 63 \times 7158.6 = 450972.9439$ kJ

Now from 1st law of Thermodynamics

$$Q_2 + W_w = Q_1$$

$$\Rightarrow W_w = Q_1 - Q_2 = 32.97 \text{ MJ } \textbf{Ans}$$

Problem 8.8 : An adiabatic turbine receives a gas (C_p = 1.09 and c_v = 0.838 kJ/kg.K) at bar and 1000°C and discharges at 1.5 bar and 665°C. Determine the second law and isentropic efficiencies of the turbine. Take T_0 = 298 K.

Solution : Given data :

p_1 = 7 bar = inlet pressure
T_1 = 1000°C = 1000 + 273 = 1273 K = inlet temperature

$p_2 = 1.5$ bar
= outlet pressure
$T_2 = 665°C = 665 + 273 = 938$ K
$T_0 = 298$ K

From general equation of maximum work obtainable in open system with no accumulation of energy in the control volume

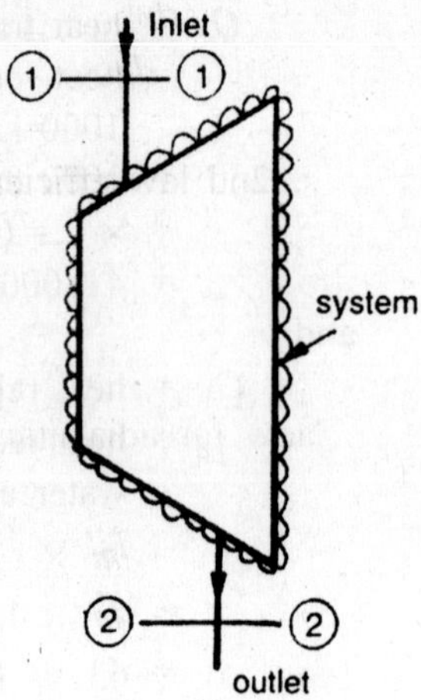

Fig. P. 8.8 (a)

$$dw = dm_1 \left[h_1 - T_0 S_1 + \frac{V_1^2}{2} + Z_1 g\right]$$

$$- dm_2 \left[h_2 - T_0 S_2 + \frac{V_2^2}{2} + Z_2 g\right]$$

Neglecting ΔK.E. & ΔP.E. is zero and $dm_1 = dm_2 = m$

$$dw_{max} = m\,[(h_1 - h_2) + T_0\,(S_2 - S_1)]$$
$$= m\,[C_P\,(T_1 - T_2) + T_0\,(S_2 - S_1)]$$

$S_2 - S_1$ = change of entropy of system

$$= \int_{T_1}^{T_2} \frac{dh}{T} - R\int_{p_1}^{p_2} \frac{dp}{p}$$

$$\therefore \quad dw_{max} = m\left[C_P(T_1 - T_2) + T_0\left(C_P \ln\frac{T_2}{T_1} - R\ln\frac{p_2}{p_1}\right)\right]$$

$$= m\left[1.09\,(1273 - 938) + 298\left(1.09 \ln\frac{938}{1273} - 0.252\ln\frac{1.5}{7}\right)\right]$$

= 381.64 kJ/kg

From S.F.E.E. equation

$$\frac{dQ}{dm} + \left(h_1 + \frac{v_1^2}{2} + Z_1 g\right)$$

$$= \frac{dw}{dm} + \left(h_2 + \frac{v_2^2}{2} + Z_2 g\right)$$

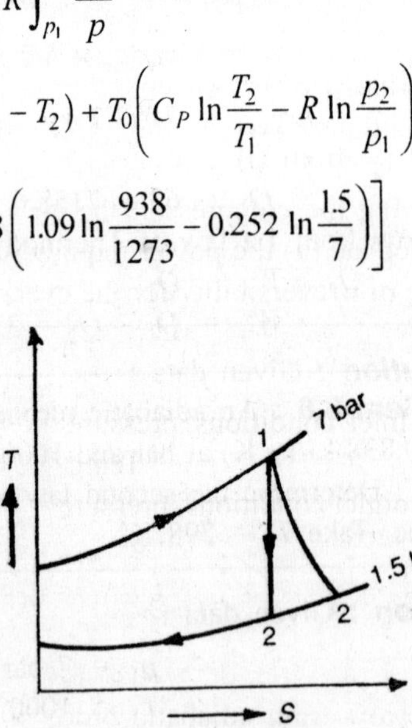

Fig. P. 8.8 (b)

Neglecting $\frac{dQ}{dm}$, ΔP.E. and ΔK.E., we get

$$\frac{dw_{act}}{dm} = h_1 - h_2 = C_P (T_1 - T_2)$$

$$= 1.09\ (1000 - 665) = 365.15 \text{ kJ/kg}$$

∴ 2nd law efficiency

$$= n_{II}$$

$$= \frac{w_{act}}{w_{max}} = 0.956 \textbf{ Ans}$$

Now for adiabatic process,

$$\frac{T_2'}{T_1} = \left(\frac{P_2}{P_1}\right)^{\frac{\gamma-1}{\gamma}}$$

where $\gamma = \dfrac{1.09}{.838} = 1.3$

$$T_2' = 892.16 \text{ K}$$

∴ isentropic efficiency

$$= \frac{T_1 - T_2'}{T_1 - T_2'} = 0.879 \textbf{ Ans}$$

Problem 8.9 : Air enters an adiabatic compressors at atmospheric conditions of 1 bar and 15°C and leaves at 5.5 bar. The mass flow rate is 0.01 kg/s and the efficiency of the compressor is 75%. After leaving the compressor, the air is cooled to 40°C in an aftercooler. Calculate (a) the power required to drive the compressor, and (b) the rate of irreversibility for the overall process (compressor and cooler).

Solution : Given data :

Inlet conditions, pressure = 1 bar = p_1

Temp = 15°C = 15 + 273 = 288 K = T_1

outlet conditions, pressure = 5.5 bar = p_2

$\dot{m}$ = mass flow ret = 0.01 kg/s

n = efficiency of compressor = 0.75

T_3 = tempr after aftercooler = 40°C = 40 + 273 = 313 K

Now from adiabatic process

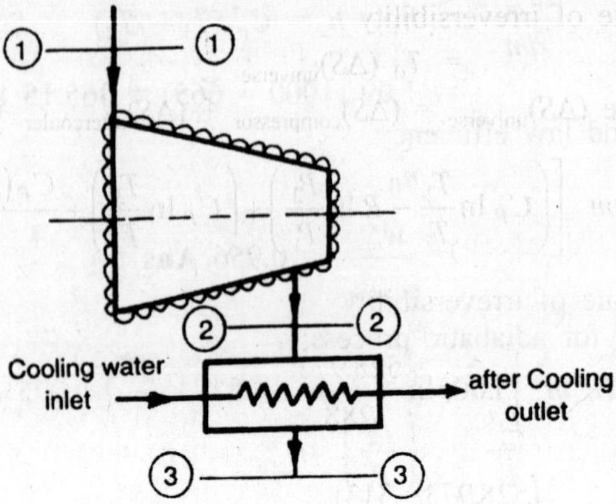

Fig. P. 8.9 (a)

$$T_2' = T_1 \left(\frac{P_2}{P_1}\right)^{\frac{\gamma-1}{\gamma}} = 288\left(\frac{5.5}{1}\right)^{\frac{.4}{1.4}} = 468.73 \text{ K}$$

$$n = 0.75 = \frac{T_2' - T_1}{T_2 - T_1}$$

$$\therefore \quad T_2 - T_1 = \frac{(T_2' - T_1)}{0.75} = 240.973 \text{ K}$$

$$\therefore \quad T_2 = 528.973 \text{ K}$$

(a) ∴ Power required to drive the compressor $= \dot{m}\,(h_2 - h_1)$

$= 0.01\,(T_2 - T_1)\,C_P$

$= 0.01 \times 240.973 \times 1.005$

$= 2.42$ kW **Ans**

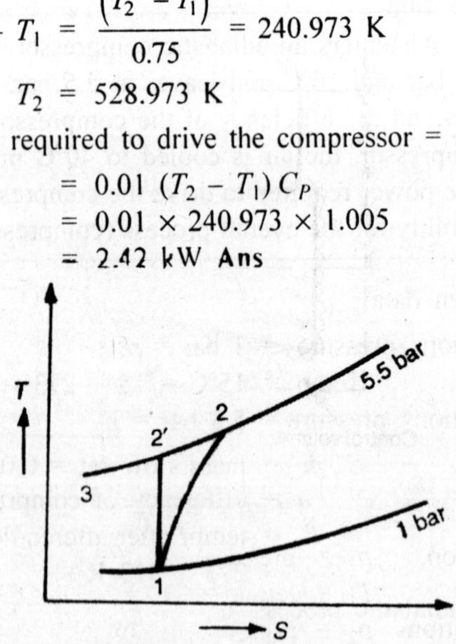

Fig. P. 8.9 (b)

(b) Rate of irreversibility

$$= T_0\,(\Delta S)_{universe}$$

where $(\Delta S)_{universe} = (\Delta S)_{compressor} + (\Delta S)_{aftercooler} + (\Delta S)_{atmosphere}$

$$= \dot{m}\left[\left(C_P \ln\frac{T_2}{T_1} - R\ln\frac{P_2}{P_1}\right) + \left(C_P \ln\frac{T_3}{T_2}\right) + \frac{C_P(528.973-313)}{288}\right]$$

$\therefore$ Rate of irreversibility

$$= T_0\ \dot{m}\left[1.005\ln\frac{528.973}{288} - 0.287\ln\frac{5.5}{1} + 1.005\ln\frac{313}{528.973}\right.$$

$$\left. + C_P\,\frac{(528.973-313)}{288}\right] = 1.0 \text{ kW } \textbf{Ans}$$

Problem 8.10 : In a rotary compressor, air enters at 1.1 bar, 21°C. Where it is compressed adiabatically to 6.6 bar, 250°C. Calculate the irreversibility and the entropy production for unit mass flow rate. The atmosphere is at 1.03 bar, 20°C. Neglect the K.E. change.

Solution: Given data :

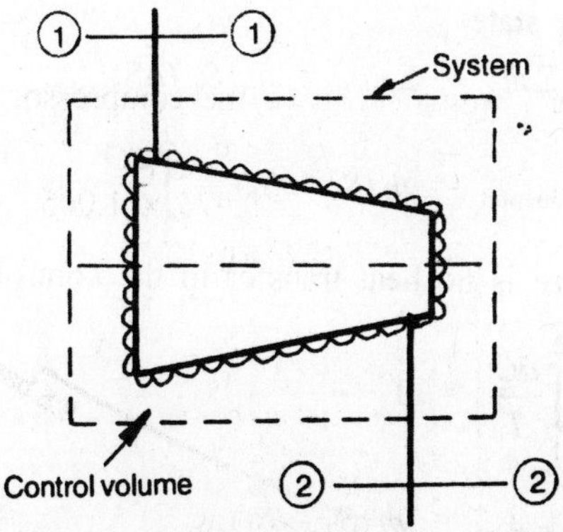

Fig. P. 8.10

Inlet condition, p_1 = pressure = 1.1 bar

T_1 = tempr = 21°C = 21 + 273 = 294 K

Outlet condition p_2 = pressure = 6.6 bar

T_2 = tempr = 250°C = 250 + 273 = 523 K

$$T_0 = \text{atmospheric tempr}$$
$$= 20°C = 20 + 273 = 293 \text{ K}$$
$$P_0 = \text{atmospheric pressure} = 1.03 \text{ bar}$$

$$\text{Irreversibility} = T_0\,(\Delta S)_{universe} \text{ kJ/kg}$$

$$(\Delta S)_{universe} = (\Delta S)_{compressor} + (\Delta S)_{atmosphere}$$

$$= \left(C_P \ln\frac{T_2}{T_1} - R\ln\frac{P_2}{P_1}\right) + 0$$

$$\therefore \text{Irreversibility} = T_0\left(C_P \ln\frac{T_2}{T_1} - R\ln\frac{P_2}{P_1}\right) \text{ kJ/kg}$$

$$= 293\left[1.005 \ln\frac{523}{294} - 0.287 \ln\frac{6.6}{1.1}\right] = 19.04 \text{ kJ/kg}$$

$\dot{S}_{production}$ = rate of entropy production = rate of entropy change of system – rate of entropy transfer + $\Sigma \dot{m}_e s_e - \Sigma \dot{m}_i s_i$

$$\dot{S}_{production} = \text{rate of entropy production}$$

$$= (\Delta S)_{system} - \int_1^2 \frac{d\dot{Q}}{T} + \Sigma \dot{m}_e s_e - \Sigma \dot{m}_i s_i$$

for steady state

$$\Sigma \dot{m}_e = \Sigma \dot{m}_i = \dot{m} \ \& \ (\Delta S)_{system} = 0$$

$$\therefore (\Delta S)_{production} = \dot{m}(S_2 - S_1) - \int_1^2 \frac{d\dot{Q}}{T}$$

Since there is no heat transfer to the control volume

$$\therefore \int_1^2 \frac{d\dot{Q}}{T} = 0$$

$$\therefore \dot{S}_{production} = \dot{m}(S_2 - S_1)$$

$$\Rightarrow \frac{S_{production}}{\dot{m}} = S_2 - S_1 = C_P \ln\frac{T_2}{T_1} - R\ln\frac{P_2}{P_1}$$

$$= 1.005 \ln\frac{523}{294} - .0287 \ln\frac{6.6}{1.1}$$

$$= 0.065 \text{ kJ/kg.K} \ \textbf{Ans}$$

> **Problem 8.11 :** An economizer, a gas-to-water finned tube heat exchanger, receives 67.5 kg/s of gas, C_P = 1.0046 kJ/kg.K, and 51.1 kg/s of water, C_P = 4.186 kJ/kg.K. The water rises in temperature from 402 to 469 K, where the gas falls in temperature from 682 to 470 K. There are no changes of kinetic energy, and P_0 = 1.03 bar and T_0 = 289 K. Determine (a) rate of change of availability of the water, (b) the rate of change of availability of the gas, and (c) the rate of entropy generation.

Solution : Given data :

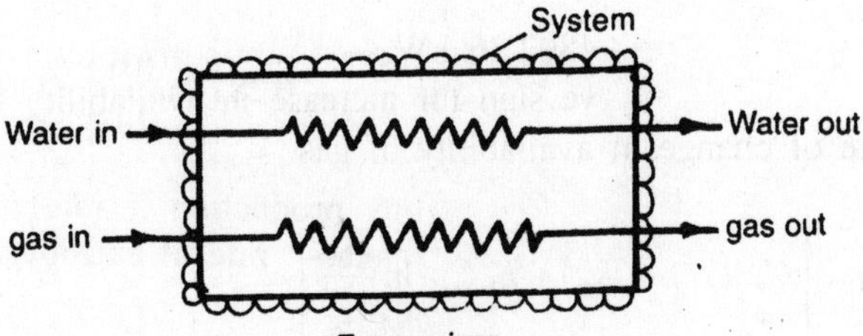

Fig. P. 8.11

$\dot{m}_g$ = rate of gas flow = 67.5 kg/s

C_{P_g} = 1.0046 kJ/kg.K

$\dot{m}_w$ = rate of water flow = 51.1 kg/s

C_{P_w} = 4.186 kJ/kg.K

T_{1w} = water inlet tempr = 402 K

T_{2w} = water outlet tempr = 469 K

T_{1g} = gas inlet tempr = 682 K

T_{2g} = gas outlet tempr = 470 K

P_0 = atmospheric pressure = 1.03 bar

T_0 = atmospheric tempr = 289 K

General equation of maximum work obtainable in an open system, with no accumulation of mass and energy

$$đ\,\dot{w} = d\dot{m}_1\left[h_1 - T_0S_1 + \frac{V_1^2}{2} + Z_1g\right] - d\dot{m}_2\left[h_2 - T_0S_2 + \frac{V_2^2}{2} + Z_2g\right]$$

For steady flow

$d\dot{m}_1 = d\dot{m}_2 = d\dot{m}$ and neglecting ΔK.E. and ΔP.E.

$d\dot{w}$ = rate of change in availability

$= d\dot{m}\ [(h_1 - h_2) + T_0\ (S_2 - S_1)]$

(a) rate of change in availability of water

$$= \dot{m}_w \left[C_{P_w} l \left(T_{1_w} - T_{2_w}\right) + T_0\, C_{P_w} \ln \frac{T_{2w}}{T_{1w}} \right]$$

$$= 51.1 \left[4.186(402 - 469) + 289 \times 4.186 \ln \frac{469}{402} \right]$$

= –4802.26 kW

= –ve sign for increase in availability **Ans**

(b) rate of change in availability of gas

$$= \dot{m}_g \left[C_{P_g} \left(T_{1g} - T_{2g}\right) + T_0\, C_{P_g} \ln \frac{T_{2g}}{T_{1g}} \right]$$

$$= 67.5 \left[1.0046\,(682 - 470) + 289 \times 1.0046 \times \ln \frac{470}{682} \right]$$

= 7079.84 kW

= +ve, i.e. the decrease in availability **Ans**

(c) rate of entropy generation

$$= d\left(\Delta\dot{S}\right) + \Sigma\dot{m}_e s_e - \Sigma\dot{m}_i s_i - \int_1^2 \frac{d\dot{Q}}{T}$$

For steady flow $d\left(\Delta\dot{S}\right) = 0$, & $\Sigma\dot{m}_e = \Sigma\dot{m}_i = \Sigma\dot{m}$

with no heat transfer, $\int_1^2 \frac{d\dot{Q}}{T} = 0$

∴ Rate of entropy generation $= \Sigma\dot{m}(S_e - S_i)$

$$= \dot{m}_g\ (S_{2g} - S_{1g}) + \dot{m}_w\ (S_{2w} - S_{1w})$$

$$= 67.5 \times 1.0046 \ln \frac{470}{682} + 51.1 \times 4.186 \ln \frac{469}{402}$$

= 7.73 kW/K **Ans**

Problem 8.12 : The exhaust from a gas turbine at 1.12 bar, 800 K flows steadily into a heat exchanger which cools the gas to 700 K without significant pressure drop. The heat transfer from the gas heats an air flow at constant pressure, which enters the heat exchanger at 470 K. The mass flow rate of air is twice that of the gas and the surroundings are at 1.03 bar, 20°C. Determine: (a) the decrease in availability of the exhaust gases, and (b) the total entropy production per kg of gas. (c) What arrangement would be necessary to make the heat transfer reversible and how much would this increase the power output of the plant per kg of turbine gas? Take C_P for exhaust gas as 1.08 and for air as 1.05 kJ/kg.K. Neglect heat transfer to the surroundings and the changes in kinetic and potential energy.

Solution : Given data :

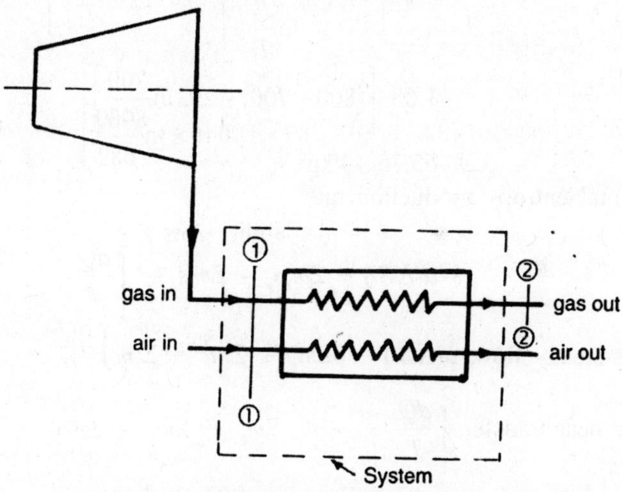

Fig. P. 8.12

p_{1g} = pressure of gas inlet = 1.12 bar

T_{1g} = tempr of gas inlet = 800 K

T_{2g} = tempr of gas outlet = 700 K

T_{1a} = tempr of air inlet = 470 K

C_{P_g} = exhaust gas = 1.08 kJ/kg.K

C_{P_a} = 1.05 kJ/kg.K = air

$\dot{m}_a$ = mass flow rate of air

$\dot{m}_g$ = mass flow rate of gas

$\Rightarrow \quad \dot{m}_a = 2\ \dot{m}_g$

T_0 = surrounding tempr = 20°C = 293 K

P_0 = surrounding pressure = 1.03 bar

Applying heat balance for the system

heat lost by gas = heat gained by air

$$\Rightarrow \dot{m}_g\ C_{P_g}\ (T_{1g} - T_{2g}) = \dot{m}_a\ C_{P_a}\ (T_{2a} - T_{1a})$$

$\therefore \quad T_{2a}$ = air outlet tempr = 521.43 K

(a) Decrease in availability of exhaust gas

$$= \dot{m}_g \left[C_{P_g}\left(T_{1g} - T_{2g}\right) + T_0 C_{P_g} \ln \frac{T_{2g}}{T_{1g}} \right]$$

$$= 1.08 \left[(800 - 700) + 293 \ln \frac{700}{800} \right]$$

$$= 65.75 \text{ kJ/kg}$$

(b) total entropy production rate

$$= d(\Delta \dot{S}) + \Sigma m_e s_e - \Sigma m_i s_i - \int_1^2 \frac{d\dot{Q}}{T}$$

for steady flow, $d(\Delta \dot{S}) = 0 \Sigma m_e = \Sigma m_i = \dot{\Sigma} m$

for heat transfer, $\int_1^2 \frac{d\dot{Q}}{T} = 0$

$\therefore$ total entropy production rate $= \Sigma \dot{m}\ (S_e - S_i)$

$$= \dot{m}_g\ [S_{2g} - S_{1g}] + \dot{m}_a\ [S_{2a} - S_{1a}]$$

$$= \dot{m}_g \left[C_{P_g} \ln \frac{T_{2g}}{T_{1g}} + 2 C_{P_a} \ln \frac{T_{2a}}{T_{1a}} \right]$$

$\therefore$ total entropy production per kg gas

$$= 1.08 \ln \frac{700}{800} + 2 \times 1.05 \times \ln \frac{521.43}{470}$$

$$= 0.0738 \text{ kJ/kg.K}$$

(c) Heat supplied to air in heat exchanger

$$= \dot{m}_g \, C_{P_g} \, (T_{1g} - T_{2g})$$

$$= 1.08 \times (800 - 700) = 108 \text{ kJ/kg of gas}$$

Now for reversible heat transfer

$$\Delta S_{\text{universe}} = 0$$

$$\Rightarrow \quad (\Delta S)_{\text{gas}} + (\Delta S)_{\text{air}} = 0$$

$$\Rightarrow \quad -(\Delta S)_{\text{gas}} = +(\Delta S)_{\text{air}}$$

$$\Rightarrow \quad -\dot{m}_g \, C_{P_g} \, \ln \frac{T_{2g}}{T_{2g}} = + \, \dot{m}_a \, C_{P_a} \, \ln \frac{T_{2a}}{T_{1a}}$$

$$\Rightarrow \dot{m}_g \, C_{P_g} \ln \frac{T_{1g}}{T_{2g}} - \dot{m}_g \, C_{P_g} \frac{(T_{2g} - T_0)}{T_0}$$

$$= 2 \dot{m}_g \, C_{P_a} \ln \frac{T_{2a}}{T_{1a}}$$

$$\Rightarrow 1.08 \ln \frac{800}{700} = 2 \times 1.05 \ln \frac{T_{2a}}{470}$$

$$\therefore \quad T_{2a} = 503.41 \text{ K}$$

$\therefore$ Now heat supplied to air in heat exchanger

$$= \dot{m}_a \, C_{P_a} \, (T_{2a} - T_{1a})$$

$$= 2 \dot{m}_g \times 1.05 \, (503.41 - 470) = 70.16$$

Problem 8.13 : For flow of an ideal gas through an insulated pipeline, the pressure drops from 100 bar to 95 bar. If the gas flows at the rate of 1.5 kg/s and has $C_P = 1.005$ and $C_v = 0.718$ kJ/kg.K and if $T_0 = 300$ K, find the rate of entropy generation and rate of loss of exergy.

Solution : Given data :

p_1 = inlet pressure
= 100 bar

p_2 = outlet pressure
= 95 bar

$\dot{m}$ = gas flow rate

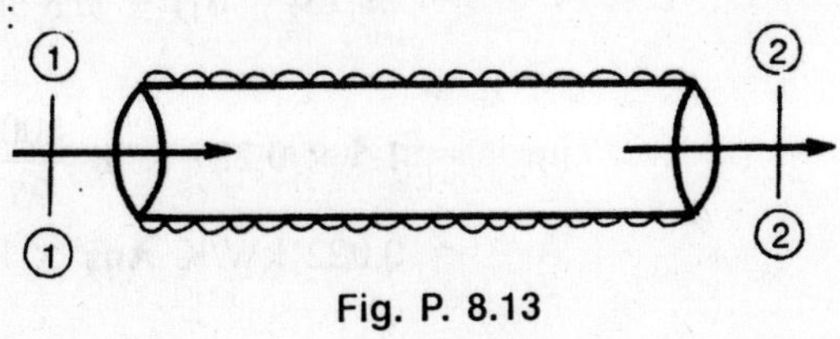

Fig. P. 8.13

$$= 1.5 \text{ kg/s}$$

$$C_P = 1.005 \text{ kJ/kg.K}$$

$$w = 0.718 \text{ kJ/kg.K}$$

$$T_0 = 300 \text{ K}$$

(a) Rate of loss of energy = $T_0\ (\Delta S)_{universe}$

$$(\Delta S)_{universe} = \dot{m}\left[C_P \ln\frac{T_2}{T_1} - R\ln\frac{P_2}{P_1}\right]$$

due to insulation,

$$T_2 = T_1$$

$$\therefore \quad (\Delta S)_{universe} = +\ \dot{m} R \ln \frac{P_1}{P_2}$$

$$\therefore \text{ Rate loss of energy} = T_0 \times \dot{m} R \ln \frac{P_1}{P_2}$$

$$= 300 \times 1.5 \times 0.287 \ln \frac{100}{95}$$

$$= 6.62 \text{ kW } \textbf{Ans}$$

(b) Rate of entropy production

$$= d(\Delta\dot{s}) + \Sigma m_e s_e - \Sigma m_i s_i - \int_1^2 \frac{d\dot{Q}}{T}$$

for steady flow

$$d(\Delta\dot{s}) = 0,\ \Sigma m_e = \Sigma m_i = \dot{m}$$

due to insulation

$$= \int_1^2 \frac{d\dot{Q}}{T} = 0$$

$\therefore$ Rate of entropy production

$$= \dot{m}\ (S_2 - S_1) = \dot{m} R \ln \frac{P_1}{P_2}$$

$$= 1.5 \times 0.287 \times \ln \frac{100}{95}$$

$$= 0.022 \text{ kW/K } \textbf{Ans}$$

Problem 8.14 : The cylinder of an internal combustion engine contains gases at 2500°C, 58 bar. Expansion takes place through a volume ratio of 9 according to $pv^{1.38}$ = const. The surroundings are at 20°C, 1.1 bar. Determine the loss of availability, the work transfer and the heat transfer per unit mass. Treat the gases as ideal having R = 0.26 kJ/kg.K and C_v = 0.82 kJ/kg.K.

Solution : Given data :

P_1 = initial pressure
= 58 bar

T_1 = initial temp
= 2500°C
= 2500 + 273
= 2773 K

$\frac{V_2}{V_1}$ = 9 following $PV^{1.38}$
= constant

T_0 = 20°C
= 20 + 273 = 293 K

P_0 = 1.1 bar,

R = 0.26 kJ/kg.K

C_v = 0.82 kJ/kg.K

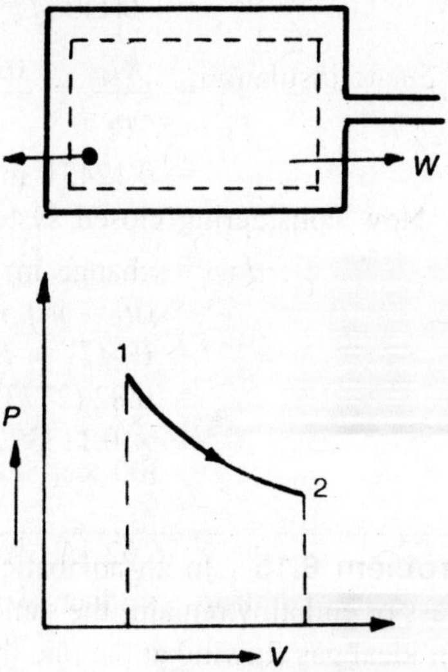

Fig. P. 8.14

Now $\quad \frac{T_2}{T_1} = \left(\frac{V_1}{V_2}\right)^{n-1} = \left(\frac{1}{9}\right)^{0.38}$

$\Rightarrow \quad T_2 = 120.32$ K

From 1st law of thermodynamics

$\therefore$ heat transferred to the surrounding $đQ = du + đw$

$$đw = \text{work transfer}$$

$$= \frac{p_1V_1 - p_2V_2}{n-1} = \frac{R(T_1 - T_2)}{n-1}$$

$$= \frac{0.26(2773 - 1203.2)}{0.38}$$

$$= 1074.07 \text{ kJ/kg } \textbf{Ans}$$

$$\therefore \quad đQ = C_v (T_2 - T_1) + đw$$
$$= -213.16 \text{ kW/kg } \textbf{Ans}$$

$$(\Delta S)_{system} = C_v \ln \frac{T_2}{T_1} + R \ln \frac{V_2}{V_1}$$

$$= 0.82 \ln \frac{1203.2}{2773} + 0.26 \times \ln 9$$

$$= -0.113377 \text{ kJ/kg.K}$$

$$V_1 = \frac{RT_1}{p_1} = \frac{.026 \times 2773}{5800}$$

$$= 0.12431 \text{ m}^3\text{/kg}$$

Now considering closed system

$$đw = \text{change in availability}$$
$$= (u_1 - u_2) + P_0 (V_1 - V_2) + T_0 (S_2 - S_1)$$
$$= C_v (T_1 - T_2) + p_0 (V_1 - gV_1) + T_0 (S_2 - S_1)$$
$$= 0.82 (2773 - 1203.2) - 110 \times 8 \times 0.12431 - 293 \times 0.113377 = 1144.5 \text{ kJ/kg } \textbf{Ans}$$

Problem 8.15 : In an adiabatic throttling process, energy per unit mass or enthalpy remains the same. However, there is a loss of exergy. An ideal gas flowing at the rate throttled from pressure p_1 to pressure p_2 when the environment is at temperature. What is the rate of exergy loss due to throttling?

Solution : Given :

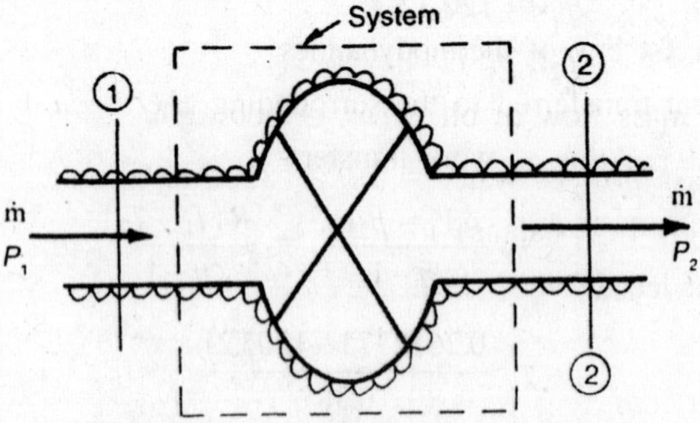

Fig. P. 8.15

$\dot{m}$ = mass flow rate
p_1 = inlet pressure
p_2 = outlet pressure
T_0 = atmospheric tempr

Rate of energy loss

$$= T_0\ (\Delta S_{universe})$$

$$\Delta S_{universe} = \Delta S_{system} + \Delta S_{surrounding}$$

$$= \dot{m}\left[R \ln \frac{P_1}{P_2}\right] + 0 = \dot{m} R \ln \frac{P_1}{P_2}$$

∴ Rate of exergy loss

$$= \dot{m} R\ T_0 \ln \frac{P_1}{P_2} \quad \textbf{Ans}$$

Problem 8.16 : In a counterflow heat exchanger, oil (C_P = 2.1 kJ/kg.K) is cooled from 440 to 320 K, while water (C_P = 4.2 kJ/kg.K) is heated from 290 K to temperature T. The respective mass flow rates of oil and water are 800 and 3200 kg/h. Neglecting pressure drop, KE and PE effects and heat loss, determine (a) the temperature T, (b) the rate of energy destruction, (c) the second law efficiency. Take T_0 = 17°C and P_0 = 1 atm.

Solution : Given data :

For oil,

T_1^0 = 440 K
T_2^0 = 320 K

For water

T_1^w = 290 K
T_2^w = T

mass flow of oil = $\dot{m}_0$ = 800 kg/h

mass flow of water = $\dot{m}_w$ = 3200 kg/h

(i) From heat balance

heat lost by oil = heat gained by water

$$\Rightarrow \quad 800 \times 2.1 \times (440 - 320)$$
$$= 3200 \times 4.2 \times (T - 290)$$

∴ T = 305 K **Ans**

(ii) From property relation, at constant pressure

$$(\Delta S)_0 = \int_{T_Y}^{T_f} \frac{dh}{T} = mC_P \int_{T_i}^{T_f} \frac{dT}{T}$$

$$\therefore \quad (\Delta S)_{oil} = \text{Change in entropy of oil} = mC_P \ln \frac{T_2^{\,0}}{T_1^{\,0}}$$

$$= 800 \times 2.1 \ln \frac{320}{440} = -535 \text{ kJ/h.K}$$

and $(\Delta S)_{water}$ = Change in entropy of water

$$= mC_P \ln \frac{T_2^{\,w}}{T_1^{\,w}} = 3200 \times 4.2 \ln \frac{305}{290}$$

$$= 677.8 \text{ kJ/h.K}$$

$$(\Delta S)_{surrounding} = 0$$

$$\therefore \quad (\Delta S)_{suniverse} = (\Delta S)_{oil} + (\Delta S)_{water} + (\Delta S)_{surrounding}$$

$$= -535 + 677.8 + 0$$

$$142.8 \text{ kJ/h.K}$$

Rate of energy destruction $= T_0\,(\Delta S)_{universe}$

$$= 290 \times 142.8 = 41.41 \text{ MJ/h}$$

$$= 41.41 \text{ MJ/h} \quad \textbf{Ans}$$

Problem 8.17 : Energy is received by a solar collector at the rate of 300 kW from a temperature of 2400 K. If 60 kW of this energy is lost to the surroundings at steady state and if the user temperature remains constant at 600 K, what are the first law and the second law efficiencies? Take T_0 = 300 K.

Solution : Given :

T_r = Source temperature = 2400 K

$\dot{Q}_r$ = solar energy available at source = 300 kW

T_0 = 300 K

$\dot{Q}_{lost}$ = energy lost to the surrounding = 60 kW

T_a = user temperature = 600 K

$\therefore$ $\dot{Q}_a$ = heat transfer by solar collector (source)

$= 300 - 60 = 240$ kW

$\therefore$ n_I = first law efficiency

$$= \frac{\dot{Q}_a}{\dot{Q}_r} = \frac{240}{300} = 0.8$$

and $\quad n_{II}$ = 2nd law efficiency

$$= n_I \frac{\left(1-\frac{T_0}{T_a}\right)}{\left(1-\frac{T_0}{T_r}\right)} = \frac{0.8\left(1-\frac{300}{600}\right)}{\left(1-\frac{300}{2400}\right)}$$

$$= 0.457 \textbf{ Ans}$$

Problem 8.18 : A carnot cycle engine receives and rejects heat with a 20°C temperature differential between itself and the thermal energy reservoirs. The expansion and compression processes have a pressure ratio of 50. For 1 kg of air as the working substance, cycle temperature limits of 1000 K and 300 K and T_0 = 280 K, determines the second law efficiency.

Solution :

n_{cornot} = efficiency of carnot cycle

$$= 1 - \frac{T_0}{T_{reservoir}}$$

$$= \left(1-\frac{280}{1020}\right)$$

and n_I = first law efficiency of cycle (actual)

$$= 1 \frac{T_2}{T_1}$$

$$= \left(1-\frac{300}{1000}\right)$$

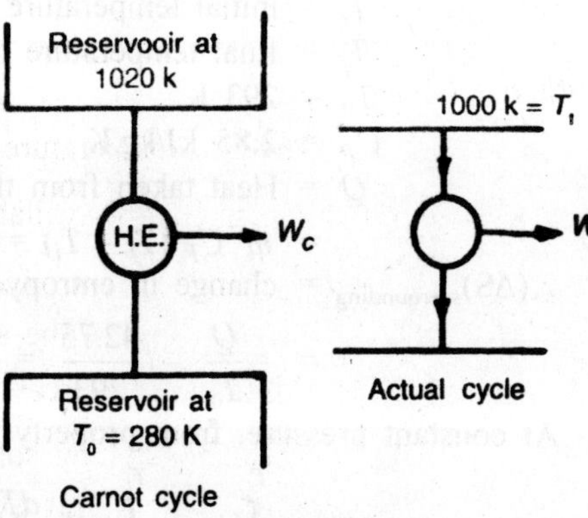

Fig. P. 8.18

$\therefore \quad n_{II}$ = 2nd law efficiency

$$= \frac{n_I}{n_{carnot}} = \frac{\left(1 - \frac{300}{1000}\right)}{\left(1 - \frac{280}{1020}\right)} = 0.965 \textbf{ Ans}$$

> **Problem 8.19** : A pipe carries a stream of brine with a mass flow rate of 5 kg/s. because poor thermal insulation the brine temperature increases from 250 K at the pipe inlet to 253 K at the exit. Neglecting pressure losses, calculate the irreversibility rate (or rate of energy degradation) associated with the heat leakage. Take T_0 = 293 K and C_P = 2.85 kJ/kg.K.

Solution : Given data :

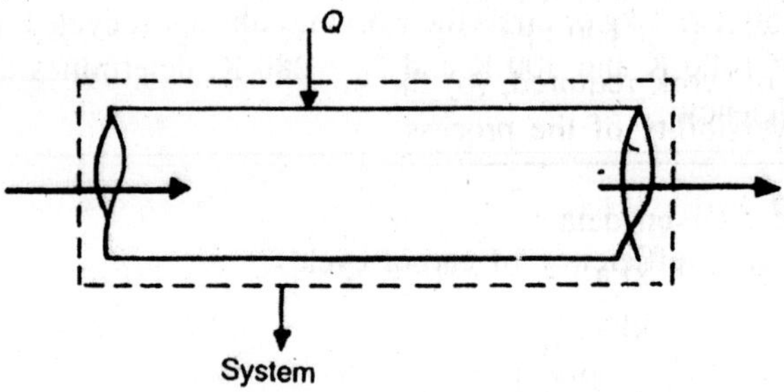

Fig. P. 8.19

$\dot{m}$ = mass flow rate of brine = 5 kg/s

T_i = initial temperature of brine = 250 K

T_f = final temperature of brine = 253 K

T_0 = 293 K

C_P = 2.85 kJ/kg.K

Q = Heat taken from the surrounding

$= \dot{m}\ C_P\ (T_f - T_i)$ = 42.75 kW

$\therefore (\Delta S)_{surrounding}$ = change in entropy of surrounding

$$= -\frac{Q}{T_0} - \frac{42.75}{293} = -0.1459 \text{ kW/K}$$

At constant pressure, from property relation

$$\Delta S = \int_{T_i}^{T_f} ds = \int_{T_i}^{T_f} mC_p \frac{dT}{T}$$

$\therefore \quad (\Delta S)_{system}$ = change in entropy of system

$$= \int_{T_i}^{T_f} mC_p \frac{dT}{T} = mC_P \ln \frac{T_f}{T_i}$$

$$= 5 \times 2.85 \times \ln \frac{253}{250} = 0.169982 \text{ kW/K}$$

$$\therefore \quad (\Delta S)_{universe} \doteq (\Delta S)_{system} + (\Delta S)_{surrounding}$$

$$= 0.024082 \text{ kW/K}$$

$\therefore$ Rate of irreversibility

$$= T_0(\Delta S)_{universe}$$

$$= 7.056 \text{ kW } \textbf{Ans}$$

Problem 8.20 : Air enters a compressor in steady flow at 140 kPa, 17°C and 70 m/s and leaves it at 350 kPa, 127°C and 110 m/s. The environment is at 100 kPa, 7°C calculate per kg of air (a) the actual amount of work required, (b) the minimum work required, and (c) the irreversibility of the process.

***Solution* :** Given data :

P_1 = 140 kPa

P_2 = 350 kPa

T_1 = 17°C = 290 K

T_2 = 127°C = 400 K

V_1 = 70 m/s

V_2 = 110 m/s

T_0 = 7°C = 280 K

P_0 = 100 kPa

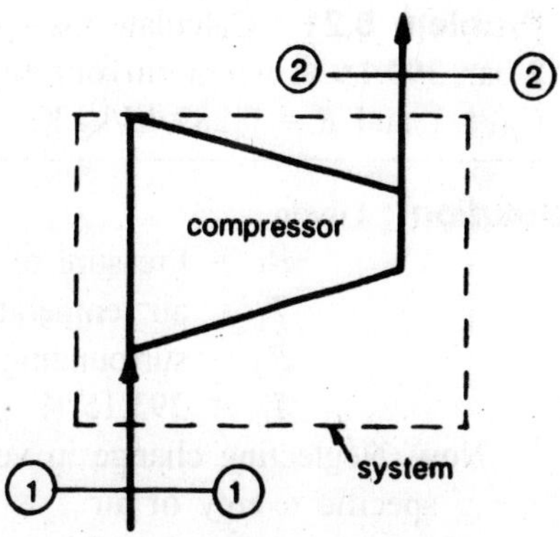

Fig. P. 8.20

(a) from 1st law of thermodynamics, neglecting change in potential energy and heat transfer

$$\frac{dw}{dm} = (h_1 - h_2) + \frac{V_1^2 - V_2^2}{2}$$

$$= C_P (T_1 - T_2) + \frac{V_1^2 - V_2^2}{2}$$

$$= 1.005\,(290 - 400) + \frac{70^2 - 110^2}{2 \times 1000}$$

$= -114.2$ kJ/kg **Ans**

–ve sign show that work is input to the system

(ii) Minimum work required = change of availability

$$= (h_1 - h_2) + \frac{V_1^2 - V_2^2}{2} - T_0\,(S_1 - S_2)$$

$$= C_P\,(T_1 - T_2) + \frac{V_1^2 - V_2^2}{2} - T_0\left[C_P \ln\frac{T_1}{T_2} - R\ln\frac{P_1}{P_2}\right]$$

$$1.005\,(290 - 400) + \frac{70^2 - 110^2}{2 \times 1000} - T_0\left[1.005 \ln\frac{290}{400} - 0.287 \ln\frac{140}{350}\right]$$

$= -97.3$ kJ/kg

= –ve sign for work is input to the system

(iii) Irreversibility of the process

$= W_{act} - W_{min} = 114.2 - 97.3$

$= 16.9$ kJ/kg **Ans**

Problem 8.21 : Calculate the specific exergy of air for a state at 2 bar, 393.15 K when surroundings are at 1 bar, 293.15 K. Take $C_P = 1$ and $R = 0.287$ kJ/kg.K.

Solution : Given :

P_1 = Pressure of air = 2 bar

T_1 = air temperature = 393.15 K

P_0 = surrounding pressure = 1 bar

T_0 = 293.15 K

Now, Neglecting change in velocity and potential,

∴ specific exergy of air

$= (h_1 - h_0) - T_0\,(S_1 - S_0)$

$$= C_P\,(T_1 - T_0) - T_0\left[c_P \ln\frac{T_1}{T_0} - R\ln\frac{P_1}{P_0}\right]$$

$$= 1.005\,(393.15 - 293.15) - 293.15\left[1.005 \ln\frac{393.15}{293.15} - 0.287 \ln\frac{2}{1}\right]$$

$= 72.35$ kJ/kg **Ans**

Problem 8.22 : Calculate the specific energy of CO_2 (C_P = 0.8659 and R = 0.1889 kJ/kg,K for a state at 0.7 bar, 268.15 K and for the environment at 1.0 bar and 293.15 K.

Solution : Given data :

$$P_1 = 0.7 \text{ bar}$$
$$T_1 = 268.15 \text{ K}$$
$$P_0 = \text{surrounding pressure} = 1 \text{ bar}$$
$$T_0 = \text{surrounding temperature} = 293.15 \text{ K}$$

Neglecting change in potential and kinetic energy.

Specific energy of CO_2

$$= (h_1 - h_0) - T_0\,(S_1 - S_0)$$

$$= C_P\,(T_1 - T_0) - T_0\left[C_P \ln\frac{T_1}{T_0} - R \ln\frac{P_1}{P_0}\right]$$

$$= 0.8659\,(268.15 - 293.15) - 293.15 \times$$

$$\left[0.8659 \ln\frac{268.15}{293.15} - 0.1889 \times \ln\frac{0.7}{1}\right] = -18.77 \text{ kJ/kg } \textbf{Ans}$$

Problem 8.23 : Air enters the compressor of a gas turbine at 1 bar, 30°C and leaves the compressor at 4 bar. The compressor has an efficiency of 82%. Calculate per kg of air (a) the work of compression, (b) the reversible work of compression, and (c) the irreversibility. [For air, use

$$\frac{T_{2s}}{T_1} = \left(\frac{P_2}{P_1}\right)^{\gamma - 1/\gamma}$$

Where T_{2s} is the temperature of air after isentropic compression and γ = 1.4. The compressor efficiency is defined as $(T_{2s} - T_1)/(T_2 - T_1)$, where T_2 is the actual temperature of air after compression.]

Solution : Given :

$$P_1 = 1 \text{ bar}$$
$$T_1 = 30°\text{C} = 303 \text{ K}$$
$$P_2 = 4 \text{ bar}$$
$$n_c = \text{compressor efficiency} = 0.82$$

For isentropic process,

$$\frac{T_{2s}}{T_1} = \left(\frac{P_2}{P_1}\right)^{\frac{\gamma-1}{\gamma}} = (4)^{\frac{0.4}{1.4}}$$

$\therefore T_{2s} = 450.43$ K

$$n_c = 0.82 = \frac{T_{2s} - T_1}{T_2 - T}$$

$$= \frac{450.43 - 303}{T_2 - 303}$$

$\therefore T_2 = 482.79$ K

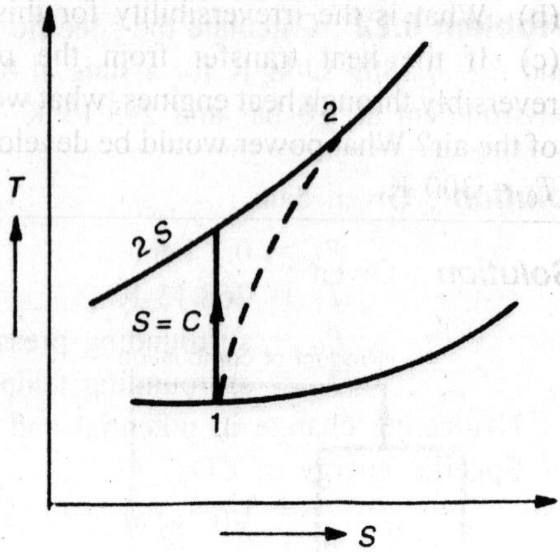

Fig. P. 8.23

(a) Work of compression

$= C_P\,(T_2 - T_1)$

$= 1.005\,(482.79 - 303)$

$= 180.6$ kJ/kg **Ans**

(b) Minimum work required = reversible work of compression

Now neglecting the change in kinetic and potential energy reversible work of compression = change in availability

$= (h_1 - h_2) - T_0\,(S_1 - S_2)$

$$= C_P\,[T_1 - T_2] - T_0 \left[C_P \ln\frac{T_1}{T_2} - R\ln\frac{P_1}{P_2}\right]$$

$$= 1.005\,[303 - 482.79] - 303\left[1.005\ln\frac{303}{482.79} - 0.287\ln\frac{1}{4}\right]$$

$= -159.4$ kJ/kg **Ans**

[–ve sign shows that work is input to the system]

$\therefore$ (c) Irreversibility

$= W_{\text{Actual}} - W_{\text{min}}$

$= 180.6 - 159.4 = 21.2$ kJ/kg **Ans**

Problem 8.24 : An air preheater is used to heat up the air used for combustion by cooling the outgoing products of combustion from a furnace. The rate of flow of the products is 10 kg/s, and the products are cooled from 300°C to 200°C, and for the products at this temperature $C_P = 1.09$ kJ/kg.K. The rate of air flow is 9 kg/s, the initial air temperature is 40°C and for the air $C_P = 1.005$ kJ/kg.K.
(a) What is the initial and final availability of the products?

(b) What is the irreversibility for this process?
(c) If the heat transfer from the products were to take place reversibly through heat engines, what would be the finally temperature of the air? What power would be developed by the heat engines? Take T_0 = 300 K.

Solution : Given :

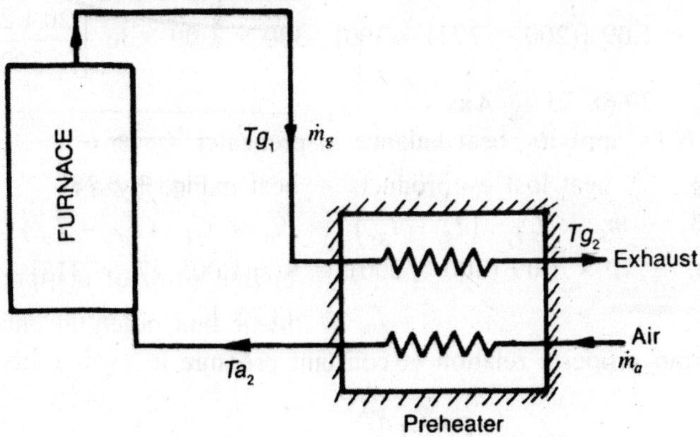

Fig. P. 8.24

$$\dot{m}_g = 10 \text{ kg/s}$$

$$T_{g_1} = 300°C$$

$$T_{g_2} = 200°C$$

$$\dot{m}_a = 9 \text{ kg/s}$$

$$T_{a_1} = 40°C$$

$$T_0 = 300 \text{ K}$$

(a) Neglecting the change in kinetic and potential energy and pressure loss

ϕ_1 = initial availability of product

$$= (h_1 - h_0) - T_0 (S_1 - S_0)$$

$$= C_{P_g} \left(T_{g_1} - T_0\right) - T_0 \left[C_{P_g} \ln \frac{T_{g_1}}{T_0}\right]$$

$$= 1.09\ [(300 + 273) - 300] - 300\left[1.09 \ln \frac{300+273}{300}\right]$$

$= 85.97$ kJ/kg **Ans**

ϕ_2 = final availability of product

$= (h_2 - h_0) - T_0\,(S_2 - S_0)$

$$= C_{P_g}\left(T_{g_2} - T_0\right) - T_0\left[1.09 \ln \frac{T_{g_2}}{T_0}\right]$$

$$= 1.09\ [(200 + 273) - 300] - 300 \times 1.09 \times \ln\left[\frac{20+273}{300}\right]$$

$= 39.68$ kJ/kg **Ans**

(b) Now applying heat balance in preheater

$\Rightarrow$ heat lost by products = heat gained by air

$$\Rightarrow \quad \dot{m}_g \times C_{P_g}\left(T_{g_1} - T_{g_2}\right) = \dot{m}_a \times C_{P_a}\,(T_{a2} - T_{a1})$$

or, $$10 \times 1.09\ (300 - 200) = 9 \times 1.005\left(T_{a_2} - 313\right)$$

$$\therefore \quad T_{a_2} = 433.5 \text{ K}$$

From property relation at constant pressure

$$\Delta S = \int_{T_i}^{T_f} \frac{m\,C_P dT}{T}$$

$\therefore$ $(\Delta S)_{product}$ = change in entropy of combustion

$$= \dot{m}_g\, C_{P_g} \ln \frac{T_{g_2}}{T_{g_1}}$$

$$= 10 \times 1.005 \ln \frac{473}{573} \text{ kJ/K}$$

and $(\Delta S)_{air}$ = change in entropy of air

$$= \dot{m}_a\, C_{P_a} \ln \frac{T_{a_2}}{T_{a_1}}$$

$$= 9 \times 1.005 \ln \frac{433.5}{313} \text{ kJ/K}$$

$$(\Delta S)_{universe} = (\Delta S)_{product} + (\Delta S)_{air} + (\Delta S)_{surrounding}$$

$$= 10 \times 1.09 \ln \frac{473}{573} + 9 \times 1.005 \ln \frac{433.5}{313} + 0$$

$$= 0.85534 \text{ kJ/K}$$

∴ Irreversibility of this process

$$= T_0 \times (\Delta S)_{surrounding}$$

$$= 300 \times 0.85534 = 256.6 \text{ kW } \textbf{Ans}$$

(c) Let T_{a_2} be the final temperature of air

Now For reversible process,

$$(\Delta S)_{universe} = 0$$

or $$(\Delta S)_{product} = -(\Delta S)_{air}$$

or, $$\dot{m}_g \times C_{P_g} \ln \frac{T_{g_2}}{T_{g_1}} = -\dot{m}_a C_{P_a} \ln \frac{T_{a_2}}{T_{a_1}}$$

or $$\ln \frac{T_{a_2}}{T_{a_1}} = -\left(10 \times 1.09 \times \ln \frac{473}{573}\right) / 9 \times 1.005$$

∴ $$T_{a_2} = 394.385 \text{ K}$$

∴ $$Q_1 = \text{heat transfer by product of combustion}$$

$$= \dot{m}_g C_{P_g} \left(T_{g_2} - T_{g_1}\right)$$

$$= 10 \times 1.09 (300 - 200)$$

$$= 1090 \text{ kW}$$

∴ $$Q_2 = \text{heat given to the air preheater}$$

$$= \dot{m}_a C_{P_a} \left(T_{a_2} - T_{a_1}\right)$$

$$= 9 \times 1.005 (394.385 - 313)$$

$$= 736.13 \text{ kW}$$

∴ from 1st law of thermodynamics

$$Q_1 = W + Q_2$$

∴ $$W = Q_1 - Q_2$$

$$= 1090 - 736.13 = 353.87 \text{ kW } \textbf{Ans}$$

Problem 8.25 : A pressure vessel has a volume of 1 m^3 and contains air at 1.4 MPa, 175°C. The air is cooled to 25°C by heat transfer to the surroundings at 25°C. Calculate the availability in the initial and final states and the irreversibility of this process.

Solution : Given :

Initial state of air

$$p_1 = 1.4 \text{ MPa}$$

$T_1 = 175°C$
$V = 1\ m^3$

Final state of air,

$T_2 = 25°C$
T_0 = Surrounding tempr = 25°C

Let P_0 = Surrounding pressure = 0.1 MPa

Taking volume of container constant

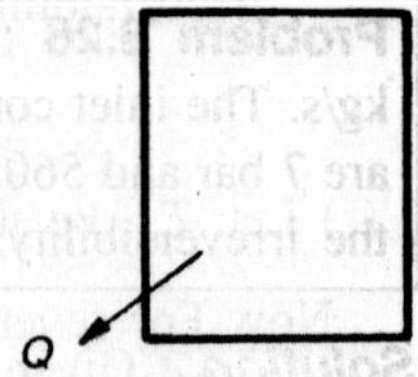

Fig. P. 8.25

$$\therefore \quad P_2 = \frac{P_1}{T_1} \times T_2 = \frac{1.4}{(175+273)} \times (25 + 273)$$

$$= 0.93125 \text{ MPa}$$

Now m = mass of air in container

$$= \frac{p_1 V_1}{RT_1} = \frac{1400 \times 1}{0.287 \times 448} = 10.89 \text{ kg}$$

Considering closed system

∴ Initial availability

$$= (u_1 - u_0) + p_0 (V_1 \overset{0}{\nearrow} - V_0) - T_0 (S_1 - S_0)$$

$$= C_v (T_1 - T_0) + p_0 (V_1 \overset{0}{\nearrow} - V_0) - T_0 \left[C_P \ln \frac{T_1}{T_0} - R \ln \frac{P_1}{P_0} \right]$$

$$= 0.718 (175 - 25) - 298 \left[1.005 \times \ln \frac{448}{298} - 0.287 \ln \frac{1.4}{0.1} \right]$$

= 211.31 kJ/kg **Ans**

final availability

$$= (u_2 - u_0) + p_0 (V_2 \overset{0}{\nearrow} - V_0) - T_0 (S_2 - S_0)$$

$$= C_p (T_2 - T_0) + T_0 \left[C_P \ln \frac{T_2}{T_0} - R \ln \frac{P_2}{P_0} \right]$$

$$= 1.005 (25 - 25) - 298 \left[C_P \ln 1 - 0.287 \ln \frac{9.3/25}{1} \right]$$

= 190.82 kJ/kg **Ans**

∴ Availability = 211.3 − 190.82

= 20.5 kJ/kg

Since work done is nil due to constant volume process

∴ irreversibility = 20.5 × 10.88 = 223 kJ **Ans**

Problem 8.26 : Air flows through an adiabatic compressor at 2 kg/s. The inlet condition are 1 bar and 310 K and the exit conditions are 7 bar and 560 K. Compute the net rate of availability transfer and the irreversibility. Take T_0 = 298 K.

Solution : Given :

$\dot{m}$ = mass flow rate = 2 kg/s

P_1 = 1 bar

P_2 = 7 bar

T_1 = 310 K

T_2 = 560 K

T_0 = 298 K

Neglecting the ΔK.E. and ΔP.E.

(i) Rate of availability transfer

$= \phi_2 - \phi_1$

$= m(h_2 - h_1) - mT_0(S_2 - S_1)$

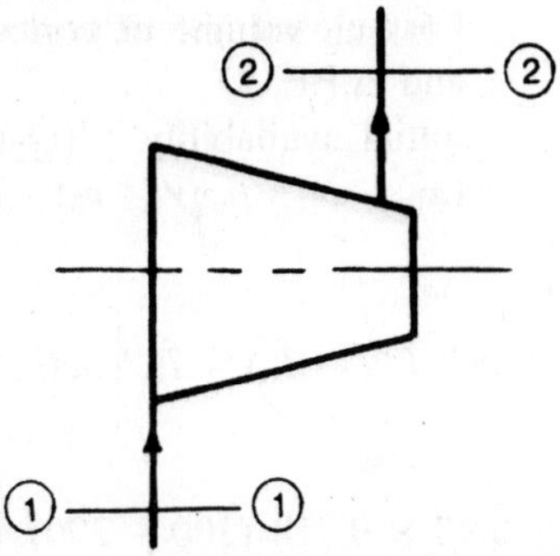

Fig. P. 8.26

$$= mC_p\,(T_2 - T_1) - m\,T_0\left[C_p \ln\frac{T_2}{T_1} - R\ln\frac{P_2}{P_1}\right]$$

$$= 2 \times 1.005 \times (560 - 310) - 2 \times 298\left[1.005 \times \ln\frac{560}{310} - 0.287 \ln\frac{7}{1}\right]$$

= 481.14 kW **Ans**

(ii) Now, $W_{actual} = m\,(h_2 - h_1) = m\,C_p\,(T_2 - T_1)$

$= 2 \times 1.005\,(560 - 310)$

= 502.5 kW

∴ Irreversibility = W_{act} − rate of availability transfer

= 502.5 − 481.14 = 21.36 kW **Ans**

Problem 8.27 : A mass of 6.98 kg of air is in a vessel at 200 kPa, 27°C. Heat is transferred to the air from a reservoir at 727°C until the temperature of air rises to 327°C. The environment is at 100 kPa, 17°C Determine (a) the initial and final availability of a (b) the maximum useful work associated with the process.

Solution : Given :

Initial Condition of vessel

m = mass of air = 6.98 kg

T_1 = 27°C = 300 K

P_1 = 200 kPa

T_3 = 727°C = 1000 K

P_0 = Surrounding pressure = 100 kPa

T_0 = surrounding temperature = 290 K

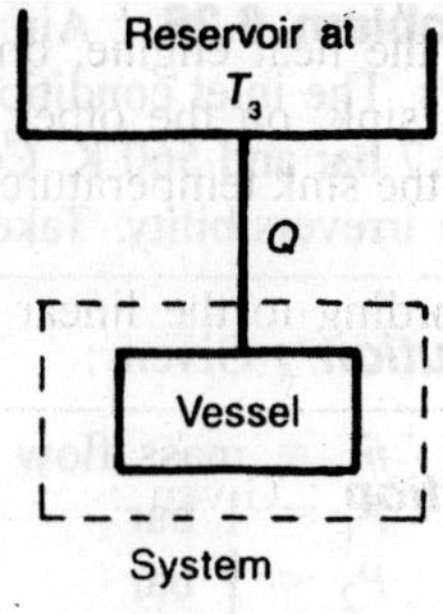

Fig. P. 8.27

For a closed system, neglecting ΔK.E. and Δ.P.E.

initial availability = (Q_1)

$(u_1 - u_0) + p_0(v_1 - v_0) - T_0(S_1 - S_0)$ (with $(v_1 - v_0) \to 0$)

$$= mC_v(T_1 - T_0) - T_0\left[m C_P \ln\frac{T_1}{T_0} - m R \ln\frac{P_1}{P_2}\right]$$

$$= 6.98 \times 0.718\,(300 - 290)\left[6.98 \times 1.005 \ln\frac{300}{290} - 6.98 \times 0.287 \ln\frac{200}{100}\right]$$

= 383.83 kJ **Ans**

considering constant volume process

$$p_2 = \frac{P_1}{T_1} \times T_2$$

$$= \frac{200}{300} \times 600 = 400 \text{ kPa}$$

∴ final availability (Q_2)

$$= (u_2 - u_0) - T_0(S_2 - S_0)$$

$$= m \times C_v(T_2 - T_0) - T_0\left[mC_P \ln\frac{T_2}{T_0} - mR \ln\frac{P_2}{P_0}\right]$$

$$= 6.98 \times 0.718\,(600 - 300) - T_0$$

$$\left[6.98 \times 1.005 \ln\frac{600}{290} - 6.98 \times .287 \ln\frac{400}{100}\right] = 829.80 \text{ kJ}$$

(b) Maximum useful work = change in availability

$$= \phi_2 - \phi_1 = 445.97 \text{ kJ} \; \textbf{Ans}$$

Problem 8.28 : What is the maximum useful work which can be obtained when 100 kJ are abstracted from a heat reservoir at 675 K in an environment at 288 K? What is the loss of useful work if (a) a temperature drop of 50°C is introduced between the heat source and the heat engine, on the one hand, and the heat engine and the heat sink, on the other, (b) the source temperature drops by 50°C and the sink temperature rises by 50°C during the heat transfer process according to the linear law $\frac{dQ}{dT} = \pm$ constant?

Solution : Given :

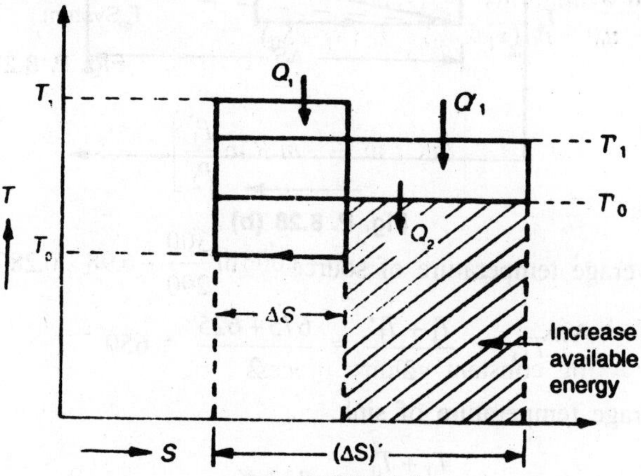

Fig. P. 8.28 (a)

$$Q_1 = 100 \text{ kJ}$$
$$T_1 = 675 \text{ K}$$
$$T_0 = 288 \text{ K}$$

(a) Availability $= \left(1 - \frac{T_0}{T_1}\right) Q_1 = \left(1 - \frac{288}{675}\right) \times 100$

$= 57.33$ kJ

From question

$$T_1' = 625 \text{ K}$$
$$T_0' = 338 \text{ K}$$

Now $\Delta S' = \frac{Q_1}{T_1'} = \frac{100}{625} = 0.16$ kJ/K

$\therefore$ $w' = \Delta S' (T_1' - T_0')$

$= 0.16 [625 - 338] = 45.92$ kJ

$\therefore$ Loss of availability

$= 57.33 - 45.92 = 11.41$ kJ **Ans**

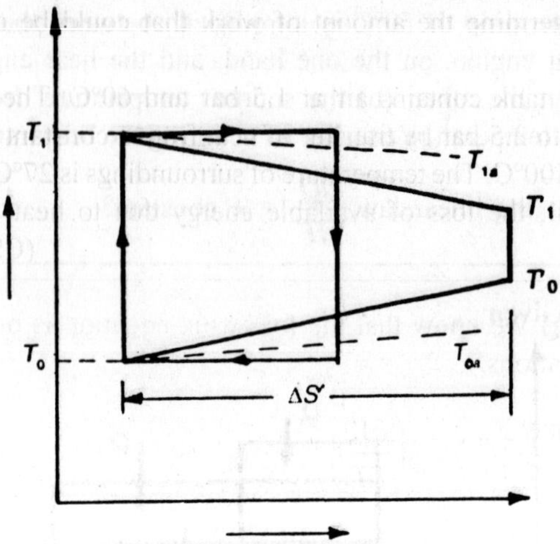

Fig. P. 8.28 (b)

(b) Average temperature of source

$$T_{1a} = \frac{T_1 + T_1'}{2} = \frac{675 + 625}{2} = 650$$

Average temperature of sink,

$$T_{0a} = \frac{T_0 + T_0'}{2} = 313 \text{ K}$$

$$(\Delta S)' = \frac{Q_1}{T_{1a}} = \frac{100}{650} = 0.1538 \text{ kJ/K}$$

$$\therefore \quad w' = \Delta S' (T_{1a} - T_{0a})$$
$$= 0.1538 (650 - 313) = 51.83 \text{ kJ}$$

$\therefore$ loss in availability

$$= 57.33 - 51.83 = 5.5 \text{ kJ } \textbf{Ans}$$

Problem 8.29 :

(a) Using the Maxwell relation derive the following *TdS* equation

$$TdS = C_P \, dT - T \left(\frac{\partial V}{\partial T}\right)_P dP$$

(b) A compressed air bottle of volume 0.15 m^3 contains air at 40 bar and 27°C. It is used to drive a turbvine which exhausts to atmosphere at 1 bar. If the pressure in the bottle is allowed to fall to 2 bar, determine the amount of work that could be delivered by the turbine.

(c) A rigid tank contains air at 1.5 bar and 60°C. The pressure of air is raised to 2.5 bar by transfer of heat from a constant temperature reservoir at 400°C. The temperature of surroundings is 27°C. Determine per kg of air, the loss of available energy due to heat transfer.

(CSE – 1998)

Solution : (a) We know that the following equation is one of the four Maxwell equations.

$$\left(\frac{\partial S}{\partial p}\right)_T = -\left(\frac{\partial V}{\partial T}\right)_P \qquad ...(1)$$

Let the entropy S be assumed as a function of T and P.

i.e., $S = S(T, p)$

i.e., $$dS = \left(\frac{\partial S}{\partial T}\right)_P dT + \left(\frac{\partial S}{\partial p}\right)_T dp$$

$$TdS = T\left(\frac{\partial S}{\partial T}\right)_P dT + T\left(\frac{\partial S}{\partial p}\right)_T dp$$

We know that,

$$T\left(\frac{\partial S}{\partial T}\right)_P = C_P$$

and using equation (1), then

$$TdS = C_P dT - T\left(\frac{\partial V}{\partial T}\right)_P dp$$

(b) $P_1 = 40$ bar; $T_1 = 300$ K; $V_1 = 0.15$ m^3

$$m_1 = \frac{P_1V_1}{RT_1} = \frac{40 \times 1.03 \times 10^5 \times 0.15}{0.287 \times 10^3 \times 300}$$

$$= 7.18 \text{ kg}$$

$p_2 = 2$ bar; T_2 ?; $V_2 = 0.15$ m^3

Assuming adiabatic expansion through the turbine.

$$\frac{T_2}{T_1} = \left(\frac{p_2}{p_1}\right)^{\frac{\gamma-1}{\gamma}}$$

$$T_2 = 300\left(\frac{2}{40}\right)^{\frac{0.4}{1.4}} = 127.47 \text{ K.}$$

$$m_2 = \frac{p_2 V_2}{RT_2} = \frac{2 \times 1.03 \times 10^5 \times 0.15}{0.287 \times 10^3 \times 127.47}$$

$$= 0.845 \text{ kg}$$

Assuming friction losses in turbine is negligible.
Output of turbine = K.E. obtained by expansion

$$(m_1 - m_2)\frac{V^2}{2} = (m_1 u_1 - m_2 u_2) - (m_1 - m_2)h$$

$$= (7.18 \times 0.771 \times 300 - 0.845 \times 0.771 \times 127.47) - (7.18 - 0.845)\, 1.007 \times 127.47$$

$$= 764.51 \text{ kJ}$$

(c) $p_1 = 1.5$ bar; $p_2 = 2.5$ bar

$$T_1 = 60 + 273 = 333 \text{ K};\ T_2 = \left(\frac{P_2}{P_1}\right) \times 333 = 555 \text{ K}$$

Constant volume process
Surrounding temperature = T_0 = 400°C = 673 K
Availability at state (1)

$$\phi_1 = (u_1 - u_0) - T_0 (S_1 - S_0) + p_0 (V_1 - V_0)$$

Availability at state (2)

$$\phi_2 = (u_2 - u_0) - T_0 (S_2 - S_0) + p_0 (V_2 - V_0)$$

Loss in availability

$$= \phi_1 - \phi_2$$

$$= (u_1 - u_2) - T_0 (S_1 - S_2) + p_0 (V_1 - V_2)$$

As $V_1 = V_2$

$$\phi_1 - \phi_2 = (u_1 - u_2) - T_0 (S_1 - S_2)$$

$$= C_v (T_1 - T_2) - T_0 \left(C_v \ln\frac{P_1}{P_2}\right)$$

$$= 0.717\left[333 - 555 - 673\left(\ln\frac{1.5}{2.5}\right)\right]$$

$$= 0.717\,[-222 - 673\,(-0.5108)]$$

Loss in availability = 87.31 kJ/kg

Problem 8.30 : (a) An ideal gas is heated from temperature T_1 to T_2 by keeping its volume constant. The gas is expanded back to its initial temperature according to the law pV^n = constant. If the entropy changes in the two processes are equal, find the value of n in terms of the adiabatic index γ.

(b) A cylinder contains one kg of water and steam at a pressure of 3.8 bar and 0.4 dry. Heat is supplied at constant volume until the pressure reaches to 10 bar. The steam is then expanded according to the law. pV = constant, until the pressure is 2 bar. Calculate

(i) Heat transfer during constant volume heating,

(ii) Heat transfer during pV = constant expansion and

(iii) Temperature of steam after the expansion.

(c) An adiabatic cylinder of volume 10 m^3 is divided into two compartments A and B, each of volume 6 m^3 and 4 m^3 respectively, by a thin sliding partition. Initially the compartment A is filled with air at 6 bar and 600 K, whilst there is a vacuum in the compartment B. Suddenly the partition is removed, the fluid in compartment A expands and fills both the compartments. Calculate the loss in available energy. Assume atmosphere is at 1 bar and 300 K.

(IES – 1997)

Solution : (a)

Change in entropy during constant volume process (1 – 2)

$$= S_2 - S_1$$

Change in entropy during pV^n = Constant process (2 – 3)

$$= S_3 - S_2$$

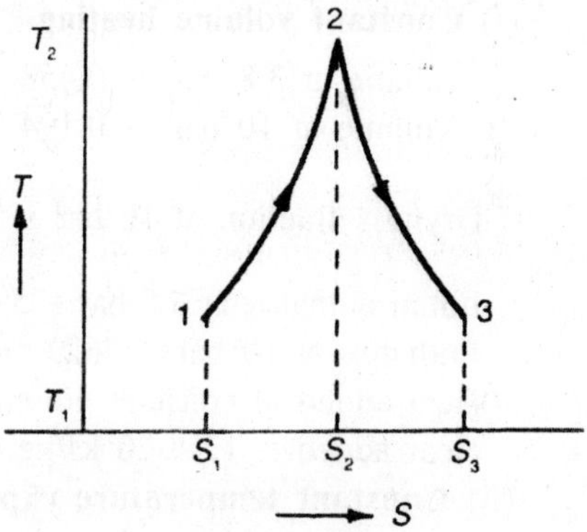

Fig. P. 8.30

Given: $S_2 - S_1 = S_3 - S_2$

$$S_2 - S_1 = C_v \ln \frac{T_2}{T_1}$$

(constant volume process) ...(1)

and $$S_3 - S_2 = C_v \ln \frac{p_3}{p_2} + C_P \ln \frac{V_3}{V_2}$$

$$= C_v \left(\frac{n}{n-1}\right) \ln \frac{T_3}{T_2} + C_P \left(\frac{1}{n-1}\right) \ln \frac{T_2}{T_3}$$

$$= \frac{C_p}{n-1} \ln \frac{T_2}{T_3} - \left(\frac{nC_v}{n-1}\right) \ln \frac{T_2}{T_3}$$

$$= \left(\frac{C_p - nC_v}{n-1}\right) \ln \frac{T_2}{T_3} \quad \ldots(2)$$

Equating (1) and (2),

$$C_v = \frac{C_p - nC_v}{n-1}$$

$$n = \frac{C_p + C_v}{2\,C_v} = \frac{1}{2}\left(\frac{C_p}{C_v} + 1\right) = \frac{1}{2}(\gamma + 1)$$

(b) From steam tables,

P (bar)	t_s (°C)	v_g (m³/kg)	h_f (kJ/kg)	h_{fg} (kJ/kg)	h_g (kJ/kg)
2	120.2	0.885	504.7	2201.6	2706.3
3.8	141.8	0.458	596.8	2138.6	2735.3
10	179.9	0.194	762.6	2013.6	2776.2

(i) **Constant volume heating**

Volume at 3.8 bar = $0.458 \times 0.4 = 0.1832$ m³/kg

Volume at 10 bar = $0.194 \times x = 0.1832$ m³/kg

Dryness fraction at 10 bar $x = \dfrac{0.1832}{0.194} = 0.944$

Initial enthalpy at 3.8 bar = 596.8 + 0.4 (2138.6) = 1452.24 kJ/kg

Enthalpy at 10 bar = 762.6 + 0.944 (2013.6) = 2663.44 kJ/kg

Heat added at constant volume = 2663.44 – 1452.24

Heat added = 1211.20 kJ/kg

(ii) **Constant temperature expansion**

$$pV = C$$

$$\text{Heat transfer} = p_1 V_1 \ln \frac{V_2}{V_1}$$

$$p_2 V_2 = p_1 V_1$$

$$= (10 \text{ bar}) \times 0.1832 \text{ m}^3/\text{kg} \ln \left(\frac{10}{2}\right)$$

$$\frac{V_2}{V_1} = \frac{p_1}{p_2}$$

$= 10.3 \times 10^5$ N/m² $\times$ 0.1832 m³/kg ln (5)

$=$ 0.3037 MJ/kg

Heat transferred= 303.7 kJ (mass = 1 kg)

(iii) **Condition of steam after heating at constant volume**

F_2 = 10 bar; Saturation temperature for 10 bar = 179.9°C

Expansion process follows the law $pV = C$

It is an isothermal process (constant temperature process).

Temperature at the end of expansion to 2 bar = 179.9°C

(c) In an unrestricted expansion for a perfect gas, the initial and final temperature are equal. In this case,

initial volume,

$$V_A = 6 \text{ m}^3$$

final volume,

$$(V_A + V_B) = 10 \text{ m}^3$$

State (1): $p_1 = 6$ bar, $T_1 = 600$ K

$V_1 = 6$ m³

State (2): $V_2 = (6 + 4) = 10$ m³ $T_2 = 600$ K

(free expression $T_1 = T_2$)

$$p_2 = p_1 \times \frac{V_1}{V_2} = 6 \times \frac{6}{10} = 3.6 \text{ m}^3$$

Ambient state (0) : p_0 = 1 bar, T_0 = 300 K

$$\text{Mass of gas, } m = \frac{p_1 V_1}{RT_1} = \frac{6 \times 1.03 \times 10^5 \times 6}{0.287 \times 10^3 \times 600} = 21.55 \text{ kg}$$

$(\Delta S)_{1-0}$ between state (1) and ambient state (0),

$$R \ln\left(\frac{P_1}{P_0}\right) + C_P \ln\left(\frac{T_0}{T_1}\right) \text{ kJ/kg.K}$$

$$= 0.287 \ln\left(\frac{6}{1}\right) + 1.005 \ln\left(\frac{300}{600}\right)$$

$$= -0.182 \text{ kJ/kg.K}$$

$(\Delta S)_{2-0}$ between state (2) and ambient state (0)

$$= 0.287 \ln\left(\frac{3.6}{1}\right) + 1.005 \ln\left(\frac{300}{600}\right)$$

$= -0.329$ kJ/kg.K

Availability state (1)

$= m[(u_1 - u_0) + T_0 \, \Delta S_{1-0}]$

$= 21.55 \, [C_v (T_1 - T_0) + T_0 \, \Delta S_{1-0}]$

$= 21.55 \, [0.718 \, (600 - 300)$
$+ 300 \, (-0.182)] = 3465.24$ kJ

Availability state (2)

$= m \, [(u_2 - u_0) + T_0 \, \Delta S_{2-0})]$

$= 21.55 \, [0.718 \, (600 - 300) + 300 \, (-0.329)]$

$= 2514.89$ kJ

Loss in availability

$= (3465.44 - 2514.89) = 950.55$ kJ

Problem 8.31 : Air expands steadily through a turbine from 6 bar, 800 K to 1 bar, 520 K. During the expansion, heat transfer from air to the surroundings at 300 K is 10 kJ/kg air. Neglect the changes in kinetic and potential energies and evaluate the irreversibility per kg air. Assume air to behave as an ideal gas with $C_P = 1.0$ kJ/kg.K and $R = 0.3$ kJ/kg.K.

Solution : Irreversibility

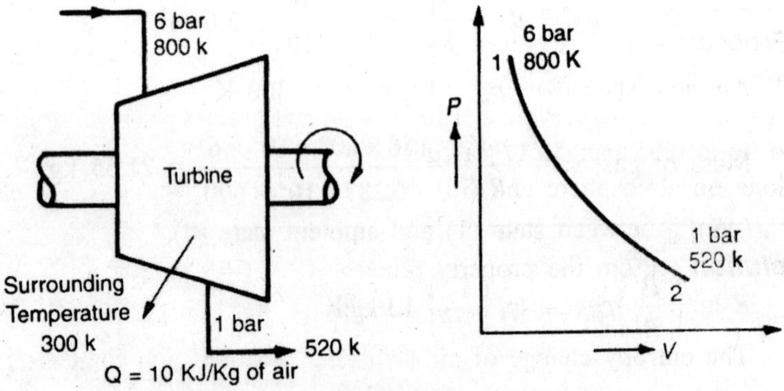

Fig. P. 8.31

$$I = T_0 \, (\Delta_{system} + \Delta_{surrounding}) + Q$$

$$= 300 \left[C_p \ln \frac{520}{800} - R \ln \frac{1}{6} \right] + 10$$

$$= 300 \, [1.005 \, (-0.43) - 0.3 \, (-1.79)] + 10$$

$$= 41.38 \text{ kJ/kg of air}$$

Problem 8.32 : In problem 8.31, find the actual work and maximum work per kg of air.

Solution :

Maximum work = Change in availability

$$\psi_1 - \psi_2 = (H_1 - T_0S_1) - (H_2 - T_0S_2)$$

$$= (H_1 - H_2) - T_0 (S_1 - S_2)$$

$$= C_p (T_1 - T_2) - T_0 \left(R \ln \frac{P_2}{P_1} - Cp \ln \frac{T_2}{T_1} \right)$$

$$= 1.005 (800 - 520) - 300 \left(0.3 \ln \frac{1}{6} - 1.005 \ln \frac{520}{800} \right)$$

$$= 281.4 - (-31.37) = 312.78 \text{ kJ/kg}$$

Actual work:

$$Q + H_1 = W + H_2$$

$$W = (H_1 - H_2) + Q = C_P (T_1 - T_2) + Q$$

$$= 1.005 (800 - 520) - 10$$

$$= 271.4 \text{ kJ/kg}$$

Problem 8.33 : A closed system contains 0.5 kg of air. It expands form 2 bar, 60°C to 1 bar, 40°C. During expansion it receives 2 kJ of heat from a reservoir at 100°C. Assuming atmospheric conditions to be at 0.95 bar, 30°C, calculate (i) the maximum work, (ii) work done on atmosphere and (iii) change in availability. (CES – 96)

Solution : From the property relation,

$$TdS = dV + pdu$$

The entropy change of air between the initial and final states is

$$\int_1^2 ds = \int_1^2 mC_v \frac{dT}{T} + \int_1^2 mR \frac{dV}{V}$$

or $$s_2 - s_1 = C_v \ln \frac{T_2}{T_1} + m R \ln \frac{V_2}{V_1}$$

$$W_{max} = (U_1 - U_2) - T_0 (s_1 - s_2)$$

$$= m\left[C_v(T_1 - T_2) + T_0\left(C_v \ln\frac{T_2}{T_1} + R\ln\frac{V_2}{V_1}\right)\right]$$

$$= 0.5 \left[0.718\,(60 - 40) + 303\left(0.718\ln\frac{313}{333} + 0.287\ln\frac{V_2}{V_1}\right)\right]$$

$$\frac{P_1V_1}{RT_1} = m = 0.5 = \frac{2\times 01^5 V_1}{287(273+60)}$$

$$V_1 = 0.239 \text{ m}^3$$

Similarly, $m = 0.5 = \dfrac{p_2v_2}{RT_2} = \dfrac{1\times 10^5}{287\ (273+40)}$

$$V_2 = 0.45 \text{ m}^3$$

(i) $W_{max} = 0.5\left[14.36 + 303\left(-0.042 + 0.287\ln\dfrac{0.45}{0.239}\right)\right]$

$$= 0.5\ [14.36 + 42] = 28.18 \text{ kJ}$$

(ii) Work done on atmosphere

$$= p_0\ (V_1 - V_2)$$

$$= 0.95\ (0.5 \times 0.287)\left[\frac{333}{2} - 313\right]$$

$$= -0.95 \times 0.5 \times 0.287\left[\frac{293}{2}\right] = -20 \text{ kJ}$$

(Negative sign indicates that work is done on the atmosphere)

(iii) Change in availability

$$= W_{max} + p_0\ (V_1 - V_2)$$

$$= 28.18 - 20 = 8.18 \text{ kJ}$$

Problem 8.34 : 500 kJ of heat is removed from a constant temperature heat reservoir maintained at 835 K. The heat is received by a system at constant temperature of 720 K. The temperature of the surroundings, the lowest available temperature, is 280 K. Illustrate the problem by *T–S* diagram and calculate the net loss of available energy as a result of this irreversible heat transfer. (CES – 92)

Solution :

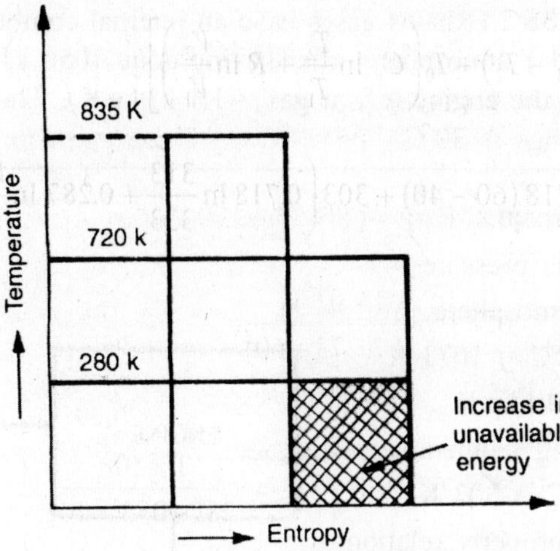

Fig. P. 8.34

Change of entropy of the source during heat transfer

$$= \frac{Q}{T_1} = \frac{-500}{835} \text{ kJ/K}$$

Change in entropy of system during heat transfer

$$= \frac{Q}{T_2} = \frac{500}{720} \text{ kJ/K}$$

The net change in entropy

$$= \frac{-500}{835} + \frac{500}{720} = 0.0956 \text{ kJ/K}$$

Available energy with source at 835 K

$$= (835 - 280)\left(\frac{500}{835}\right) = 333.2 \text{ kJ}$$

Available energy with system at 720 K

$$= (720 - 280)\left(\frac{500}{720}\right) = 305.6 \text{ kJ}$$

$\therefore$ Change in available energy

$$= 333.2 - 305.6 = 27.6 \text{ kJ}$$

Increase in unavailable energy

$$= (\Delta\phi)_{net} \times 280$$

$$= (0.0956) \times (280) = 26.7 \text{ kJ}$$

Problem 8.35 : Exhaust gases have an internal combustion engine at 800°C and 1 atmosphere, after having done 1050 kJ of work per kg of gas in the engine (C_p of gas = 1.1 kJ/kg.K). The temperature of surroundings is 30°C.

Solution : Given :

exhaust gas pressure

P_1 = 1 atmosphere

T_1 = 800°C = 1073 K

C_P = 1.1 kJ/kg.K

Surrounding temperature,

T_0 = 30°C = 303 K

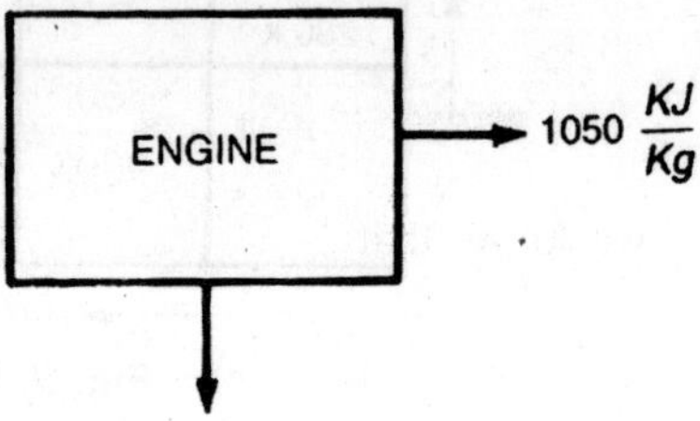

(a) From property relation

$Tds = C_p dt - vdp$

or $$\Delta S = \int_{T_1}^{T_2} C_P \frac{dT}{T} - R \int_{p_1}^{p_2} \frac{dp}{p}$$

Now if $p_1 = p_2$

$$\therefore \quad \Delta s = \int_{T_1}^{T_2} C_P \frac{dT}{T}$$

∴ loss in available energy

$$= \phi_1 - \phi_0 = C_P (T_1 - T_0) - T_0 (S_1 - S_0)$$

$$= 1.1 \times (1073 - 303) - 303 \left[c_P \ln \frac{1073}{303} \right]$$

$$= 425.55 \text{ kJ/kg } \textbf{Ans}$$

(b) ∴ ratio of lost available energy to engine work = 0.41 **Ans**

Problem 8.36 : A heat sowice at 627°C transfers heat at the rate of 300 kJ/min to a system maintained at 287°C. A heat sink is available at 27°C. Assuming these temperatures to remain constant. Find :

(i) Change in entropy of sowice

(ii) Entropy porduction accompanying heat transfer

(iii) The original available energy

(iv) The available energy after heat transfer

Solution : Given :

$$\dot{Q}_2 = 3000 \text{ kJ/min}$$
$$T_s = 900 \text{ K}$$
$$T_{sys} = 560 \text{ K}$$
$$T_{surr} = 300 \text{ K}$$

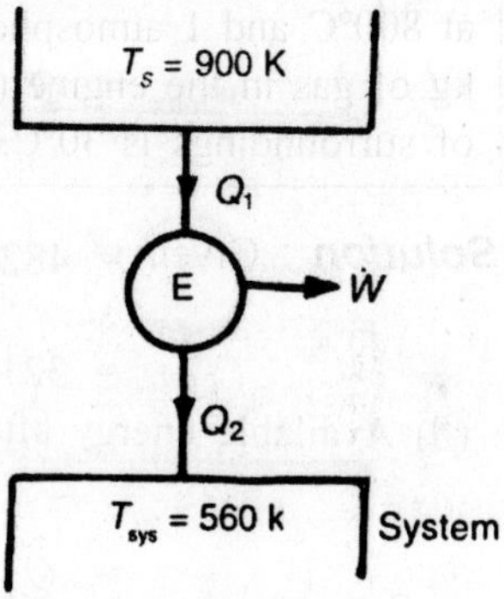

Fig. P. 8.36

Assuming every process reversible

$$\text{Cycle efficiency} = n = 1 - \frac{T_{sys}}{T_s}$$

$$= 1 - \frac{560}{900} = 0.38$$

we know that

$$\Rightarrow \quad n = 1 - \frac{Q_R}{Q_A} = 1 - \frac{\dot{Q}_2}{\dot{Q}_1}$$

$$\Rightarrow \quad \frac{\dot{Q}_2}{\dot{Q}_1} = 1 - n$$

$$\Rightarrow \quad \dot{Q}_1 = \frac{\dot{Q}_2}{1-n}$$

$$= \frac{3000}{1-0.38} = 4821.43 \text{ kJ/min}$$

(a) Change in entropy of source

$$(\Delta S)_{source} = -\frac{\dot{Q}_1}{T_s} = -\frac{4821.43}{900}$$

$$= -5.357 \text{ kJ/min K} \quad \textbf{Ans}$$

(b) Entropy production, ΔS_{gen}

$$(\Delta \dot{S})_{syst} = +\frac{\dot{Q}_2}{T_{sys}} = \frac{3000}{560} = 5.357 \text{ kJ/min K}$$

$\therefore$ From Goudy Stodola theorem =

$$\Delta \dot{S}_{gen} = (\Delta \dot{S})_{univ} = (\Delta \dot{S})_{source} + (\Delta \dot{S})_{system} + (\Delta \dot{S})_{Engine}$$

$$= -5.357 + 5.357 + 0 = 0$$

Hence the entropy production is zero.
Hence the assumption of reversible process is valid.

(c) Original available energy

$$w_{max} = \dot{Q}_1 \left(1 - \frac{T_0}{T_s}\right)$$

$$= 4821.43 \times \left(1 - \frac{300}{900}\right)$$

$$= 3214.28 \text{ kJ/min } \textbf{Ans}$$

(d) Available energy after heat transfer

$$= \dot{Q}_2 \left[1 - \frac{T_0}{T_{sys}}\right] = 3000 \times \left[1 - \frac{300}{560}\right]$$

$$= 1392.85 \text{ kJ/min } \textbf{Ans}$$

Problem 8.37 : Steam flows through an adiabatic flow turbine. The enthalpy at entrance is 4142 kJ/kg and at exit 2585 kJ/kg. The values of flow availability of steam at entrance and exit are 1787 kJ/kg and 140 kJ/kg respectively. If the dead state temperature T_0 is 300 K, determine per kg of steam. Neglect changes in kinetic and potential energy.

Solution : Given :

$$h_1 = 4142 \text{ kJ/kg}$$
$$h_2 = 2585 \text{ kJ/kg}$$
$$T_0 = 300 \text{ K}$$

flow availability at entrance = 1787 kJ/kg
flow availability at exit = 140 kJ/kg

∴ From first law of thermodynamics, for a steady flow process neglecting ΔK.E. and ΔP.E.

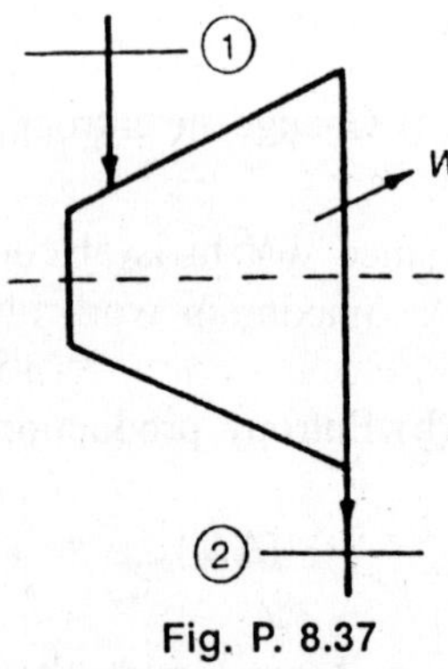

Fig. P. 8.37

$$\frac{dw}{dm} = h_1 - h_2 = 4142 - 2585$$

$$= 1557 \text{ kJ/kg } \textbf{Ans}$$

Maximum possible work = change in availability = $\phi_1 - \phi_2$

$$\therefore \quad = (h_1 - h_2) - T_0 (S_1 - S_2),$$

neglecting change in ΔK.E. and ΔP.E.

$$d\, w_{max} = (h_1 - T_0 S_1) - (h_2 - T_0 S_2)$$

= flow availability at entrance – flow availability at exit

$$= 1787 - 140 = 1647 \text{ kJ/kg } \textbf{Ans}$$

Now $\quad d\!\!\!{}^{-} w_{max} = 1647 \text{ kJ/kg} = (h_1 - h_2) - T_0 (S_1 - S_2)$

or, $\quad 1647 = 1557 - 300 (S_1 - S_2)$

or $\quad S_1 - S_2 = -0.333$ kJ/kg.K

∴ change in entropy

$$= S_2 - S_1 = +0.333 \text{ kJ/kg.K } \textbf{Ans}$$

Problem 8.38 : A system maintained at constant volume is initially at temperature T_1. If a heat resrervoir at temperature T_0, which is less than T_1 is available determine the maximum work obtainable as the system is cooled to T_0 in terms of T_0, jT_1 and C_v.

Solution : Given :

System is cooled from T_1 to T_0 at constant volume and $T_0 < T_1$

Considering the system as a closed system availability

$$= \phi - \phi_0$$

$$= (u - u_0) - T_0 (S - S_0)$$

$$= (u - u_0) + T_0 (S_0 - S)$$

Now change in entropy of system when cooled to T_0

$$= S_0 - S = C_v \ln \frac{T_0}{T_1},$$

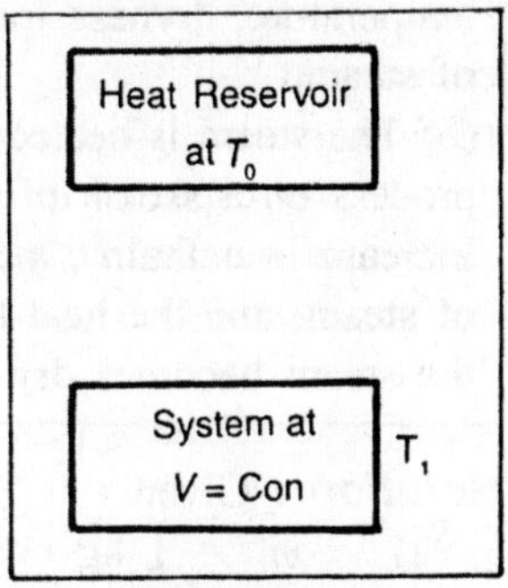

Fig. P. 8.38

since system is at constant volume

∴ maximum work obtainable = availability of the system

$$= (u - u_0) + T_0 (S_0 - S)$$

$$= C_v (T_1 - T_0) + T_0 C_v \ln \frac{T_0}{T_1} \text{ kJ/kg}$$

$$= C_v (T_1 - T_0) + T_0 \ln \frac{T_0}{T_1}, \text{ kJ/kg } \textbf{Ans}$$

Chapter 9

Properties of Pure Substances

Problem 9.1 : (a) A rigid vessel of volume 0.86 m³ contains 1 kg of steam at a pressure of 2 bar. Evaluate the specific volume temperature, dryness fraction, internal energy, enthalpy, and entropy of steam.

(b) The steam is heated to raise its temperature to 150°C. Show the process on a sketch of the p–v diagram, and evaluate the pressure, increase in enthalpy, increase in internal energy, increase in entropy of steam, and the heat transfer. Evaluate also the pressure at which the steam becomes dry saturated.

***Solution* :** Given :

(a)
$$m_s = 1 \text{ kg}$$
$$p_1 = 2 \text{ bar}$$
$$V = 0.86 \text{ m}^3$$

From steam table at 2 bar

$$v_f = 1.061 \times 10^{-3} \text{ m}^3/\text{kg}$$
$$V_g = 0.8857 \text{ m}^3/\text{kg}$$
$$h_f = 504.7 \text{ kJ/kg}$$
$$h_{f_g} = 2201.9 \text{ kJ/kg}$$
$$S_f = 1.5301 \text{ kJ/kg.K}$$
$$S_g = 7.1271 \text{ kJ/kg.K}$$
$$t_{sat} = 120.23°\text{C}$$

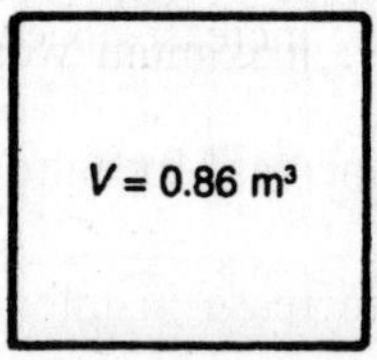

Fig. P. 9.1 (a)

$$\therefore \quad v = \text{Specific volume} = \frac{V}{m} = 0.86 \text{ m}^3/\text{kg}$$

or
$$v = 0.86 = V_f + x\,(V_g - V_f)$$

$$= 0.001061 + x\ (0.8857 - 0.001061)$$

$\therefore \quad x = 0.971$ **Ans**

$\therefore \quad h_1 = \text{enthalpy} = h_f + 0.971\ h_{fg}$

$= 2642.7$ kJ/kg **Ans**

$S_1 = \text{entropy} = S_f + 0.971 \times (s_g - s_f)$

$= 6.9648$ kg/kg.K

$u_1 = \text{internal energy} = h_1 - pv$

$= 2642.7 - 2 \times 100 \times 0.86$

$= 2470.7$ kJ/kg

(b) Assuming constant volume process

At = 150°C

from steam table,

$V_f = 0.001091\ \text{m}^3/\text{kg}$

$V_g = 0.3928\ \text{m}^3/\text{kg}$

Since $V_g < v$

$\therefore$ hence point 2 will be in superheated region.

Now at 150°C and

$v_d = 0.86\ \text{m}^3/\text{kg}$ from superheated steam table,

By interpolation

$p_2 = 2.3$ bar **Ans**

$h_2 = 2769.4$ kJ/kg

$s_2 = 7.2215$ kJ/kg.K

$u_2 = h_2 - p_2 v$

$= 2769.4 - 2.3 \times 100 \times 0.86$

$= 2571.6$ kJ/kg

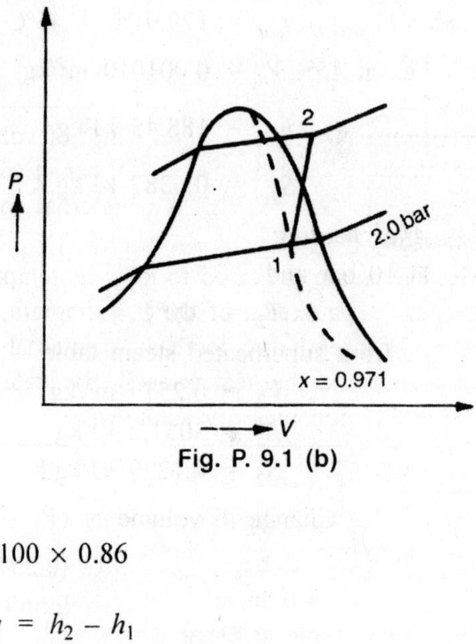

Fig. P. 9.1 (b)

$\therefore$ increase in enthalpy $= h_2 - h_1$

$= 126.7$ kJ/kg **Ans**

increase in entropy $= S_2 - S_1$

$= 0.2567$ kJ/kg **Ans**

increase in internal energy

$= u_2 - u_1$

$= 100.9$ kJ/kg **Ans**

Since process is constant volume

$\therefore \quad đ\omega = 0$

$\therefore$ from 1st law of thermodynamics

$đQ = du = 100.9$ kJ/ks **Ans**

Problem 9.2 : Ten kg of water at 45°C is heated at a constant pressure of 10 bar until it becomes superheated vapour at 300°C. Find the changes in volume, enthalpy, internal energy and entropy.

Solution : Given :

$$m = 10 \text{ kg}$$
$$T_1 = 45°C = 45 + 273 = 318 \text{ K}$$
$$p_1 = 10 \text{ bar} = p_2$$
$$T_2 = 300°C = 573 \text{ K at superheated state}$$

Now from steam table at 10 bar

$$t_{sat} = 179.91°C > 45°C$$

i.e., at 45°C $V_{f_1} = 0.001010 \text{ m}^3/\text{kg}$

$$h_{f_1} = 188.45 \text{ kJ/kg}$$
$$S_{f_1} = 0.6387 \text{ kJ/kg.K}$$

After heating
at 10 bar and

$$T_2 = 300°C,$$

∴ from superheated steam table

$$V_2 = 0.2579 \text{ m}^3/\text{kg}$$
$$h_2 = 3051.2 \text{ kJ/kg}$$
$$S_2 = 7.1229 \text{ kJ/kg.K}$$

Change in volume $= (V_2 - V_{f_1})\, m$

∴ $= (0.2579 - 0.001010) \times 10$

$= 2.569 \text{ m}^3$ **Ans**

Change in enthalpy $= (h_2 - h_{f_1})\, m$

$= (3051.2 - 188.45) \times 10$

$= 28627.5 \text{ kJ}$ **Ans**

Change in entropy $= (s_2 - s_{f1})\, m$

$= (17.1229 - 0.6387) \times 10$

$= 64.842 \text{ kJ/K}$ **Ans**

Change in internal energy $= m\,(u_2 - U_1)$

$= m\,[(h_2 - h_1) - (p_2V_2 - p_1V_1)]$

$= [28627.5 - 10 \times 100\,(2.569)]$

$= 26058.5 \text{ kJ}$ **Ans**

Problem 9.3 : Water at 40°C is continuously sprayed into a pipeline carrying 5 tones of steam at 5 bar, 300°C per hour. At a section downstream where the pressure is 3 bar, the quality is to be 95%. Find the rate of water spray in kg/h.

Solution : Given

$$\dot{m}_1 = \text{mass of steam} = 5000 \text{ kJ/hr}$$

initial pressure = P_1 = 5 bar
initial temperature T_1 = 300°C = 573 K
At down steam,

$$p_2 = 3 \text{ bar}$$

$$\text{quality} = x_2 = 0.95$$

$$T_3 = \text{water spraying temperature} = 40°\text{C}$$

From steam tables
at 5 bar and 300°C,

$$h_1 = 3064.2 \text{ kJ/kg}$$

at 3 bar

$$h_{f_2} = 561.47 \text{ kJ/kg}$$

$$h_{fg_2} = 2163.8 \text{ kJ/kg}$$

$$\therefore \quad h_2 = h_f + x_2 h_{f_g}$$

$$= 2617.31 \text{ kJ/kg}$$

at 40°C

$$h_{f_3} = 167.57 \text{ kJ/kg}$$

Let $\dot{m}_w$ be the rate of water spray

∴ Applying 1st law of thermodynamics

$$\dot{m}_1 h_1 + \dot{m}_w h_{f_3} = \dot{m}_2 h_2 + \left[\dot{m}_w + (\dot{m}_1 - \dot{m}_2)\right] h_{f_2}$$

where $\dot{m}_2$ = mass of steam at downstream section

$$\Rightarrow \quad 5000 \times 3064.2 + \dot{m}_w h_{f_3}$$

$$= 5000 \times 0.95 \times 2617.31 + \dot{m}_w h_{f_2} + 0.05 \times 5000 \times 561.47$$

$$\therefore \quad \dot{m}_w = 6977.43 \text{ kg/hr}$$ **Ans**

Problem 9.4 : A rigid closed tank of volume 3 m^3 contains 5 kg of wet steam at a pressure of 200 kPa. The tank is heated until the steam becomes dry saturated. Determine the final pressure and the heat transfer to the tank.

Solution : Given :

V = volume of closed tank = 3 m^3

m_1 = mass of wet steam = 5 kg

p_1 = 200 kPa

After heating = condition of steam becomes dry saturated

For initial condition

v_1 = specific volume

$$= \frac{V}{m_1} = \frac{3}{5} = 0.6 \text{ m}^3/\text{kg}$$

from steam at 200 KPa.

h_{f_1} = 504.7 kJ/kg

h_{fg_2} = 2201.9 kJ/kg

V_{f_1} = 1.061×10^{-3} m^3/kg

V_{g_1} = 0.8857 m^3/kg

Now $V_1 = 0.6 = V_{f_1} + x_1 (0.8857 - 0.001061)$

quality = x_1 = 0.677

$\therefore$ $h_1 = h_{f_1} + x_1 h_{fg_1}$

$= 504.7 + 0.677 \times 2201.9$

$= 1995.48$ kJ/kg

For dry saturated steam

at v_1 = 0.6 m^3/kg

from steam table

by interpolation,

P_2 = 304 bar

and h_2 = 2725.87 kJ/kg

$\therefore$ heat transfer to the tank = $m (h_2 - h_1)$ = 3651.94 kJ **Ans**

Problem 9.5 : Steam flows through a small turbine at the rate of 5000 kg/h entering at 15 bar, 300°C and leaving at 0.1 bar with 4% moisture. The steam enters at 80 m/s at a point 2 m above the discharge and leaves at 40 m/s. Compute the shaft power assuming that the device is adiabatic but considering kinetic and potential energy changes. How much error would be made if these terms were neglected? Calculate the diameters of the inlet and discharge tubes.

Solution : Given :

$$\dot{m} = \text{steam flow rate} = 5000 \text{ kg/hr}$$
$$P_1 = 15 \text{ bar}$$
$$P_2 = 0.1 \text{ bar}$$
$$T_1 = 300°C$$
$$x_2 = 0.96$$
$$V_1 = 80 \text{ m/s}$$
$$V_2 = 40 \text{ m/s}$$
$$(Z_1 - Z_2) = 2 \text{ m}$$

Now from steam table

at 15 bar and 300°C, from interpolation

$$\Rightarrow \quad h_1 = 3037.4$$
$$V_1 = 0.17045 \text{ m}^3\text{/kg}$$

at 0.1 bar $h_{f_2} = 191.83$ kJ/kg

$$h_{fg_2} = 2392.8 \text{ kJ/kg}$$
$$V_{f_2} = 0.001010 \text{ m}^3\text{/kg}$$
$$V_{g_2} = 14.67 \text{ m}^3\text{/kg}$$
$$h_2 = h_{f_2} + 0.96\ h_{fg_2} = 2488.92 \text{ kJ/kg}$$
$$V_2 = 0.0010 + 0.96\ (14.67 - 0.001010)$$
$$= 14.0832 \text{ m}^3\text{/kg}$$

Now from 1st law of thermodynamics

$$\dot{m}\left(h_1 + \frac{V_1^2}{2} + Z_1\right) + \dot{Q} = \dot{m}\left(h_2 + \frac{V_2^2}{2} + Z_2\right) + \dot{w}$$

$$\therefore \quad \dot{w} = \dot{m}\left[(h_1 - h_2) + \frac{V_1^2 - V_2^2}{2} + (Z_1 - Z_2)\right] \quad [\because \dot{Q} = 0]$$

$$= \frac{5000}{3600}\left[(3037.4 - 2488.92) + \frac{80^2 - 40^2}{2 \times 1000} + \frac{2}{1000}\right]$$

$$= 765.12 \text{ kW } \textbf{Ans}$$

Now neglecting ΔK.E. and ΔP.E.

$$\dot{w} = (h_1 - h_2)\frac{5000}{3600}$$

$$= 761.78 \text{ kW}$$

$\therefore$ % error = 0.435% **Ans**

$$\dot{m} = \text{mass flow rate}$$

$$= \frac{A_1V_1}{v_1} = \frac{A_2V_2}{v_2}$$

$$\therefore \quad A_1 = \frac{v_1 \times \dot{m}}{V_1}$$

$$= \frac{0.17045 \times 5000}{3600 \times 80}$$

or inlet diameter $= d_1 = \sqrt{\dfrac{4A_1}{\pi}} = 6.14$ cm **Ans**

Similarly d_2 = outlet diameter

$$= \sqrt{\frac{4A_2}{\pi}} = \sqrt{\frac{4 \times v_2 \times \dot{m}}{\pi V_2}}$$

$$= \sqrt{\frac{4 \times 14.0832 \times 5000}{\pi \times 40 \times 3600}} = 78.91 \text{ cm } \textbf{Ans}$$

Problem 9.6 : A sample of steam from a boiler drum at 3 MPa is put through a throttling calorimeter in which the pressure and temperature are found to be 0.1 MPa, 120°C. Find the quality of the sample taken from the boiler.

Solution : Given :

p_1 = pressure in boiler drum = 3 MPa

and p_2 = pressure in calorimeter = 0.1 MPa

and T_2 = 120°C

Now let x be the quality of steam taken from the boiler drum then from steam table,

at 3 MPa,

$$h_{f_1} = 1008.42 \text{ kJ/kg}$$

h_{fg_1} = 1795.7 kJ/kg

at 0.1 MPa and 120°C

by interpolation,

h_2 = 2716.28 kJ/kg

Since,

$h_1 = h_2 = h_{f1} + x\, h_{fg_1}$

$\Rightarrow$ 2716.28

= 1008.42 + $x \times$ 1795.7

$\therefore$ x = 0.951 **Ans**

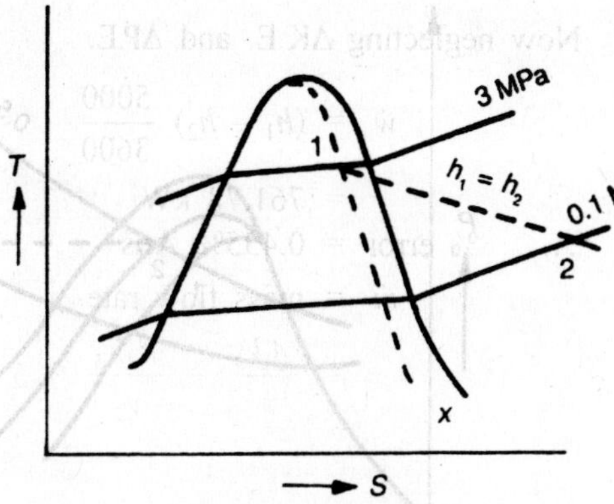

Fig. P. 9.6

Problem 9.7 : It is desired to measure the quality of wet steam at 0.5 MPa. The quality of steam is expected to be not more than 0.9.
(a) Explain why a throttling calorimeter to atmospheric pressure will not serve the purpose.
(b) Will the use of a separating calorimeter, ahead of the throttling calorimeter, serve the purpose, if at best 5°C degree of superheat is desirable at the end of throttling? What is the minimum dryness fraction required at the exit of the separating calorimeter to satisfy this condition?

Solution : Given :

p_1 = pressure of wet steam = 0.5 MPa

x_1 = quality = 0.9

from steam table at

p_1 = 0.5 MPa

h_{f_1} = 640.23 kJ/kg

h_{fg_1} = 2108.5 kJ/kg

$\therefore$ $h_1 = h_{f_1} + x_1 h_{fg_1}$ = 640.23 + 0.9 × 2108.5 = 2537.8 kJ/kg

Now at atmospheric pressure = p_2 = 1 bar = 0.1 MPa

h_{f_2} = 417.46 kJ/kg

h_{fg_2} = 2258.6 kg/kg

t_{sat} = 99.63°C

Since $h_1 - h_2 \Rightarrow$ 2537.8 = $h_{f_2} + x_2 h_{fg_2}$

$\therefore$ x_2 = 0.94

Hence at atmospheric pressure, the final condition is not superheated, but wet. And for measuring quality from throttling calorimeter outlet

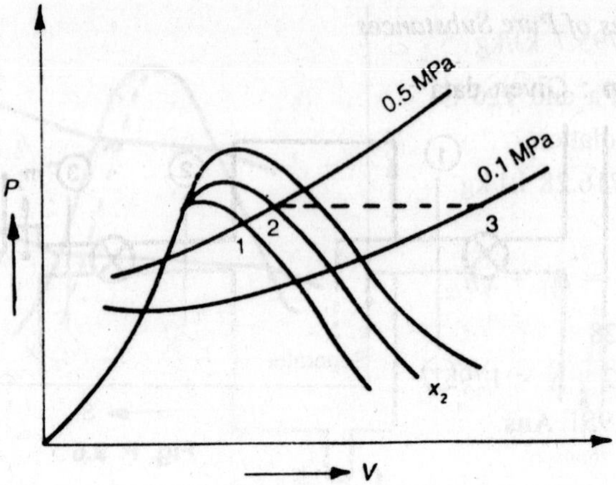

Fig. P. 9.7

condition must be superheated or minimum saturated. Therefore for the given condition, this calorimeter will not serve the purpose. **Ans**

(b) Yes, separating calorimeter ahead of throttling calorimeter will serve our purpose.

Now exit condition is 0.1 MPa & $T_3 = t_{sat} + 5°C = 104.63°C$ from steam table

$$h_3 = 2685.48 \text{ kJ/kg}$$

Point 2 indicates the exit of separating calorimeter at 0.5 MPa

Hence from steam table at 0.5 MPa

$$h_{f_2} = 640.23 \text{ kJ/kg,}$$

$$h_{fg_2} = 2108.5 \text{ kJ/kg}$$

Since $h_2 = h_3$

or, $2685.48 = h_{f_2} + x_2 h_{fg_2} = 640.23 + x_2 \times 2108.5$

$\therefore \quad x_2 = 0.97$

= minimum dryness fraction required at the exit of the separating calorimeter. **Ans**

Problem 9.8 : The following observations were recorded in an experiment with a combined separating and throttling calorimeter:
Pressure in the steam main = 15 bar
Mass of water drained from the separator = 0.55 kg
Mass of steam condensed after passing through the throttle valve = 4.20 kg
Pressure and temperature after throttling = 1 bar, 120°C
Evaluate the dryness fraction of the steam in the main, and state with reason whether the throttling calorimeter alone could have been used for this test.

Solution : Given data :

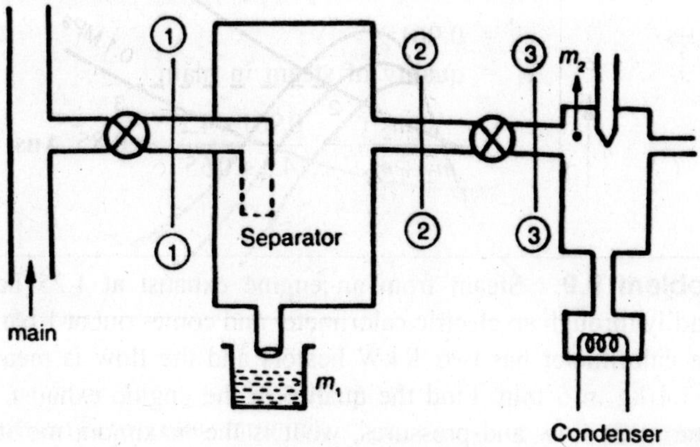

Fig. P. 9.8 (a)

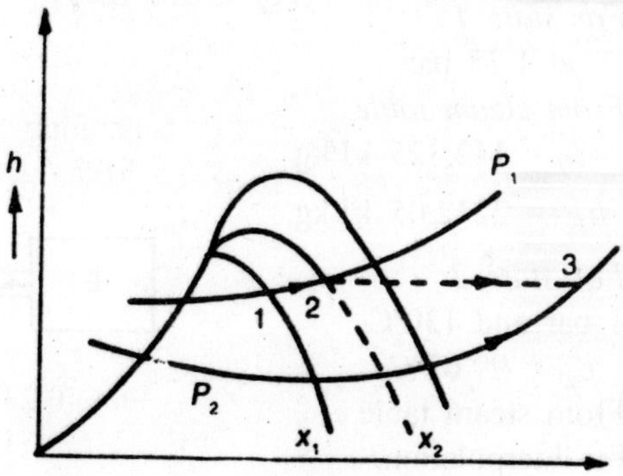

Fig. P. 9.8 (b)

p_1 = 15 bar

m_1 = 0.55 kg

m_2 = 4.2 kg

p_3 = 1 bar,

T_3 = 120°C
= 120 + 273
= 393 K

Now steam table at p_3 and

at 100°C

h_a = 2676.2 kJ/kg

at 150°C

h_b = 2776.4 kJ/kg

∴ by interpolation, $\frac{h - h_a}{h_b - h_a}$

$$= \frac{T - T_a}{T_b - T_a} = \frac{120 - 100}{150 - 100} = \frac{20}{50} = 0.4$$

∴ h_3 = 2716.28 kJ/kg

$h_3 = h_2$ = 2716.28

from steam table, at 15 bar

h_f = 844.89 kJ/kg

h_{fg} = 1947.3 kJ/kg

$$\therefore \quad h_2 = 2716.28 = h_f + x_2 h_{f_g}$$

$$\Rightarrow \quad x_2 = 0.961$$

$$\therefore \quad x_1 = \text{quality of steam in main}$$

$$= \frac{x_2 m_2}{m_2 + m_2} = \frac{0.961 \times 4.2}{4.2 + 0.55} = 0.85 \textbf{ Ans}$$

Problem 9.9 : Steam from an engine exhaust at 1.25 bar flows steadily through an electric calorimeter and comes out at 1 bar, 130°C. The calorimeter has two 1 kW heaters and the flow is measured to be 3.4 kg in 5 min. Find the quality in the engine exhaust. For the same mass flow and pressures, what is the maximum moisture that can be determined if the outlet temperature is at least 105°C?

Solution :

For state 1

at 1.25 bar

From steam table

$h_f = 442.325$ kJ/kg

$h_{f_g} = 2242.05$ kJ/kg

For state 2

1 bar and 130°C

$t_{sat} = 99.63$°C

From steam table

by interpolation,

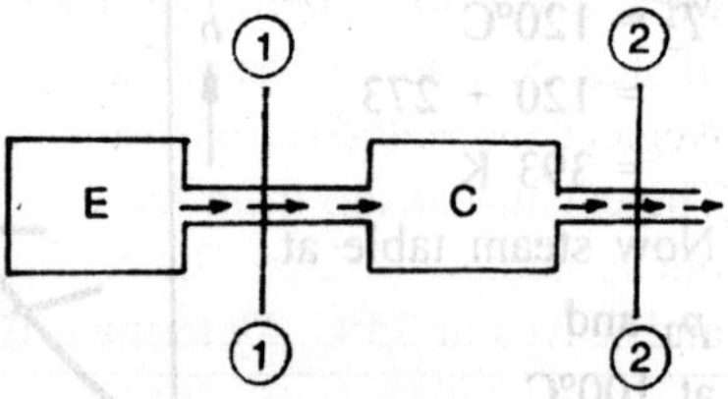

Fig. P. 9.9

$$h_2 = 2736.32 \text{ kJ/kg}$$

$$\dot{m} = \text{mass rate of flow}$$

$$= \frac{3.4}{5} \text{ kJ/min} = \frac{3.4}{300} \text{ kg/sec.}$$

$\therefore$ from S.F.E.E. neglecting ΔK.E. ΔP.E. and $\frac{dw}{d\tau}$, we get,

$$h_1 = -\frac{Q}{m} + h_2 = -\frac{2 \text{ kJ/s}}{\frac{304}{300} \text{ kg/s}} + 2736.32$$

$$= 2559.85 \text{ kJ/kg}$$

(a) *Now from state 1*

$$h_1 = 2559.85$$

$$= h_{f_1} + x_1 h_{fg_1} = 442.32 + x_1 \times 2242.05$$

$$\Rightarrow \quad x_1 = 0.944 \textbf{ Ans}$$

(b) *Now at 1 bar at 105°C*

by interpolation

$$h_2 = 2686.21 \text{ kJ/kg}$$

$$h_2 = h_2 - \frac{Q}{m_1}$$

$$= 2686.21 - \frac{2.0}{\frac{3.4}{300}} = 2509.74 \text{ kJ/kg}$$

$$\text{Hence} \Rightarrow \quad h_2 = 2509.74 = 442.32 + x_1 \times 2242.05$$

$$\therefore \quad x_1 = 0.922 \textbf{ Ans}$$

Problem 9.10 : Steam expands isentropically in a nozzle from 1 MPa, 250°C to 10 kPa. The steam flow rate is 1 kg/s. Find the velocity of steam at the exit from the nozzle, and the exit area of the nozzle. Neglect the velocity of steam at the inlet to the nozzle.
The exhaust steam from the nozzle flows into a condenser and flows out as saturated water. The cooling water enters the condenser at 25°C and leaves at 35°C. Determine the mass flow rate of cooling water.

Solution :

For state 1

1 MPa, 250°C

From steam table

$$V_1 = 2327 \text{ m}^3\text{/kg}$$

$$h_1 = 2942.6 \text{ kJ/kg}$$

$$S_1 = 6.9247 \text{ kJ/kg.K}$$

State 2

10 kPa

from steam table

$$V_f = 0.00101 \text{ m}^3\text{/kg}$$

$$V_g = 14.67 \text{ m}^3\text{/kg}$$

$$h_f = 191.83 \text{ kJ/kg,}$$

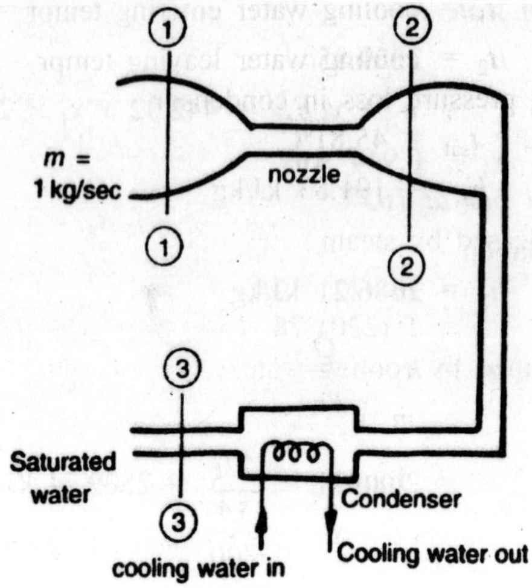

Fig. P. 9.10

$$h_{f_g} = 2392.8 \text{ kJ/kg}$$

$$S_f = 0.6493 \text{ kJ/kg.K},$$

$$S_g = 8.1502 \text{ kJ/kg.K}$$

here $S_1 = S_2 = 6.9247$

$$= 0.6493 + x_2 (8.1502 - 0.6493)$$

$$x_2 = 0.84$$

$$\therefore \quad v_2 = 0.00101 + 0.84 (14.67 - 0.00101)$$

$$= 12.32 \text{ m}^3\text{/kg}$$

$$h_2 = 2201.78 \text{ kJ/kg}$$

Now applying S.F.E.E.

$$\Rightarrow \quad h_1 = h_2 + \frac{V_2^2}{2}$$

$$\Rightarrow \quad V_2 = \sqrt{2(h_1 - h_2)}$$

$$= \sqrt{2(2942.6 - 2201.78) \times 1000}$$

$$= 1217.23 \text{ m/s } \textbf{Ans}$$

Now $\dot{m} = \rho AV$

$$\Rightarrow \quad A = \frac{\dot{m}}{\rho V} = \frac{12.32}{1217.23} = 0.0101 \text{ m}^2$$

(b) t_1 = cooling water entering tempr = 25°C

t_2 = cooling water leaving tempr = 35°C

Neglecting pressure loss in condensor

⇒ 10 kPa, t_{sat} = 45.81°C

h_{f_3} = 191.83 kJ/kg

∴ heat released by steam

$$= \dot{m}\,(h_2 - h_3)$$

$$= 1\,(2201.78 - 191.83) = 2009.95 \text{ kW/s}$$

∴ heat gained by cooling water

$$= \dot{m}_w\, C_{p_w}\, \Delta t$$

$$\therefore \quad \dot{m}_w = \frac{2009.95}{C_{p_w}\,\Delta t}$$

$$= \frac{2009.95}{4.187 \times 10} = 48 \text{ kg/s } \textbf{Ans}$$

Problem 9.11 : Two insulated tanks *A* and *B*, are connected by a valve. Tank A has a volume of 0.70 m³ and contains steam at 1.5 bar, 200°C. Tank B has a volume of 0.35 m³ and contains steam at 6 bar with a quality of 90%. The valve is then opened, and the two tanks come to a uniform state. If there is no heat transfer during the process, what is the final pressure? Compute the entropy change of the universe.

Solution : Given :

Tank is insulated i.e.

$dQ = 0$

For Tank A

$V_A = 0.7 \text{ m}^3$

$P_A = 1.5 \text{ bar}$

$T_A = 200°C$

$= 473 \text{ K}$

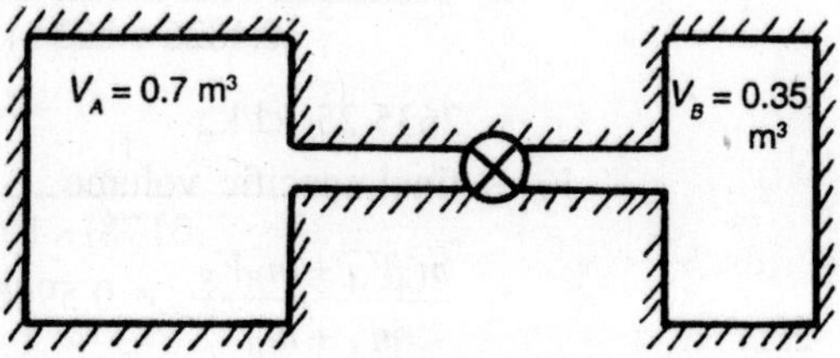

Fig. P. 9.11

For Tank B

$V_B = 0.35 \text{ m}^3$

$P_B = 6 \text{ bar}$

$x_B = 90\%$

Now from steam table

At 1.5 bar and 200°C,

$$S_A = 7.6464 \text{ kJ/kg.K}$$

$$h_A = 2873.1 \text{ kJ/kg}$$

$$v_A = \text{sp. volume} = 1.4505 \text{ m}^3\text{/kg}$$

$$\therefore \quad m_A = \text{mass of steam in tank A}$$

$$= \frac{V_A}{v_A} = \frac{0.7}{1.4505} = 0.4826 \text{ kg}$$

At $P_B = 6$ bar, and $x = 0.9$

$$V_f = 0.001101 \text{ m}^3\text{/kg}$$

$$S_f = 1.9312 \text{ kJ/kg.K}$$

$$V_g = 0.3157 \text{ kJ/kg.K}$$

$$S_g = 6.76 \text{ kJ/kg.K}$$

$$h_f = 670.56 \text{ kJ/kg}$$

$$h_{f_g} = 2086.3 \text{ kJ/kg}$$

$$\therefore \quad V_B = V_f + x\,(V_g - V_f) = 0.28424 \text{ m}^3\text{/kg}$$

$$h_B = V_f + x\ h_{f_g} = 2548.23 \text{ kJ/kg}$$

$$S_B = S_f + x\,(S_g - S_f) = 6.27712 \text{ kJ/kg.K}$$

$$\therefore \quad M_B = \text{mass of steam in Tank } B$$

$$= \frac{V_B}{v_B} = 1.2314 \text{ kg}$$

After mixing

$$h_f = \text{final enthalpy} = \frac{m_A h_A + m_B h_B}{m_A + m_B}$$

$$= \frac{0.4826 \times 2873.1 + 1.2314 \times 2548.23}{(0.4826 + 1.2314)}$$

$$= 2635.25 \text{ kJ/kg}$$

$$V_f = \text{final specific volume}$$

$$= \frac{m_A V_A + m_B V_B}{m_A + m_B} = 0.5966 = 0.6 \text{ m}^3\text{/kg}$$

Hence from Mollin diagram

ath_f and V_f, P_f = final pressure = 322.6 kPa **Ans**

$$S_f = 6.78 \text{ kJ/kg.K}$$

$\therefore$ change in entropy of universe $= (\Delta S)_A + (\Delta S)_B$

$= m_A (S_f - S_A) + m_B (S_f - S_B)$

$= 0.4826 (6.78 - 7.6464) + 1.2314 (6.78 - 6.27714)$

$= 0.2010$ kJ/kg **Ans**

Problem 9.12 : A sample of wet steam from a steam main flows steadily through a partially open valve into a pipeline in which is fitted an electric coil. The valve and the pipeline are well insulated. The steam mass flow rate is .008 kg/s while the coil takes 3.91 amperes at 230 volts. The main pressure is 4 bar, and the pressure and temperature of the steam downstream of the coil are 2 bar and 160°C respectively. Steam velocities may be assumed to be negligible.
(a) Evaluate the quality of steam in the main.
(b) State, with reasons, whether an insulated throttling calorimeter could be used for this test.

Solution : Given data :

$\dot{m}$ = steam flow rate
= 0.008 kg/s
I = current = 3.91 A
V = 230 V
P_0 = Main pressure = 4 bar
P_1 = pressure at downstream of coil
= 2 bar
T_1 = temperature of steam at downstream of coil
= 160°C
= 160 + 273 = 433 K
Q = Heat supplied by coil
$= V.I = 3.91 \times 250 = 0.8993$ kJ

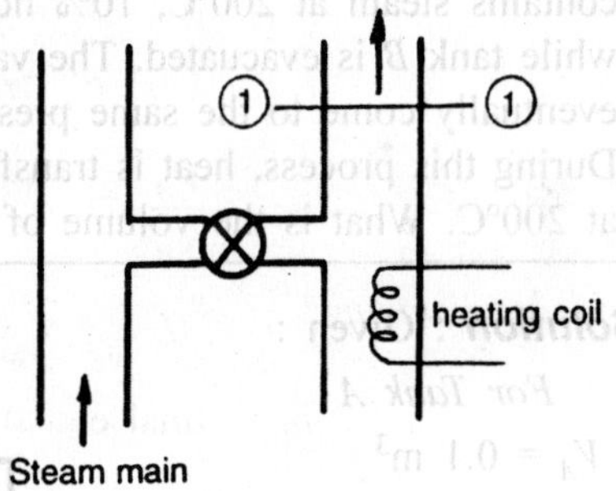

Fig. P. 9.12

From steam table
at 4 bar,

$h_{f_4} = 604.74$ kJ/kg,

$h_{fg_4} = 2133.8$ kJ/kg

at 2 bar
at 150°C, $h_{150} = 2768.8$ kJ/kg
at 200°C, $h_{200} = 2870.5$ kJ/kg

By interpolation,

$$\frac{h_{160} - h_{150}}{h_{200} - h_{150}} = \frac{160-150}{200-150} = \frac{10}{50} = \frac{1}{5}$$

$$\therefore \quad h_{160} = 2789.14 \text{ kJ/kg}$$

Now energy balance,

$$Q = h_{160} - h_{main}$$

or $$h_{main} = h_{100} - Q$$

$$= 2789.14 - 0.8993$$

$$= 2676.73 \text{ kJ/kg}$$

OR $$h_{main} = 2676.73 = h_{f_4} + x_{main}\, h_{fg_4}$$

or, $$x_{main} = \frac{(2676.73 - 604.74)}{2133.8} = 0.97 \textbf{ Ans}$$

(b) Since quality of steam in main are 0.97, so, there is no use of emulated calorimeter for this test.

Problem 9.13 : Tank *A* (fig. P 9.13) has a volume of 0.1 m³ and contains steam at 200°C, 10% liquid and 90% vapour by volume, while tank *B* is evacuated. The valve is then opened, and the tanks eventually come to the same pressure, which is found to be 4 bar. During this process, heat is transferred such that the steam remains at 200°C. What is the volume of tank *B*?

Solution : Given :

For Tank A

$V_A = 0.1$ m³

$T_A = 200$°C,

$x_A = 0.9$

From steam table

at 200°C,

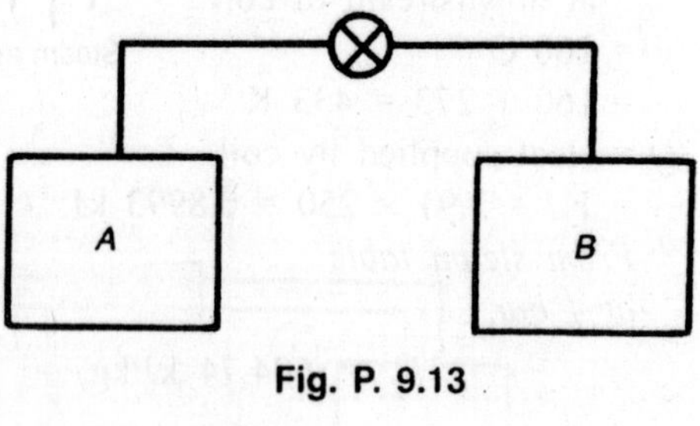

Fig. P. 9.13

$V_f = 1.157 \times 10^{-3}$ m³/kg,

$V_g = 0.12736$ m³/kg

$\therefore v_A$ = specific volume of steam in tank A

$= V_f + 0.9\,(V_g - V_f)$

$= 0.001157 + 0.9\,(0.12736 - 0.001157)$

$= 0.11474$ m³/kg

$$\therefore \quad m_A = \frac{V_A}{v_A} = \frac{0.1}{0.11474} = 0.872 \text{ kg}$$

After opening valve and healing

$$T_f = 200°C$$

and $$P_f = 4 \text{ bar}$$

∴ from steam table,

$$V_f = 0.6548 \text{ m}^3/\text{kg}$$

Now by mass balance

mass of steam in both tank before mixing = mass in both tank after mixing

$$\Rightarrow \quad m_A + 0 = m_A' + m_B'$$

or $$0.872 = \frac{V_A}{V_f} + \frac{V_B}{V_f}$$

or $$V_A + V_B = 0.872 \times V_f = 0.872 \times 0.6548 = 0.571 \text{ m}^3$$

$$\therefore \quad V_B = 0.571 - 0.01 = 0.471 \text{ m}^3 \textbf{ Ans}$$

Problem 9.14 : Boiler steam at 8 bar, 250°C, reaches the engine control valve through a pipeline at 7 bar, 200°C. It is throttled to 5 bar before expanding in the engine to 0.1 bar, 0.9 dry. Determine per kg of steam (a) the heat loss in the pipeline, (b) the temperature drop in passing through the throttle valve, (c) the work output of the engine, (d) the entropy change due to throttling and (e) the entropy change due to throttling and (e) the entropy change in passing through the engine.

Solution :

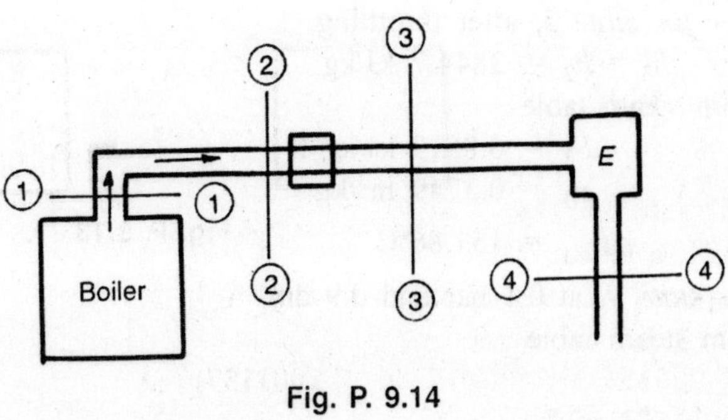

Fig. P. 9.14

For state 1

$$p_1 = 8 \text{ bar}$$
$$T_1 = 250°C = 250 + 273 = 523 \text{ K}$$

From steam table $v_1 = 0.2931 \text{ m}^3/\text{kg}$

$$h_1 = 2950.0 \text{ kJ/kg}$$
$$S_1 = 7.0384 \text{ kJ/kg.K}$$

For state 2

$$p_2 = 7 \text{ bar}$$
$$T_2 = 200°C = 200 + 273 = 473 \text{ K}$$

from steam table

at 0.6 MPa at 200°C	at 0.7 MPa and 200°C
$V_1 = 0.3520 \text{ m}^3/\text{kg}$	$V_2 = 0.2608 \text{ m}^3/\text{kg}$
$h_1 = 2850.1 \text{ kJ/kg}$	$h_2 = 2839.3 \text{ kJ/kg}$
$S_1 = 6.9665 \text{ kJ/kg.K}$	$S_2 = 6.8158 \text{ kg/kg.K}$

by interpolation,
at 7 bar or 0.7 MPa

$$\frac{v - v_1}{v_2 - v_1} = \frac{\frac{1}{p} - \frac{1}{p_1}}{\frac{1}{p_2} - \frac{1}{p_2}} = \frac{\frac{1}{.7} - \frac{1}{.6}}{\frac{1}{.8} - \frac{1}{.6}} = 0.57143$$

$$\& \quad \frac{S - S_1}{S_2 - S_1} = \frac{\ln\frac{p}{p_1}}{\ln\frac{p_2}{p_1}} = \frac{\ln\frac{.07}{0.6}}{\ln\frac{0.8}{0.6}} = 0.53584$$

$$\& \quad \frac{h - h_1}{h_2 - h_1} = \frac{p - p_1}{p_2 - p_1} = \frac{0.7 - 0.6}{0.8 - 0.6} = \frac{0.1}{0.2} = 0.5$$

$\therefore$ $v_2 = 0.2999 \text{ m}^3/\text{kg}$, $S_2 = 6.8857$ kJ/kg.K, $h_2 = 2844.7$ kJ/kg

Now for state 3, after throttling,

$$h_3 = h_2 = 2844.7 \text{ kJ/kg}$$

From steam table

$$S_3 = 6.8513 \text{ kJ/kg K}$$
$$v_3 = 0.3749 \text{ m}^3/\text{kg}$$
$$t_{sat_3} = 151.86°C$$

For state 4, at 0.1 bar and 0.9 dry
from steam table

$$v_f = 1.010 \times 10^{-3} \text{ m}^3/\text{kg}$$
$$v_g = 14.67 \text{ m}^3/\text{kg}$$
$$h_g = 191.83 \text{ kJ/kg}$$
$$h_{f_g} = 2392.8 \text{ kJ/kg}$$
$$S_t = 0.6493 \text{ kJ/kg.K}$$
$$S_g = 8.1502 \text{ kJ/kg.K}$$
$$\therefore \quad v_4 = 13.20 \text{ m}^3/\text{kg},$$
$$h_4 = 2345.35 \text{ kJ/kg}$$
$$S_4 = 7.4011 \text{ kJ/kg.K}$$

(a) heat loss in pipe line

$$= h_1 - h_2 = 2950.0 - 2844.7$$
$$= 105.3 \text{ kJ/kg}$$

(b) the tempr in passing through the throttle valve:
from mollier diagram, tempr drop = 5°C

(c) Work out of engine $= h_3 - h_4$

$$= 2844.7 - 2345.35 = 499.35 \text{ kJ/kg}$$

(d) entropy change due to throttling

$$= S_3 - S_2$$
$$= 6.8213 - 6.8857 = -0.0644 \text{ kJ/kg.K}$$

= decrease in entropy

(e) entropy change in passing through the engine

$$= S_4 - S_3 = -6.8213 + 7.40011$$
$$= 0.57881 \text{ kJ/kg.K}$$

= increase in entropy **Ans**

Problem 9.15 : Two streams of steam, one at 2 MPa, 300°C and the other at 2 MPa, 400°C, mix in a steady flow adiabatic process. The rates of flow of the two streams are 3 kg/min and 2 kg/min respectively. Evaluate the final temperature of the emerging stream, if there is no pressure drop due to the mixing process. What would be the rate of increase in the entropy of the universe? This stream with a negligible velocity now expands adiabatically in a nozzle to a pressure of 1 kPa. Determine the exit velocity of the stream and the exit area of the nozzle.

Solution :

For stream 1

$p_1 = 2$ MPa,

$\dot{m}_1 = 3$ kg/min

$T_1 = 300°C$
$= 300 + 273$
$= 573$ K

From steam table

$h_1 = 3023.5$ kJ/kg

$S_1 = 6.7664$ kJ/kg.K

$v_1 = 0.12547$ m^3/kg

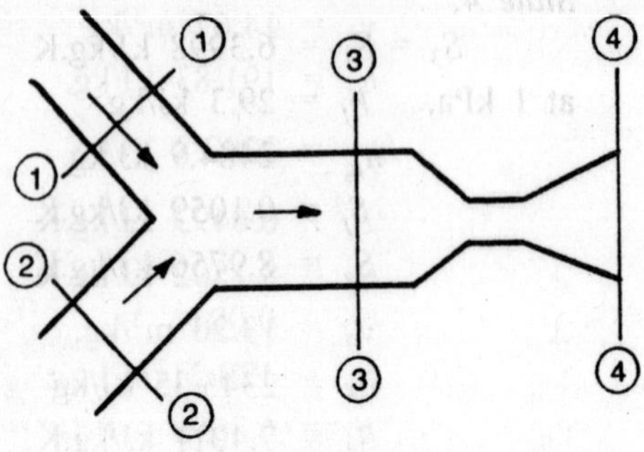

Fig. P. 9.15

For stream 2

$p_2 = 2$ MPa,

$T_2 = 400°C = 400 + 273 = 673$ K

$\dot{m}_2 = 2$ kg/min

From steam table

$h_2 = 3247.6$ kJ/kg

$S_2 = 7.1271$ kJ/kg.K

$v_2 = 0.15120$ m^3/kg

For stream 3
after mixing

$$h_3 = \frac{m_1 h_1 + m_2 h_2}{(m_1 + m_2)} = \frac{3 \times 3023.5 + 2 \times 3247.6}{(3+2)}$$

$$= 3113.14 \text{ kJ/kg}$$

$$v_3 = \frac{m_1 v_1 + m_2 v_2}{m_1 + m_2} = \frac{3 \times .12547 + 2 \times .15120}{5}$$

$$= 0.135762 \text{ m}^3/\text{kg}$$

Hence steam will be in the superheated region because

h_g at 2 MPa $= 2799.5$ kJ/kg and $t_{sat} = 212.42°C$

Again at 2 MPa and $t = 350°C$,

$h = 3137.0$ kJ/kg

Hence by interpolation $\dfrac{h_3 - h_{2\,\text{MPa}}}{h_{2\,\text{MPa}}\,350°\text{C} - h_{2\,\text{MPa}}} = \dfrac{t_3 - t_{sat}}{350 - t_{sat}}$

$\therefore \quad t_3 = 340.27°C$ **Ans**

Similarly by interpolation,

$S_3 = 6.3998$ kJ/kg.K

State 4,

$$S_3 = S_4 = 6.3998 \text{ kJ/kg.K}$$

at 1 kPa,

$$h_f = 29.3 \text{ kJ/kg}$$
$$h_{fg} = 2484.9 \text{ kJ/kg}$$
$$S_f = 0.1059 \text{ kJ/kg.K}$$
$$S_g = 8.9756 \text{ kJ/kg.K}$$
$$v_f = 0.001$$
$$v_g = 129.21 \text{ m}^3\text{kg}$$

$$\therefore \quad S_4 = 6.3998 = 0.1059 + x_4 (8.9756 - 0.1059)$$

$$\Rightarrow \quad x_4 = 0.71$$

$$\therefore \quad h_4 = 29.3 + 0.71 \times 2484.9 = 1793.6 \text{ kJ/kg}$$

$$v_4 = 0.001 + 0.71 (129.21 - 0.001)$$
$$= 91.74 \text{ m}^3/\text{kg}$$

(b) rate of entropy increase of the universe

$$= \dot{m}_1 (S_3 - S_1) + \dot{m}_2 (S_3 - S_2) + (\dot{m}_1 + \dot{m}_2)(S_4 - S_3)$$

$$= \frac{3}{60}(6.3998 - 6.7664) + \frac{2}{60}(6.3998 - 7.1271) + 0$$

$$= -0.0426 \text{ kJ/kg/min } \textbf{Ans}$$

(c) Applying S.F.E.E in nozzle neglecting the inlet velocity and ΔP.E.

$$h_3 = h_4 + \frac{V_4^2}{2}$$

$$\Rightarrow \quad V_4 = \sqrt{2(h_3 - h_4)}$$

$$= \sqrt{2(3113.14 - 1793.6) \times 1000}$$

$$= 1624.5 \text{ m/sec}$$

Problem 9.16 : A spherical aluminium vessel has an inside diameter of 0.3 m and a 0.62 cm thick wall. The vessel contains water at 25°C with a quality of 1%. The vessel is then heated until the water inside is saturated vapour. Considering the vessel and water together as a system, calculate the heat transfer during this process. The density of aluminium is 2.7 g/cm^3 and its specific heat is 0.896 kJ/kg.K.

Solution : Given data

d_i = inside diameter = 0.3 m

t = thickness of vessel = 0.62 cm = 0.0062 m

∴ d_0 = outside diameter = 0.3 + 2 × 0.0062

= 0.3124 m

(i) *Initial condition*

water at 25°C and quality = 1% = 0.01

(ii) *final condition*, – saturated vapour

ρ_a = density of A1 = 2.7 g/cm³ = 2.7 × 10³ kg/m³

C_a = specific heat capacity = 0.896 kJ/kg.K

Now from steam table at 25°C,

$$v_{f_1} = 1.003 \times 10^{-3} \text{ m}^3/\text{kg},$$

$$v_{g_1} = 43.36 \text{ m}^3/\text{kg}$$

$$h_{f_1} = 104.89 \text{ kJ/kg}$$

$$h_{fg_1} = 2442.3 \text{ m}^3/\text{kg}$$

$$\therefore \quad h_1 = h_{f_1} + 0.01\ h_{fg_1} = 129.313 \text{ kJ/kg}$$

Specific volume

$$v_1 = v_{f_1} + 0.01\ (v_{g_1} + v_{f_1})$$

$$= 0.43459 \text{ m}^3/\text{kg}$$

Volume of aluminium container

$$= \frac{\pi}{6}(d_0^3 - d_i^3) = \frac{\pi}{6}[(0.3124)^3 - (0.3)^3]$$

$$= 1.826 \times 10^{-3} \text{ m}^3$$

∴ mass of vessel

$$= \text{volume of container} \times \rho_a$$

$$= 2.7 \times 10^3 \times 1.826 \times 10^{-3} = 4.93 \text{ kg}$$

Again, mass of fluid

$$= \frac{\text{Volume of container}}{\text{specific volume}} = \frac{\frac{\pi}{6} d_i^3}{0.43459}$$

$$= \frac{\pi}{6} \times \frac{(0.3)^3}{0.43459} = 0.3253 \text{ kg}$$

Now for saturated vapour at 100°C, h_2 = 2676.1 kJ/kg

∴ Heat taken by aluminium container = mass of fluid × C × ΔT

$= 4.93 \times .0896 \times (100 - 25)$

$= 331.296$ kJ

and heat taken by water

= mass of fluid $(h_2 - h_1)$

$= 0.03253\ (2676.1 - 129.313) = 82.85$ kJ

∴ Total heat received during the process

$= 331.296 + 82.85 = 414.15$ kJ **Ans**

Problem 9.17 : Steam of 65 bar, 400°C leaves the boiler to enter a steam turbine fitted with a throttle governor. At a reduced load, as the governor takes action, the pressure of stream is reduced to 59 bar by throttling before it is admitted to the turbine. Evaluate the availabilities of steam before and after the throttling process and the irreversibility due to it.

Solution : Given data

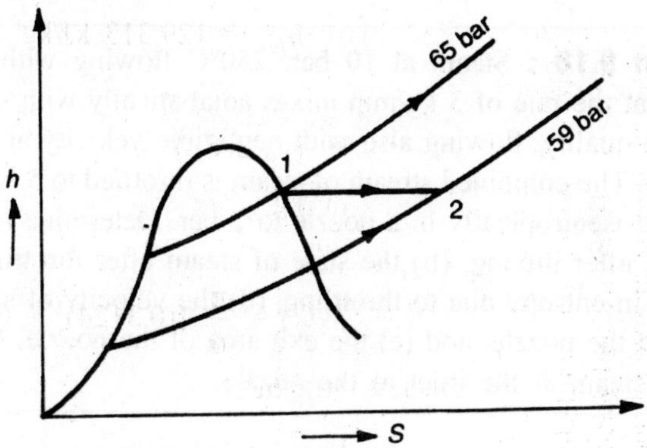

Fig. P. 9.17

Before throttling

$p_1 = 65$ bar

$T_1 = 400$°C

From steam table by interpolation

$h_1 = 3167.65$ kJ/kg

$S_1 = 6.4930$ kJ/kg

After throttling

$p_2 = 59$ bar

$$h_2 = h_1 = 3167.65 \text{ kJ/kg}$$

from Mollier diagram,

$$S_2 = 6.54 \text{ kJ/kg.K}$$

Let the surrounding temperature is 20°C

From steam table,

$$h_0 = 83.96 \text{ kJ/kg}$$

$$S_0 = 0.2966 \text{ kJ/kg.K}$$

∴ (i) Availability of steam before the throttling

$$= (h_1 - h_0) - T_0 (S_1 - S_0)$$

$$= (3167.65 - 83.96) - 293 (6.4930 - 0.2966)$$

$$= 1268.152 \text{ kJ/kg } \textbf{Ans}$$

Availability of steam after throttling

$$= (h_2 - h_0) - T_0 (S_2 - S_0)$$

$$= (3167.65 - 83.96) - 293 (6.54 - 0.2966)$$

$$= 1254.3738 \text{ kJ/kg } \textbf{Ans}$$

(ii) Irreversibility $= T_0 (\Delta S)_{universe}$

$$= T_0 (S_2 - S_1)$$

$$= 293 (6.54 - 6.493) = 13.771 \text{ kJ/kg } \textbf{Ans}$$

Problem 9.18 : Steam at 10 bar, 250°C flowing with negligible velocity at the rate of 3 kg/min mixes adiabatically with steam at 10 bar, 0.75 quality, flowing also with negligible velocity at the rate of 5 kg/min. The combined stream of steam is throttled to 5 bar and then expanded isentropically in a nozzle to 2 bar. Determine (a) the state of steam after mixing, (b) the state of steam after throttling, (c) the increase in entropy due to throttling, (d) the velocity of steam at the exit from the nozzle, and (e) the exit area of the nozzle. Neglect the K.E. of steam at the inlet to the nozzle.

Solution :

For state 1

$$P_1 = 10 \text{ bar}$$

$$T_1 = 250°\text{C}$$

$$m_1 = 3 \text{ kg/min}$$

∴ from steam table,

$$h_1 = 2942.6 \text{ kJ/kg}$$

At state 2

$$P_2 = 10 \text{ bar}$$

$$x_2 = 0.75$$

$$m_2 = 5 \text{ kg/min}$$

from steam table

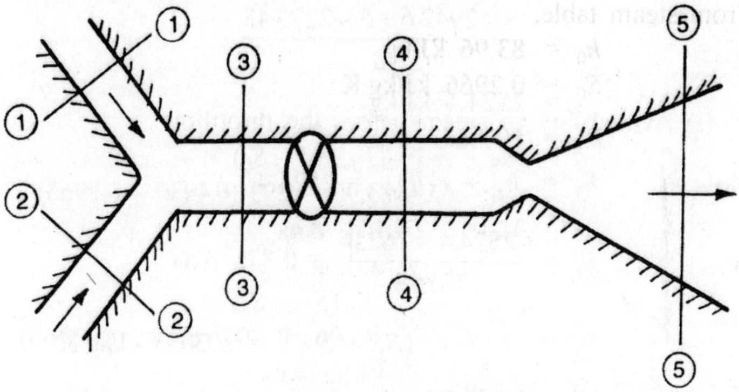

Fig. P. 9.18 (a)

$$h_{f_2} = 762 \text{ kJ/kg}$$
$$h_{fg_2} = 2015.3 \text{ kJ/kg}$$
$$S_{f_2} = 2.1387 \text{ kJ/kg.K}$$
$$S_{g_2} = 6.5865 \text{ kJ/kg.K}$$
$$\therefore \quad h_2 = h_{f_2} + x_2 \, h_{fg_2} = 762 + 0.75 \times (2015.3) = 2273.48 \text{ kJ/kg}$$

At state 3

After mixing,

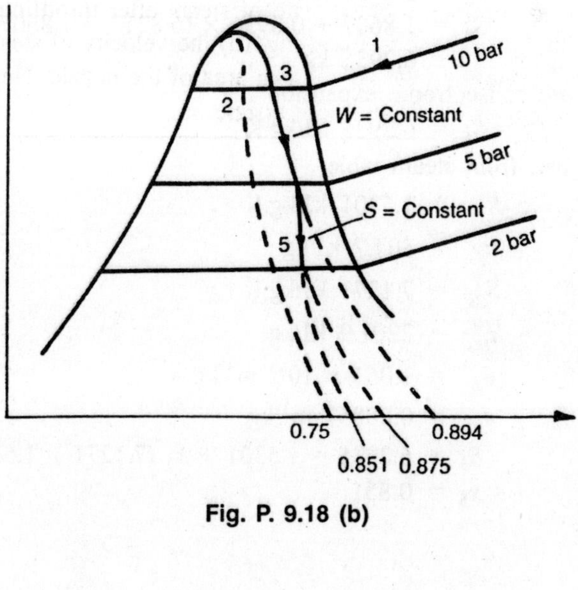

Fig. P. 9.18 (b)

$$h_3 = h_m = \frac{m_1 h_1 + m_2 h_2}{(m_1 + m_2)}$$

$$= \frac{3 \times 2942.6 + 5 \times 2273.48}{8}$$

$$= 2524.4 \text{ kJ/kg}$$

$$P_3 = 10 \text{ bar}$$

Now $h_3 = h_{f_2} + x_3 h_{fg_2} = 2524.4$

or, $x_3 = \dfrac{2524.4 - 762.0}{2015.3} = 0.875$ **Ans**

$$S_3 = S_{f_2} + x_3 \left(S_{g_2} - S_{f_2}\right) = 6.03053 \text{ kJ/kg.K}$$

At state 4, After throttling

$h_3 = h_4 = 2524.4$ kJ/kg

Now at 5 bar, from steam table,

$$h_{f_4} = 640.23 \text{ kJ/kg,}$$

$$h_{fg_4} = 2108.5 \text{ kJ/kg}$$

$$S_{f_4} = 1.8607 \text{ kJ/kg.K}$$

$$S_{g_4} = 6.8213 \text{ kJ/kg.K}$$

or $h_3 = 2524.4 = 640.23 + x_4 \times 2108.5$

$\therefore$ $x_4 = 0.8936$ **Ans**

$\therefore$ $S_4 = 1.8607 + 0.8936 \times (6.8213 - 1.8607)$

$= 6.2935$ kg

At state 5, isentropic expansion

$S_4 = S_5 = 6.2935$ kJ/kg.K

at 2 bar, from steam table

$$S_{f_5} = 1.5301 \text{ kJ/kg.K}$$

$$h_{f_5} = 504.7 \text{ kJ/kg}$$

$$S_{g_5} = 7.1271 \text{ kJ/kg.K}$$

$$h_{fg_5} = 2201.9 \text{ kJ/kg}$$

$$v_{f_5} = 1.061 \times 10^{-3} \text{ m}^3\text{/kg}$$

$$v_{g_5} = 0.08857 \text{ m}^3\text{/kg}$$

or $S_5 = 6.2935 = 1.5301 + x_5 (7.1271 - 1.5301)$

$\therefore$ $x_5 = 0.851$

Specific volume

$$= v_5 = 0.001061 + 0.851\,(0.8857 - 0.001061)$$
$$= 0.754 \text{ m}^3/\text{kg}$$
$$h_5 = 504.7 + 2201.9 \times 0.851$$
$$= 2378.52 \text{ kJ/kg}$$

∴ (a) at 10 bar and 0.875 dry **Ans**

(b) at 5 bar and 0.894 dry **Ans**

(c) Increase in entropy due to throttling $= S_4 - S_3$

$$= 6.2935 - 6.03053 = 0.26297 \text{ kJ/kg.K}$$

(d) Applying S.F.E.E. neglecting ΔK.E. and ΔP.E. for the nozzle

Here $\frac{dQ}{dm} = 0$ and $\frac{dw}{dm} = 0$

V_4 = incoming velocity to the nozzle = 0

$$h_4 = h_5 + \frac{V_5^2}{2}$$

$$\therefore \quad V_5 = \sqrt{2(h_4 - h_5)}$$
$$= \sqrt{2 \times 1000(2524.4 - 2378.52)}$$
$$= 540.15 \text{ m/s } \textbf{Ans}$$

(e) $\dot{Q}$ = Volume flow rate

$$= \frac{V_5 \times \dot{m}}{60} = \frac{0.754 \times (5+3)}{60}$$
$$= 0.100533 \text{ m}^3/\text{sec}$$

$$\therefore \text{ Area} = \frac{\dot{Q}}{V_5} = \frac{0.100533}{540.15} = 0.1861 \times 10^{-3} \text{ m}^2$$
$$= 1.861 \text{ m}^2 \textbf{ Ans}$$

Problem 9.19 : A mass of wet steam at temperature 165°C is expanded at constant quality 0.8 to pressure 3 bar. It is then heated at constant pressure to a degree of superheat of 66.5°C. Find the enthalpy and entropy changes during expansion and during heating. Draw the *T*–s and *h*–s diagrams.

Solution : Given data :

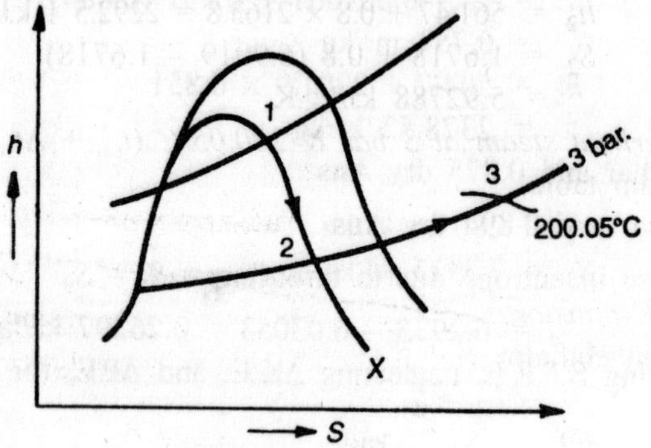

Fig. P. 9.19 (a)

For wet steam at

$$T_1 = 165°C, x = 0.8$$

∴ from steam table,

$$h_1 = h_f + 0.8\, h_{f_g}$$
$$= 697.38 + 0.8 \times 2066.05 = 2350.22 \text{ kJ/kg}$$
$$S_1 = S_f + 0.8\,(S_g - S_f)$$
$$= 1.9923 + 0.8\,(6.70825 - 1.9923) = 5.7651 \text{ kJ/kg.K}$$

For wet steam at 3 bar and $x = 0.8$

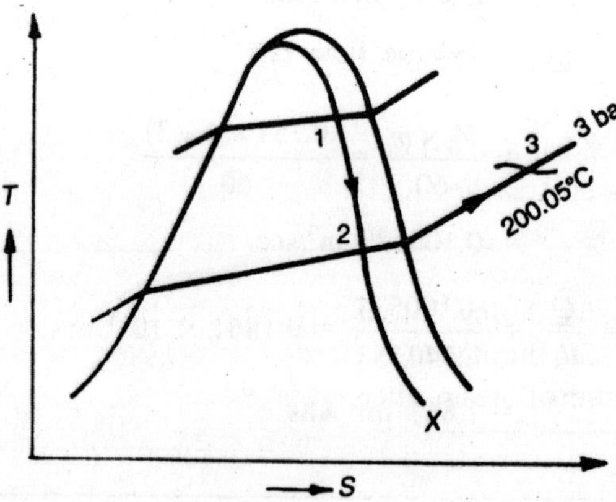

Fig. P. 9.19 (b)

from steam table

$$t_{sat} = 133.55°C$$
$$h_f = 561.47 \text{ kJ/kg}$$
$$h_{f_g} = 2163.8 \text{ kJ/kg}$$
$$S_f = 1.6718 \text{ kJ/kg.K}$$

$$S_g = 6.9919 \text{ kJ/kg.K}$$

$$\therefore \quad h_2 = 561.47 + 0.8 \times 2163.8 = 2292.5\ 1 \text{ kJ/kg}$$

$$S_2 = 1.6718 + 0.8\ (6.9919 - 1.6718)$$

$$= 5.92788 \text{ kJ/kg.K}$$

For superheat steam at 3 bar & 200.05°C ($t_{sat} + \Delta t_{superheat}$) from steam table,

$$h_3 = 2865.6 \text{ kJ/kg}$$

$$S_3 = 7.3115 \text{ kJ/kg.K}$$

During Expansion,

Change in enthalpy

$$= h_2 - h_1 = (-2350.22 + 2292.5)$$

$$= -57.71 \text{ kJ/kg } \textbf{Ans}$$

Change in entropy

$$= S_2 - S_1 = 5.92788 - 5.7651$$

$$= 0.1628 \text{ kJ/kg.K } \textbf{Ans}$$

During heating

Change in enthalpy

$$= h_3 - h_2$$

$$= 2865.6 - 2292.51 = 573.09 \text{ kJ/kg } \textbf{Ans}$$

Change in entropy

$$= S_3 - S_2 = 7.3115 - 5.9278.8$$

$$= 1.38362 \text{ kJ/kg.K } \textbf{Ans}$$

Problem 9.20 : A cylinder contains one kg of water and steam at a pressure of 3.8 bar and 0.4 dry. Heat is supplied at constant volume untill the pressure reaches to 10 bar. The steam is then expanded according to the law pv = constant. Untill the pressure is 2 bar. Calculate

(i) Heat transfer during constant volume heating

(ii) Heat transfer during pv = constant expansion

(iii) Temperature of steam after the expansion

Solution :

Initial condition

$$m = \text{mass of water + mass of steam}$$

$$= 1 \text{ kg mass of steam}$$

$$p_1 = 3.8 \text{ bar}, x = 0.4$$

$\therefore$ mass of water = m_w = 0.6 kg & m_s = mass of steam = 0.4 kg

After heat addition

$p_2 = 10$ bar

final condition after expansion at

$pv =$ Const.

$p_3 = 2$ bar

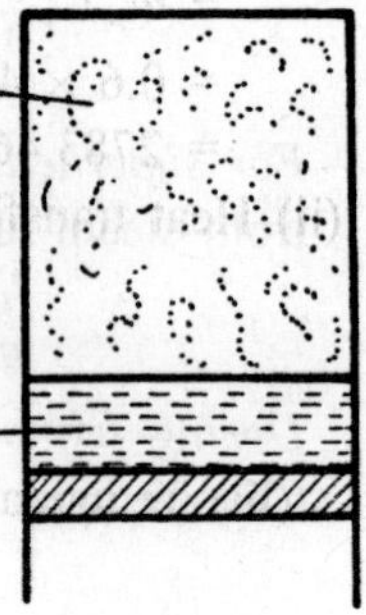

Fig. P. 9.20

From steam table *at 3.8 bar*

v_f = sp. volume of water

$= 0.001082 \text{ m}^3/\text{kg}$

v_g = sp. volume of steam

$= 0.485 \text{ m}^3/\text{kg}$

$h_f = 596.8 \text{ kJ/kg}$

$h_g = 2736.2 \text{ kJ/kg}$

$t_{sat} = T_1 = 141.8°\text{C}$

$= 414.8 \text{ K}$

$\therefore \quad v$ = volume of mixture

$= v_f + xv_g$

$= 0.001082 + 0.4 \times 0.485$

$= 0.195082 \text{ m}^3/\text{kg}$

$\therefore$ initial volume of cylinder

$= V_1 = v \times m = 0.195082 \text{ m}^3$

$h_1 = 596.8 + 0.4 \times (2736.2 - 596.8)$

$= 1452.56 \text{ kJ/kg}$

After heating at constant volume

$$T_2 = \left(\frac{T_1}{P_1}\right) P_2$$

$$= \frac{414.8}{3.8} \times 10 = 1091.58 \text{ K} = 818.58°\text{C}$$

$V_2 = V_1$

From steam table at 10 bar and at T_2°C, $h_2 = 4160.7$ kJ/kg

Final Condition at isothermal process

$p_3 = 2$ bar

$$v_3 = \frac{p_2 v_2}{p_3}$$

$$= \frac{10 \times 0.195082}{2} = 0.97541 \text{ m}^3/\text{kg}$$

from steam table at 2 bar and T_2°C, $h_3 = 4165.2$ kJ/kg

(i) heat transfer during constant volume heating

$= m_w C_{p_w} (T_2 - T_1) + m_s(h_2 - h_1)$

$= 0.6 \times 4.187\ (1091.58 - 414.8) + 0.4\ (4160.7 - 1452.56)$

$= 2783.46$ kJ **Ans**

(ii) Heat transfer during *pv*

= constant expansion

$= m\ (h_3 - h_2)$

$= 1 \times (4165.2 - 4160.7) = 4.5$ kJ **Ans**

(iii) Temperature of steam after the expansion

$= T_3 = T_2$

$= 818.58°C$ **Ans**

Chapter 10

Properties of Gases and Gas Mixtures

Problem 10.1 : The usual cooking gas (mostly methane) cylinder is about 25 cm in diameter and 80 cm in height. It is charged 12 MPa at room temperature (27°C). (a) Assuming the ideal gas law, find the mass of gas filled in the cylinder. (b) Explain how the actual cylinder contains nearly 15 kg of gas. (c) If the cylinder is to be protected against excessive pressure by means of a fusible plug, at what temperature should the plug melt to limit the maximum pressure to 15 MPa ?

Solution : Given data :

D = Cylinder dia = 25 cm

h = Cylinder height = 80 cm

P = Pressure in the cylinder = 12 MPa

T = Temperature in the cylinder

= 27°C = 27 + 273 = 300 K.

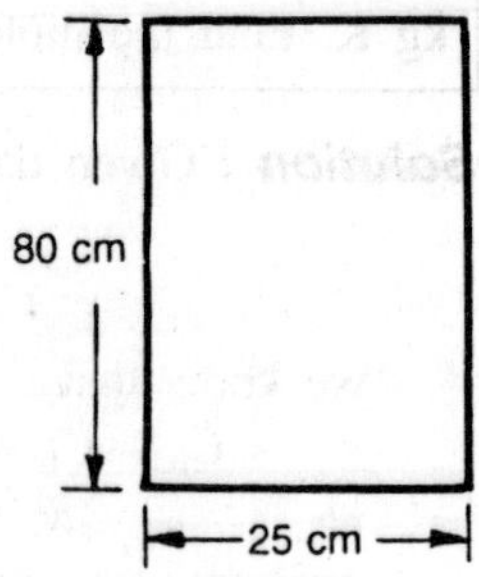

Fig. P. 10.1

(a) Volume of cylinder

$$= \frac{\pi}{4} D^2 \times h$$

$$= \frac{\pi}{4} (0.25)^2 \times 0.80 = 0.0393 \text{ m}^3$$

from ideal gas law

$$m = \frac{PV}{RT}$$

where $$R = \frac{\bar{R}}{M} = \frac{8.314}{16} \text{ kJ/kg.K}$$

$$M = \text{molcular wt} = 16 \text{ for } CH_4$$

$$= \frac{12 \times 10^3 \times 0.0393 \times 16}{8.314 \times 300} = 3.03 \text{ kg } \textbf{Ans}$$

(b) for mass to be 15 kg

from ideal gas law

$$P \times V = MRT$$

$$\Rightarrow \quad P = \frac{MRT}{V} = \frac{15 \times 8.314 \times 300}{16 \times 0.0393} = 59.5 \text{ MPa}$$

If pressure within the cylinder becomes 59.5 MPa, mass will be 15 kg at the same temp.

(c) Pressure max^m = 15 MPa

mass = 3.03 kg

from ideal gas law

$$T = \frac{PV}{MR} = \frac{15 \times 10^3 \times 0.0393 \times 16}{3.03 \times 8.314}$$

$$= 374.41 \text{ K} = 101.41°\text{C}$$

at this temp fusible plug will meltdown.

Problem 10.2 : A certain gas has $C_p = 0.913$ and $C_v = 0.653$ kJ/kg K. Find the molecular weight and the gas constant R of gas

Solution : Given data :

$$C_p = 0.913 \text{ kJ/kg.K}$$

$$C_v = 0.653 \text{ kJ/kg.K}$$

we know that

$$C_p - C_v = R$$

$$\Rightarrow \quad R = 0.913 - 0.653 = 0.26 \text{ kJ/kg.K}$$

we know gas constant

$$R = \frac{\text{Universal gas constant } (\bar{R})}{\text{Molecular weight } (M)}$$

$$\therefore \quad M = \frac{\overline{R}}{R} = \frac{8.314}{0.26} = 31.98 \approx 32 \textbf{ Ans.}$$

Problem 10.3 : From an experimental determination the specific heat ratio for acetylene (C_2H_2) is found to 1.26. Find the two specific heats.

Solution : Given data :

$$\frac{C_p}{C_v} = 1.26 \qquad ...(1)$$

for C_2H_2,

$$M = \text{molecular wt} = 2 \times 12 + 2 \times 1 = 26$$

$$\therefore \quad R = \text{Gas constant} = \frac{\overline{R}}{M} = \frac{8.314}{26}$$

$$= 0.31977 \text{ kJ/kg.K}$$

$$\therefore \quad C_p - C_v = 0.31977 \text{ kJ/kg.K}$$

$$\text{or } 1.26\, C_v - C_v = 0.31977$$

$$\text{or,} \quad C_v = \frac{0.31977}{0.26} = 1.23 \text{ kJ/kg.K}$$

$$\& \quad C_p = 1.55 \text{ kJ/kg.K}$$

Problem 10.4 : A supply of natural gas is required on a site 800 m above storage level. The gas at –150°C, 1.1 bar from storage is pumped steadily to a point on the site where its pressure is 1.2 bar, its temperature 15°C, and its flow rate 1000 m³/hr. If the work transfer to the gas at the pump is 15 kW, find the heat transfer to the gas between the two points. Neglect the change in K.E. and assume that the gas has the properties of methane (CH_4) which may be treated as an ideal gas having $\gamma = 1.33$ ($g = 9.75$ m/s²)

Solution : Given data :

At site 1 $\quad P_1 = 1.1$ bar

$$T_1 = -150°C = -150 + 273 = 123 \text{ K}$$

At site 2 $\quad P_2 = 1.2$ bar

$T_2 = 15°C = 15 + 273$
$= 288$ K

$\dot{m}$ = mass flow rate
$= 1000$ m³/hr

$= \frac{1000}{3600}$ m³/sec

$\gamma = 1.33,$
$g = 9.75$ m/s²,
M = molecular weight
$= 16$

$d w$ = ware tranfer
$= 15$ kW

$d Q = ?$

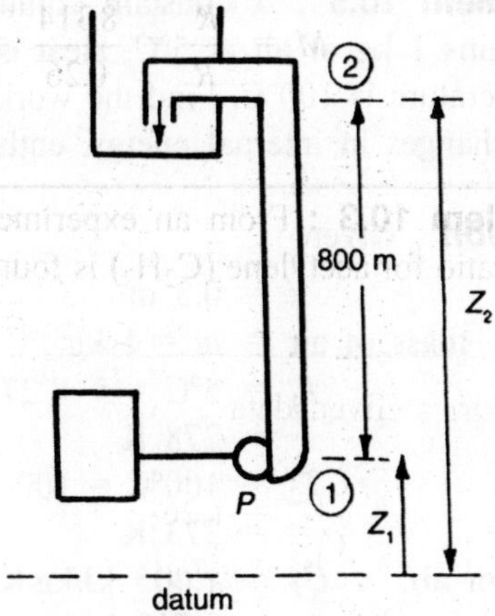

Fig. P. 10.4

Applying S.F.E.E. with ΔK.E $= 0$ and $\dot{m}_1 = \dot{m}_2 = \dot{m}$

$$\Rightarrow \qquad \dot{m}\ (h_1 + Z_1 g) + \frac{dQ}{d\tau}$$

$$= \dot{m}\ (h_1 + Z_2 g) + \frac{d\omega}{dt}$$

$$\therefore \qquad \frac{dQ}{d\tau} = \dot{m}\ \left[(h_1 - h_2) + (Z_2 - Z_1)g\right] + \frac{d\omega}{dt} \quad \ldots(1)$$

$$\text{Now} \qquad R = \frac{\overline{R}}{M} = \frac{8.314}{16} \text{ kJ/kg.K}$$

$$C_p = R + w$$

$$\Rightarrow C_P - C_P/\gamma = R$$

$$C_p = \frac{R\gamma}{\gamma - 1}$$

$$\Rightarrow \quad \frac{8.314}{16} \times \frac{1.33}{33} = 2.094 \text{ kJ/kg.K}$$

$$\therefore \qquad h_2 - h_1 = C_p[T_2 - T_1] = 345.55 \text{ kJ/kg}$$

$\therefore$ from (1)

$$\frac{dQ}{dt} = \frac{1000}{3600}\left[345.55 + \frac{800 \times 9.75}{1000}\right] - 15$$

$$= 83.15 \text{ kW}$$

Problem 10.5 : A constant volume chamber of 0.3 m^3 capacity contains 1 kg of air at 5°C. Heat is transferred to the air until the temperature is 100°C. Find the work done, the heat transferred, and the changes in internal energy, enthalpy and entropy.

Solution : Given :

$$V = 0.3 \text{ m}^3$$
$$\text{mass of air} = m = 1 \text{ kg}$$
$$T_1 = 5°\text{C} = 5 + 273 = 278 \text{ K}$$
$$T_2 = 100°\text{C} = 100 + 273 = 373 \text{ K}$$

For air, $C_P = 1.005$ kJ/kg.K

$C_v = 0.718$ kJ/kg.K

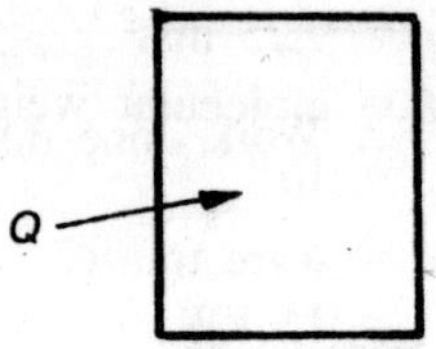

Fig. P. 10.5

(i) heat transfer $= mC_v\Delta T$

$$= 1 \times 0.718\ (373 - 278) = 68.21 \text{ kJ}$$

(ii) work done = 0. Since change in volume = 0

(iii) du = change in internal energy $= mC_v(\Delta T)$

$$= 1 \times .718 \times 95 = 68.21 \text{ kJ}$$

(iv) dh = change in enthalpy $= mC_p\Delta T$

$$= 1 \times 1.005 \times 95 = 95.475 \text{ kJ}$$

(v) Change in entropy

$$= \Delta S = m\ C_v \ln \frac{T_2}{T_1}$$

$$= 1 \times 0.718 \times \ln \frac{373}{278} = 0.211 \text{ kJ/K} \textbf{ Ans}$$

Problem 10.6 : The indicator diagram for a certain water-cooled cylinder and piston air compressor shows that during compression $pv^{1.3}$ = constant. The compression starts at 100 kPa, 25°C and ends at 600 kPa. If the process is reversible, how much heat is transferred per kg of air ?

Solution : Given data :

$$pv^{1.3} = \text{Const.}$$

Before compression,

$$p_1 = 100 \text{ kPa}$$
$$T_1 = 25°C = 25 + 273 = 298 \text{ K}$$

After compression,

$$P_2 = 600 \text{ kPa}$$

from given law,

$$\frac{T_2}{T_1} = \left(\frac{P_2}{P_1}\right)^{\frac{1.3-1}{1.3}} \Rightarrow T_2 = 450.6 \text{ K}$$

$$\therefore \text{ Work done during compression } = \frac{P_1V_1 - P_2V_2}{n-1}$$

$$= \frac{MR(T_1 - T_2)}{n-1}$$

$$= \frac{1 \times 0.287 \times (298 - 450.6)}{1.3 - 1}$$

$$= -145.99 \text{ kJ/Kg} = -146 \text{ kJ/kg}$$

Now from 1st law of thermodynamics

$$đQ = du + đw$$
$$= (u_2 - u_1) + đw$$
$$= 0.718\ (450.6 - 298\ ) - 146$$
$$= -36.4332 \text{ kJ/kg}$$

$-v_e$ sign indicates that heat is relaxed from the system.

Problem 10.7 : An ideal gas cycle of three processes uses Argon (Mol. wt. 40) as a working substance. Process 1-2 is a reversible adiabatic expansion from 0.014 m^3, 700 kPa, 280°C to 0.056 m^3. Process 2-3 is a reversible isothermal process. Process 3-1 is a constant pressure process in which heat transfer is zero. Sketch the cycle in the *p-v* and *T-s* planes, and find (a) the work transfer in process 1-2, (b) the work transfer in process 2-3, and (c) the net work of the cycle. Take $\gamma = 1.67$.

***Solution* :** Given data :

$$M = \text{molecular weight} = 40$$
$$p_1 = 700 \text{ kPa},$$
$$V_1 = 0.014 \text{ m}^3$$

$$T_1 = 280°C = 280 + 273 = 553 \text{ K}$$
$$V_2 = 0.056 \text{ m}^3$$

expansion follows

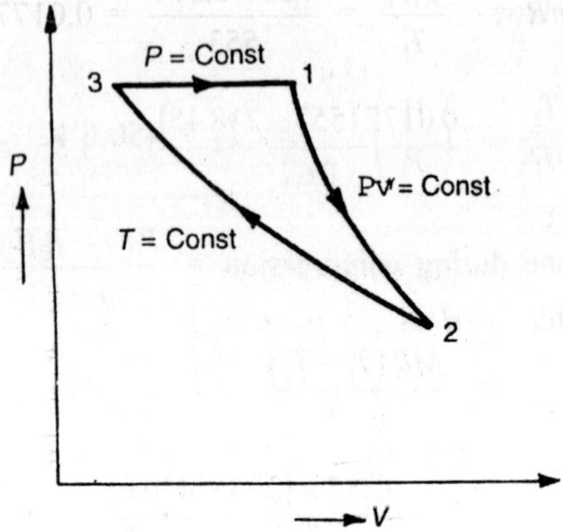

Fig. P. 10.7 (a)

$$pv^{1.67} = \text{const.}$$

$$\therefore \quad \frac{T_2}{T_1} = \left(\frac{P_2}{P_1}\right)^{\frac{1.67-1}{1.67}} = \left(\frac{V_1}{V_2}\right)^{1.67-1}$$

$$\therefore \quad T_2 = T_1 = \left(\frac{0.014}{0.056}\right)^{0.67} = 218.45 \text{ K}$$

$$P_2 = p_1 \left(\frac{V_1}{V_2}\right)^{1.67}$$

$$= 700 \left(\frac{0.014}{0.056}\right)^{1.67}$$

$$= 69.13 \text{ kPa}$$

∴ (a) work transfer in 1-2

$$đw_{1-2} = \frac{p_1 v_1 - p_2 v_2}{n-1}$$

$$= \frac{mR(T_1 - T_2)}{n-1}$$

from ideal gas eq.

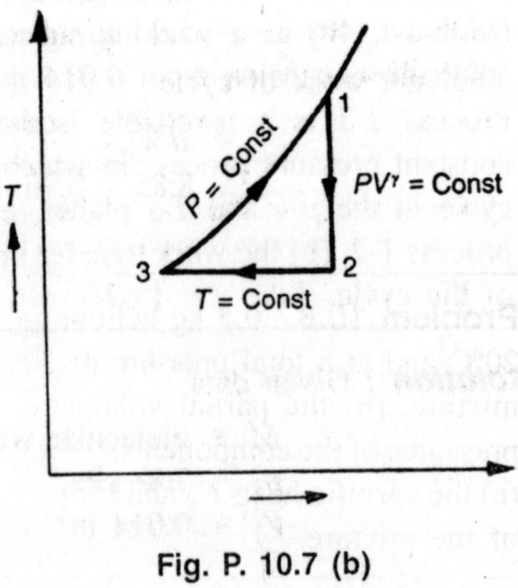

Fig. P. 10.7 (b)

$$\Rightarrow \qquad pV = m\frac{\bar{R}}{M}T = mRT$$

$$\Rightarrow \qquad mR = \frac{p_1 v_1}{T_1} = \frac{700 \times 0.014}{553} = 0.0177$$

$$\therefore \qquad d\,w_{1-2} = \frac{0.0177\,(553 - 218.45)}{0.67} = 8.85 \text{ kJ/kg.}$$

(b) Process 2-3

$$\text{Work transfer} = d\,w_{2-3} = p_2 v_2 \ln \frac{V_3}{V_2}$$

$$= p_2 v_2 \ln \frac{p_2}{p_3} = p_2 v_2 \ln \frac{p_2}{p_1}$$

$$= 69.13 \times 0.056 \ln \frac{69.13}{700} = -\ 8.96 \text{ kJ/kg}$$

= i.e. it indicates that work in done on system

(c) In process 2-3, for constant temperature pressure

$$\frac{V_3}{V_2} = \frac{P_2}{P_3} = \frac{69.13}{700}$$

$$\Rightarrow \qquad V_3 = 0.00553 \text{ m}^3$$

Now in process 1-3,

$$d\,w_{1-3} = p_1\,(V_1 - V_3) = 700\,(0.014 - 0.00553)$$
$$= 5.93 \text{ kJ/kg}$$

$\therefore$ net work of cycle

$$= d\,w_{1-2} + d\,w_{2-3} + d\,w_{3-1}$$
$$= 8.85 - 8.96 + 5.93 = 5.82 \text{ kJ/kg}$$

Problem 10.8 : 0.5 kg helium and 0.5 kg of nitrogen are mixed at 20°C and at a total pressure of 100 kPa. Find (a) the volume of the mixture, (b) the partial volumes of the components, (c) the partial pressures of the components, (d) the mole fractions of the components, (e) the specific heats C_p and C_v of the mixture, and (f) the gas constant of the mixture.

Solution : Given data :

mass of helium = m_1 = 0.5 kg

mass of nitrogen

= m_2 = 0.5 kg

temperature of mixing

= T = 20°C = 20 + 273 = 293 K

Total pressure = P = 100 kPa

Let p_1, v_1 and p_2, v_2 are partial pressure and partial volume of helium and nitrogen

M_1 = molecular wt of helium = 4

M_2 = molecular wt of nitrogen = 28

$$\therefore \quad n_1 = \text{no. of moles of helium} = \frac{m_1}{M_1} = \frac{0.5}{4} = 0.125$$

$$n_2 = \text{no. of moles of nitrogen} = \frac{m_2}{M_2} = \frac{0.5}{28} = 0.0179$$

(i) $$v_1 = \frac{n_1 \bar{R} T}{p} = \frac{0.25 \times 8.314 \times 293}{100} = 3.045 \text{ m}^3$$

$$\therefore \quad v_2 = \frac{n_2 \bar{R} T}{p} = \frac{0.0179 \times 8.314 \times 293}{100} = 0.436 \text{ m}^3$$

∴ Volume of the mixture

$= v_1 + v_2 = 3.481 \text{ m}^3$

(ii) $v_1 = 3.095 \text{ m}^3$

$v_2 = 0.436 \text{ m}^3$

(iii) $$p_1 = \frac{n_1}{n_1 + n_2}$$

$$p = \frac{0.125}{0.125 + 0.175} \times 100 = 87.47 \text{ kPa}$$

$$p_2 = 100 - 87.47 = 12.53 \text{ kPa}$$

(iv) $$x_1 = \frac{n_1}{n_1 + n_2} = \text{mole fraction of helium} = 0.875$$

$$x_2 = \frac{n_2}{n_1 + n_2} = \text{mole fraction of nitrogen}$$

$= 0.125$

(v) C_p of mixture $= \dfrac{m_1 C_{p_1} + m_2 C_{p_2}}{m_1 + m_2}$

For helium, $R_1 = \dfrac{\overline{R}}{M_1} = \dfrac{8.314}{4} = 2.0785$ kJ/kg.K

and $r = 1.67$ for the for monoatomic

$$\therefore \quad C_{p_1} = \frac{\gamma \cdot R}{\gamma - 1}$$

$$= \frac{2.0785 \times 1.67}{0.67} = 5.18 \text{ kJ/kg.K}$$

$$C_{v_1} = \frac{R}{\gamma - 1} = \frac{2.0785}{.67} = 3.10 \text{ kJ/kg.K}$$

For nitrogen

$$R = \frac{8.314}{28} = 0.297 \text{ kJ/kg.K}$$

and $\gamma = 1.4$ for diatomic

$$\therefore \quad C_{p_2} = \frac{R\gamma}{\gamma - 1} = 1.039 \text{ kJ/kg.K}$$

$$C_{v_2} = \frac{R}{\gamma - 1} = 0.742 \text{ kJ/kg.K}$$

$$\therefore \quad C_{p_m} = \frac{m_1 C_{p_1} + m_2 C_{p_2}}{m_1 + m_2} = 3.1095 \text{ kJ/kg.K}$$

$$C_{v_m} = \frac{m_1 C_{v_1} + m_2 C_{v_2}}{m_1 + m_2} = 1.921 \text{ kJ/kg.K}$$

(iv) $$R_m = \frac{m_1 R_1 + m_2 R_2}{m_1 + m_2} = 0.1485 \text{ kJ/kg.K}$$

Problem 10.9 : A Carnot cycle uses 1 kg of air as the working fluid. The maximum and minimum temperatures of the cycle are 600 K and 300 K. The maximum pressure of the cycle is 1 MPa and the volume of the gas doubles during the isothermal heating process. Show by calculation of net work and heat supplied that the effciency is the maximum possible for the given maximum and minimum temperatures.

Solution : Given data :

m = mass of air = 1 kg

T_1 = maximum temperature = 600 K

T_2 = minimum temperature = 300 K

P_4 = maximum pressure = 1 MPa

$V_1 = 2\ V_4$

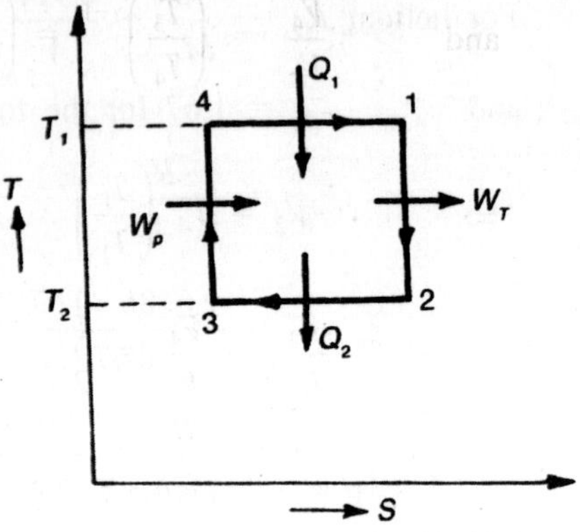

Fig. P. 10.9 (a)

For process 4-1, constant temp. reversible heat additon process.

$$đQ_{4-1} = đw_{4-1}$$

$$= p_1v_1 \ln \frac{V_1}{V_4} = p_4v_4 \ln \frac{V_1}{V_4}$$

$$= mRT_1 \ln \frac{V_1}{V_4}$$

$$= 1 \times 0.287 \times 600\ \text{m}^2$$

$$= 119.36\ \text{kJ}$$

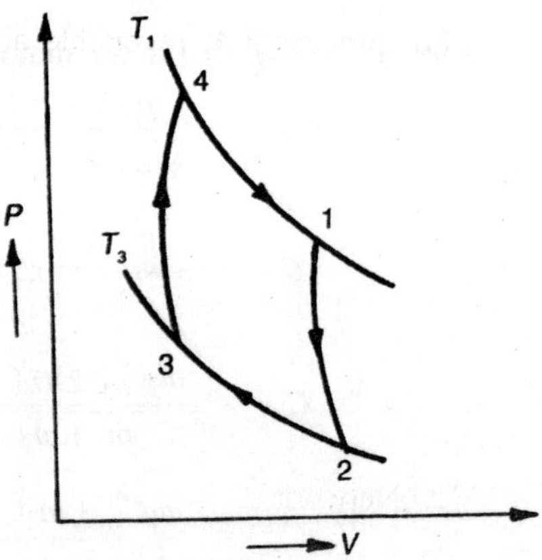

Fig. P. 10.9 (b)

For process 1-2, reversible adiabatic process

$$dw_{1-2} = \frac{p_1v_1 - p_2v_2}{\gamma - 1}$$

$$= \frac{mR(T_1 - T_2)}{\gamma - 1}$$

$$= \frac{1 \times 0.287(600 - 300)}{0.4} = 215.25\ \text{kJ}$$

For process 3-2, reversible adiabatic process

$$đQ_{3-2} = đw_{3-2}$$

$$= p_2v_2 \ln \frac{V_3}{V_2} = mRT_2 \ln \frac{V_3}{V_2}$$

Now $\frac{V_2}{V_1} = \left(\frac{T_1}{T_2}\right)^{\frac{1}{\gamma-1}}$

and $\frac{V_4}{V_3} = \left(\frac{T_3}{T_4}\right)^{\frac{1}{\gamma-1}} = \left(\frac{T_2}{T_1}\right)^{\frac{1}{\gamma-1}}$

$$\Rightarrow \quad V_2 = V_1\left(\frac{T_1}{T_2}\right)^{\frac{1}{\gamma-1}} \;\&\; V_3 = V_4\left(\frac{T_1}{T_2}\right)^{\frac{1}{\gamma-1}}$$

$$\therefore \quad \frac{V_3}{V_2} = \frac{V_4}{V_1} = \frac{1}{2}$$

$$\therefore \quad đw_{3\text{-}2} = dQ_{3\text{-}2} = mRT_2 \ln \frac{V_3}{V_2}$$

$$= 1 \times 0.287 \times 300 \ln \frac{1}{2} = 59.68 \text{ kW}$$

For process 3-4, reversible adiabatic process

$$đw_{3-4} = \frac{p_3V_3 - p_4V_4}{\gamma - 1}$$

$$= mR\frac{(T_3 - T_4)}{\gamma - 1} = \frac{mR(T_2 - T_1)}{\gamma - 1}$$

$$= \frac{1 \times 0.287\,(300 - 600)}{0.4} = -215.25 \text{ kJ}$$

$\therefore$ Network

$$= dw_{1-2} + dw_{3-4} + dw_{2-3} + dw_{4-1}$$
$$= 215.25 - 215.25 + 119.36 - 59.68$$
$$= 59.68 \text{ kJ}$$

that supplied $= đQ_{4-1} = 119.36$ kJ

$\therefore \quad \eta_{th}$ = efficiency (thermal)

$$= \frac{\text{network}}{\text{heat supplied}} = \frac{59.68}{119.36} = 0.5 \textbf{ Ans}$$

Problem 10.10 : An ideal gas cycle consists of three reversible processes in the following sequence: (a) constant volume pressure rise, (b) isentropic expansion to r times the initial volume, and (c) constant pressure decreased in volume. Sketch the cycle on the p-v and T-s diagrams. Show that the efficiency of the cycle is

$$\eta_{cycle} = \frac{r^{\gamma} - 1 - \gamma(r-1)}{r^{\gamma} - 1}$$

Evaluate the cycle efficiency when $\gamma = \frac{4}{3}$ and $r = 8$.

Solution : Given $= \frac{V_3}{V_1} = r$

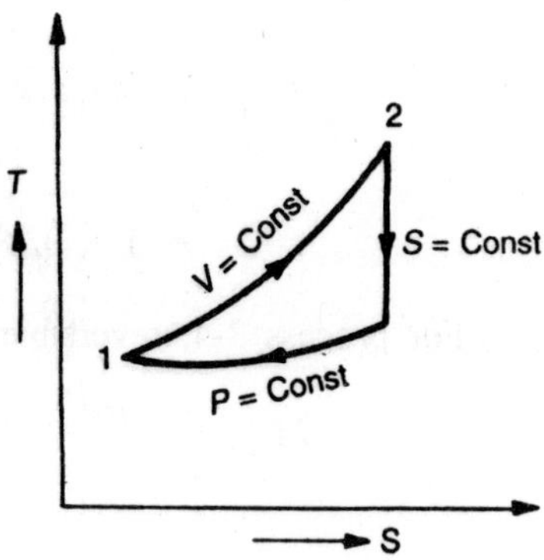

Fig. P. 10.10 (a)

Expansion follow

$= Pv^{\gamma}$

$=$ constant

Process 1-2,

Constant volume process,

$$\frac{P_2}{P_1} = \frac{T_2}{T_1} = dw_{1-2} = 0$$

and $dQ_{1-2} = du + dw_{12}$

$= u_2 - u_1$

$= 1 \times C_v (T_2 - T_1)$ kJ/kg

Process 2-3, insentropic process.

$$dw_{2-3} = \frac{p_2V_2 - p_3V_3}{\gamma - 1}$$

$$= \frac{mR(T_2 - T_3)}{\gamma - 1} = \frac{R(T_2 - T_3)}{\gamma - 1} \text{ kJ/kg}$$

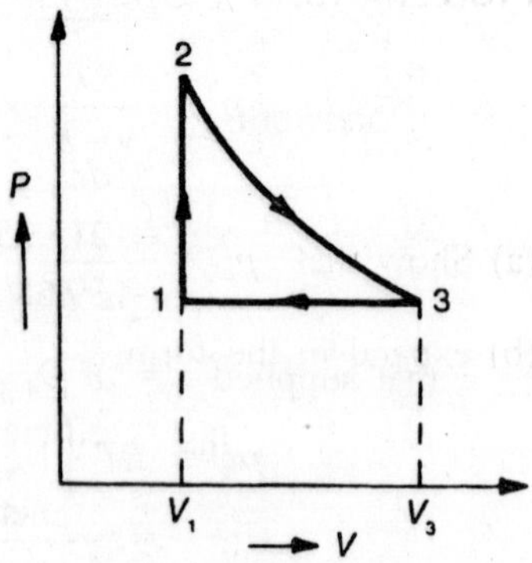

Fig. P. 10.10 (b)

and $\frac{P_2}{P_3} = \left(\frac{V_3}{V_2}\right)^{\gamma} = \left(\frac{V_3}{V_1}\right)^{\gamma}$

$= r^{\gamma} = \frac{P_2}{P_1}$

Process 3-1, constant pressure process

$$dQ_{3-1} = m\ C_p\ (T_1 - T_3)$$
$$= -C_p\ (T_3 - T_1)\ \text{kJ/kg}$$

$$\Rightarrow \quad \frac{T_3}{T_1} = \frac{V_3}{V_1} = r$$

$\therefore$ Net work $=$ area enclosed by the curves

$$= C_v(T_2 - T_1) - C_p(T_3 - T_1)$$

heat suppled, $= dQ_{1-2} = C_v(T_2 - T_1)$

$$\therefore \quad \eta_{th} = \frac{\text{Net work}}{\text{heat supplied}}$$

$$= \frac{C_v(T_2 - T_1) - C_p(T_3 - T_1)}{C_v(T_2 - T_1)}$$

$$= 1 - \frac{\gamma(T_3 - T_1)}{(T_2 - T_1)} = 1 - \frac{\gamma\left(\frac{T_3}{T_1} - 1\right)}{\left(\frac{T_2}{T_1} - 1\right)}$$

$$= \frac{1 - \gamma(r - 1)}{r^\gamma - 1} = \frac{r^\gamma - 1 - \gamma(r - 1)}{r^\gamma - 1}$$

Problem 10.11 : Using the Dieterici equation of state

$$p = \frac{RT}{v - b} \times \exp\left(-\frac{a}{RT_v}\right)$$

(a) Show that $p_c = \dfrac{a}{4e^2b^2}$, $v_c = 2b$, $T_c = \dfrac{a}{4Rb}$

(b) expand in the form

$$pv = RT\left(1 + \frac{B}{v} + \frac{C}{v^2} + \ldots\right)$$

(c) show that $T_B = \dfrac{a}{bR}$

Solution : Given data :

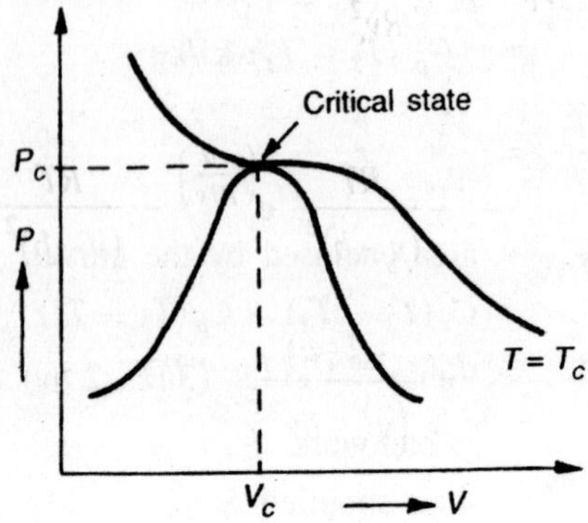

Fig. P. 10.11

$$p = \frac{RT}{v-b} \times e\left(\frac{-a}{RT_v}\right) \quad \text{...(i)}$$

At critical state,

$$\left(\frac{\partial p}{\partial v}\right)_{T=T_c} = 0 \text{ and } \left(\frac{\partial^2 p}{\partial v^2}\right)_{T=T_c} = 0$$

Diffentiation eq (i) w.r.to volume.

Now $\left(\frac{\partial p}{\partial v}\right)_{T=T_c}$

$$= -\frac{RT}{(v-b)^2} e^{-\frac{a}{RTv}} + \frac{RT}{v-b} \times \left(\frac{+a}{RTv^2}\right) e^{\frac{-a}{RTv}} = 0$$

$$\Rightarrow \quad e^{\frac{-a}{RTv}} \left(\frac{-RT}{(v-b)^2} + \frac{RT}{(v-b)} \times \frac{a}{RTv^2}\right) = 0$$

$$\Rightarrow \quad \frac{RT}{(v-b)^2} = \frac{RT}{(v-b)} \times \frac{a}{RTv^2}$$

$$\Rightarrow \quad \frac{1}{v-b} = \frac{a}{RTv^2} \text{ or } T = \frac{a(v-b)}{Rv^2}$$

For $\quad T = T_c, v = v_c$

$$\Rightarrow \qquad T_c = \frac{a(v_c - b)}{Rv_c^2} \qquad ...(a)$$

Again

$$\left(\frac{\partial^2 p}{\partial v^2}\right)_{T=T_c} = -\frac{RT}{(v-b)^2}\, e^{\left(\frac{-a}{RTv}\right)} - \frac{RT}{(v-b)^2}\left(\frac{a}{RTv^2}\right) \times e^{\left(\frac{-a}{RTv}\right)}$$

$$+ \frac{(-a)}{\left(v^3 - bv^2\right)^3}(3v^2 - 2\,bv)\, e^{\left(\frac{-a}{RTv}\right)} + \frac{a}{v^3 - bv^2}$$

$$\left(\frac{a}{RTv^2}\right) e^{\left(\frac{-a}{RTv}\right)} = 0$$

or

$$e^{\left(\frac{-a}{RTv}\right)}\left[\frac{2RT}{(v-b)^3} - \frac{a}{(v-b)^2 v^2} - \frac{a\left(3v^2 - 2bv\right)}{\left(v^3 - bv^2\right)^2}\right.$$

$$+ \frac{a^2}{\left(v^3 - bv^2\right)RTv^2} = 0$$

or, at $\quad T = T_c \; v = v_c$

$$\Rightarrow \qquad \frac{2RT_c}{(v_c - b)^3} - \frac{a}{(v_c - b)^2 v_c^2} - \frac{a\left(3v_c^2 - 2bv_c\right)}{\left(v_c^3 - bv_c^2\right)^2}$$

$$+ \frac{a^2}{\left(v_c^3 - bv_c^2\right)RT_c v_c^2} = 0$$

from (a)

$$\Rightarrow \qquad \frac{2Ra(v_c - b)}{Rv_c^2(v_c - b)^3} - \frac{a}{(v_{c-b})^2 v_c^2} - \frac{a(3v_c - 2b)v_c}{v_c^4(v_c - b)^2}$$

$$+ \frac{a^2 R V_c^2}{V_c^2(v_c - b)R\, v_c^2 \times a(v_{c-b})} = 0$$

$$\Rightarrow 2a - a - \frac{a(3v_c - 2b)}{v_c} + a = 0$$

$$\Rightarrow \quad 2v_c = 3v_c - 2b$$

$$\Rightarrow \quad v_c = 2b \qquad \text{...(b)}$$

∴ from (a)

$$T_c = \frac{a(v_c - b)}{R(v_c^2)} = \frac{a\,b}{R \times 4b^2} = \frac{a}{4bR} \qquad \text{...(c)}$$

Now

$$p_c = \frac{RT_c}{v_{c-b}}\, e^{\frac{-a}{RT_c v_c}}$$

$$= \frac{a}{v_c^2}\, e^{-\frac{aRv_c^2}{Ra(v_c - b) \times v_c}}$$

$$= \frac{a}{4b^2}\, e^{-\frac{2b}{b}} = \frac{a}{4e^2b^2} \qquad \text{...(c)}$$

Hence

$$P_c = \frac{a}{4e^2b^2},\; v_c = 2b,\; T_c = \frac{a}{4bR}$$

(b)

$$P = \frac{RT}{v - b}\, e^{-\left(\frac{a}{RT_v}\right)}$$

$$p = \left[1 - \frac{a}{RTv} + \frac{a^2}{2R^2T^2v^2} - \frac{a^3}{6R^3T^3v^3} + \ldots\ldots\right]\frac{RT}{v - b}$$

$$= RT\left[(v - b)^{-1} - \frac{a}{RTv}(v - b)^{-1} + \frac{a^2(v - b)^{-1}}{2R^2T^2v^2} - \ldots\ldots\right]$$

$$= RT\left[v^{-1}\left(1 - \frac{b}{v}\right)^{-1} - \frac{a}{RTv}v^{-1}\left(1 - \frac{b}{v}\right)^{-1} + \ldots\ldots\right]$$

$$= RT$$

$$\left[v^{-1}\left(1 + \frac{b}{v} + \frac{b^2}{v^2} + \frac{b^3}{v^3} + \ldots\ldots\right) - \frac{a}{RTv^2}\left(1 + \frac{b}{v} + \frac{b^2}{v^2} + \frac{b^3}{v^3} + \ldots\ldots\right) + \ldots\ldots\right]$$

$$\Rightarrow p = \frac{RT}{v}\left[\left(1+\frac{b}{v}+\frac{b^2}{v^2}+\frac{b^3}{v^3}\right)-\frac{a}{Rv}\left(1+\frac{b}{v}+\frac{b^2}{v^2}+\frac{b^3}{v^3}+\ldots\right)+\ldots\right]$$

$$\Rightarrow pv = RT\left[1+\frac{1}{v}\left(b-\frac{a}{RT}\right)+\ldots\right]$$

$$\Rightarrow pv = RT\left[1+\frac{B}{v}+\frac{c}{v^2}+\ldots\right] \quad \textbf{Ans} \qquad \ldots\text{(ii)}$$

When $B = b - \dfrac{a}{RT}$ etc....

(iii) we know from above eq. (ii)
2nd virial coeficient

$$= B = b - \frac{a}{RT}$$

Eq (ii) is

$$Pv = RT\left[1+\frac{Bp}{\overline{R}T}+\frac{cp^2}{\overline{R}^2T^2}+\ldots\right]$$

Toditmine boyle temperature T_B

$$\Rightarrow \left[\frac{d(pv)}{dp}\right]_{\text{at } p=0} = 0 = RT\left[0+\frac{B}{\overline{R}T}+0\right]$$

$$\Rightarrow \qquad \frac{B}{RT} = 0 \Rightarrow B = 0$$

$$\therefore \quad b - \frac{a}{RT_B} = 0 \Rightarrow T_B = \frac{a}{bR} \quad \textbf{Ans}$$

Problem 10.12 : An ideal gas at temperature T_1 is heated at constant pressure to T_2 and then expanded reversibly, according to the law pv^n = constant, until temperature is once again T_1. What is the required value of n, if the changes of entropy during the separate processes are equal ?

Solution : Given :

$$T_3 = T_1$$

and $\Delta S_{2-1} = \Delta S_{3-2}$

Now from property relation

$$ds = \frac{dh}{T} - R\frac{dP}{P}$$

$$= C_p \frac{dT}{T} - R\frac{dP}{P}$$

$$\Delta S = C_p \ln\frac{T_2}{T_1} - R\ln\frac{P_2}{P_1}$$

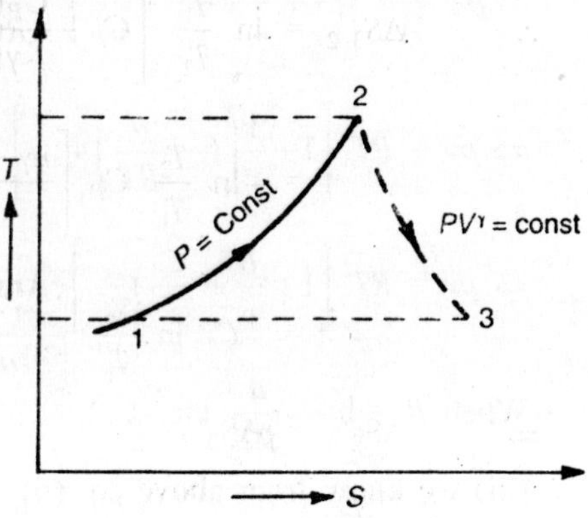

Fig. P. 10.12 (a)

For process 1-2,

$$P_2 = P_1$$

$$\therefore \quad \Delta S_{3-2} = C_p \ln\frac{T_2}{T_1}$$

For process 2-3,

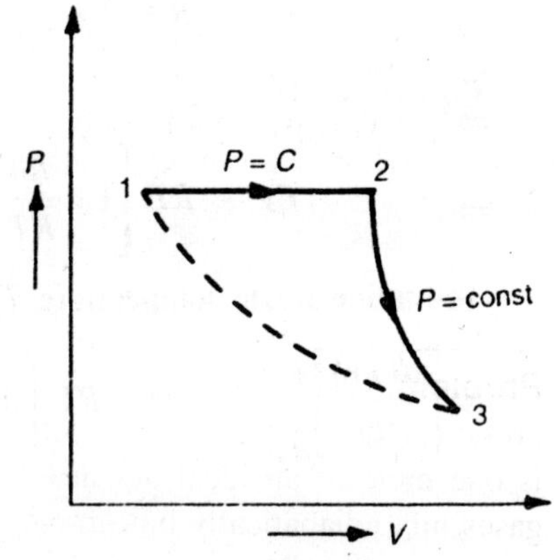

Fig. P. 10.12 (b)

$$\Delta S_{3-2} = C_p \ln\frac{T_3}{T_2} - R\ln\frac{P_3}{P_2}$$

$$= C_p \ln\frac{T_1}{T_2} - R\ln\frac{P_3}{P_2}$$

Now $\dfrac{P_3}{P_2} = \left(\dfrac{T_3}{T_2}\right)^{\frac{n}{n-1}} = \left(\dfrac{T_1}{T_2}\right)^{\frac{n}{n-1}}$

$$\therefore \quad \Delta S_{3-2} = C_P \ln\frac{T_1}{T_2} - R\frac{n}{n-1}\ln\frac{T_1}{T_2}$$

$$= \ln\frac{T_1}{T_2}\left[C_P - \frac{Rn}{n-1}\right]$$

Now $\quad C_p = \dfrac{R\gamma}{\gamma - 1} \Rightarrow$

$$R = \frac{C_p(\gamma - 1)}{\gamma}$$

$$\therefore \quad \Delta S_{3-2} = \ln \frac{T_1}{T_2}\left[C_P - \frac{C_P(\gamma-1)\,n}{\gamma(n-1)}\right]$$

$$= -\ln \frac{T_2}{T_1}\, C_P \left[\frac{n\gamma-\gamma-n\gamma+n}{\gamma(n-1)}\right]$$

$$= -C_P \ln \frac{T_2}{T_1}\left[\frac{n-\gamma}{\gamma(n-1)}\right]$$

$$\Rightarrow \quad \Delta S_{2-1} = \Delta S_{3-2}$$

$$\Rightarrow \quad C_p \ln \frac{T_2}{T_1} = -C_P \ln \frac{T_2}{T_1}\left[\frac{n-\gamma}{\gamma(n-1)}\right]$$

$$\Rightarrow \quad r\,(n-1) = -[n-\gamma]$$

$$\Rightarrow \quad n\gamma - \gamma = -n + \gamma$$

$$\Rightarrow \quad n\,(\gamma+1) = 2\gamma$$

$$\Rightarrow \quad n = \frac{2\gamma}{\gamma+1}$$

Problem 10.13 : Two moles of an ideal gas at temperature T and pressure P are contained in a compartment. In an adjacent compartment is one mole of an ideal gas at temperature $2T$ and pressure P. The gases mix adiabatically but do not react chemically when a partition separating the compartments is withdrawn. Show that the entropy increase due to the mixing process is given by

$$R\left(\ln\frac{27}{4} + \frac{\gamma}{\gamma-1}\ln\frac{32}{27}\right)$$

Provided that the gases are different and that the ratio of specific heat γ is the same for both gases and remains constant.
What would the entropy change be if the mixing gases were of the same species?

Solution : Given data :

Compartmint A

$$n_1 = \text{no of mole} = 2$$

T_1 = T = Temperature

P_1 = P = Pressure

Compoutmit B

$n_2 = 1$

P_2 = Pressure = P

T_2 = Temperature = 2 T

For mixing adiabatically

T_f = final temperature

$$= \frac{n_1 T_1 + n_2 T_2}{n_1 + n_2}$$

$$= \frac{2T + 2T}{3} = \frac{4T}{3}$$

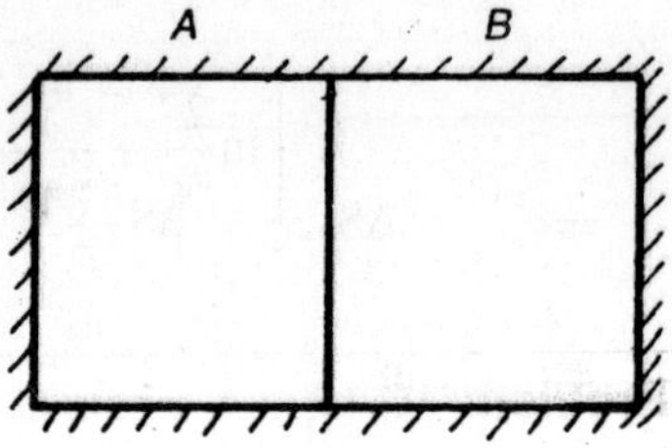

Fig. P. 10.13

$$V_f = V_1 + V_2$$

$$= \frac{n_1 \bar{R} T}{P} + \frac{n_2 \bar{R} T \times 2}{p} = \frac{4 \bar{R} T}{P}$$

$\therefore$ P_f, final pressure

$$= \frac{(n_1 + n_2)\bar{R} T_f}{V_f}$$

$$= \frac{3 \bar{R} T_f}{4 \bar{R} T / P} = \frac{3P \times 4T}{4T \quad 3} = P$$

$\therefore$ P_{f1}, for Ist ideal gas

$$= x_1 P = \frac{2}{3} P$$

P_{f2} for 2nd ideal gas

$$= x_2 P = \frac{1}{3} P$$

$\therefore$ from property relation

$$\Delta S = C_P \ln \frac{T_2}{T_1} - \text{R} \ln \frac{P_2}{P_1}$$

$\therefore$ change in entropy due to mixing

$$= \left(2\, C_P \ln \frac{4T}{3T} - 2R \ln \frac{2P}{3P}\right) + \left(C_P \ln \frac{4T}{3 \times 2T} - R \ln \frac{P}{3P}\right)$$

$$= C_p \ln \frac{16}{9} - R \ln \frac{4}{9} + C_P \ln \frac{2}{3} - R \ln \frac{1}{3}$$

$$= C_p \ln \frac{32}{27} - R \ln \frac{4}{27}$$

$$= \frac{\gamma R}{\gamma - 1} \ln \frac{32}{27} + R \ln \frac{27}{4}$$

$$= \text{R} \left[\ln \frac{27}{4} + \frac{\gamma}{\gamma - 1} \ln \frac{32}{27} \right]$$

Problem 10.14 : A single-acting air compressor has a cylinder of bore 15 cm and the piston stroke is 25 cm. The crank speed is 600 rpm. Air taken from the atmosphere (1 atm, 27°C) is delivered at 11 bar. Assuming polytropic compression $pv^{1.25}$ = const., find the power required to drive the compressor, when its mechanical efficiency is 80% . The compressor has a clearance volume which is 1/20th of the stroke volume. How long will it take to deliver 1 m³ of air at the compressor outlet conditions ? Find the volumetric efficiency of the compressor.

Solution : Given data :

d = bone of cylinder
= 15 cm = 0.15 m

L = Piston stroke
= 25 cm. = 0.25 m

N = 600 rpm

P_1 = Suction pressure
= 1 atm = 1.013 bar

T_1 = Suction temperature
= 27°C = 300 K

P_2 = Dilivery presure
= 11 bar

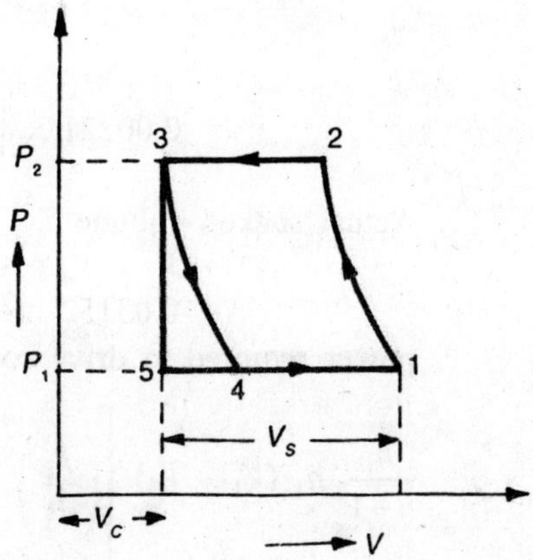

Fig. P. 10.14

Polytropic comprssion law $pv^{1.25}$ = conot

n_{mech} = mechanical efficincy = 0.8

V_c = Clearance volume = $\frac{V_s}{20}$ (stroke volume)

From polytropic law

$$T_2 = \text{temperature after compression}$$

$$= \left(\frac{p_2}{p_1}\right)^{\frac{n-1}{n}} = \left(\frac{11}{1.013}\right)^{\frac{0.25}{1.25}} = 483.37 \text{ K}$$

$$V_s = \frac{\pi}{4} d^2 \text{ Ln} = \frac{\pi}{4} \times (0.15)^2 \times 25 \times \frac{600}{60}$$

$$= 0.0442 \text{ m}^3\text{/sec}$$

$$V_c = \frac{V_s}{20} = 0.00221 \text{ m}^3\text{/sec}$$

$$\therefore \quad V_1 = V_s + V_c = 0.0442 + 0.00221$$

$$= 0.04641 \text{ m}^3\text{/sec}$$

Now for expansion process, Assuning following the same law $pv^{1.25} = C$

$$\frac{V_4}{V_3} = \left(\frac{p_3}{p_4}\right)^{\frac{1}{n}}$$

or

$$V_4 = V_c \times \left(\frac{p_2}{p_1}\right)^{\frac{1}{n}}$$

$$= 0.00221 \times \left(\frac{11}{1.013}\right)^{\frac{1}{1.25}} = 0.01489 \text{ m}^3\text{/sec}$$

$\therefore$ Actual sucked volume

$$= V_1 - V_4 = 0.04641 - 0.01489$$

$$= 0.03152 \text{ m}^3\text{/sec}$$

$\therefore$ power required to drive compressor

$$= \frac{n}{n-1} p_1 (V_1 - V_4) \left[\left(\frac{P_2}{P_1}\right)^{\frac{n-1}{n}} - 1\right] \times \frac{1}{n_{mech}}$$

$$= \frac{1.25}{0.25} \times 1.013 \times 100 \times 0.03152 \left[\left(\frac{11}{1.013}\right)^{\frac{0.25}{1.25}} - 1\right] \times \frac{1}{0.8} = 12.2 \text{ kW}$$

$$n_v = \text{Volumetric efficincy}$$

$$= 1 + k - k\left(\frac{P_2}{P_1}\right)^{\frac{1}{n}}, \text{ where } k = \frac{V_c}{V_S}$$

$$= 1 + \frac{1}{20}\left[1 - \left(\frac{11}{1.013}\right)^{\frac{1}{1.25}}\right] = 71.3\ \%$$

Now from polytropic law

$$V_2 = (V_1 - V_4) \times \left(\frac{P_1}{P_2}\right)^{\frac{1}{n}} = 0.03152\left(\frac{1.013}{11}\right)^{\frac{1}{1.25}} \text{ m}^3\text{/sec}$$

$$\therefore \text{ time taken to deliver 1 m}^3 \text{ of air} = \frac{1\text{m}^3}{v_2 \dfrac{\text{m}^3}{\text{sec}}}$$

$= 213.816$ sec $= 3.56$ min **Ans**

Problem 10.15 : A multistage air compressor compresses air from 1 bar to 40 bar. The maximum temperature in any stage is not to exceed 400 K. (a) If the law of compression for all the stages is $pv^{1.3}$ = const., find the number of stages for the minimum power input (b) Find the intermediate pressures for optimum compression was well as the power needed, (c) What is the heat transfer in each of the intercooler?

Solution : Given data :

P_n = 40 bar = delivery pressure
P_1 = 1 bar = inital pressure.
maximum temperature in any stage = 400 K
Law of compression $pv^{1.3}$ = const.
Let initial temperature = T_1 = 300 K.
(a) For minimum work output,

$$\frac{P_2}{P_1} = \frac{P_3}{P_2} = \frac{P_4}{P_3} = \ldots\ldots = \frac{P_n}{P_{n-1}}$$

or $$\left(\frac{P_2}{P_1}\right) = \left(\frac{P_3}{P_2}\right) = \ldots\ldots = \left(\frac{P_n}{P_1}\right)^{\frac{1}{n}} \quad \ldots(1)$$

where n = no. of stages

Now from polythropic law

$$\left(\frac{P_2}{P_1}\right) = \left(\frac{T_2}{T_1}\right)^{\frac{n}{n-1}}$$

$$= \left(\frac{400}{300}\right)^{\frac{1.3}{0.3}}$$

$$= 3.48$$

∴ from (1)

$$\left(\frac{P_2}{P_1}\right) = \left(\frac{P_n}{P_1}\right)^{\frac{1}{n}}$$

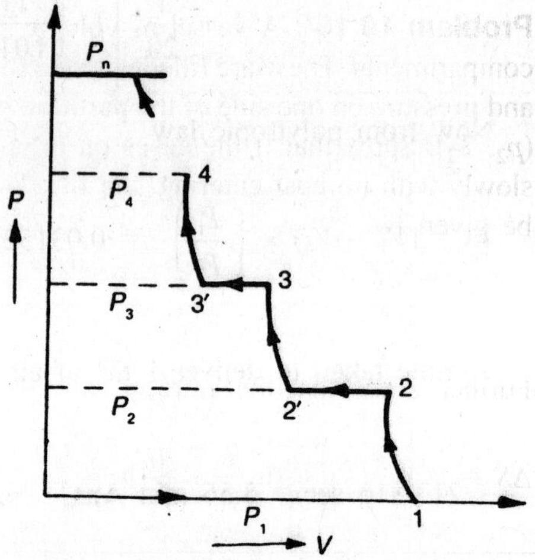

Fig. P. 10.15

or

taking log both the side

$$\Rightarrow \ln\left(\frac{P_2}{P_1}\right) = \frac{1}{n}\ln\left(\frac{P_n}{P_1}\right)$$

or $$n = \frac{\ln(P_n/P_1)}{\ln(P_2/P_1)} = \frac{\ln 40}{\ln 3.480} = 2.96 \approx 3$$

(b) $$\frac{P_2}{P_1} = \frac{P_3}{P_2} = 3.48$$

∴ $$P_2 = 3.48 \times P_1 = 3.48 \text{ bar}$$

$$P_3 = 3.48 \times P_2 = 12.1 \text{ bar}$$

∴ power needed to drike the compressure

$$= \frac{3n}{n-1}\, mRT_1 \left\{\left(\frac{P_2}{P_1}\right)^{\frac{n-1}{n}} - 1\right\}$$

$$= \frac{3 \times 1.3}{1.3-1} \times 1 \times 0.28 \times 300 \left\{(3.48)^{\frac{0.3}{1.3}} - 1\right\}$$

$= 373.24$ kJ/kg

(c) heat transfer in each inter cooler $= mC_p\,(\Delta T)$

$= 1 \times 1.005\,(400 - 300)$

$= 100.5$ kJ/kg

Problem 10.16 : A vessel of volume 2V is divided into two equal compartments. These are filled with the same ideal gas, the temperature and pressure on one side of the partition being (p_1, T_1) and on the other (p_2, T_2). Show that if the gases on the two sides are allowed to mix slowly with no heat entering, the final pressure and temperature will be given by

$$P = \frac{p_1 + p_2}{2}, \quad T = \frac{T_1 T_2 (p_1 + p_2)}{p_1 T_2 + p_2 T_1}$$

Further, show that the entropy gain is

$$\Delta S = V\left[\left(\frac{c_p}{R}\right)\left\{\frac{p_1}{T_1}\ln\frac{T}{T_1} + \frac{p_2}{T_2}\ln\frac{T}{T_2}\right\} - \frac{p_1}{T_1}\ln\frac{p}{p_1} - \frac{p_2}{T_2}\ln\frac{p}{p_2}\right]$$

Solution : Given data :

Before mixing

m_1 = mass of gas in compartment (i)

$$= \frac{p_1 V}{RT_1}$$

m_2 = mass of gas in compartment (ii)

$$= \frac{p_2 V}{RT_2}$$

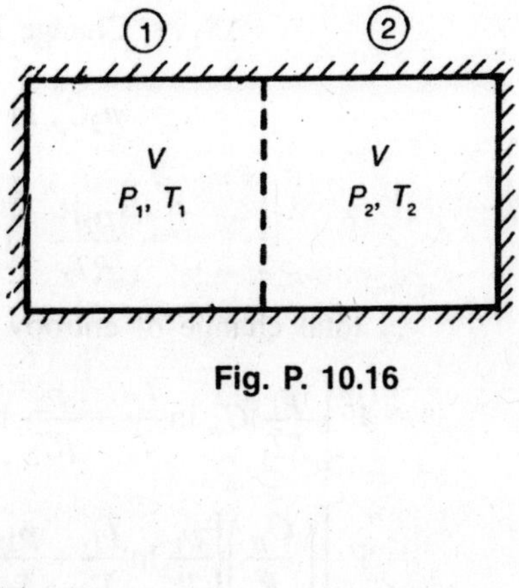

Fig. P. 10.16

During mixing no temperature and work done taken place.

$\therefore \quad U = \text{constant}$

$\Rightarrow \quad U_{\text{before mixing}} = U_{\text{after mixing}}$

$\Rightarrow \quad m_1 C_v T_1 + m_2 C_v T_2$

$= (m_1 + m_2)\, C_v T_f,$

where $\quad T_f$ = final temperature

$$\therefore \quad T_f = \frac{m_1T_1 + m_2T_2}{m_1 + m_2} = \frac{\frac{p_1V}{RT_1} \times T_1 + \frac{p_2V}{RT_2} T_2}{\frac{p_1V}{RT_1} + \frac{p_2V}{RT_2}}$$

$$T_f = \frac{p_1 + p_2}{\frac{p_1}{T_1} + \frac{p_2}{T_2}} = \frac{(p_1 + p_2)\, T_1\, T_2}{p_1T_1 + p_2T_2} \quad \textbf{Ans}$$

Now from ideal gas law : – $P_f\, V_f = mRT_f$

$$\Rightarrow \quad p_f = \frac{(m_1 + m_2)\, R\, T_f}{2V}$$

$$= \frac{(m_1 + m_2)}{2V} \times R \times \frac{(m_1\, T_1 + m_2\, T_2)}{m_1 + m_2}$$

$$= \frac{m_1T_1 + m_2T_2}{2V} \times R = \frac{p_1 + p_2}{2}$$

(c) Δ S_1 = change in intropy of molucules of compartment 1

$$= m_1C_p \ln \frac{T_f}{T_1} - m_1R \ln \frac{p_f}{p_1}$$

$$= \frac{p_1V}{RT_1} \left[C_p \ln \frac{T_f}{T_2} - R \ln \frac{p_f}{p_2} \right]$$

ΔS_2 = Change in intropy of compartment 2

$$= m_2C_p \ln \frac{T_f}{T_2} - m_2R \ln \frac{P_f}{P_2}$$

$$= \frac{p_2V}{RT_2} \left[C_p \ln \frac{T_f}{T_2} - R \ln \frac{P_f}{P_2} \right]$$

$\therefore$ total change of entropy = $\Delta S_2 + \Delta S_2$

$$= V \left[\frac{p_1}{RT_1} C_p \ln \frac{T_f}{T_1} + \frac{p_2}{RT_2} \ln \frac{T_f}{T_2} - \frac{p_1}{T_1} \ln \frac{p_f}{p_1} - \frac{p_2}{T_2} \ln \frac{p_f}{p_2} \right]$$

$$= V \left[\left(\frac{C_p}{R} \right) \left\{ \frac{p_1}{T_1} \ln \frac{T_f}{T_1} + \frac{p_2}{T_2} \ln \frac{T_f}{T_2} \right\} - \frac{p_1}{T_1} \ln \frac{p_f}{p_1} - \frac{p_2}{T_2} \ln \frac{p_f}{p_2} \right]$$

> **Problem 10.17 :** In a heat engine cycle, air isothermally compressed. Heat is then added at constant pressure, after which the expands isentropically to its original state. Draw the cycle on p–v and T–s coordinates. Show that the cycle efficiency can be expressed in the following form
>
> $$n = 1 - \frac{(\gamma - 1)\ln r}{\gamma\left[r^{\gamma-1/\gamma} - 1\right]}$$
>
> where r is the pressure ratio, p_2/p_1.Determine the pressure ratio and the cycle efficiency if the initial temperature is 27°C and the maximum temperature is 327°C .

Solution : Given data :

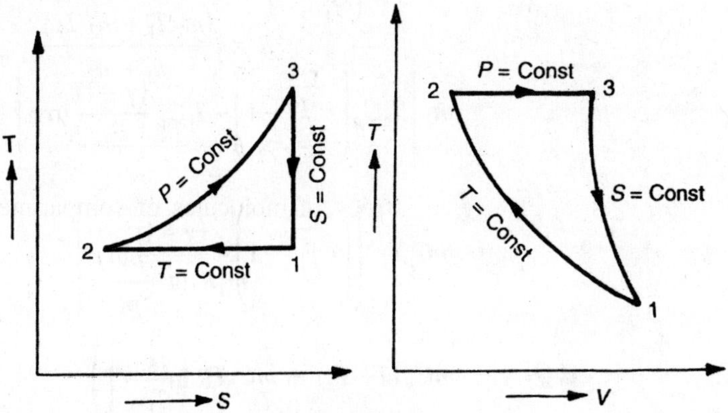

Fig. P. 10.17

$$r = \frac{p_2}{p_1} = \text{pressure ratio}$$

Assume mass of air = m kg

In process 2-3

Constant pressure process

heat added = $m \times C_P\,(T_3 - T_2) = C_P\,(T_3 - T_2)$

$$\frac{T_3}{T_2} = \frac{V_3}{V_2}$$

In process 3-1, isentropic process

$$đQ_{3-1} = 0,\; đw_{3-1} = mC_P\,(T_3 - T_1) = mC_PT_1\left[\frac{T_3}{T_1} - 1\right]$$

$$= m\ C_P\ T_1 \left[r^{\frac{\gamma-1}{\gamma}} - 1 \right] \left[\because \frac{T_3}{T_1} = \left(\frac{P_3}{P_1}\right)^{\frac{\gamma-1}{\gamma}} = \left(\frac{p_2}{p_1}\right)^{\frac{\gamma-1}{\gamma}} = r^{\frac{\gamma-1}{\gamma}} \right]$$

In process 1-2,

$$đw_{2-1} = p_1 V_1 \ln \frac{V_2}{V_1} = mRT_1 \ln \frac{p_1}{p_2}$$

$$= -mRT_1 \ln \frac{p_2}{p_1} = -m\ RT_1 \ln r$$

$$\therefore \quad \text{net work} = đw_{3-1} + đw_{2-1}$$

$$= T_1 m C_p \left(\frac{T_3}{T_1} - 1\right) - mRT_1 \ln r$$

$$= m \left[T_1 C_p \left(r^{\frac{\gamma-1}{\gamma}} - 1 \right) - T_1 C_p \frac{(\gamma-1)}{\gamma} \ln r \right]$$

$$= mC_p T_1 \left[\left(r^{\frac{\gamma-1}{\gamma}} - 1 \right) - \frac{(\gamma-1)}{\gamma} \ln r \right]$$

$$đQ_{2-3} = mC_p(T_3 - T_2) = mC_p T_1 \left(\frac{T_3}{T_1} - 1\right)$$

$$= mC_p\ T_1 \left(r^{\frac{\gamma-1}{\gamma}} - 1 \right)$$

$$\therefore \quad n_{th} = \frac{\text{net work}}{\text{heat supplied}}$$

$$= \frac{mC_p T_1 \left[\left(r^{\frac{\gamma-1}{\gamma}} - 1 \right) - \frac{(\gamma-1)}{\gamma} \ln r \right]}{m\,C_p T_1 \left(r^{\frac{\gamma-1}{\gamma}} - 1 \right)}$$

$$= 1 - \frac{(\gamma - 1)\ln r}{\gamma\left(r^{\frac{\gamma-1}{\gamma}} - 1\right)}$$

2nd part
given

$$T_1 = 27°C = 27 + 273 = 300 \text{ K}$$
$$T_3 = 327°C = 327 + 273 = 600 \text{ K}$$

In process 3-1

$$\frac{p_3}{p_1} = \left(\frac{T_3}{T_1}\right)^{\frac{\gamma}{\gamma-1}}$$

$$\Rightarrow \quad \frac{p_2}{p_1} = \left(\frac{600}{300}\right)^{\frac{1.4}{0.4}} = 11.3$$

$$n = 1 - \frac{0.4 \times \ln 11.31}{1.4\left[11.3^{\frac{0.4}{1.4}} - 1\right]} = 30.67$$

Problem 10.18 : An ideal gas cycle consists of the following reversible process: (i) isentropic compression, (ii) constant volume heat addition, (iii) isentropic expansion, and (iv) consant pressure heat rejection. Show that the efficiency of this cycle is given by

$$n = 1 - \frac{1}{r_k^{\gamma-1}}\left[\frac{\gamma\left(a^{1/\gamma} - 1\right)}{a - 1}\right]$$

where r_k is the compression ratio and a is the ratio of pressures after and before heat addition.

An engine operating on the above cycle with a compression ratio of 6 starts the compression with air at 1 bar, 300 K. If the ratio of pressures after and before heat addition is 2.5, calculate the efficiency and the m.e.p of the cycle. Take g = 1.4 and C_v = 0.718 kJ/kg.K

Solution : Given :

In process 2-3.

constant volume heat addition process,

$$\frac{T_3}{T_2} = \frac{P_3}{P_2} = a$$

from

Ist law of thermodymaics

$$đQ_1 = du + dw$$
$$dw = du + pdv = du$$
$$= mC_v (T_3 - T_2)$$

In process 4-1.

Constant pressure heat rejection process,

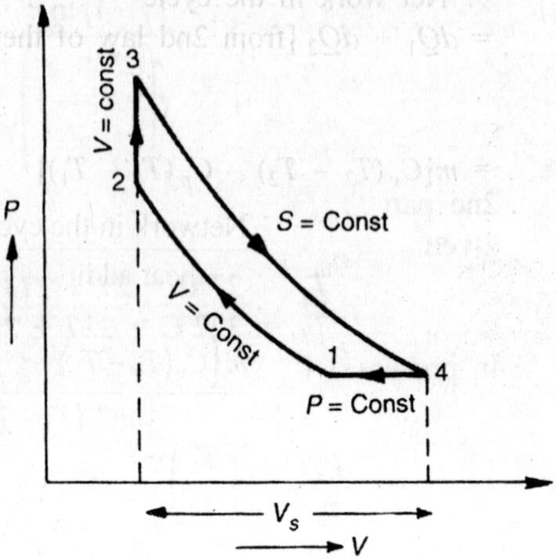

Fig. P. 10.18 (a)

$$\frac{T_4}{T_1} = \frac{V_4}{V_1} = \frac{\left(\frac{T_3}{T_4}\right)^{\frac{1}{\gamma-1}} \times V_3}{\left(\frac{T_2}{T_1}\right)^{\frac{1}{\gamma-1}} \times V_2}$$

From 1st law of thermodynamics

$$dQ = dh - V\,dp = dh$$
$$= mC_p (T_1 - T_4)$$
$$dQ_2 = -mC_p (T_4 - T_1)$$

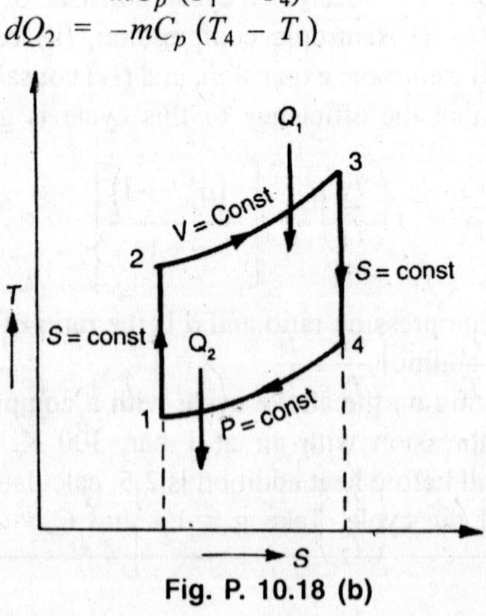

Fig. P. 10.18 (b)

∴ Net work in the cycle

$= dQ_1 + dQ_2$ [from 2nd law of thermodynamics

$$\oint Qdw = \oint dQ]$$

$= m[C_v(T_3 - T_2) - C_p(T_4 - T_1)]$

$$\therefore \quad n_{th} = \frac{\text{Network in the cycle}}{\text{heat addition}}$$

$$= \frac{m\left[C_v(T_3 - T_2) - C_p(T_4 - T_1)\right]}{mC_v(T_3 - T_2)}$$

$$\therefore \quad 1 - \frac{C_p}{C_v}\frac{(T_4 - T_1)}{(T_3 - T_2)} = 1 - \frac{\gamma\, T_1\left(\frac{T_4}{T_1} - 1\right)}{T_2\left(\frac{T_3}{T_2} - 1\right)}$$

Now in process 1-2

$$\left(\frac{V_1}{V_2}\right)^{\gamma-1} = \frac{T_2}{T_1} = \left(\frac{P_2}{P_1}\right)^{\frac{\gamma-1}{\gamma}} = r_K^{\frac{\gamma-1}{\gamma}}$$

$$\Rightarrow \quad \frac{T_1}{T_2} = \frac{1}{r_k^{\frac{\gamma-1}{\gamma}}}$$

in process 3-4

$$\frac{V_4}{V_3} = \left(\frac{T_3}{T_4}\right)^{\frac{1}{\gamma-1}}$$

$$\text{Hence} \quad \frac{T_4}{T_1} = \left(\frac{T_3}{T_4} \times \frac{T_1}{T_2}\right)^{\frac{1}{\gamma-1}} = \left(\frac{T_3 / T_2}{T_4 / T_1}\right)^{\frac{1}{\gamma-1}}$$

$$\Rightarrow \quad \left(\frac{T_4}{T_1}\right)^{\frac{\gamma}{\gamma-1}} = \left(\frac{T_3}{T_2}\right)^{\frac{1}{\gamma-1}} = a^{\frac{1}{\gamma-1}} \left[\because \frac{T_4}{T_1} = a^{\frac{1}{\gamma}}\right]$$

$$\therefore \quad n_{th} = 1 - \frac{\gamma\left(a^{\frac{1}{\gamma}} - 1\right)}{r_K^{\frac{\gamma-1}{\gamma}}(a-1)}$$

2nd Part $r_K = 6$

$a = 2.5$

$r = 1.4$

$$n_{th} = 1 - \frac{1}{6^{1.4-1}}\left[\frac{1.4\left(2.5^{\frac{1}{1.4}} - 1\right)}{(2.5-1)}\right] = 0.579 \textbf{ Ans}$$

$$\text{m.e.p.} = \frac{\text{Net work in cycle}}{\text{stroke volume}}$$

$$= \frac{m\left[C_v(T_3 - T_2) - C_p(T_4 - T_1)\right]}{V_4 - V_2}$$

given $P_1 = 1$ bar,

$T_1 = 300\text{K}$, $C_v = 0.718$ kJ/kg.K

$C_p = 1.005$ kJ/kg.K

$T_2 = T_1 r_K^{\gamma-1} = 300 \times 6^{1.4} = 611.3$ K

$T_3 = a\,T_2 = 611.3 \times 2.5 = 1535.75$ K

$T_4 = a^{1/\gamma}\,T_1 = 577.25$ K

$$V_1 = \frac{RT_1}{P_1} = \frac{0.287 \times 300}{1 \times 100} = 0.861 \text{ m}^3/\text{kg}$$

Now $$\frac{V_1}{V_2} = 6 = \frac{P_2}{P_1}$$

$$\Rightarrow \quad V_2 = \frac{V_1}{6} = \frac{.0861}{6} = 0.1435 \text{ m}^3/\text{kg}$$

Again $$\frac{V_4}{V_1} = \frac{T_4}{T_1}$$

$$\Rightarrow \quad V_4 = \left(\frac{577.25}{300}\right) \times .861 = 1.6567 \text{ m}^3\text{/kg}$$

$$\therefore \quad \text{m.e.p.} = \frac{[0.718(1535.75 - 611.3) - 1.005(577.25 - 300)]}{1.6567 - 0.1535}$$

$$= 2.5322$$

Problem 10.19 : The relation between u, p and v for many gases is of the form $u = a + bpv$ where a and b are constants. Show that for a reversible adiabatic process pv^γ = constant, where $\gamma = (b + 1)/b$.

Solution : Given data :

$$u = a + bpv$$

Now from 1st law of thermodynamics

$$dQ = du + pdv$$

for adiabatic process

$$du + pdv = 0$$

$$\Rightarrow \quad Vbdp + bpdv + pdv = 0$$

$$\Rightarrow \quad pav\ (b+1) + b\ vdp = 0$$

$$\Rightarrow \quad \frac{(b+1)}{b}\ p\ dv + v\ dp = 0$$

$$\Rightarrow \quad \frac{b+1}{b} \times RT\ \frac{dv}{v} + RT\ \frac{dp}{p} = 0$$

$$\Rightarrow \quad \frac{dp}{p} + \frac{b+1}{b}\ \frac{dv}{v} = 0$$

Integrating, we get,

$$\ln \quad p + \frac{b+1}{b} \ln v = \ln C$$

$$\Rightarrow \quad \ln p\ v^{\frac{b+1}{b}} = \ln C$$

$$\Rightarrow \quad p\,v^{\frac{b+1}{b}} = C$$

Comparing with pv^γ = constant, we get $\gamma = \dfrac{b+1}{b}$

Problem 10.20 : (a) Show that the slope of a reversible adiabatic process on p–v coordinates is

$$\frac{dp}{dv} = -\frac{1}{kv}\frac{C_p}{C_v} \text{ where } k = -\frac{1}{v}\left(\frac{\partial v}{\partial p}\right)_T$$

(b) Hence, show that for an ideal gas, pv^γ = constant, for a reversible adiabatic process.

Solution :

(a) for a reverrible adiabatic process, pv^γ = const = C

⇒ differentiating, we get

$p\gamma v^{\gamma-1} + v^\gamma\, dp = 0$

$$\Rightarrow \quad \frac{dp}{dv} = -\gamma\frac{p}{v} \qquad \text{...(i)}$$

Now isothermal compressibility

$$= K_T = -\frac{1}{v}\left(\frac{dv}{dp}\right)_T$$

again for isothermal process

$$pv = RT$$

differentiating w.r.to p

$$\Rightarrow \quad p\left(\frac{\partial v}{\partial p}\right)_T + v = 0$$

$$\Rightarrow \therefore \quad -\frac{p}{v} = +\frac{1}{\left(\frac{\partial v}{\partial p}\right)_T}$$

∴ from (i)

$$\frac{dP}{dv} = -\gamma\,\frac{p}{v}$$

$$= \gamma \times \frac{1}{(\partial v/\partial p)_T} = -\frac{C_p}{C_v} \times \frac{1}{K_T v} \quad \textbf{Ans}$$

(b) For a reversible adiabatic process

$$Tds = du + pdv = 0$$

$$\Rightarrow \quad -du = pdv \Rightarrow -C_v\, dT = p\, dv$$

from ideal gas eqn,

$$pv = pdv$$

$$\Rightarrow \quad pdv + vdp = RdT$$

$$\Rightarrow \quad dT = \frac{1}{R}(pdv + vdp)$$

$$\Rightarrow \quad -\frac{C_v}{R} = (pdv + Vdp) = pdv$$

$$\Rightarrow -\frac{C_v}{R} \quad Vdp = pdv\left(1+\frac{C_v}{R}\right)$$

$$= pdv\left(\frac{R+C_v}{R}\right) = pdv\ \frac{C_p}{R}$$

$$\Rightarrow \quad \frac{C_p}{C_v} pdv + vdp = 0$$

$$\Rightarrow \quad \gamma\frac{dv}{v} + \frac{dp}{p} = 0$$

$\Rightarrow$ integrating, we get

$$\Rightarrow \quad \gamma \ln v + \ln p = \ln c$$

$$\Rightarrow \quad pv^{\gamma} = \text{cont}$$

Problem 10.21 : A certain gas obeys the clausius equation of state $p\,(v - b) = RT$ and has its internal energy given by $u = C_v\,T$. Show that the equation for a reversible adiabatic process is $p(v - b)^{\gamma}$ = constant, where $\gamma = C_p/C_v$.

Solution : Given :

Clausius equation of state,

$$p\,(v - b) = RT$$

& $\quad u = C_vT$

From property relation

$$TdS = du + pdv$$

From clausius equation

$$p\,(v - b) = RT$$

on diffentiationg we get

$$\Rightarrow \quad dp\,(v - b) + pdv = RdT$$

from (i), $$dT = -\frac{pdv}{C_v}$$

$$\therefore \quad dp(v-b) + pdv = -\frac{pdv\,R}{C_v}$$

or $$dp\,(v-b) = -\left(1+\frac{R}{C_v}\right) pdv = -\frac{(C_v+R)}{C_v}\,\text{pdv}$$

or $$dp(v-b) = -\frac{C_p}{C_v}\,pdv = -\gamma pdv \quad [\therefore C_p = C_v + R]$$

or $$\frac{dp}{p} + \gamma\frac{dv}{v-b} = 0$$

by intigration, we get.

$$\ln P + \gamma \ln (v-b) = c$$

or $$\ln [p\,(v-b)^{\gamma}] = c$$

$$\Rightarrow \quad P\,(v-b)^{\gamma} = e^c = \text{const}$$

Problem 10.22 : (a) Two curves, one representing a reversible adiabatic process undergone by an ideal gas and the other an isothermal process by the same gas, intersect at the same point on the p–v diagram. Show that the ratio of the slope of the adiabatic curve to the slope of the isothermal curve is equal to γ. (b) Determine the ratio of work done during a reversible adiabatic process to the work done during an isothermal process for a gas having $\gamma = 1.6$. Both processes have a pressure ratio of 6.

***Solution* :** Given data :

For isothermal process

$$pv = c$$

On differentiation

$$vdp + pdv = 0$$

or $$\left(\frac{dp}{dv}\right)_{T=c} = -\frac{p}{v} \qquad \text{...(i)}$$

for reversible adiabatic process

$$Pv^{\gamma} = C$$

or, $V^{\gamma}dp + p\gamma\, v^{\gamma-1}dv = 0$

or $\left(\dfrac{dp}{dv}\right)_{S=c} = -\dfrac{p\gamma}{v}$ ...(ii)

$\therefore$ from equation (i) & (ii)

$$\left(\frac{dp}{dv}\right)_{S=c} = \gamma\left(\frac{dp}{dv}\right)_{T=c}$$

Hence slope of reversible process = $\gamma \times$ slope of isothermal process.

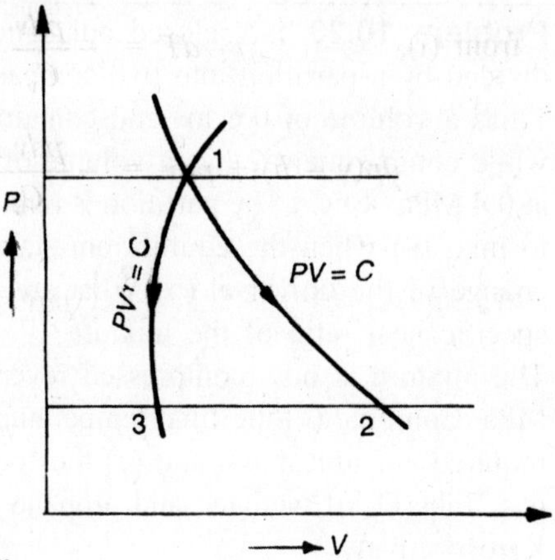

Fig. P. 10.22

(b) For isothermal process

$$w_{1-2} = p_1 v_1 \ln\frac{p_1}{p_2}$$

$$= mRT_1 \ln\frac{p_1}{p_2}$$

Given $\dfrac{p_1}{p_2} = 6$

$\therefore$ $w_{1-2} = mRT_1 \ln 6$

For isentropic process

$$w_{1-3} = \frac{p_1v_1 - p_2v_2}{\gamma - 1} = \frac{mRT_1\left[1 - \dfrac{T_2}{T_1}\right]}{\gamma - 1}$$

$$= \frac{mRT_1}{\gamma - 1}\left[1 - \left(\frac{P_2}{P_1}\right)^{\frac{\gamma-1}{\gamma}}\right]$$

or, $$w_{1-3} = \frac{mRT_1}{1.6 - 1} \times \left[1 - \left(\frac{1}{6}\right)^{\frac{1.6-1}{1.6}}\right]$$

Now $$\frac{w_{1-3}}{w_{1-2}} = \frac{\left[1 - \left(\dfrac{1}{6}\right)^{\frac{1.6-1}{1.6}}\right]}{0.6 \times \ln 6} = 0.46 \textbf{ Ans}$$

Problem 10.23 : A closed adiabatic cylinder of volume 1 m^{3}is divided by a partition into two comparments 1 and 2. Compartment 1 has a volume of 0.6 m^3 and contains methane at 0.4 MPa, 40°C, while compartment 2 has a volume of 0.4 m^3 and contains propane at 0.4 MPa, 40°C . The partition is removed and the gases are allowed to mix. (a) when the equilibrium state is reached, find the entropy change of the universe. (b) What are the molecular weight and the specific heat ratio of the mixture?
The mixture is now compressed reversibly and adiabatically to 1.2 MPa. Compute (c) the final temperature of the mixture, (d) the work required per unit mass, and (e) the specific entropy change for each gas. Take C_p of methane and propane as 35.72 and 74.56 kJ/kg mol K respectively.

Solution : Given data :

For compartment 1 (methane)

$V_1 = 0.6$ m^3

$p_1 = 0.4$ MPa

$T_1 = 40$°C

Compartment 2 (propane)

$V_2 = 0.4$ m^3

$P_2 = 0.4$ MPa

$T_2 = 40$°C

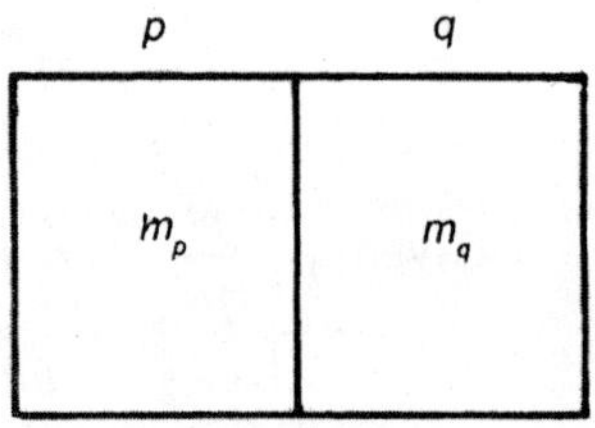

Fig. P. 10.23

For methane, molecular wt = 16

$$\therefore R_1 = \frac{\overline{R}}{16} = \frac{8.314}{16} \text{ kJ/kg.K}$$

$$\therefore \quad m_1 = \frac{p_1 v_1}{R_1 T1} = \frac{0.4 \times 10^3 \times 0.6 \times 16}{8.314 \times (273 + 40)} = 1.48 \text{ kg}$$

$$n_1 = \text{mole} = \frac{m_1}{M_1} = \frac{148}{16} = 0.0925$$

For propane, molecular wt = 44

$$\therefore \quad R_2 = \frac{\overline{R}}{44} = \frac{8.314}{44} \text{ kJ/kg.K}$$

$$\therefore \quad m_2 = \frac{p_2 v_2}{R_2 T_2} = \frac{0.4 \times 10^3 \times 0.4 \times 44}{8.314 \times (273 + 40)} = 2.71 \text{ kg.}$$

$$n_2 = \text{mole} = \frac{m_2}{M_2} = \frac{2.71}{44} = 0.0616$$

final temperature

$$T_{m_1} = 40°C$$

and pressure remains the same.

$$\therefore \quad R_m = \text{gass constant of the mixture}$$

$$= \frac{m_1 R_1 + m_2 R_2}{m_1 + m_2}$$

$$= \frac{1.48 \times \frac{8.314}{16} + 2.71 \times \frac{8.314}{44}}{(1.48 + 2.71)}$$

$$= 0.3058 \text{ kJ/kg.K}$$

$$C_{p_1} = 35.72 \text{ kJ/kg mole K}$$

$$\therefore \quad C_{v_1} = C_{p_1} - \bar{R} = 35.72 - 8.314$$

$$= 27.41 \text{ kJ/kg mole K}$$

$$C_{p_2} = 74.56 \text{ kJ/kg mole K}$$

$$\therefore \quad C_{v_2} = C_{p_2} - \bar{R} = 74.56 - 8.314$$

$$= 66.25 \text{ kJ/kg mole.K.}$$

$$\therefore \quad C_{p_m} = \frac{\frac{m_1 C_{p_1}}{M_1} + \frac{m_2 C_{p_2}}{M_2}}{m_1 + m_2}$$

$$= \frac{1.48 \times \frac{35.72}{16} + 2.71 \times \frac{74.56}{44}}{(1.48 + 2.71)}$$

$$C_{v_m} = \frac{m_1 \frac{C_{v_1}}{M_1} + m_2 \frac{C_{v_2}}{M_2}}{m_1 + m_2}$$

$$= \frac{\frac{1.48 \times 27.41}{16} + \frac{2.71 \times 66.25}{44}}{1.48 + 2.71}$$

$$= 1.57896 \text{ kJ/kg mole K}$$

$$\therefore \quad \gamma_m = \frac{C_{p_m}}{C_{v_m}} = 1.193$$

$$M = \text{molucular wt of mixture}$$

$$= \frac{n_1 M_1 + n_2 M_2}{n_1 + n_2}$$

$$= \frac{0.0925 \times 16 + 0.0616 \times 44}{0.0925 + 0.0616} = 27.2 \quad \textbf{Ans}$$

From property relation

$$\Delta S = \int d_s = \int mC_v \frac{dT}{T} + \int mR \frac{dV}{V}$$

since temperature is same

$$\therefore \quad \Delta S = mR \int \frac{dV}{V}$$

$$\therefore \quad (\Delta S)_1 = \text{change in entropy to methane}$$

$$= m_1 R_1 \ln \frac{V}{V_1}$$

$$= 1.48 \times \frac{8.314}{16} \ln \left(\frac{1}{0.6}\right) = 0.39285 \text{ kJ/K}$$

$(\Delta S)_2$ = change in entropy of propane

$$= m_2 R_2 \, l_n \frac{V}{V_2}$$

$$= 2.71 \times \frac{8.314}{44} \times \ln \left(\frac{1}{0.4}\right)$$

$$= 0.4692 \text{ kJ/K}$$

$$\therefore \quad (\Delta S)_{\text{universe}} = (\Delta S)_1 + (\Delta S)_2$$

$$= 0.86205 \text{ kJ/K}$$

For compressing reversibly and adibatically,

(c) $$\frac{T_{m_2}}{T_{m_1}} = \left(\frac{p_{m_2}}{p_{m_1}}\right)^{\frac{\gamma-1}{\gamma}}$$

$$T_{m_2} = T_{m_1}\left(\frac{1.2}{0.4}\right)^{\frac{0.193}{1.193}}$$

$$= 373.88 \text{ K} = 100.88°\text{C}$$

(d) work required per unit mass $= \dfrac{R_m\left(T_{2_m} - T_{m_1}\right)}{\gamma - 1}$

$$= \frac{0.3058\,(100.88 - 40)}{0.193} = 96.5 \text{ kJ/kg } \textbf{Ans}$$

Problem 10.24 : (a) A closed system of 2 kg of air initially at pressure 5 atm and temperature 227°C, expands reversibly to pressure 2 atm following the law $pv^{1.25}$ = const. Assuming air as an ideal gas, determine the work done and the heat transferred.
(b) If the system does the same expansion in a steady flow process, what is the work done by the system?

Solution : Given data :

m = mass of air = 2 kg
P_1 = initial pressure = 5 atm
T_1 = initial temperature = 227°C = 500 K
P_2 = final pressure = 2 atm.

from the expansion law $pv^{1.25} = c$

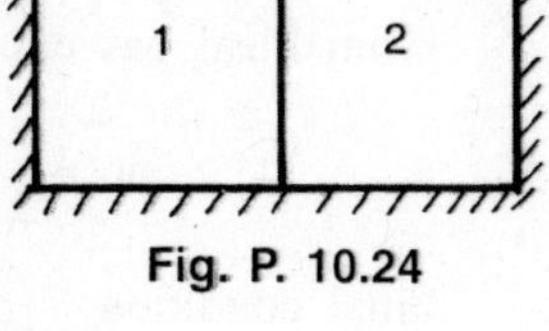

Fig. P. 10.24

$$T_2 = T_1\left(\frac{p_2}{p_1}\right)^{\frac{1.25-1}{1.25}}$$

$$= 500\left(\frac{2}{5}\right)^{\frac{0.25}{1.25}} = 416.28 \text{ K.}$$

$$\therefore \text{ work done} = \frac{MR\,(T_1 - T_2)}{n - 1}$$

$$= \frac{2 \times 0.287 \times (500 - 416.28)}{0.25} = 192.22 \text{ kJ}$$

Now from first law of thermodynamics

$$dQ = du + dw$$

$$\therefore \quad Q = mC_v (T_2 - T_1) + dw$$

$$= 2 \times 0.718 (416.28 - 500) + 192.22$$

$$= 71.998$$

$$= 72 \text{ kJ}$$

(b) for steady flow system,

Applying S.F.E.E neglecting Δ K.E. and Δ P.E.

$$mh_1 + dQ = mh_2 + dw$$

$$\therefore \quad w = m(h_1 - h_2) + dQ$$

$$= mC_p(T_1 - T_2) + Q$$

$$= 2 \times 1.005 (500 - 416.28) + 72$$

$$= 240.27 \text{ kJ}$$

Problem 10.25 : Air is contained in cylinder fitted with a frictionless piston. Initially the cylinder contains 0.5 m³ of air at 1.5 bar, 20°C.The air is then compressed reversibly according to the law pv^n = constant until the final pressure is 6 bar, at which point the temperature is 120°C .Determine: (a) the polytropic index *n*, (b) the final volume of air, (c) the work done on the air and the heat transfer, and (d) the net change in entropy.

Solution : Initial condition

$$V_1 = 0.5 \text{ m}^3$$

$$P_1 = 1.5 \text{ bar}$$

$$T_1 = 20°\text{C} = 293 \text{ K}$$

From ideal gas equation,

$$m = \frac{P_1V_1}{RT_1} = \frac{1.5 \times 100 \times 0.5}{0.287 \times 293} = 0.892 \text{ kg}$$

Final condition

$$P_2 = 6 \text{ bar}$$

$$T_2 = 120°\text{C} = 393 \text{ K}$$

Law of compression follows pv^n = const.

$$\therefore \quad \left(\frac{P_2}{P_1}\right)^{\frac{n-1}{n}} = \frac{T_2}{T_1}$$

$$\text{or,} \quad \left(\frac{6}{1.5}\right)^{\left(1-\frac{1}{n}\right)} = \frac{393}{293}$$

$$\text{or} \quad \left(1-\frac{1}{n}\right) = \frac{\ln(393/293)}{\ln(6/1.5)}$$

(a) $\therefore$ $n = 1.2687$

(b) $$V_2 = V_1\left(\frac{P_1}{P_2}\right)^{\frac{1}{n}} = 0.5 \times \left(\frac{1.5}{6}\right)^{\frac{1}{1.2687}}$$

$$= 0.1676 \text{ m}^3$$

(c) work done on air

$$= \frac{p_1v_1 - p_2v_2}{n-1} = \frac{mR(T_1 - T_2)}{n-1}$$

$$= \frac{0.287(293-393)\times 0.892}{0.2687} = -95.26 \text{ kJ}$$

From 1st law of thermodynamics

$$đQ = đu + đw$$

$$\therefore \quad Q = mc_v(T_2 - T_1) + đw$$

$$= 0.892 \times 0.718(120 - 20) - 95.26$$

$$= -31.21 \text{ kJ}$$ **Ans**

Let surrounding temperature be 20°C

(d) Net change in intropy

$= (\Delta S)_{system} + (\Delta S)_{surrounding}$

$$= m\left[c_p \ln\frac{T_2}{T_1} - R\ln\frac{P_2}{P_1}\right] + \frac{Q}{293}$$

$$= 0.892\left[1.005\ln\frac{393}{293} - 0.287\ln\frac{0.6}{1.5}\right] + \frac{31.21}{293} = 0.01485 \text{ kJ/K}$$

Problem 10.26 : The specific heat a constant pressure for air is given by

$C_p = 0.9169 + 2.577 \times 10^{-4}T - 3.974 \times 10^{-8}\, T^2$ kJ/kg.K

Determine the change in internal energy and that in entropy of air when it undergoes a change of state from 1 atm and 298 K to a temperature of 2000 K at the same pressure.

Solution : Given data :

Initial condition

$$P_1 = 1 \text{ atm}$$
$$T_1 = 298 \text{ K}$$

final condition

$$P_2 = 1 \text{ atm.}$$
$$T_2 = 2000 \text{ K}$$

we know that,

$$C_p - C_v = R$$

$$\therefore \quad C_v = C_p - R$$
$$= 0.9169 + 2.577 \times 10^{-4}T - 3.974 \times 10^{-8}\, T^2 - 0.287$$
$$= 0.6299 + 2.577 \times 10^{-4}\, T - 3.974 \times 10^{-8}\, T^2 \text{ kJ/kg.K}$$

$$\therefore \text{ Change in internal energy} = \Delta U = \int_{T_1}^{T_2} mC_v dT$$

$$\therefore \quad \Delta u = \int_{T_1=298k}^{T_2=2000k} [.6299 + 2.577 \times 10^{-4}\, T - 3.974 \times 10^{-8}\, T^2]\, dT$$

$$= \left[0.6299\, T + \frac{2.577 \times 10^{-4}}{2} T^2 - \frac{3.974 \times 10^{-8}}{3} T^3\right]_{T_1=298K}^{T_2=2000K}$$

$$= 0.06299\,[2000 - 298] + \frac{2.577 \times 10^{-4}}{2}[(2000)^2 - (298)^2]$$

$$- \frac{3.974 \times 10^{-8}}{3}[(2000)^3 - (298)^3]$$

$$= 1470.42 \text{ kJ/kg}$$

From property relation

$$\Delta s = \int C_p \frac{dT}{T} - R \int \frac{dP}{P}$$

at constant pressure, $dp = 0$

$$\therefore \quad \Delta s = \int_{T_1}^{T_2} C_p \frac{dT}{T}$$

$\therefore$ change in entropy $= \Delta s$

$$= \int_{T_1=298\text{K}}^{T_2=2000\text{K}} \left[\frac{0.9169}{T} + 2.577 \times 10^{-4} - 3.974 \times 10^{-8} T\right] dT$$

$$= \left[0.9169 \ln T + 2.577 \times 10^{-4}\ T - \frac{3.974 \times 10^{-8}}{2} T^2\right]_{T_1=298\text{K}}^{T_2=2000\text{k}}$$

$$= 0.9169 \ln \frac{2000}{298} + 2.577 \times 10^{-4} [2000 - 298] - \frac{3.974 \times 10^{-8}}{2} [(2000)^2 - (298)^2]$$

$= 2.1065$ kJ/kg.K

Problem 10.27 : A closed system allows nitrogen to expand reversibly from a volume of 0.25 m^3 along the path $pv^{1.32}$ = const. The original pressure of the gas is 250 kPa and its initial temperature is 100°C.(a) Draw the *p-v* and *T-s* diagrams. (b) What are the final temperature and the final pressure of the gas? (c) How much work is done and how much heat is transferred? (d) What is the entropy change of nitrogen?

Solution : Given data :

$$V_1 = 0.25\text{m}^3$$
$$P_1 = 250 \text{ kPa}$$
$$T_1 = 100°\text{C} = 373 \text{ K}$$
$$V_2 = 0.75 \text{ m}^3$$

(b) From

$$Pv^{1.32} = C$$

$$\frac{T_2}{T_1} = \left(\frac{V_1}{V_2}\right)^{1.32-1}$$

or,
$$T_2 = 373 \left(\frac{0.25}{0.75}\right)^{0.32} = 262.44 \text{ K}$$

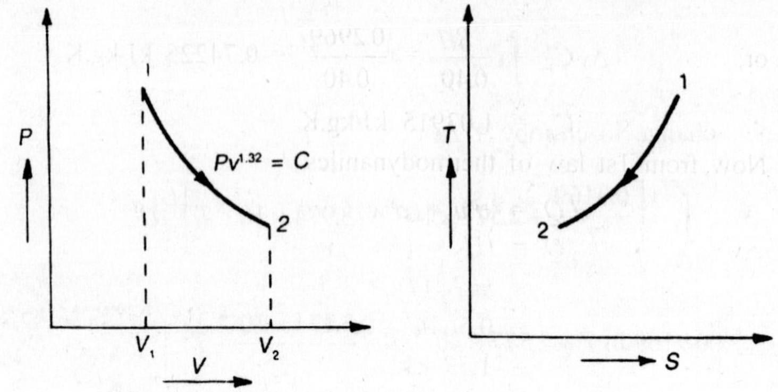

Fig. P. 10.27

$\therefore$ Final temperature = T_2 = 262.44 **Ans**

$$\text{Similarly, } \frac{P_2}{P_1} = \left(\frac{V_1}{V_2}\right)^{1.32}$$

$$\therefore \text{ Final pressure} = P_2 = P_1\left(\frac{0.25}{0.75}\right)^{1.32} = 58.63 \text{ kPa}$$

(c) work done $= \dfrac{p_1v_1 - p_2v_2}{(1.32-1)}$

$$= \frac{250 \times 0.25 - 58.63 \times 0.75}{0.32} = 57.89 \text{ kJ}$$

Now gas constant for $N_2 = R = \dfrac{\bar{R}}{28} = \dfrac{8.314}{28} = 0.2969$ kJ/kg.K.

$\therefore$ From ideal gas equation,

$$m = \frac{p_1v_1}{RT_1} = \frac{250 \times 0.25}{0.2969 \times 373} = 0.5644 \text{ kg}$$

$$\frac{C_p}{C_v} = 1.40$$

or, $C_p = 1.40\ C_v$ [since N_2 is diaatomic gas]

Now $C_p - C_v = R$

or, $1.40\ C_v - C_v = R$

or, $C_v = \frac{R}{0.40} = \frac{0.2969}{0.40} = 0.74225$ kJ/kg.K

$\therefore$ $C_p = 1.03915$ kJ/kg.K

Now from 1st law of thermodynamics

$$dQ = du + dw$$

$$\therefore \quad Q = (U_2 - U_1) + w$$

$$= mCv\,(T_2 - T_1) + w$$

$$= 0.5644 \times 0.74225\,(262.44 - 373) + 57.89$$

$$= 1.57 \text{ kJ}$$

(d) change in entropy

$$= mC_v\, l_n \frac{T_2}{T_1} + mR \ln \frac{V_2}{V_1}$$

$$= 0.5644 \times 0.74225 \times \ln \frac{262.44}{373} + 0.5644 \times 0.2969 \ln \frac{0.75}{0.25}$$

$$= 0.03682 \text{ kJ/K}$$

Problem 10.28 : Air is compressed reversibly according to the law $pv^{1.25}$ = const. from an initial pressure of 1 bar and volume of 0.9 m^3 to a final volume of 0.6 m^3. Determine the final pressure and the change of entropy per kg of air.

Solution : Given data :

Initial condition

$$P_1 = 1 \text{ bar}$$

$$V_1 = 0.9 \text{ m}^3$$

Final condition

$$V_2 = 0.6 \text{ m}^3$$

From law of compression $pv^{1.25}$ = const

$$\frac{p_2}{p_1} = \left(\frac{V_1}{V_2}\right)^{1.25}$$

$\therefore$ final pressure $= P_2$

$$= p_1 \left(\frac{V_1}{V_2}\right)^{1.25}$$

$$= 1 \times \left(\frac{0.9}{0.6}\right)^{1.25}$$

$$= 1.66 \text{ bar } \textbf{Ans}$$

change of entropy

$$= C_p \ln \frac{T_2}{T_1} - R \ln \frac{P_2}{P_1}$$

$$= C_p \frac{(n-1)}{n} \ln \frac{P_2}{P_1} - R \ln \frac{P_2}{P_1}$$

$$= \left[C_P \frac{(n-1)}{n} - R\right] \ln \frac{P_2}{P_1}$$

$$= \left[\frac{1.005 \times 0.25}{1.25} - 0.287\right] \ln \left(\frac{1.66}{1}\right)$$

$$= -0.0436 \text{ kJ/kg.K.}$$

> **Problem 10.29** : n_1 moles of an ideal gas at pressure P_1 and temperature T are in one a partition of an insulated container. In an adjoining compartment, separated by a partition, are n_2 moles of an ideal gas at pressure p_2 and temperature T. When the partition is removed, calculate (a) the final pressure of the mixture, (b) the entropy change when the gases are identical, and (c) the entropy change when the gases are different. Prove that the entropy change in (c) is the same as that produced by two independent free expansions.

Solution : Given :

Compartment 1

n_1 = no. of moles.
p_1 = initial pressure
T = initial temperature

Compartment 2

n_2 = no. of moles
p_2 = initial pressure
T = initial temperature

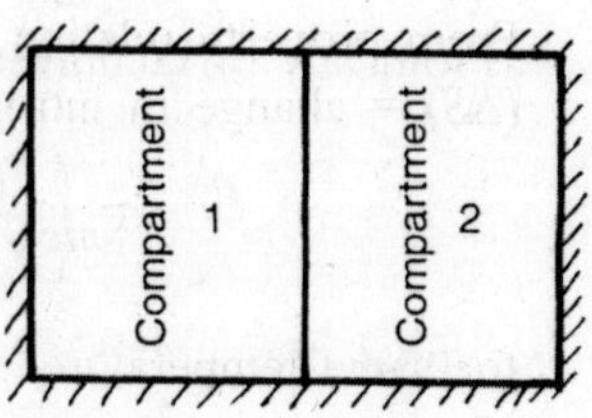

Fig. P. 10.29

From ideal gas equation,

Now V_1 = volume of compartment 1 = $\dfrac{n_1 \bar{R} T}{P_1}$

V_2 = volume of compartment 2 = $\dfrac{n_2 \bar{R} T}{P_2}$

$\therefore$ V = final volume = $V_1 + V_2 = \left(\dfrac{n_1}{p_1} + \dfrac{n_2}{p_2}\right) \bar{R}\, T$

(i) $\therefore$ P = final pressure = $\dfrac{(n_1 + n_2)\bar{R} T}{V}$

$$= \frac{n_1 + n_2}{\dfrac{n_1}{p_1} + \dfrac{n_2}{p_2}} = \frac{(n_1 + n_2) p_1 p_2}{n_1 p_2 + n_2 p_1}$$

(ii) *when two different gases are mixed,*

P_{01} = partial pressure of gas in compartment 1

$$= \Sigma x_1 p = \frac{n_1}{n_1 + n_2} \times \frac{(n_1 + n_2) p_1 p_2}{n_1 p_2 + n_2 p_1}$$

$$= \frac{n_1 p_1 p_2}{n_1 p_2 + n_2 p_1}$$

p_{02} = parital pressure of gas in compartment 2

$$= \Sigma x_2 p = \frac{n_2}{n_1 + n_2} \times \frac{(n_1 + n_2) p_1 p_2}{n_1 p_2 + n_2 p_1}$$

$$= \frac{n_2 p_1 p_2}{n_1 p_2 + n_2 p_1}$$

From property relation
(ΔS) = change in intropy

$$= mc_p \ln \frac{T_2}{T_1} - n\ \bar{R}\ ln\ \frac{p_2}{p_1}$$

for same temperature

$$\Delta S = -\, n \bar{R}\ \ln \frac{p_2}{p_1}$$

Now $(\Delta S)_1$ = change in entropy of gas which was in compartment 1

$$= -n_1 \bar{R} \ln \frac{p_{o_1}}{p_1}$$

$$= -n_1 \bar{R} \ln \frac{n_1 p_2}{n_1 p_2 + n_2 p_1}$$

Now $(\Delta S)_2$ = change in entropy of gas which was in compartment 1

$$= -n_2 \bar{R} \ln \frac{p_{o_2}}{p_2}$$

$$= -n_2 \bar{R} \ln \frac{n_2 p_2}{n_1 p_2 + n_2 p_1}$$

∴ total entropy change

$$= \Delta S_1 + \Delta S_2$$

$$= +n_1 \bar{R} \ln \left(\frac{n_1 p_2}{n_2 p_1} + 1\right) + n_2 \bar{R} \ln \left(\frac{n_2 p_1}{n_1 p_2} + 1\right)$$

(iii) *when same gases are mixed*

$$p = \text{final pressure} = \frac{(n_1 + n_2) p_1 p_2}{n_1 p_2 + n_2 p_1}$$

∴ change in entropy

$$= -n_1 \bar{R}_1 \ln \left\{\frac{(n_1 p_1 + n_2 p_1)}{(n_1 + n_2) p_2}\right\} + n_2 \bar{R}_2 \ln \left\{\frac{(n_1 p_1 + n_2 p_1)}{(n_1 + n_2) p_1}\right\}$$

Problem 10.30 : Assume that 20kg of steam are required at a pressure of 600 bar and a temperature of 750°C in order to conduct a particular experiment. A 140-litre heavy duty tank is available for storage.

Predict if this is an adequate storage capacity using :

(a) ideal gas theory,

(b) steam tables. the error in each.

(c) Mollier chart

Solution : Given data :

P = Pressure of steam = 600 bar
T = Temperature of steam
= 750°C = 750 + 273 = 1023 K
m = mass = 20 kg

(a) From ideal gas theroy

At 600 bar and 750°C, steam (superheated) is considered as an ideal gas.

$$\therefore \quad R = \frac{\bar{R}}{18} = \frac{8.317}{18} \text{ kJ/kg.K}$$

$$\therefore \quad V = \frac{mRT}{p} = \frac{20 \times 8.314}{18} \times \frac{1023}{600 \times 100}$$

$$= 0.157504 \text{ m}^3 = 157.5 \text{ ltrs}$$

since capacity of tank is only = 140 ltrs
It is not adequate for storage.

(b) From steam table.

at 750°C and 600 bar,
specific volume = 0.6865×10^{-2} m³/kg
∴ volume of steam = $m \times$ sp. volume
= 13.73×10^{-2}
= 0.1373 m³
= 137.3 ltrs.

since capacity is more than the volume of steam. Hence it is adequate for storage

(c) From Mollier chart,

at 600 bar and 750 K,

$$V = \text{specific volume} \approx 0.01 \text{ m}^3/\text{kg}$$

∴ Volume of steam = 0.01 × 20
= 0.2 m³
= 200 Lts. > 140 lts.

idence it is not adequate for storage.

Problem 10.31 : Measurements of pressure and temperature at various stages in an adiabatic air turbine show that the states of air lie of the line $pv^{1.25}$ = constant. If kinetic and gravitational potential energy are neglected, prove that the shaft work per kg as a function of pressure is given by the following relation.

$$W = 3.5\, p_1 v_1 \left[1 - \left(\frac{P_2}{P_1}\right)^{1/5}\right]$$

Take γ for air as 1.4.

Solution : Given data :

Law of expansion follows $pv^{1.25}$ = constant

$$\therefore \quad \frac{T_2}{T_1} = \left(\frac{P_2}{P_1}\right)^{\frac{n-1}{n}} = \left(\frac{P_2}{P_1}\right)^{\frac{0.25}{1.25}} = \left(\frac{P_2}{P_1}\right)^{1/5}$$

Consider the flow through turbine is steady then from S.F.E.E neglecting ΔK.E and ΔP.E

$$\dot{m}(h_1) + \frac{dQ}{dt} = \dot{m}(h_2) + \frac{dw}{d\tau}$$

Since preocess is adiabatic,

$$\frac{dQ}{d\tau} = 0$$

$$\text{or,} \quad \frac{dw}{dm} = h_1 - h_2 = C_p (T_1 - T_2)$$

$$= C_p T_1 \left[1 - \frac{T_2}{T_1}\right]$$

$$= C_p \frac{P_1 v_1}{R} \left[1 - \left(\frac{P_2}{P_1}\right)^{\gamma_s}\right] \left[\because T_1 = \frac{p_1 v_1}{R}\right]$$

Now $C_p - C_v = R$

or, $C_p - \dfrac{C_v}{\gamma} = R$

or $\dfrac{C_p}{R} = \dfrac{\gamma}{\gamma - 1} = \dfrac{1.4}{0.4} = 3.5$

$$\therefore \quad \frac{dw}{dm} = 3.5\, p_1 v_1 \left[1 - \left(\frac{p_2}{p_1}\right)^{\frac{1}{5}}\right]$$

Problem 10.32 : Air flows steadily into a compressor at a temperature of 17°C and a pressure of 1.05 bar and leves at a temperature of 247°C and a pressure of 6.3 bar. There is no heat transfer to or from the air as it flows through the compressor; changes in elevation and velocity are negligible. Evaluate the external work done per kg of air, assuming air as an ideal gas for which $R = 0.287$ kJ/kg.K and $\gamma = 1.4$. Evaluate the minimum external work required to compress the air adiabatically from the same initial state to the same final pressure and the isentropic efficiency of the compressor.

Solution : Given data :

$T_1 = 17°C = 290$ K

$P_1 = 1.05$ bar

$T_2 = 247°C = 520$ K

$P_2 = 6.3$ bar

Considuning flow through compressor is steady.

Applying S.F.E.E for the system neglecting Δ K.E Δ P.E

$$\Rightarrow m\,(h_1) + \frac{dQ}{d\tau} = \dot{m}\,(h_2) + \frac{dw}{d\tau}$$

here $\dfrac{dQ}{d\tau} = 0$

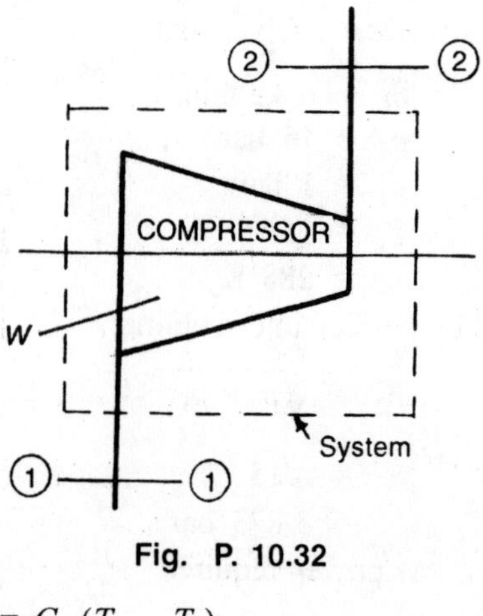

Fig. P. 10.32

$$\therefore \quad \frac{dw}{dm} = h_1 - h_2 = C_p\,(T_1 - T_2)$$

$$= 1.005\,(290 - 520) = -231.15 \text{ kJ/kg } \textbf{Ans}$$

Minimum work required to compress the air adiabatically

$$= \frac{\gamma}{\gamma - 1} RT_1 \left[1 - \left(\frac{p_2}{p_1}\right)^{\frac{\gamma-1}{\gamma}}\right]$$

$$= \frac{1.4}{0.4} \times 0.287 \times 290 \left[1 - \left(\frac{6.3}{1.05}\right)^{\frac{0.4}{1.4}}\right]$$

$$= -\ 194.98 \text{ kJ/kg}$$

Now isentropic efficiency

$$= \frac{\text{minimum work (adiabatic work) required}}{\text{Actual work}} = \frac{194.98}{231.15} = 84.4\%$$

Problem 10.33 : A single-acting two-stage reciprocating air compressure with complete intercooling delivers 6 kg/min at 15 bar pressure. Assume an intake condition of 1 bar and 15°C and that the ocmpression and expansion process are polytropic with. $n = 1.3$. Calculate; (a) the power required, (b) the isothermal efficiency.

Solution : Given data :

$$\dot{m} = 6 \text{ kg/min.}$$
$$p_3 = 15 \text{ bar}$$
$$p_1 = 1 \text{ bar}$$
$$T_1 = 15^\circ = 288 \text{ K}$$

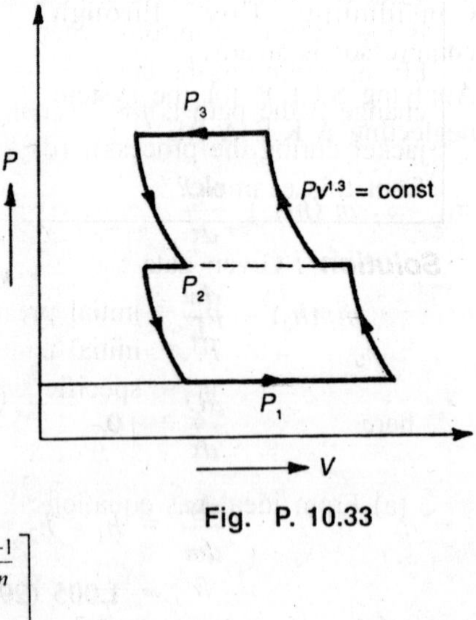

Fig. P. 10.33

For perfect inter-cooling,

$$p_2 = \sqrt{p_3 \times p_1}$$
$$= \sqrt{15}$$
$$= 3.873 \text{ bar}$$

(a) power required

$$= \frac{2n}{n-1} mR\, T_1 \left[1 - \left(\frac{p_3}{p_1}\right)^{\frac{n-1}{2n}}\right]$$

$$= \frac{2 \times 1.3}{0.3} \times \frac{6}{60} \times 0.287 \times 288 \left[1 - \left(\frac{15}{1}\right)^{\frac{0.3}{2.6}}\right]$$

$$= -\ 26.2 \text{ kW}$$

= –ve sign shows that power is fed into the compressor

(b) Now isothermal power

$$= p_1 v_1 \ln \frac{p_1}{p_3} = mRT_1 \ln \frac{p_1}{p_3}$$

$$= \frac{6}{60} \times 0.287 \times 288 \ln \left(\frac{1}{15}\right)$$

$$= -\ 22.38 \text{ kW}$$

$$\therefore \text{ isothermal efficincy} = \frac{\text{Isothermal power}}{\text{Adiabatic power}} = \frac{22.38}{26.2}$$

$$= 85.4\% \textbf{ Ans}$$

> **Problem 10.34 :** A mass of an ideal gas exists initially at a pressure of 200 kPa, temperature 300 K, and specific volume 0.5 m^3/kg. The value of γ is 1.4. (a) Determine the specific heats of the gas. (b) What is the change in entropy when the gas is expanded to pressure 100 kPa according to the law $pv^{1.3}$ = const? (c) What will be the entropy change if the path is $pv^{1.5}$ = const. (by the application of a cooling jacket during the process)? (d) What is the inference you can draw from this example?

Solution : Given data :

p_1 = initial pressure = 200 kPa
T_1 = initial temperature = 300 K
v_1 = specific volume = 0.5 m^3/kg
γ = 1.4

(a) From ideal gas equation

$$R = \frac{p_1 v_1}{T_1} = \frac{200 \times 0.5}{300} = 0.333 \text{ kJ/kg.K}$$

Now $\frac{C_p}{C_v} = \gamma = 1.4$

and $C_p - C_v = R$

or, $C_v (\gamma - 1) = R$

$$\therefore \quad C_v = \frac{R}{\gamma - 1} = \frac{0.333}{0.4} = 0.833 \text{ kJ/kg.K}$$

and $C_p = 1.4\, C_v = 1.167$ kJ/kg.K

(b) Final pressure = 100 kPa

law of expansion,

$$pv^{1.3} = \text{constant}$$

(ΔS) = Change in enterpy of gas

$$= C_p \ln \frac{T_2}{T_1} - R \ln \frac{p_2}{p_1}$$

$$= C_p \ln \left(\frac{p_2}{p_2}\right)^{\frac{n-1}{n}} - R \ln \frac{p_2}{p_1}$$

$$= 1.167 \ln \left(\frac{100}{200}\right)^{\frac{0.3}{1.3}} - 0.333 \ln \frac{100}{200}$$

$$= 0.0441 \text{ kJ/kg.K}$$

(c) For expansion law $pv^{1.5} = c$

(ΔS) = change in entropy

$$= C_p \ln \frac{T_2}{T_1} - R \ln \frac{p_2}{p_1}$$

$$= C_p \ln \left(\frac{p_2}{p_1}\right)^{\frac{n-1}{n}} - R \ln \frac{p_2}{p_1}$$

$$= 1.167 \ln \left(\frac{100}{200}\right)^{\frac{0.05}{1.5}} - 0.333 \ln \frac{100}{200}$$

$$= -\,0.0388 \text{ kJ/kg.K}$$

(d) when $n < \gamma$, there will be increase in intropy and when $n > \gamma$ these will be deanaure in entropy.

Problem 10.35 : Two vessels, A and B, each of volume 3 m^3 may be connected together by a tube of negligible volume. Vessel *A* contains air at 7 bar, 95°C while *B* contains air at 3.5 bar, 205°C. Find the change of entropy when *A* is connected to *B*. Assume the mixing to be complete and adiabatic.

Solution : Given data :

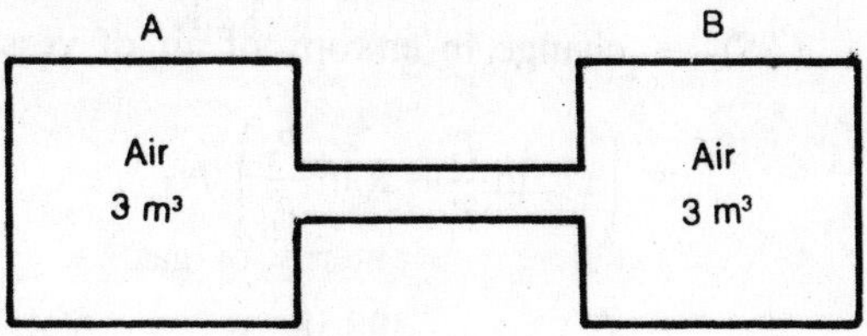

Fig. P. 10.35

For vissel A

$$P_1 = 7 \text{ bar}$$
$$T_1 = 95°\text{C}$$

For Vessel B

$$P_2 = 3.5 \text{ bar}$$
$$T_2 = 205°\text{C}$$

Mixing is complete and adiabatic

Now mass of air in vessel A is given by

$$m_1 = \frac{p_1 V_1}{RT_1} = \frac{7 \times 100 \times 3}{0.287\,(95+273)} = 19.88 \text{ kg}$$

and mass of air in vessel B =

$$m_2 = \frac{p_2 V_2}{RT_2} = \frac{3.5 \times 100 \times 3}{0.287 \times (205+273)} = 7.654 \text{ kg}$$

since mixing is adiabatic

∴ internal energy before mixing = internal energy after mixing

or $m_1 C_v T_1 + m_2 C_v T_2 = (m_1 + m_2)\, C_v T_3$

where T_3 is final temperature

$$\therefore \quad T_3 = \frac{m_1 T_1 + m_2 T_2}{m_1 + m_2}$$

$$= \frac{19.88 \times 368 + 7.654 \times 478}{19.88 + 7.654} = 398.58 \text{ K}$$

Let p_3 be final pressure

$$\therefore \quad p_3 = \frac{(m_1 + m_2)RT_3}{V_3}$$

where V_3 = final volume = 6 m³

$$\frac{(19.88 + 7.654) \times 0.287 \times 398.58}{6} = 524.95 \text{ kPa}$$

$\therefore$ $(\Delta S)_1$ = change in entropy of air of vessel A

$$= \left(C_p \ln\frac{T_3}{T_1} - R\ln\frac{P_3}{P_1}\right) m_1$$

$$= \left(1.005 \times \ln\frac{398.58}{368} - 0.287 \ln\frac{524.95}{700}\right) \times 19.88$$

$= 3.25182$ kJ/K

$(\Delta S)_2$ = change in entropy of air of vessel B

$$= \left(C_p \ln\frac{T_3}{T_2} - R\ln\frac{P_3}{P_2}\right) m_2$$

$$= \left(1.005 \ln\frac{398.88}{478} - 0.287 \ln\frac{524.95}{350}\right) \times 7.654$$

$= -2.28239$ kJ/K

$\therefore$ Total change in entropy

$= \Delta S_1 + \Delta S_2$

$= 0.969$ kJ/K

Problem 10.36 : A certain mass of sulphur dioxide (SO_2) is contained in a vessel of 0.142 m³ capacity, at a pressure and temperature of 23.1 bar and 18°C respectively. A value is opened momentarily and the pressure falls immediately to 6.9 bar. Sometime later the temperature is again 18°C and the pressure is observed to be 9.1 bar. Estimate the value of specific heat ratio.

Solution : Given data :

V = Volume of container

$= 0.142$ m³

p_1 = initial pressure = 23.1 bar

T_1 = initial temperature = 18°C

After valve opening

$p_2 = 6.9$ bar

After mixing $T_3 = 18°C$

$p_3 = 9.1$ bar

Molecular wt of SO_2 = 64

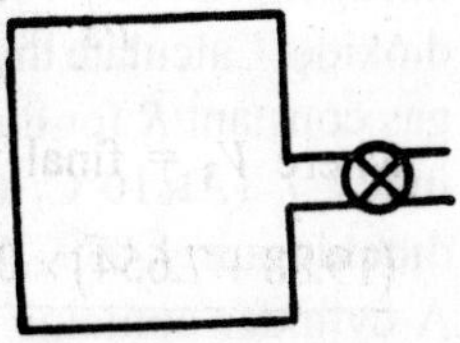

Fig. P. 10.36

$$\therefore \text{gas constant} = R = \frac{\overline{R}}{M} = \frac{8.314}{64} = 0.1299 \text{ kJ/kg.K}$$

Let m_1 be initial mass of SO_2 in vessel

$$\therefore \quad m_1 = \frac{p_1 V}{RT_1} = \frac{0.142 \times 23.1 \times 100}{0.1299 \times 291} = 8.678 \text{ kg.}$$

$$p_3 = 9.1 \text{ bar}$$

$$T_3 = 291 \text{ K}$$

$$\therefore \quad m_3 = \frac{p_3 V}{RT_3} = \frac{9.1 \times 100 \times 0.142}{0.1299 \times 291} = 3.418 \text{ kg}$$

$$m_3 = m_2 = 3.418 \text{ kg}$$

$$\therefore \quad T_2 = \frac{p_2 V}{Rm_2} = \frac{6.9 \times 0.142 \times 100}{0.1299 \times 3.418} = 220.676 \text{ K}$$

Now considering the momentarily valve opening to be reversible adiabatic process then,

$$\left(\frac{p_2}{p_1}\right)^{\frac{\gamma-1}{\gamma}} = \frac{T_2}{T_1}$$

$$\text{or,} \quad \left(1 - \frac{1}{\gamma}\right) \ln \frac{p_2}{p_1} = \ln \frac{T_2}{T_2}$$

$$\text{or} \quad \gamma = 1 / \left[1 - \frac{\ln (T_2 / T_1)}{\ln (p_2 / p_1)}\right]$$

$$= 1 / \left[1 - \frac{\ln (220.676 / 291)}{\ln (6.9 / 23.1)}\right] = 1.297 \textbf{ Ans}$$

Problem 10.37 : A gaseous mixture contains 21% by volume of nitrogen, 50% by volume of hydrogen, and 29% by volume of carbon-dioxide. Calculate the molecular weight of the mixture, the characteristic gas constant R for the mixture and the value of the reversible adiabatic index γ. (At 10°C, the C_p value of nitrogen, hydrogen, and carbon diocide are 1.039, 14.235, and 0.828 kJ/kg.K respectively.)
A cylinder contains 0.085 m^3 of the mixture at 1 bar and 10°C. The gas undergoes a reversible non-flow process during which its volume is reduced to one-fifth of its original value. If the law of compression is $pv^{1.2}$ = constant, determine the work and heat transfers in magnitude and sense and the change in entropy.

Solution : Given data :

$$V_{N_2} = 21\% \text{ of V}$$

$$V_{H_2} = 50\% \text{ of V}$$

$$V_{CO_2} = 29\% \text{ of V}$$

where V be the volume of mixture.

$$x_{N_2} = \text{mole fraction of } N_2$$

$$= \frac{V_{N_2}}{V_{N_2} + V_{H_2} + V_{CO_2}} = 0.21$$

$$x_{H_2} = \text{mole fraction of } H_2$$

$$= \frac{V_{H_2}}{V_{N_2} + V_{H_2} + V_{CO_2}} = 0.50$$

$$x_{CO_2} = \text{mole fraction of } CO_2$$

$$= \frac{V_{CO_2}}{V_{N_2} + V_{H_2} + V_{CO_2}} = 0.29$$

$$\therefore \quad M = \text{Molucular wt of mixture}$$

$$= x_{N_2} M_{N_2} + x_{H_2} M_{H_2} + x_{CO_2} M_{CO_2}$$

$$= 0.21 \times 28 + 0.50 \times 2 + 0.29 \times 44$$

$$= 19.64 \text{ kg/kg mol } \textbf{Ans}$$

$$R_m = \text{Characteristic gas constant}$$

$$= \frac{\bar{R}}{M} = \frac{8.314}{19.64} = 0.4233 \text{ kJ/kg.K } \textbf{Ans}$$

$$Cp_m = \frac{m_{H_2} C_{P_{H_2}} + m_{N_2} \times C_{P_{N_2}} + m_{CO_2} C_{P_{CO_2}}}{m_{H_2} + m_{N_2} + m_{CO_2}}$$

$$= \frac{x_{N_2} \times M_{N_2} \times C_{P_{N_2}} + x_{H_2} M_{H_2} C_{P_{H_2}} + x_{CO_2} M_{CO_2} C_{P_{CO_2}}}{x_{N_2} M_{N_2} + x_{H_2} + x_{CO_2} M_{CO_2}}$$

$$= \frac{0.21 \times 28 \times 1.039 + 0.5 \times 2 \times 14.235 + 0.29 \times 44 \times 0.828}{21 \times 28 + 0.5 \times 2 + 0.29 \times 44}$$

$$= 1.574 \text{ kJ/kg.K}$$

$$C_{P_m} = \frac{R_m \gamma}{\gamma - 1}$$

or
$$\frac{\gamma - 1}{\gamma} = \frac{R_m}{C_{P_m}}$$

or
$$\left(1 - \frac{1}{\gamma}\right) = \frac{R_m}{C_{P_m}}$$

or
$$\gamma = \frac{1}{\left(1 - \frac{R_m}{C_{P_m}}\right)}$$

$$= \frac{1}{\left(1 - \frac{0.4233}{1.574}\right)} = 1.367 \textbf{ Ans}$$

$$\therefore \quad C_{v_m} = \frac{C_{P_m}}{\gamma} = 1.1514 \text{ kJ/kg.K}$$

initial condition

$$V_1 = \text{O}.085 \text{ m}^3$$
$$p_1 = 1 \text{ bar}$$
$$T_1 = 10°\text{C}$$

final condition,

$$V_2 = \frac{1}{5} \times 0.085 = 0.017 \text{ m}^3$$

Law of compression follow $pv^{1.2}$ = constant

$$\therefore \quad p_2 = p_1\left(\frac{V_1}{V_2}\right)^{1.2}$$

$$= 1 \times (5)^{1.2} = 6.9 \text{ bar}$$

$$T_2 = T_1\left(\frac{V_1}{V_2}\right)^{1.2-1}$$

$$= 283 \times (5)^{0.2} = 390.5 \text{ K}$$

and mass of mixture

$$m = \frac{p_1V_1}{R_mT_1} = \frac{1 \times 100 \times 0.085}{0.4233 \times 283} = 0.071 \text{ kg}$$

$$w_{1\text{-}2} = \text{work transfer} = \frac{mR_m(T_1 - T_2)}{n-1}$$

$$= \frac{0.071 \times 0.4233(283 - 390.5)}{0.2}$$

$$= -16.14 \text{ kJ } \textbf{Ans}$$

Now from Ist law of thermodynamics

$$đQ_{1-2} = du + đw_{1-2}$$

$$\therefore \quad Q_{1-2} = mC_v(T_2 - T_1) + w_{1-2}$$

$$= 0.071 \times 1.1514 \times (3905 - 283) - 16.14$$

$$= -7.35 \text{ kJ}$$

ΔS = Change in entropy

$$= C_p \ln \frac{T_2}{T_1} - R \ln \frac{p_2}{p_1}$$

$$= 1.574 \ln \frac{390.5}{283} - 0.4233 \ln \frac{6.9}{1}$$

$$= -0.311 \text{ kJ/kg.K}$$

Problem 10.38 : A closed rigid cylinder is divided by a diaphragm into two equal compartments, each of volume 0.1 m^3. Each compartment contains air at a temperature of 20°C. The pressure in one compartment is 2.5 MPa and in the other compartment is 1 MPa. The diaphragm is ruptured so that the air in both the compartments mixes to bring the pressure to a uniform value througout the cylinder which is insulated. Find the net change of entropy for the mixing process.

Solution : Given data :

Compartment 1

$p_1 = 2.5$ MPa
$V_1 = 0.1\ m^3$
$T_1 = 20°C$

Compartment 2

$p_2 = 1$ MPa
$V_2 = 0.1\ m^3$
$T_2 = 20°C$

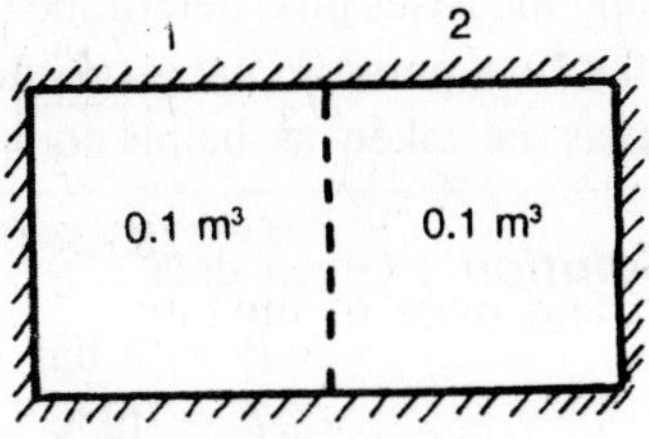

Fig. P. 10.38

$\therefore\ m_1$ = mass of air in compartment 1

$$= \frac{p_1V_1}{RT_1} = \frac{2.5 \times 10^3 \times 0.1}{0.287 \times 293} = 2.973 \text{ kg}$$

m_2 = mass of air in compartment 2

$$= \frac{p_2V_2}{RT_2} = \frac{1 \times 10^3 \times 0.1}{0.287 \times 293} = 1.189 \text{ kg}$$

Now after mixing ($T_1 = T_2 = T_f$)

$$p_f = \text{final pressure} = \frac{(m_1 + m_2)RT_f}{V_f}$$

$$= \frac{(2.973 + 1.189) \times 0.287 \times 293}{0.2} = 1749.93 \text{ kPa} = 1.75 \text{ MPa}$$

$\therefore\ \Delta S$ = change in entropy for the mixing process

$$= (\Delta S)_{\text{compact 1}} + (\Delta S)_{\text{compact 2}}$$

$$= m_1\left[C_p \ln\frac{T_f}{T_1} - R\ln\frac{p_f}{p_1}\right] + m_2\left[C_p \ln\frac{T_f}{T_2} - R\ln\frac{p_f}{p_2}\right]$$

$$= -m_1R \ln\frac{p_f}{p_1} - m_2R \ln\frac{p_f}{p_2}$$

$$= -2.973 \times 0.287 \ln\frac{1.75}{2.5} - 1.189 \times 0.287 \ln\frac{1.75}{1}$$

$$= 0.1134 \text{ kJ/K}$$

Problem 10.39 : A vessel is divided into three compartments (a), (b), and (c) by two partitions. Part (a) contains oxygen and has a volume of 0.1 m^3, (b) has a volume of 0.2 m^3 and contains nitrogen, while (c) is 0.05 $m^3$and holds CO_2. All three parts are at a pressure of 2 bar and a temperature of 13°C. When the partitions are removed and the gases mix, determine the change of entropy of each constituent, the final pressure in vessel and partial pressure of each gas. The vessel may be taken as being completely isolated from its surroundings.

Solution : Given data :

$$P = 2 \text{ bar}$$

$$T = 13°C = 13 + 273 = 286 \text{ K}$$

After mixing, final pressure

$$= 2 \text{ bar}$$

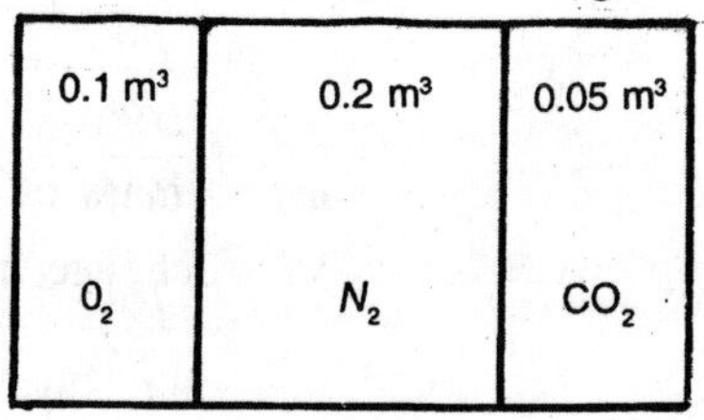

Fig. P. 10.39

For compartment a

$$R_a = \frac{\bar{R}}{M} = \frac{8.314}{32} = 0.2598 \text{ kJ/kg.K},$$

$$m_a = \frac{pV_a}{R_a T} = \frac{2 \times 100 \times 0.1}{0.2598 \times 286} = 0.269 \text{ kg}$$

n_a = no. of mole of

For compartment b

$$R_b = \frac{\bar{R}}{M} = \frac{8.314}{28} = 0.297 \text{ kJ/kg.K}$$

$$m_b = \frac{pV_b}{R_b \times T} = \frac{2 \times 100 \times 0.2}{0.297 \times 286} = 0.471 \text{ kg}$$

For compartment c

$$R_c = \frac{\bar{R}}{M} = \frac{8.314}{44} = 0.189 \text{ kJ/kg.K}$$

$$m_c = \frac{p \times V_c}{R_b \times T} = \frac{200 \times 0.05}{0.189 \times 286} = 0.185 \text{ kg}$$

Now partial pressure of

$$O_2 = p_a = \frac{V_a}{V_a + V_b + V_c} \times p$$

$$= \frac{0.1}{0.35} \times 2 = 0.5714 \text{ bar}$$

paritial pressure of

$$N_2 = p_b = \frac{V_b}{V_a + V_b + V_c} p$$

$$= \frac{0.2}{0.35} \times 2 = 1.142 \text{ bar}$$

paritial pressure of

$$CO_2 = p_c = \frac{V_c}{V_a + V_b + V_c} p$$

$$= \frac{0.05}{0.35} \times 2 = 0.286 \text{ bar}$$

ΔS = change in entropy at constant temperature

$$= mC_p \ln \frac{T_2}{T_1} - mR \ln \frac{p_2}{p_1}$$

$$\Delta S = -mR \ln \frac{p_2}{p_1}$$

$\therefore$ $(\Delta S)_a$ = change in entropy

$$= -m_a R_a \ln \frac{p_a}{p}$$

$$= -0.269 \times 0.2598 \ln \frac{0.5714}{2} = 0.0875 \text{ kJ/K}$$

$$(\Delta S)_b = -m_b R_b \ln \frac{p_b}{p}$$

$$= -0.471 \times 0.297 \ln \frac{1.143}{2} = 0.0783 \text{ kJ/kg.K}$$

$$(\Delta S)_c = -m_c R_c \ln \frac{p_c}{p}$$

$$= -0.185 \times 0.189 \ln \frac{0.2857}{2} = 0.068 \text{ kJ/kg.K}$$

Problem 10.40 : A gas occupies 0.024 m³ at 700 kPa 95°C. It is exppanded in the nonflow process according to the law $pv^{1.2}$= constant to a pressure of 70 kPa after which it is heated at constant pressure back to its original temperature. Sketch the process on the *p-v* and *T–s* diagrams, and calculate for the whole process the work done, the heat transferred, and the change of entropy. Take C_p = 1.047 and C_v = 0.775 kJ/kg.K for the gas.

Solution : Given data :

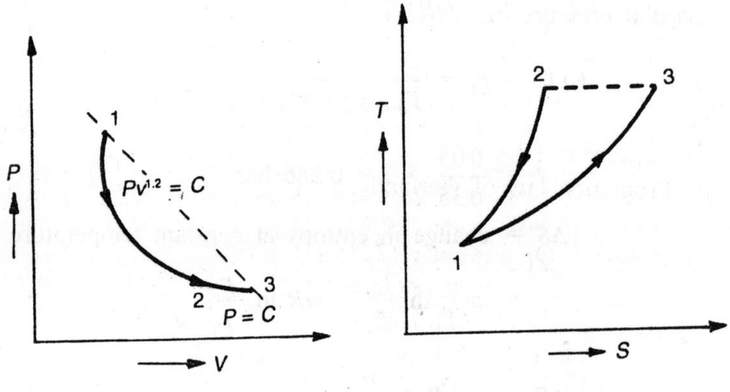

Fig. P. 10.40

Initial conditions,

$$V_1 = 0.024\text{m}^3$$

$$p_1 = 700 \text{ kPa}$$

$$T_1 = 95°\text{C} = 95 + 273 = 368 \text{ K}$$

After expansion

$$P_2 = 70 \text{ kPa}$$

$$P_3 = P_2$$

$$T_3 = 368 \text{ K}$$

For process 1-2

$$pv^{1.2} = C$$

$$\therefore \quad T_2 = T_1 \left(\frac{P_2}{P_1}\right)^{\frac{1.2-1}{1.2}}$$

$$= 368 \left(\frac{70}{700}\right)^{\frac{0.2}{1.2}} = 250.72 \text{ K}$$

$$V_2 = V_1 \left(\frac{P_1}{P_2}\right)^{\frac{1}{\gamma}} = 0.024 \times (10)^{\frac{1}{1.2}} = 0.164 \text{ m}^3$$

$\because \quad C_p - C_v = R$

or $\quad R = 1.047 - 0.775 = 0.272$ kJ/kg.K

$$\text{Mass of gas} = m = \frac{p_1 V_1}{RT_1} = \frac{700 \times 0.024}{0.272 \times 368} = 0.168 \text{ kg}$$

$$\therefore \quad w_{1-2} = \frac{mR(T_1 - T_2)}{n-1}$$

$$= \frac{0.168 \times 0.272(368 - 250.72)}{0.2} = 26.77 \text{ kJ}$$

From first law of thermodynamics

$$dQ = du + dw$$

$$\therefore \quad Q_{1-2} = mCv\,(T_2 - T_1) + w_{1-2}$$

$$= 0.168 \times 0.775\,(250.72 - 368) + 26.77$$

$$= 11.5 \text{ kJ}$$

$$(\Delta S)_{1-2} = \text{Change in entropy}$$

$$= mC_p \ln \frac{T_2}{T_1} - mR \ln \frac{p_2}{p_1}$$

$$= 0.168 \left[1.047 \ln \frac{250.72}{368} - 0.272 \ln \frac{70}{700}\right]$$

$$= 0.03772 \text{ kJ/K}$$

For process 2-3

At constant process

$$\Rightarrow \quad V_3 = V_2 \frac{T_3}{T_2} = 0.164 \times \frac{368}{250.72} = 0.247 \text{ m}^3$$

$$\therefore \quad w_{2-3} = p_2\,(V_3 - V_2)$$

$$= 70\,(0.247 - 0.164) = 5.81 \text{ kJ}$$

$\therefore$ From first law of thermodynamics,

$$dQ = dh - Vdp$$

at constant pressure

$$dQ = dh$$

$$\therefore \quad Q_{2-3} = mCp\ (T_3 - T_2)$$

$$= 0.168 \times 1.047\ (368 - 250.72)$$

$$= 20.63 \text{ kJ}$$

$\therefore \quad (\Delta S_{2-3})$ = Change in entropy at constant pressure

$$= mC_p \ln \frac{T_3}{T_2}$$

$$= 0.168 \times 1.\ 047 \ln \frac{368}{250.72}$$

$$= 0.168 \times 1.047 \ln = 0.0675 \text{ kJ/K.}$$

$\therefore$ Work transfer for whole procss

$$= w_{1-2} + w_{2-3} = 26.77 + 5.81$$

$$= 32.58 \text{ kJ}$$

Heat transfer for whole process

$$= Q_{1-2} + Q_{2-3}$$

$$= 11.5 + 20.63 = 32.13 \text{ kJ}$$

Change in entropy of whole process

$$= (\Delta S)_{1-2} + (\Delta S)_{2-3} = 0.10522 \text{ kJ/K}$$

Problem 10.41 : The specific heats of a gas are of the form $C_p = a + KT$ and $C_v = b + kT$ where a, b and k are constants and T is in K. Derive the formula $T^b\ V^{a-b}\ e^{KT}$ = Constant for the adiabatic expanision of the gas. (CES – 1996)

Solution : Given data :

The heat transfer during any process is given by

$$dQ = W + \Delta U$$

$$= pdv + mC_v dT$$

The change in entropy is given by

$$ds = \frac{dQ}{T} = \frac{pdv}{T} + \frac{mC_v dT}{T}$$

$$= \frac{mRdv}{v} + m\frac{C_v\ dT}{T} \quad \text{as } pv = mRT$$

where the value of R is heat units.

$$\therefore \quad s_2 - s_1 = mR \int_1^2 \frac{dv}{v} + mC_v \int_1^2 \frac{dT}{T}$$

$$= m\,(C_p - C_v) \int_1^2 \frac{dv}{v} + mC_v \int_1^2 \frac{dT}{T}$$

$$= m\,(a - b) \int_1^2 \frac{dv}{v} + m \int_1^2 (b + kT) \frac{dT}{T}$$

$$= m\,(a - b)\,\log_e \left(\frac{v_2}{v_1}\right) + m \left[b \log_e \left(\frac{T_2}{T_1}\right) + k \cdot (T_2 - T_1) \right]$$

For adiabatic expansion,

$$s_2 - s_1 = 0 = (a - b)\,\log_e \left(\frac{v_2}{v_1}\right) + b\,\log_e \left(\frac{T_2}{T_1}\right) + K\,(T_2 - T_1)$$

i.e. $v^{(a-b)} \times T^b \times e^{kT} = \text{Constant}$

Problem 10.42 : A gas obeys the clasius equation of state $P\,(V - b) = RT$, where b is a constant. The internal energy of this gas obeys the law $U = C_v T + \text{constant}$. For a reversible adiabatic, process executed by this gas show that $p(v - b)^{\gamma} = \text{constant}$, where $\gamma = \left(\frac{C_p}{C_v}\right)$. (CES – 1995)

Solution : Given data :

$$P(V - b) = RT \qquad \ldots(1)$$

$$U = C_v\,T + \text{constant} \qquad \ldots(2)$$

$$V - b = \frac{RT}{P}$$

Differentiating

$$dV = \frac{P\,(RdT) - RT\,(dP)}{P^2} \qquad \ldots(3)$$

Differentiating equation (2), we get

$$dU = C_v\,(dT) \qquad ...(4)$$

Since the process is reversible, adiabatic heat transfer is equal to zero. According to first law of thermodynamics

$$dQ = dU + dW = 0 \qquad ...(5)$$

$$du = C_v\,(dT)$$

$$dW = pdV$$

$$= p\left[\frac{P\,(RdT) - RT\,(dP)}{P^2}\right]$$

$$= RdT - RT\left(\frac{dP}{P}\right)$$

$$0 = C_v\,(dT) + R\,(dT) - RT\left(\frac{dP}{P}\right)$$

Dividing by RT

$$\frac{C_v}{R}\left(\frac{dT}{T}\right) + \left(\frac{dT}{T}\right) - \left(\frac{dP}{P}\right) = 0 \qquad ...(6)$$

$$P\,(V - b) = RT$$

Differentiating

$$P\,(dV) + (V - b)\,dP = R\,(dT)$$

Dividing by

$$P\,(V - b) = RT$$

$$\frac{dV}{(V-b)} + \left(\frac{dP}{P}\right) = \frac{dT}{T} \qquad ...(7)$$

Substituting this in equation (6)

$$\frac{C_v}{R}\left(\frac{dT}{T}\right) + \frac{dV}{(V-b)} + \left(\frac{dP}{P}\right) - \left(\frac{dP}{P}\right) = 0$$

$$\frac{dV}{(V-b)} + \frac{C_v}{R}\left[\frac{dV}{(V-b)} + \left(\frac{dP}{P}\right)\right] = 0$$

$$\frac{dV}{(V-b)}\left[\frac{R}{C_v}+1\right]+\frac{dP}{P} = 0$$

$$\left(\because R = C_p - C_v \text{ and } \gamma = \frac{C_p}{C_v}\right)$$

$$\gamma\frac{dV}{(V-b)}+\frac{dP}{P} = 0$$

Integrating this, we get

$$\gamma \log (V-b) + \log P = 0$$

$$P(V-b)^{\gamma} = 0$$

Chapter 11

Thermodynamic Relations

Problem 11.1 : Derive the following equations

(a) $$U = F - T\left(\frac{\partial F}{\partial T}\right)_v = -T^2\left(\frac{\partial F/T}{\partial T}\right)_v$$

(b) $$C_v = -T\left(\frac{\partial^2 F}{\partial T^2}\right)_v$$

(c) $$H = G - T\left(\frac{\partial G}{\partial T}\right)_p = -T^2\left(\frac{\partial G/T}{\partial T}\right)_p$$

(d) $$C_p = -T\left(\frac{\partial^2 G}{\partial T^2}\right)_p$$

Solution: We know that Helmholte function

$$F = u - Ts$$

or, $$dF = du - Tds - sdT$$

or $$dF = -pdv - sdT \qquad [\because du = Tds - pdv]$$

or $$dF = -pdv - sdT \qquad \ldots(i)$$

Now $$F = f(T, v)$$

or, $$dF = dV\left(\frac{\partial F}{\partial v}\right)_T + dT\left(\frac{\partial F}{\partial T}\right)_v \qquad \ldots(ii)$$

Equating the coefficients of dV and dT, we get

$$S = -\left(\frac{\partial F}{\partial T}\right)_v$$

$$\therefore \quad u = F + TS = F - T\left(\frac{\partial F}{\partial T}\right)_v$$

$$= -T^2\left[\frac{1}{T}\left(\frac{\partial F}{\partial T}\right)_v - \frac{F}{T^2}\right]$$

$$= -T^2\left[\frac{1}{T}\left(\frac{\partial F}{\partial T}\right)_v + F\frac{\partial}{\partial T}\left(\frac{1}{T}\right)\right]$$

$$= -T^2\left[\frac{1}{T}\frac{\partial}{\partial T}(F) + F\frac{\partial}{\partial T}\left(\frac{1}{T}\right)\right]_v$$

$$= -T^2\left[\cdot\frac{\partial}{\partial T}(F/T)\right]_v \quad \textbf{Ans}$$

(b) We know $C_v = T\left(\frac{\partial S}{\partial T}\right)_v$

and, $S = -\left(\frac{\partial F}{\partial T}\right)_v$

differentiating with respect to T we get,

$$\therefore \quad \left(\frac{\partial S}{\partial T}\right)_v = -\left(\frac{\partial^2 F}{\partial T^2}\right)_v$$

$$\therefore \quad C_v = -T\left(\frac{\partial^2 F}{\partial T^2}\right)_v \quad \textbf{Ans}$$

(c) Gibbs function,

$$G = H - Ts$$

or, $dG = dH - Tds - sdT$

$$= Vdp - SdT \qquad \ldots\text{(i)}$$

Now $G = f(T.P.)$

or $$dG = dP\left(\frac{\partial G}{\partial P}\right)_T + dT\left(\frac{\partial G}{\partial T}\right)_p \qquad \text{...(ii)}$$

Equating coefficients of dT from eqn (i) & (ii)

$$S = -\left(\frac{\partial G}{\partial T}\right)_p$$

$$H = G + TS = G - T\left(\frac{\partial G}{\partial T}\right)_p$$

$$= -T^2\left[\frac{1}{T}\left(\frac{\partial G}{\partial T}\right)_p - \frac{G}{T^2}\right]$$

$$= -T^2\left[\frac{1}{T}\frac{\partial}{\partial T}(G) + G\frac{\partial}{\partial T}\left(\frac{1}{T}\right)\right]_p$$

$$= -T^2\left[\frac{\partial}{\partial T}(G/T)\right]_p \quad \textbf{Ans}$$

(d) We know that,

$$C_p = T \times \left(\frac{\partial S}{\partial T}\right)_p$$

and $$S = -\left(\frac{\partial G}{\partial T}\right)_p$$

defferentiating with respect to T

$$-\left(\frac{\partial S}{\partial T}\right)_p = -\left(\frac{\partial^2 G}{\partial T^2}\right)_p$$

differentiating with respect to T

$$\left(\frac{\partial S}{\partial T}\right)_p = -\left(\frac{\partial^2 G}{\partial T^2}\right)_p$$

$$\therefore \qquad C_p = -T\left(\frac{\partial^2 G}{\partial T^2}\right)_p \quad \textbf{Ans}$$

Problem 11.2 : (a) Derive the equation

$$\left(\frac{\partial C_v}{\partial v}\right)_T = T \times \left(\frac{\partial^2 p}{\partial T^2}\right)_v$$

(b) Prove that C_v of an ideal gas is a function of T only.

(c) In the case of a gas obeying the equation of state

$$\frac{pv}{RT} = 1 + \frac{B''}{v}$$

where B'' is a function of T only, show that

$$C_v = -\frac{RT}{v}\frac{d^2}{dT^2}(B''T) + (c_v)_0$$

where $(c_v)_0$ is the value at very large volumes.

Solution: (a) We know

$$C_v = T\left(\frac{\partial S}{\partial T}\right)_v$$

differentiating with respect to v

$$\text{or} \quad \left(\frac{\partial C_v}{\partial v}\right)_T = T\frac{\partial^2 S}{\partial T \partial v}$$

$$\text{Now} \quad \left(\frac{\partial S}{\partial v}\right)_T = +\left(\frac{\partial p}{\partial T}\right)_v$$

$$\therefore \quad \frac{\partial^2 S}{\partial T \partial v} = +\left(\frac{\partial^2 p}{\partial T^2}\right)_v$$

$$\therefore \quad \left(\frac{\partial c_v}{\partial v}\right)_T = +T\left(\frac{\partial^2 p}{\partial T^2}\right)_v \quad \textbf{Ans}$$

(b) For an ideal gas

$$P = \frac{n\bar{R}T}{v}$$

$$\therefore \quad \left(\frac{\partial p}{\partial T}\right)_v = \frac{n\bar{R}}{v}$$

and $\left(\frac{\partial^2 p}{\partial T^2}\right)_v = 0$

$\therefore \quad \left(\frac{\partial c_v}{\partial v}\right)_T = 0$

i.e. C_P is a function of T alone

(c) $$\frac{pv}{RT} = 1 + \frac{B''}{v}$$

or $$\frac{B''}{v} = \frac{pv}{RT} - 1$$

or $$B''T = \frac{pv^2}{R} - vT$$

or $$\left(\frac{\partial (B''T)}{\partial T}\right)_v = \frac{v^2}{R}\left(\frac{\partial p}{\partial T}\right)_v - v$$

or $$\left[\frac{\partial^2 (B''T)}{\partial T^2}\right]_v = \frac{v^2}{R}\left(\frac{\partial^2 p}{\partial T^2}\right)_v$$

$$= \frac{v^2}{TR}\left(\frac{\partial c_v}{\partial v}\right)_T \qquad \left[\because \left(\frac{\partial C_v}{\partial v}\right)_T = T\left(\frac{\partial^2 p}{\partial T^2}\right)_v\right]$$

or $$\left(\frac{\partial c_v}{\partial v}\right)_T = \frac{RT}{v^2}\left[\frac{\partial^2 (B''T)}{\partial T^2}\right]_v$$

on integrating, we get

$$c_v = -\frac{RT}{v}\frac{d^2}{dT^2}(B''T) + C_{v0}$$

where C_{v0} = Constant of integration **Ans**

Problem 11.3 : Derive the third *TdS* equation

$$TdS = C_v\left(\frac{\partial T}{\partial p}\right)_v dp + C_p\left(\frac{\partial T}{\partial V}\right)_p dV$$

and show that the three TdS equations may be written as

(a) $$TdS = C_v dT + \frac{\beta T}{k}\, dV$$

(b) $$TdS = C_p dT - V\beta\, Tdp$$

(c) $$TdS = \frac{C_v}{\beta}\, kdp + \frac{C_p}{\beta V}\, dV$$

Solution : Let

$$S = f(P, v)$$

or, $$ds = \left(\frac{\partial S}{\partial p}\right)_v dP + \left(\frac{\partial S}{\partial v}\right)_p dv$$

or $$TdS = T\left(\frac{\partial S}{\partial p}\right)_v dp + T\left(\frac{\partial S}{\partial v}\right)_p dV \quad \text{...(i)}$$

Now $$C_v = T\left(\frac{\partial S}{\partial T}\right)_v$$

or $$C_v\left(\frac{\partial T}{\partial p}\right)_v = T\left(\frac{\partial s}{\partial T}\right)_v \left(\frac{\partial T}{\partial p}\right)_v$$

$$= T\left(\frac{\partial s}{\partial p}\right)_v \quad \text{...(ii)}$$

$$\left[\because \left(\frac{\partial S}{\partial T}\right)_v \left(\frac{\partial T}{\partial p}\right)_v \left(\frac{\partial p}{\partial S}\right)_v = 1\right]$$

Similarly $$C_p = T\left(\frac{\partial S}{\partial T}\right)_p$$

or $$C_p\left(\frac{\partial T}{\partial v}\right)_p = T\left(\frac{\partial S}{\partial T}\right)_p \left(\frac{\partial T}{\partial v}\right)_p$$

$$= T\left(\frac{\partial S}{\partial v}\right)_p \quad \text{...(iii)}$$

$$\left[\because \left(\frac{\partial S}{\partial T}\right)_p \left(\frac{\partial T}{\partial v}\right)_p \left(\frac{\partial u}{\partial S}\right)_p = 1\right]$$

$\therefore$ from (i), (ii) and (iii)

$$TdS = C_v \left(\frac{\partial T}{\partial p}\right)_v dp + C_p \left(\frac{\partial T}{\partial v}\right)_p dV$$

(a) 1st *TdS* equation

$$TdS = C_v dT + T\left(\frac{\partial p}{\partial T}\right)_v dV$$

We know

Volume of expansivity $\beta = \frac{1}{V}\left(\frac{\partial V}{\partial T}\right)_p$

isothermal compressibility, $K = -\frac{1}{V}\left(\frac{\partial V}{\partial p}\right)_T$

$$\therefore \quad \frac{\beta}{k} = -\left(\frac{\partial v}{\partial T}\right)_p \times \left(\frac{\partial p}{\partial v}\right)_T$$

$$= +\left(\frac{\partial p}{\partial T}\right)_v$$

$$\therefore \quad TdS = C_v dT \times \frac{\beta T}{k} dV \textbf{ Ans}$$

(b) 2nd *TdS* equation

$$TdS = C_p dT - T\left(\frac{\partial v}{\partial T}\right)_p dp$$

We know $\beta = \frac{1}{V}\left(\frac{\partial v}{\partial T}\right)_p$

or $\left(\frac{\partial v}{\partial T}\right)_p = \beta v$

$$\therefore \quad TdS = C_p dT - V\beta\, Tdp \textbf{ Ans}$$

(c) $$TdS = C_v \left(\frac{\partial T}{\partial p}\right)_v dp + C_p \left(\frac{\partial T}{\partial v}\right)_p dV$$

we know $$\beta = \frac{1}{V}\left(\frac{\partial v}{\partial T}\right)_p$$

$$\kappa = -\frac{1}{V}\left(\frac{\partial v}{\partial p}\right)_T$$

$$\frac{\kappa}{\beta} = -\left(\frac{\partial u}{\partial p}\right)_T \left(\frac{\partial T}{\partial v}\right)_p = \left(\frac{\partial T}{\partial p}\right)_v$$

$$\frac{1}{\beta} = v\left(\frac{\partial T}{\partial v}\right)_p$$

$\therefore$ $$TdS = C_v \frac{\kappa}{\beta} dp + \frac{C_p}{\beta v} dv \textbf{ Ans}$$

Problem 11.4 : Derive the equations

(a) $$C_p = T\left(\frac{\partial V}{\partial T}\right)_p \left(\frac{\partial P}{\partial T}\right)_s$$

(b) $$\left(\frac{\partial P}{\partial T}\right)_s = \frac{C_p}{V\beta T}$$

(c) $$\frac{(\partial p/\partial T)_s}{(\partial p/\partial T)_v} = \frac{\gamma}{\gamma - 1}$$

Solution: We know that

$$C_p = T\left(\frac{\partial S}{\partial T}\right)_p$$

$\because$ $$\left(\frac{\partial S}{\partial T}\right)_p \left(\frac{\partial T}{\partial p}\right)_s \left(\frac{\partial p}{\partial S}\right)_T = -1$$

$$\therefore \quad \left(\frac{\partial S}{\partial T}\right)_p = -\left(\frac{\partial S}{\partial p}\right)_T \left(\frac{\partial p}{\partial T}\right)_s$$

$$= \left(\frac{\partial v}{\partial T}\right)_p \times \left(\frac{\partial p}{\partial T}\right)_s$$

$\because$ from Maxwell's equation

$$\left(\frac{\partial v}{\partial T}\right)_p = -\left(\frac{\partial S}{\partial p}\right)_T$$

$$\therefore \quad C_P = T\left(\frac{\partial v}{\partial T}\right)_p \times \left(\frac{\partial p}{\partial T}\right)_s \quad \textbf{Ans}$$

(b) We know that

$$\left(\frac{\partial p}{\partial T}\right)_s \times \left(\frac{\partial T}{\partial S}\right)_p \times \left(\frac{\partial S}{\partial p}\right)_T = -1$$

or

$$\left(\frac{\partial p}{\partial T}\right)_s = -\left(\frac{\partial S}{\partial T}\right)_p \times \left(\frac{\partial p}{\partial S}\right)_T \quad \text{...(i)}$$

Now

$$C_p = T\left(\frac{\partial S}{\partial T}\right)_p$$

$$\beta = \frac{1}{V}\left(\frac{\partial V}{\partial T}\right)_p$$

From Maxwell's eqn., $\left(\frac{\partial v}{\partial T}\right)_p = -\left(\frac{\partial S}{\partial p}\right)_T$

putting above values in equation (i)

$$\therefore \quad \left(\frac{\partial p}{\partial T}\right)_s = \frac{C_p}{TV\beta} \quad \textbf{Ans}$$

(c)

$$\frac{\gamma}{\gamma - 1} = \frac{\dfrac{C_p}{C_v}}{\dfrac{C_p}{C_v} - 1} = \frac{\dfrac{C_p}{C_v}}{(C_p - C_v)/C_v}$$

$$\text{or} \qquad \frac{C_p}{C_p - C_v} = \frac{T\left(\frac{\partial v}{\partial T}\right)_p \times \left(\frac{\partial p}{\partial T}\right)_s}{T\left(\frac{\partial p}{\partial T}\right)_v \times \left(\frac{\partial v}{\partial T}\right)_p}$$

$\left[\because \text{ From equating } TdS \text{ equation } C_P - C_v = T\left(\frac{\partial p}{\partial T}\right)_v \times \left(\frac{\partial v}{\partial T}\right)_p\right]$

$$\therefore \qquad \frac{\gamma}{\gamma - 1} = \frac{(\partial p / \partial T)_s}{(\partial p / \partial T)_v} \quad \textbf{Ans}$$

Problem 11.5 : Derive the equations

(a) $$C_v = -T\left(\frac{\partial p}{\partial T}\right)_v \left(\frac{\partial V}{\partial T}\right)_s$$

(b) $$\left(\frac{\partial V}{\partial T}\right)_s = -\frac{C_v k}{\beta T}$$

(c) $$\frac{(\partial V / \partial T)_s}{(\partial V / \partial T)_p} = \frac{1}{1-\gamma}$$

Solution : We know that

$$C_v = T\left(\frac{\partial S}{\partial T}\right)_v$$

$$\because \quad \left(\frac{\partial S}{\partial T}\right)_v \times \left(\frac{\partial T}{\partial V}\right)_s \times \left(\frac{\partial V}{\partial S}\right)_T = -1$$

$$\therefore \quad \left(\frac{\partial S}{\partial T}\right)_v = -\left(\frac{\partial V}{\partial T}\right)_s \times \left(\frac{\partial S}{\partial V}\right)_T$$

$$\therefore \quad C_v = -T\left(\frac{\partial V}{\partial T}\right)_s \times \left(\frac{\partial S}{\partial V}\right)_T$$

From Maxwell's eqn.

$$\left(\frac{\partial S}{\partial V}\right)_T = \left(\frac{\partial p}{\partial T}\right)_v$$

$$\therefore \quad C_v = -T\left(\frac{\partial V}{\partial T}\right)_s \times \left(\frac{\partial p}{\partial T}\right)_v \quad \textbf{Ans}$$

(b) $\because$

$$C_v = -T\left(\frac{\partial v}{\partial T}\right)_s \times \left(\frac{\partial p}{\partial T}\right)_v$$

$$\therefore \quad \left(\frac{\partial v}{\partial T}\right)_S = -\frac{C_v}{T\left(\frac{\partial p}{\partial T}\right)_v}$$

Now

$$K = \frac{1}{V}\left(\frac{\partial V}{\partial p}\right)_T$$

$$\beta = \frac{1}{V}\left(\frac{\partial V}{\partial T}\right)_p$$

$$\Rightarrow \quad \frac{k}{\beta} = -\left(\frac{\partial V}{\partial P}\right)_T \left(\frac{\partial T}{\partial V}\right)_p = \left(\frac{\partial P}{\partial T}\right)_v$$

$$\therefore \quad \left(\frac{\partial V}{\partial T}\right)_s = -\frac{C_v K}{T\,\beta} \quad \textbf{Ans}$$

(c)

$$\frac{1}{1-\gamma} = \frac{1}{1-\frac{C_p}{C_v}} = -\frac{C_v}{C_p - C_v}$$

$$= +\frac{T\left(\frac{\partial p}{\partial T}\right)_v \times \left(\frac{\partial v}{\partial T}\right)_s}{T\left(\frac{\partial p}{\partial T}\right)_v \times \left(\frac{\partial v}{\partial T}\right)_p}$$

$$\because \quad C_v = -T\left(\frac{\partial p}{\partial T}\right)_v \left(\frac{\partial V}{\partial T}\right)_s$$

and $$C_p - C_v = T\left(\frac{\partial p}{\partial T}\right)_v \left(\frac{\partial V}{\partial T}\right)_p$$

$$\therefore \quad \frac{1}{1-\gamma} = \frac{(\partial v / \partial T)_s}{(\partial v / \partial T)_p} \quad \textbf{Ans}$$

> **Problem 11.6:** Prove that the slope of a curve on a Moiller diagram representing a reversible isothermal process is equal to
>
> $$T - \frac{1}{\beta}$$

Solution : We know

$$dw = TdS + Vdp$$

And from 2nd TdS equation, $TdS = C_p dT - T\left(\frac{\partial v}{\partial T}\right)_p dp$

$$dh = C_P\, dT - \left[T\left(\frac{\partial v}{\partial T}\right)_p - V\right] dp$$

For isothermal process, $dT = 0$

$$\therefore \quad \left(\frac{\partial h}{\partial p}\right)_T = -\left[T\left(\frac{\partial v}{\partial T}\right)_p - v = -\left(\frac{\partial v}{\partial T}\right)_p \left[T - \frac{1}{\frac{1}{v}\left(\frac{\partial v}{\partial T}\right)_p}\right]\right]$$

Now $\because$ $$\left(\frac{\partial h}{\partial p}\right)_T = \left(\frac{\partial h}{\partial S}\right)_T \times \left(\frac{\partial S}{\partial p}\right)_T$$

and from Maxwell's relation,

$$\left(\frac{\partial S}{\partial p}\right)_T = -\left(\frac{\partial V}{\partial T}\right)_p$$

$$\therefore \quad \left(\frac{\partial h}{\partial p}\right)_T = \left(\frac{\partial h}{\partial S}\right)_T \times (-) \left(\frac{\partial V}{\partial T}\right)_p$$

$$= (-) \left(\frac{\partial V}{\partial T}\right)_p \left[T - \frac{1}{\beta}\right]$$

$$\left[\because \beta = \frac{1}{V}\left(\frac{\partial V}{\mathrm{T}}\right)_p\right]$$

or $$\left(\frac{\partial h}{\partial S}\right)_T = T - \frac{1}{\beta} \quad \textbf{Proved}$$

Problem 11.7: (a) Show that

$$\mu_J C_P = T^2 \left(\frac{\partial V / T}{\partial T}\right)_p$$

For 1 mole of a gas, in the region of moderate pressures, the equation of state may be written as

$$\frac{p\bar{v}}{\bar{R}T} = 1 + B'p + C'p^2$$

where B' *and* C' *are functions of temperature only*

(b) Show that as $p \to 0$

$$\mu_J C_P \to \bar{R}T^2 \frac{dB'}{dT}$$

(c) Show that the equation of the inversion curve is

$$p = -\frac{dB'/dT}{dC'/dT}$$

Solution: From property relation

$$dh = TdS + VdP$$

and $$TdS = C_p dT - T\left(\frac{\partial V}{\partial T}\right)_p dp$$

$$\therefore \qquad dh = C_p\, dT - \left[T\left(\frac{\partial V}{\partial T}\right)_p - V\right] dP$$

For constant enthapy process, $dh = 0$

$$\therefore \qquad \mu_j = \left(\frac{\partial T}{\partial p}\right)_h$$

$$= \frac{1}{C_p}\left[T\left(\frac{\partial V}{\partial T}\right)_p - V\right]$$

or $$C_p \times \mu_J = T^2\left[\frac{1}{T}\left(\frac{\partial V}{\partial T}\right)_p - \frac{V}{T^2}\right]$$

$$= T^2\left[\frac{1}{T}\left(\frac{\partial V}{\partial T}\right)_p + V \times \frac{\partial}{\partial T}\left(\frac{1}{T}\right)\right]$$

$$= T^2\left[\frac{\partial (V/T)}{\partial T}\right]_p$$

(b) $$\frac{p\bar{V}}{\bar{R}T} = 1 + B'p + C'p^2$$

or $$\frac{\bar{V}}{T} = \frac{\bar{R}}{p} + B'\bar{R} + C'\bar{R}\,p$$

or, $$\left[\frac{\partial\left(\bar{V}/T\right)}{\partial T}\right]_p$$

$$= 0 + \bar{R}\,\frac{\partial B'}{\partial T} + p\,\bar{R}\,\frac{\partial C'}{\partial T}$$

as $p \to 0$,

$$\left[\frac{\partial\left(\bar{V}/T\right)}{\partial T}\right]_p = \bar{R}\,\frac{\partial B'}{\partial T}$$

Hence $C_P \times \mu_J = T^2 \left[-\dfrac{\partial(\bar{V}/T)}{\partial T} \right]_p$

or $$C_P M_J \to T^2 \bar{R} \frac{\partial B'}{\partial T} \quad \textbf{Ans}$$

(c) $$C_p m_J = T^2 \left[\frac{\partial(\bar{V}/T)}{\partial T} \right]_p$$

$$= T^2 \left[\bar{R}\frac{\partial B'}{\partial T} + p\bar{R}\frac{\partial C'}{\partial T} \right]_p$$

For inversion curve,

$$\mu_J = 0$$

Hence from above eqn.,

$$\left[\bar{R}\frac{\partial B'}{\partial T} + p\bar{R}\frac{\partial C'}{\partial T} \right]_p = 0$$

or $$\frac{dB'}{dT} + p\frac{dc'}{dT} = 0$$

$\therefore$ $$p = -\frac{dB'/dT}{dc'/dT} \quad \textbf{Ans}$$

Proplem 11.8: Estimate the maximum inversion temperature of hydrogen if it is assumed to obey the equation of state

$$pV = RT + B_1 p + V_2 p^2 + B_3 p^3 + \ldots$$

For hydrogen,

$$B_1 \times 10^5 = a + 10^{-2}\, bT + 10^2\, c/T$$

where $a = 166,\ b = -7.66,\ c = -172.33$

Solution: Given :

$$pV = RT + B_1 p + B_2 p^2 + B_3 p^3 + \ldots$$

or $$V = \frac{RT}{p} + B_1 + B_2 p + B_3 p^2 + \ldots$$

$$\text{Now } \left(\frac{\partial V}{\partial T}\right)_p = \frac{R}{p} + \left(\frac{\partial B_1}{\partial T}\right)_p + p\left(\frac{\partial B_2}{\partial T}\right)_p + p^2 \times \left(\frac{\partial B_3}{\partial T}\right)_p + \dots$$

$$\therefore \qquad \mu_J = \frac{1}{C_p}\left[T\left(\frac{\partial V}{\partial T}\right)_p - V\right] = \frac{1}{C_p}$$

$$\left[\frac{RT}{p} + T\left(\frac{\partial B_1}{\partial T}\right)_p + pT\left(\frac{\partial B_2}{\partial T}\right)_p + p^2T\left(\frac{\partial B_3}{\partial T}\right)_p + \dots\right.$$

$$\left. - \frac{RT}{p} - B_1 - B_2 p - B_3 p^2 - \dots\right]$$

$$= \frac{1}{C_p}\left[T\left(\frac{\partial B_1}{\partial T}\right)_p + pT\left(\frac{\partial B_2}{\partial T}\right)_p + \dots - B_1 - B_2 p - B_3 p^2 \dots\right]$$

For inversion curve $\underset{p \to 0}{Lt}\ \mu_J = 0$

$$\text{or} \qquad \frac{1}{C_p}\left[T\left(\frac{\partial B_1}{\partial T}\right)_p - B_1\right] = 0 \text{ or } T = \frac{B_1}{\left(\frac{\partial B_1}{\partial T}\right)_p}$$

$$\text{Now } B_1 \times 10^5 = a + 10^{-2}\, bT + 10^{+2}\, \frac{C}{T}$$

$$10^5 \times \left(\frac{\partial B_1}{\partial T}\right)_p = 0 + 10^{-2}\, b - 10^2\, C/T^2$$

$$\therefore \qquad T = \frac{B_1}{\left(\frac{\partial B_1}{\partial T}\right)_p} = \frac{a + 10^{-2} bT + 10^2 C/T}{10^{-2}\, b - 10^{+2}\, \frac{C}{T^2}}$$

$$\text{or } 10^{-2}\, bT - 10^{+2}\, \frac{C}{T}$$

$$= a + 10^{-2}\, bT + 10^2\, \frac{C}{T}$$

or, $$a = -2 \times 10^2 \frac{C}{T}$$

or $$T_i = \text{inversion temperature}$$

$$= -2 \times 10^{+2} \frac{C}{a}$$

$$= \frac{-2 \times 10^{+2} \times (-)172.33}{166}$$

$$= 207.63°\text{C} \textbf{ Ans}$$

Problem 11.9: The vapour pressure of mercury at 399 K and 401 K is found to be 0.988 mm and 1.084 mm of mercury respectively. Calculate the latent heat of vaporization of liquid mercury at 400 K.

Solution: *We know that*

$$l_{vap} = \frac{\bar{R}T^2}{p} \times \frac{dp}{dT}$$

given at 399 K vapour pressure = 0.988 mm of Hg
at 401 K vapour pressure = 1.084 mm of Hg

$$\therefore \quad \frac{dp}{dT} = \frac{1.084 - 0.988}{401 - 399} \text{ mm of Hg/K}$$

$$= 0.048 \text{ mm of Hg/K}$$

∴ from interpolaiton, Vapour pressure at 400°C

$$= \left(\frac{400 - 399}{401 - 399}\right) \times (1.084 - 0.988) + 0.988$$

$$= 1.036 \text{ mm of Hg}$$

∴ Latent heat of Vapourization at 400 K

$$= \frac{\bar{R}T^2}{p} \times \frac{dp}{dT}$$

$$= \frac{8.314 \times 400^2 \times 0.048}{1.036}$$

$$= 61632.74 \text{ kJ/kg.mol}$$

Problem 11.10: In the vicinity of the triple point, the vapour pressure of liquid ammonia (in atmospheres) is represented by

$$\ln p = 15.16 - \frac{3063}{T}$$

This is the equation of the liquid–vapour boundary curve in a p–T diagram. Similarly, the vapour pressure of solid ammonia is

$$\ln p = 18.70 - \frac{3754}{T}$$

(a) What is the temperature and pressure at the triple point?
(b) What are the latent heats of sublimation and vaporization?
(c) What is the latent heat of fusion at the triple point?

Solution: (a) A the triple point, the saturated solid and saturated liquid–line meets.

∴ vapour pressure of liquid ammonia = vapour pressure of solid ammonia

$$\Rightarrow 5.16 - \frac{3063}{T} = 18.7 - \frac{3754}{T}$$

or $$\frac{691}{T} = 3.54$$

or $$T = 195.2 \text{ K } \textbf{Ans}$$

or $$\ln p = 15.16 - \frac{3063}{T}$$

$$= 15.16 - \frac{3063}{195.2}$$

∴ $$p = 0.587 \text{ atm } \textbf{Ans}$$

(b) From clausius clapeyron equation for sublimation. [$V_{\text{Vap}} \gg V_{\text{solid}}$]

$$\frac{dp}{dT} = \frac{p}{\bar{R}T^2} \times l_{\text{sub}}$$

Now for solid ammonia

$$\ln p = 18.7 - \frac{3754}{T}$$

$$\frac{1}{p}\frac{dp}{dT} = 0 + \frac{3754}{T^2}$$

or $$\frac{1}{p}\frac{dp}{dT} = \frac{l_{sub}}{\overline{R}T^2}$$

$$\therefore \quad l_{sub} = \frac{3754}{T^2} \times \overline{R}\,T^2$$
$$= 3754 \times 8.314 \text{ kJ/kg.mol}$$
$$= \frac{3754 \times 8.314}{17} \text{ kJ/kg} = 1836 \text{ kJ/kg } \textbf{Ans}$$

Similarly for liquid ammonia

$$\ln p = 15.16 - \frac{3063}{T}$$

or $$\frac{1}{p}\frac{dp}{dT} = 0 + \frac{3063}{T^2}$$

$\therefore$ from clausius clapeyron equation for vaporization

$$\Rightarrow \quad \frac{1}{p}\frac{dp}{dT} = \frac{l_{vap}}{\overline{R}T^2}$$

or $$l_{vap} = 3063 \times \overline{R}$$
$$= 3063 \times 8.314 \text{ kJ/kg mol}$$
$$= \frac{3063 \times 8.317}{17} \text{ kJ/kg } \textbf{Ans}$$

(c) At triple point,

$$l_{sub} = l_{vap} + l_{fusion}$$
$$\therefore \quad l_{fusion} = \text{latent heat of fusion} = l_{sub} - l_{vap}$$
$$= 1836 - 1498 = 338 \text{ kJ/kg } \textbf{Ans}$$

Problem 11.11: Compute μ_J for a gas whose equation of state is $p\,(v - b) = RT$

Solution : We know that

$$\mu_J = \frac{1}{C_p}\left[T\left(\frac{\partial V}{\partial T}\right)_T - V\right]$$

Now $P\,(V - b) = RT$

or $$V - b = \frac{RT}{p}$$

or $$V = \frac{RT}{p} + b$$

or $$\left(\frac{\partial v}{\partial T}\right)_p = \frac{R}{p}$$

$\therefore$ $$\mu_j = \frac{1}{C_p}\left[T\left(\frac{\partial v}{\partial T}\right) - v\right]$$

$$= \frac{1}{C_p}\left[\frac{RT}{p} - v\right]$$

$$= \frac{1}{C_p}\left[\frac{RT}{p} - \frac{RT}{p} - b\right] = -\frac{b}{C_p} \quad \textbf{Ans}$$

Problem 11.12: Show that

(a) $$\left(\frac{\partial \beta}{\partial p}\right)_T = -\left(\frac{\partial k}{\partial T}\right)_p$$

(b) $$\left(\frac{\partial u}{\partial p}\right)_T = -T\left(\frac{\partial v}{\partial T}\right)_p - p\left(\frac{\partial v}{\partial p}\right)_T$$

Solution:

Volume expansivity $= \beta = \frac{1}{v}\left(\frac{\partial v}{\partial T}\right)_p$

$\therefore$ $$\left(\frac{\partial \beta}{\partial T}\right)_T = \frac{1}{v}\left[\frac{\partial}{\partial p}\left(\frac{\partial v}{\partial T}\right)_p\right]_T + \left(\frac{\partial v}{\partial T}\right)_p\left(-\frac{1}{v^2}\right)_p\left(\frac{\partial v}{\partial p}\right)_T$$

$$= \frac{\partial}{\partial T}\left[\frac{1}{V}\left(\frac{\partial v}{\partial p}\right)_T\right]_p - \frac{1}{v^2}\left(\frac{\partial u}{\partial T}\right)_p\left(\frac{\partial v}{\partial p}\right)_T \quad \text{...(i)}$$

and $$K = -\frac{1}{v}\left(\frac{\partial v}{\partial p}\right)_T$$

or $$\left(\frac{\partial K}{\partial T}\right)_p = -\frac{1}{v}\left[\frac{\partial}{\partial T}\left(\frac{\partial v}{\partial p}\right)_T\right]_p + \left(\frac{\partial v}{\partial p}\right)_T \frac{1}{v^2}\left(\frac{\partial v}{\partial T}\right)_p \quad \text{...(ii)}$$

Hence from (i) and (ii)

$$\left(\frac{\partial \beta}{\partial p}\right)_T = -\left(\frac{\partial K}{\partial T}\right)_p \quad \textbf{Ans}$$

(b) From 1st TdS relation,

$$du = c_v dT + \left[T\left(\frac{\partial p}{\partial T}\right)_v - p\right] dv \quad \text{...(i)}$$

Let $$u = f(T, v)$$

$$du = \left(\frac{\partial u}{\partial T}\right)_v dT + \left(\frac{\partial u}{\partial v}\right)_T dv$$

again $$u = f(T, p)$$

$$du = \left(\frac{\partial u}{\partial T}\right)_p dT + \left(\frac{\partial u}{\partial p}\right)_T dp \quad \text{...(ii)}$$

At constant temperature, $dT = 0$

$$\therefore \quad du = \left(\frac{\partial u}{\partial p}\right)_T dp = \left[T\left(\frac{\partial p}{\partial T}\right)_v - p\right] dv$$

or $$\left(\frac{\partial u}{\partial p}\right)_T = \left[T\left(\frac{\partial p}{\partial T}\right)_v - p\right]\left(\frac{\partial v}{\partial p}\right)_T$$

$$= T \times \left(\frac{\partial p}{\partial T}\right)_v \left(\frac{\partial v}{\partial p}\right)_T - p\left(\frac{\partial v}{\partial p}\right)_T$$

$$= -T\left(\frac{\partial v}{\partial T}\right)_p - p\left(\frac{\partial V}{\partial p}\right)_T \quad \textbf{Ans}$$

$$\left[\because \left(\frac{\partial p}{\partial T}\right)_v \times \left(\frac{\partial v}{\partial p}\right)_T \times \left(\frac{\partial T}{\partial v}\right)_p = -1\right]$$

Problem 11.13: Two particular systems have the following equations of state

$$\frac{1}{T^{(1)}} = \frac{3}{2} \overline{R} \frac{N^{(1)}}{U^{(1)}}$$

and
$$\frac{1}{T^{(2)}} = \frac{5}{2} \overline{R} \frac{N^{(2)}}{U^{(2)}}$$

where $\overline{R}$ = 8.3143 kJ/kg mol K. The mole number of the first system is $N^{(1)} = 2$, and that of the second is $N^{(2)} = 3$. The two systems are separated by a diathermal wall, ands the total energy in the composite system is 25.120 kJ. What is the internal energy of each system in equilibrium?

Solution: Given

$$U^{(1)} = \frac{3}{2} \overline{R} N^{(1)} T^{(1)}$$

$$U^{(2)} = \frac{5}{2} \overline{R} N^{(2)} T^{(2)}$$

$$\begin{aligned} \overline{R} &= 8.3143 \text{ kJ/kg mol K} \\ N^{(1)} &= 2 \text{ mole,} \\ N^{(2)} &= 3 \text{ mole} \\ U_f &= U_{\text{Total}} = 25.120 \text{ kJ} \end{aligned}$$

Let T_f be the temperature at equillibrium

then $U_f^{(1)}$ = final energy of subsystem 1

$$= \frac{3}{2} \overline{R} N^{(1)} T_f$$

Similarly
$$U_f^{(2)} = \frac{5}{2} \overline{R} N^{(2)} T_f$$

$\therefore$ for a composite system at equillibrium

$$U_f^{(1)} + U_f^{(2)} = U_f$$

or $\frac{3}{2} \times 8.3143 \times 2 \times T_f + \frac{5}{2} \times 8.3143 \times 3\ T_f$

$$= 25.120 \text{ kJ}$$

or $T_f = 287.74$ K

$\therefore$ $U_f^{(1)} = \frac{3}{2} \times 8.3143 \times 2 \times 287.74 = 7.2$ kJ

Similarly $U_f^{(2)} = \frac{5}{2} \times 8.3143 \times 3 \times 287.74 = 17.92$ kJ **Ans**

Problem 11.14: Two systems with the equations of state given in Problem 11.13 are separated by a diathermal wall. The respective mole numbers are $N^{(1)} = 2$ and $N^{(2)} = 3$. The initial temperature are $T^{(1)} = 250$ K and $T^{(2)} = 350$ K. What are the values of $U^{(1)}$ and $U^{(2)}$ after equilibrium has been established? What is the equilibrium temperature?

Solution: Given

$$N^{(1)} = 2 \text{ mole},$$
$$N^{(2)} = 3 \text{ mole}$$
$$T^{(1)} = 250 \text{ K}$$
$$T^{(2)} = 350 \text{ K}$$

For a composit system

$$U^{(1)} + U^{(2)} = \text{constant}$$

or $\Delta U^{(1)} + \Delta U^{(2)} = 0$

or $\Delta U^{(1)} = -\Delta U^{(2)}$

Now T_f be the final equilibrium temperature

Then

$$\frac{3}{2}\ \bar{R}\ \bar{N}^{(1)}\ (T_f - T^{(1)}) = -\frac{5}{2}\ \bar{R}\ N^{(2)}\ (T_f - T^{(2)})$$

or $3 \times 2\ (T_f - 250) = 5 \times 3\ (350 - T_f)$

$\Rightarrow$ $T_f = \dfrac{5 \times 3 \times 350 + 3 \times 2 \times 250}{3 \times 2 + 5 \times 3}$

$= 321.43$ K **Ans**

$\therefore$ $U_f^{(1)}$ = final enternal energy of subsystem, i.e., at equillibrium of subsystem 1

$$= \frac{3}{2} R N^{(1)} T_f$$

$$= \frac{5}{2} \times 8.3143 \times 2 \times 321.43$$

$$= 8.02 \text{ kJ } \textbf{Ans}$$

and $V_f^{(2)}$ = equilibrium internal energy of subsystem 2

$$= \frac{5}{2} R N^{(1)} T_f$$

$$= \frac{5}{2} \times 8.3143 \times 3 \times 341.2$$

$$= 20.04 \text{ kJ } \textbf{Ans}$$

Problem 11.15: Show that the change in latent heat L with temperature is given by the following relation

$$\left(\frac{dL}{dT}\right) = \left(C_p''' - C_p''\right) + \frac{L}{T} - \frac{v'''\beta''' - v''\beta''}{v''' - v''} L$$

Solution: From Clausius – Clapeycon equation for latent heat of vaporisation

$$L = T \frac{dp}{dT} \left(v''' - v''\right), \text{ where } v''' \gg v''$$

$$\therefore \quad \frac{dL}{dT} = \frac{dp}{dT} \left(v''' - v''\right) + \left(v''' - v''\right) T \frac{d^2 p}{dT^2} + T \frac{dp}{dT} \left[\frac{dv'''}{dT} - \frac{dv'}{dT}\right]$$

Now $$C_p = T \left(\frac{\partial S}{\partial T}\right)_p$$

or $$\left(\frac{\partial C_p}{\partial V}\right)_T = T \left[\frac{\partial}{\partial T}\left(\frac{\partial S}{\partial V}\right)_T\right]_p$$

$$= T \left[\frac{\partial}{\partial T}\left(\frac{\partial p}{\partial T}\right)_p\right]_p$$

$$\left[\because \text{From Maxwell's eqn,} \left(\frac{\partial S}{\partial v}\right)_T = (-)\left(\frac{\partial p}{\partial T}\right)_p\right]$$

$$\therefore \quad \left(\frac{\partial C_p}{\partial v}\right)_T = -T\left[\frac{\partial^2 p}{\partial T^2}\right]_p$$

or

$$dC_P = -T\,\frac{\partial^2 p}{\partial T^2}\,(v''' - v'')\,dV$$

or

$$C_p''' - C_p'' = -T\,\frac{\partial^2 p}{\partial T^2}\,(V''' - V'')$$

again

$$\beta = \frac{1}{V}\left(\frac{dV}{dT}\right)_p$$

$$\therefore \quad \frac{dV'''}{dT} = \beta'''\,V''' \;\&\; \frac{dV''}{dT} = \beta''\,V''$$

$$\therefore \quad \frac{dL}{dT} = \frac{L}{T} - (C_P''' - C_P'') + \frac{L\left(V'''\,\beta''' - V''\,\beta''\right)}{\left(V''' - V''\right)}$$ **Ans**

Problem 11.16: Show that for a van der Waals gas, the Joule–Thomson coefficient is given by

$$\mu_j = \frac{v}{C_p}\left[\frac{2a(v-b)^2 - RTbv^2}{RTv^3 - 2a(v-b)^2}\right]$$

Solution: We know that Joule–Thomson coefficient

$$\mu_j = \frac{1}{C_p}\left[T\left(\frac{\partial v}{\partial T}\right)_p - v\right]$$

Now from Vander Waal's equation of state

$$\left(p + \frac{a}{v^2}\right)(v - b) = RT$$

$$pv + \frac{a}{v} - pb - \frac{ab}{v^2} = RT$$

differentiating w.r.t. to T at constant pressure, we get,

$$p\left(\frac{\partial v}{\partial T}\right)_p - \frac{a}{v^2}\left(\frac{\partial v}{\partial T}\right)_p + \frac{2ab}{V^3}\left(\frac{\partial v}{\partial T}\right)_p = \text{R}$$

or $$\left(\frac{\partial v}{\partial T}\right)_p = \frac{R}{p - \frac{a}{v^2} + \frac{2ab}{v^3}}$$

$$= \frac{R}{\left(p + \frac{a}{v^2}\right) - \frac{2a(v-b)}{v^3}}$$

or $$\left(\frac{\partial v}{\partial T}\right)_p = \frac{R}{\frac{RT}{(v-b)} - \frac{2a(v-b)}{v^3}} \quad \left[\because \left(p + \frac{a}{v^2}\right) = \frac{RT}{v-b}\right]$$

$$= \frac{Rv^3(v-b)}{RTv^3 - 2a(v-b)^2}$$

or $$T\left(\frac{\partial v}{\partial T}\right)_p - v = \frac{RTv^3(v-b)}{RTv^3 - 2a(v-b)^2} - v$$

$$= \frac{RTv^3 \times v - RTv^3b - RTv^4 + 2av(v-b)^2}{RTv^3 - 2a(v-b)^2}$$

$$= \frac{2av(v-b)^2 - RTV^3b}{RTv^3 - 2a(v-b)^2}$$

$$\therefore \quad \mu_J = \frac{v}{C_P}\left[\frac{2a(v-b)^2 - RTv^2b}{RTv^3 - 2a(v-b)^2}\right] \quad \textbf{Ans}$$

Problem 11.17: At 273.15 K the specific volumes of water and ice are 0.001 and 0.001091 m^3/kg and the latent heat of fusion of ice is 334 kJ/kg. Determine the melting point increase due to increase of pressure by 1 atm (101.325 kPa).

Solution: Given :

at 273.15 K

$$v'' = \text{specific volume of water} = 0.001 \text{ m}^3/\text{kg and}$$
$$v' = \text{specific volume of ice} = 0.001091 \text{ m}^3/\text{kg}$$
$$l_{fu} = \text{latent heat of fusion} = 334 \text{ kJ/kg}$$
$$dp = 101.325 \text{ kPa}$$
$$dT = ?$$

Now from clausius clapeyron equation for fusion

$$\frac{dp}{dT} = \frac{l_{fu}}{T(v'' - v')}$$

or

$$\frac{101.325}{dT} = \frac{334}{273.15(0.001 - 0.001091)}$$

$$\therefore \quad dT = -0.00754 \text{ K } \textbf{Ans}$$

Problem 11.18: Calculate the latent heat of vaporization of steam formed by boiling water under a pressure of 101.325 kPa. At a pressure near this, a rise of temperature of 1 K causes an increase of vapour pressure of 3.62 kPa.

Solution: Given :

$$\text{at } p = \text{pressure} = 101.325 \text{ KPa}$$
$$dT = \text{rise in teperature} = 1 \text{ K}$$
$$dp = 3.62 \text{ KPa}$$

Hence from clausius clapeyron equation for vaporisation

$$l_{\text{vap}} = \frac{\dot{R}T^2}{P}\left(\frac{dp}{dT}\right)$$

$$= \frac{8.314}{18} \times \frac{(373)^2}{101.325} \times \frac{(3.62)}{1}$$

$$= 2295.8 \text{ kJ/kg } \textbf{Ans}$$

Problem 11.19: Show that:

(a) $$\left(\frac{\partial u}{\partial v}\right)_T = T^2\left[\frac{\partial (p/T)}{\partial T}\right]_v$$

(b) $$\left(\frac{\partial h}{\partial p}\right)_T = -T^2\left[\frac{\partial (v/T)}{\partial T}\right]_v$$

Solution: (a) From property relation

$$du = TdS - pdv$$

and1st *Tds* equation

$$TdS = C_v dT + T\left(\frac{\partial p}{\partial T}\right)_v dv$$

$$\therefore \quad du = C_v dT + \left[T\left(\frac{\partial p}{\partial T}\right)_v - p\right] dv \qquad \text{...(i)}$$

Now

$$u = f(T,v)$$

$$\text{or} \quad du = \left(\frac{\partial u}{\partial T}\right)_v dT + \left(\frac{\partial u}{\partial v}\right)_T dv \qquad \text{...(ii)}$$

Equating the coefficient of *dv*, we get,

$$\left(\frac{\partial u}{\partial v}\right)_T = T\left(\frac{\partial p}{\partial T}\right)_v - p$$

$$= T^2\left[\frac{1}{T}\times\left(\frac{\partial p}{\partial T}\right)_v - \frac{p}{T^2}\right]$$

$$= T^2\left[\frac{1}{T}\times\left(\frac{\partial p}{\partial T}\right)_v + p\frac{\partial}{\partial T}(1/T)\right]$$

$$= T^2\left[\frac{\partial (p/T)}{\partial T}\right]_v \quad \textbf{Ans}$$

(b) Similarly

$$dh = C_p dT - \left[T\left(\frac{\partial v}{\partial T}\right)_p - V\right] dp$$

and
$$h = f(T, p)$$

$$dh = \left(\frac{\partial h}{\partial T}\right)_p dT + \left(\frac{\partial h}{\partial p}\right)_T dp$$

Equating the coefficient of dp, we get,

$$\left(\frac{\partial h}{\partial p}\right)_T = -\left[T\left(\frac{\partial v}{\partial T}\right)_p - v\right]$$

$$= -T^2\left[\frac{1}{T}\left(\frac{\partial v}{\partial p}\right)_p - \frac{v}{T^2}\right]$$

$$= -T^2\left[\frac{1}{T}\left(\frac{\partial v}{\partial p}\right)_p + \frac{v\partial}{\partial T}(1/T)\right]$$

$$= -T^2\left[\frac{\partial(v/T)}{\partial T}\right]_p \quad \textbf{Ans}$$

Problem 11.20: Find the change of entropy of a gas following Clausius equation of state if constant temperature

$$p(v - b) = RT$$

Solution: From property relation

$$TdS = C_v dT + pdv$$

at constant temperature,

$$dT = 0$$

or
$$ds = \frac{p}{T} dv$$

Now $p(v - b) = RT$

$$\therefore \quad \frac{p}{T} = \frac{R}{(v-b)}$$

$$\therefore \quad ds = \frac{Rdv}{(v-b)}$$

$$\therefore \quad S_2 - S_1 = \int_1^2 ds = R\int_1^2 \frac{dv}{(v-b)}$$

$$= R \ln \frac{v_2 - b}{v_1 - b} \quad \textbf{Ans}$$

Problem 11.21: (a) Show that for a van der Waals gas

$$\beta = \frac{Rv^2(v-b)}{RTv^3 - 2a(v-b)^2}$$

$$k_T = \frac{v^2(v-b)^2}{RTv^3 - 2a(v-b)^2}$$

(b) What is the value of k_T/β expressed in its simplest form?

(c) What do the above relations become when $a = 0$, $b = 0$ (ideal gas)?

Solution: (a) Vander Waals equation

$$\left(p + \frac{a}{v^2}\right)(v - b) = RT$$

differentiating w.r.t. T at constant temperature

$$\text{or,} \quad \left(\frac{\partial v}{\partial T}\right)_p \left(0 + \frac{(-2)a}{v^3}\right)(v - b) + \left(p + \frac{a}{v^2}\right)\left(\frac{\partial v}{\partial T}\right)_p = R$$

$$\text{or} \quad \left(\frac{\partial v}{\partial T}\right)_p = \frac{R}{\left(p + \frac{a}{v^2}\right) - \frac{2a(v-b)}{v^3}}$$

$$= \frac{R}{\dfrac{RT}{(v-b)} - \dfrac{2a(v-b)}{u^3}}$$

$$\left[\because \left(p + \frac{a}{v^2}\right) = \frac{RT}{v-b}\right]$$

or $$\left(\frac{\partial v}{\partial T}\right)_p = \frac{R(v-b)v^3}{RTv^3 - 2a(v-b)^2}$$

Now $$\beta = \frac{1}{v}\left(\frac{\partial v}{\partial T}\right)_p$$

$$= \frac{R(v-b)v^2}{RTv^3 - 2a(v-b)^2} \quad \textbf{Ans}$$

we know that, $$K_T = -\frac{1}{v}\left(\frac{\partial v}{\partial p}\right)_T$$

Now $$\left(p + \frac{a}{v^2}\right)(v - b) = RT$$

differentiating w.r. to p at constant temperature, we get,

$$\left(p + \frac{a}{v^2}\right)\left(\frac{\partial v}{\partial p}\right)_T + (v - b)\left\{1 + \frac{(-2a)}{v^3} \times \left(\frac{\partial v}{\partial p}\right)_T\right\} = 0$$

$$\left(\frac{\partial v}{\partial p}\right)_T \left[\left(p + \frac{a}{v^2}\right) - \frac{2a(v-b)}{v^3}\right] = -(v - b)$$

or $$\left(\frac{\partial v}{\partial p}\right)_T = +\frac{(v-b)}{\dfrac{2a(v-b)}{v^3} - \left(p + \dfrac{a}{v^2}\right)}$$

$$= \frac{(v-b)}{\dfrac{2a(v-b)}{v^3} - \dfrac{RT}{v-b}} \qquad \left[\because \left(p + \frac{a}{v^2}\right) = \frac{RT}{v-b}\right]$$

$$= \frac{(v-b)^2 v^3}{2a(v-b)^2 - RTv^3}$$

$$\therefore \quad K_T = -\frac{1}{v}\left(\frac{\partial v}{\partial p}\right)_T = \frac{(v-b)^2 v^2}{RTv^3 - 2a(v-b)^2} \quad \textbf{Ans}$$

(b) $$\frac{K_T}{\beta} = \frac{v^2(v-b^2)}{Rv^2(v-b)} = \frac{(v-b)}{R} \quad \textbf{Ans}$$

(c) when $a = 0$, $b = 0$

$$K_T = \frac{v^4}{RTv^3} = \frac{v}{RT}$$

$$\beta = \frac{Rv^3}{R+v^3} = \frac{1}{T}$$

$$\therefore \quad \frac{K_T}{\beta} = v/R \quad \textbf{Ans}$$

Problem 11.22: (a) Show that

(i) $$\left(\frac{\partial u}{\partial p}\right)_v = \frac{k}{\beta} \times C_v$$

(ii) $$\left(\frac{\partial u}{\partial v}\right)_p = \frac{C_p}{v\beta} - p$$

(b) Hence show that the slope of a reversible adiabatic process on p–v coordinates

$$\frac{dp}{dv} = -\frac{\gamma}{kv}$$

where k is the isothermal compressibility.

Solution: (a) From property relation

$$du = TdS - pdv$$

and from 3rd TdS equation

$$TdS = C_v \left(\frac{\partial T}{\partial p}\right)_v dp + C_p \left(\frac{\partial T}{\partial v}\right)_p dv$$

$$du = C_v \left(\frac{\partial T}{\partial p}\right)_v dp + \left[C_p \left(\frac{\partial T}{\partial v}\right)_p - p\right] dv \qquad \text{...(i)}$$

Now let $u = f(p, v)$

$$\therefore \quad du = \left(\frac{\partial u}{\partial p}\right)_v dp + \left(\frac{\partial u}{\partial v}\right)_p dv \qquad \text{...(ii)}$$

Equating the coefficients of dp and dv from the equation (i) & (ii), we get,

$$\left(\frac{\partial u}{\partial p}\right)_v = C_v \left(\frac{\partial T}{\partial p}\right)_v = \frac{C_v(-1)}{\left(\frac{\partial p}{\partial v}\right)_T \left(\frac{\partial v}{\partial T}\right)_p}$$

$$\left[\because \left(\frac{\partial T}{\partial p}\right)_v \times \left(\frac{\partial p}{\partial v}\right)_T \times \left(\frac{\partial v}{\partial T}\right)_p = -1\right]$$

or

$$\left(\frac{\partial u}{\partial p}\right)_v = \frac{-\frac{1}{v}\left(\frac{\partial v}{\partial p}\right)_T C_v}{\frac{1}{v}\left(\frac{\partial v}{\partial T}\right)_p}$$

or

$$\left(\frac{\partial u}{\partial p}\right)_v = \frac{K}{\beta} C_v \textbf{ Ans}$$

Similarly

$$\left(\frac{\partial u}{\partial v}\right)_p = C_p \left(\frac{\partial T}{\partial v}\right)_p - p$$

$$= \frac{C_p}{v \times \frac{1}{v}\left(\frac{\partial v}{\partial T}\right)_p} - p = \frac{C_p}{v \times \beta} - p$$

(b) For reversible adiabatic process

$$pv^{\gamma} = \text{const.}$$

on differentiating, we get,

$$\gamma v^{\gamma-1} \frac{\partial v}{\partial p} \times p + v^{r} \times 1 = 0$$

or
$$\frac{dp}{dv} = -\frac{\gamma v^{\gamma-1} p}{v^{\gamma}}$$

$$-\gamma p v^{-1} = -\frac{\gamma p}{v}$$

Now
$$pv = RT$$

or
$$p\left(\frac{\partial v}{\partial p}\right)_T + v = 0$$

∴
$$-\frac{1}{v}\left(\frac{\partial v}{\partial p}\right)_T = \frac{1}{p}$$

or
$$K = \frac{1}{p} = \text{isothermal compressibility}$$

∴
$$\frac{dp}{dv} = -\frac{\gamma}{KV} \quad \textbf{Ans}$$

Problem 11.23: The following expressions for the equation of state and the specific heatg C_p are obeyed by a certain gas:

$$v = \frac{RT}{p} + \alpha T^2$$

and
$$C_p = A + BT + C_p$$

where α, A, B, C are constants. Obtain an expression for (a) the Joule–Thomson coefficient, and (b) the specific heat C_v.

Solution: (a) We know that

Joule–Thomson coefficient = μ_j

$$= \frac{1}{C_p}\left[T\left(\frac{\partial v}{\partial T}\right)_p - v\right]$$

Given $\qquad v = \dfrac{RT}{p} + \alpha T^2$

or $\qquad \left(\dfrac{\partial v}{\partial T}\right)_p = \dfrac{R}{p} + 2\alpha\, T$

$\therefore \qquad \mu_j = \dfrac{1}{(A + BT + C_p)}\left[\dfrac{RT}{p} + 2\alpha T^2 - v\right]$

$$= \frac{1}{(A + BT + C_p)}\left[\frac{RT}{p} + 2\alpha T^2 - \frac{RT}{p} - \alpha T^2\right]$$

$(\alpha T^2)/(A + BT + C_p)$ **Ans**

(b) $\qquad C_p = A + BT + C_p$

$$V = \frac{RT}{p} + \alpha T^2$$

$\therefore \qquad \left(\dfrac{\partial v}{\partial T}\right)_p = \dfrac{R}{p} + 2\alpha T$

and $\qquad \left(\dfrac{\partial p}{\partial v}\right)_T = \dfrac{-RT}{\left(v - \alpha T^2\right)^2}$

Now $\qquad C_p - C_v = -T\left(\dfrac{\partial v}{\partial T}\right)_p^2 \left(\dfrac{\partial p}{\partial v}\right)_T$

$$C_p - C_v = -T\left(\frac{R}{p} + 2\alpha T\right)^2 (-1)\, \frac{RT}{\left(v - \alpha T^2\right)^2}$$

$$C_v = C_p - \frac{R\left(\dfrac{RT}{p} + 2\alpha T^2\right)^2}{\left(v - \alpha T^2\right)^2}$$

$$= A + BT + C_p - \frac{R\left(v + \alpha T^2\right)^2}{\left(v - \alpha T^2\right)^2}$$

$$C_v = A + BT + C\frac{RT}{v - \alpha T^2} - R\left(\frac{v + \alpha T^2}{v - \alpha T^2}\right)^2$$

Problem 11.24: (a) From the virial form of the equation of state of a gas

$$v = \frac{RT}{p} + RTB'(T)_p + RTC'(T_p) \ldots$$

show that the Joule–Thomson coefficient is

$$\mu_j = \frac{RT^2}{C_p}\left[\frac{dB'}{dT} + \frac{dC'}{dT}p + \ldots\right]$$

(b) For a van der Waals gas

$$B'(T) = \frac{bRT - a}{R^2T^2}$$

Show that the limiting value of μ_j at low pressures is

$$\mu_j = \frac{1}{C_p}\left(\frac{2a}{RT} - b\right)$$

Solution: (a) Know that Joule–Thomson coefficient

$$\mu_j = \frac{1}{C_p}\left[T\left(\frac{\partial v}{\partial T}\right)_p - v\right]$$

given $$v = \frac{RT}{p} + RT\,B'(T) + RTC'(T)_p + \ldots$$

$$\therefore \left(\frac{\partial v}{\partial T}\right)_p = \frac{R}{p} + RB'(T) + RT\frac{dB'(T)}{dT} + RC'(T)p + TRP\frac{dc'(T)}{dT}$$

$$\text{or}\; T\left(\frac{\partial v}{\partial T}\right)_p - v = RT^2\frac{dB'(T)}{dT} + RT^2p\frac{dc'(T)}{dT} + \ldots$$

$$\therefore \quad \mu_j = \frac{RT^2}{C_P}\left[\frac{dB'(T)}{dT} + p\frac{dc'(T)}{dT} + \ldots\right] \quad \textbf{Ans}$$

(b)
$$\mu_j = \frac{RT^2}{C_p}\left[\frac{dB'(T)}{dT} + p\frac{dc'(T)}{dT} + \ldots\right]$$

$$\underset{p\to 0}{Lt}\ \mu_j = \frac{RT^2}{C_p}\left[\frac{dB'(T)}{dT} + 0\right]$$

Now
$$B'(T) = \frac{b}{RT} - \frac{a}{R^2T^2}$$

$$\therefore \quad \frac{d\,B'(T)}{dT} = -\frac{b}{RT^2} + \frac{2a}{R^2T^3}$$

$$\therefore \quad \underset{p\to 0}{Lt}\ \mu_j = \frac{RT^2}{C_p}\left[\frac{2a}{R^2T^3} - \frac{b}{RT^2}\right]$$

$$= \frac{1}{C_p}\left[\frac{2a}{RT} - b\right] \quad \textbf{Ans}$$

Problem 11.25: (a) If z is a function of x and y, then prove that

$$\left(\frac{\partial z}{\partial x}\right)_y\left(\frac{\partial x}{\partial y}\right)_z\left(\frac{\partial y}{\partial z}\right)_x = -1.$$

(b) Prove that for an ideal gas, $\left(\frac{\partial H}{\partial V}\right)_T = 0 = \left(\frac{\partial H}{\partial p}\right)_T$

(c) Derive $\Delta H = \int C_p\, dT + \int [V - T\,(\partial V/\partial T)_p]\, dp$

(d) Derive $\Delta S = \int C_p\, dT/T - \int (\partial V/\partial T)_p]\, dp$

[Because the right–hand sides involve only measurable variables, these have been called the "laboratory equations".]

Solution: Since z is a function of x and y, we have

$$dz = \left(\frac{\partial z}{\partial x}\right)_y dx + \left(\frac{\partial z}{\partial y}\right)_x dy$$

If z is held constant, then $dz = 0$. Thus

$$0 = \left(\frac{\partial z}{\partial x}\right)_y dx + \left(\frac{\partial z}{\partial y}\right)_x dy$$

$$\therefore \quad \frac{\left(\dfrac{\partial z}{\partial}\right)_y}{\left(\dfrac{\partial z}{\partial x}\right)_x} = -\left(\frac{\partial y}{\partial x}\right)_z$$

$$\therefore \quad \left(\frac{\partial z}{\partial x}\right)_y \left(\frac{\partial x}{\partial y}\right)_z \left(\frac{\partial y}{\partial z}\right)_x = -1$$

(b) By definition $H = E + PV = (E + RT)$ for 1 mole of an ideal gas.

$$\therefore \quad \left(\frac{\partial H}{\partial V}\right)_T = \left(\frac{\partial E}{\partial V}\right)_T = T\left(\frac{\partial P}{\partial T}\right)_V - P$$

$$= \frac{RT}{V} - P = 0$$

$$\left(\frac{\partial H}{\partial P}\right)_T = V - T\left(\frac{\partial V}{\partial T}\right)_P$$

$$= V - \frac{T \times R}{P} = 0$$

(c) $dh = Tds + V\,dp$ implies

$$\left(\frac{\partial h}{\partial T}\right)_p = T\left(\frac{\partial s}{\partial T}\right)_p = C_p$$

and
$$\left(\frac{\partial h}{\partial P}\right)_T = T\left(\frac{\partial s}{\partial P}\right)_T + V$$

But $(\partial s/\partial p)_T = -(\partial v/\partial T)_P$;

therefore $dh = C_p\, dT + [v - T\,(\partial v/\partial T)_p]\, dp$ which is the differential form of the desired equation.

(d) Using $dh = T\, ds + Vdp$

$$ds = \frac{1}{T}\,dh - \frac{V}{T}\,dp$$

$$= \frac{1}{T}\left\{C_p dT + \left[V - T\left(\frac{\partial V}{\partial T}\right)_p\right]dp\right\} - \frac{V}{T}\,dp$$

$$= \frac{C_p}{T}\,dT - \left(\frac{\partial V}{\partial T}\right)_p dp$$ which is the differential form of the desired equaiton

Problem 11.26: (a) Derive $\Delta U = C_v\, dT + [T\,(\partial P/\partial T)_v - P]\, dV$
(b) Establish $(\partial S/\partial V)_T = (\partial p/\partial T)_v\, dV$ by using a cyclic relation.

(c) Show that $\left(\frac{\partial p}{\partial V}\right)_s = \left(\frac{\partial p}{\partial V}\right)_T - \frac{T}{C_v}\left[\left(\frac{\partial p}{\partial T}\right)_v\right]^2$

Solution: (a) From

$$du = T\, ds - p\, dv,$$

we get
$$(\partial u/\partial T)_v = (\partial s/\partial T)_v = C_v \text{ and}$$
$$(\partial u/\partial V)_T = T\,(\partial s/\partial V)_T - p$$
$$= T\,(\partial p/\partial T)_v - p$$

Hence desired differential expression is

$$du = C_v\, dT + [T\,(\partial p/\partial T)_v - p]\, dV$$

(b) From calculus, if three variables are functionally related say, $\phi(x, y, z) = 0$, then $d\phi = \phi_x\, dx + \phi_y\, dy + \phi_x\, dz = 0$ which gives

$$\left(\frac{\partial x}{\partial y}\right)_z = -\left(\frac{\phi_y}{\phi_z}\right)\left(\frac{\partial y}{\partial z}\right)_x$$

$$= -\frac{\phi_z}{\phi_y}\left(\frac{\partial z}{\partial x}\right)_y = -\frac{\phi_x}{\phi_z}$$

Taking the product of these three equations gives the general cyclic formula

$$\left(\frac{\partial x}{\partial y}\right)_z\left(\frac{\partial y}{\partial z}\right)_x\left(\frac{\partial z}{\partial x}\right)_y = -1$$

Applying (1) to the variables T, v, s and using the Maxwell relation

$$(\partial T/\partial v)_s = -(\partial p/\partial s)_v$$

$$\left(\frac{\partial p}{\partial s}\right)_v \left(\frac{\partial v}{\partial s}\right)_T \left(\frac{\partial s}{\partial T}\right)_v = 1$$

But, $(\partial p/\partial s)_v\ (\partial s/\partial T)_v = (\partial p/\partial T)_v$ and $(\partial v/\partial s)_T = 1/(\partial s/\partial v)_T$

Therefore $\left(\frac{\partial p}{\partial T}\right)_v = \left(\frac{\partial s}{\partial y}\right)_T$

This problem illustrates the cylical connection of the Maxwell relations.

(c) From $p = p(V, T)$ we get

$$\left(\frac{\partial p}{\partial V}\right)_s = \left(\frac{\partial p}{\partial V}\right)_T + \left(\frac{\partial p}{\partial V}\right)_V \left(\frac{\partial T}{\partial V}\right)_s$$

$$= \left(\frac{\partial p}{\partial V}\right)_T - \left(\frac{\partial p}{\partial T}\right)_V \left(\frac{\partial P}{\partial S}\right)_V$$

$$= \left(\frac{\partial p}{\partial V}\right)_T - \left(\frac{\partial p}{\partial T}\right)_V \left(\frac{\partial p}{\partial T}\right)_V \left(\frac{\partial T}{\partial s}\right)_V$$

But, at constant volume,

$T\,ds = du = C_v\,dT$, or $(\partial S/\partial s)_v = T/C_v$

$$\therefore \quad \left(\frac{\partial p}{\partial V}\right)_s = \left(\frac{\partial p}{\partial V}\right)_T - \frac{T}{C_v}\left(\frac{\partial p}{\partial T}\right)_V^2$$

Problem 11.27 : (a) Integrate the Clausius–Clapeyron equation to obtain an expression for the variation with temperature of the vapour pressure of a pure liquid. What three assumptions are necessary in this derivation?

(b) At 20°C, steam tables give the vapour pressure of water as 2338.87 Pa and at 40°C, as 7383.83 Pa. What value would the integrated Clausius–Clapeyron equation product for the vapour pressure at 30°C?

(c) Prove that $C_P - C_v = T\,(\partial p/\partial T)_V\ (\partial V/\partial T)_p$

Solution : (a) The first approximation made is that, far from the critical point, $v_{f_g} = V_g$

so that the Clausius–Clapeyron equation sbecomes $dp/dT = h_{f_g}/Tv_g$

The next approximation is that the vapour is ideal, giving
$v_g = RT/p$ and $dp/dT = ph_{fg}/RT^2$,
or $dp/p = (h_{fg}/R) \times (dT/T^2)$

This expression can be integrated easily if the third approximaiton is made that the enthalpy of vaporization is independent of temperature:

$$\int \frac{dp}{p} = (h_{fg}/R) \int \frac{dT}{T^2},$$

or
$$\ln p = A + \frac{B}{T}$$

(b) Using the data to evaluate the constants in $\ln p = A + \frac{B}{T}$ (with the teperature in Kelvin),

we get

$$B = -5276.8 \text{ and } A = 25.75776$$

Hence at 30°C, i.e., 303.15 K,

$$\ln p = 25.75776 - (5276.8/303.15)$$
$$= 8.35120 \text{ or } p = 4325.24 \text{ Pa}$$

(c) Let specific entropy be first considered to be a function of pressure and temperature; then

$$ds = (\partial s/\partial p)_T\, dp + (\partial s/\partial T)_p\, dT$$

But $C_P = T\,(\partial s/\partial T)_p$ and,

from Maxwell's relation,

$$(\partial s/\partial p)_T = -(\partial v/\partial T)_p$$

Hence
$$ds = \frac{C_P}{T}\,dT + \left(\frac{\partial v}{\partial T}\right)_p dp \qquad ...(1)$$

If now the specific entropy is considered to be a funciton of specific volume and temperature, then

$$ds = (\partial s/\partial p)_T\, dv + (\partial s/\partial p)_v\, dT$$

But $C_v = T\,(\partial s/\partial T)_v$ and,

from Maxwell's relation

$$(\partial s/\partial v)_T = (\partial p/\partial T)_v$$

Hence $$ds = \frac{C_v}{T} dT + \left(\frac{\partial v}{\partial T}\right)_v dv \quad ...(2)$$

Equating the two results for ds in equations (1) and (2):

$$(C_P - C_v)\, dT = T\,[(\partial v/\partial T)_p\, dp + (\partial p/\partial T)_v\, dv]$$

Now let $T = f(p, v)$, so that

$$dT = (\partial T/\partial p)_v\, dp + (\partial T/\partial v)_p\, dv$$

Then $$(C_P - C_v)\left(\frac{\partial T}{\partial p}\right)_v dp + (C_P - C_v)\left(\frac{\partial T}{\partial v}\right)_p dv$$

$$= T\left(\frac{\partial v}{\partial T}\right)_p dp + T\left(\frac{\partial p}{\partial T}\right)_v dv \quad ...(3)$$

Equating the coefficients of dp or of dv in (3) yields the desired result

$$(C_P - C_v) = T\left(\frac{\partial p}{\partial T}\right)_v \times \left(\frac{\partial v}{\partial T}\right)_p$$

> **Problem 11.28:** Calculate the temperature of melting point of ice hwen it is subjected to a pressure of 100 atmosphere. Take density of ice 0.917 gm/cm³ and latent heat of ice 336.1 kJ/kg.

Solution: Specific volume of ice $= \frac{1}{0.917} = 1.0906 \times 10^{-3}$ m³/kg

h_{sf} = 336.1 kJ/kg

Pressure = 100 atm = 10.13 MPa

Data at triple point is

T = 273.16 K

p = 631 Pa

$v_{sf} = 1 - 1.0906 = -0.0906 \times 10^{-3}$ m³/kg

$$\frac{\Delta T}{\Delta p} = \frac{T v_{sf}}{h_{sf}}$$

$$\therefore \quad \Delta T = \frac{\Delta p \times T \times v_{sf}}{h_{sf}}$$

$$= \frac{\left(10.13\times10^{6}-631\right)\times273.16\times\left(-9.06\times10^{-5}\right)}{336.1\times10^{3}}$$

$$= -0.746 \text{ K}$$

Melting point of ice at 100 atm $= T + \Delta T$

$$= 273.16 - 0.746 = 272.414 \text{ K}$$

Problem 11.29: The specific volume of liquid water is 0.0010002 m³/kg and that of ice is 0.001091 m³/kg at 0.01°C. The latent heat of fusion of ice at this temperature is 333.5 kJ/kg. Calculate the rate of change of melting point of ice with pressure in K/bar.

Solution:

$$\frac{dP}{dT} = \frac{h_f - h_i}{T\left(v_f - v_i\right)}$$

$$= \frac{(273.01)(0.0010002-0.001091)}{333.5}$$

$$= -7.433 \times 10^{-5} \text{ K/kPa}$$

$$= -7.433 \times 10^{-8} \text{ K/Pa} = -0.007433 \text{ K/bar}$$

Problem 11.30: Anice skater is able to glide over the ice because the skate blade exerts sufficient pressure on the ice that a thin layer of ice is melted. The skate blade then glides over this thin lubricating layer. Determine the pressure exerted on ice skate blade to allow smooth ice skatin gat –10°C. Use h_{fg} = 3.34 × 10⁵ J/kg; ρ_{liq} = 1.000 g/cm³ and ρ_{ice} = 0.917 g/cm³ over the range of temperatures and pressures involved.

Solution:

$$\frac{dP}{dT} = \frac{h_{fg}}{Tv_{fg}}$$

or

$$\int_{P_1}^{P_2} dP = \int_{T_1}^{T_2} \frac{h_{fg}}{v_{fg}}\frac{dT}{T}$$

$$P_2 - P_1 = \frac{h_{fg}}{v_{fg}} \log \frac{T_2}{T_1}$$

But at $P_1 = 1$ atm, we know $T_1 = 0°C$

$$P_1 = 1.013 \times 10^5 \text{ N/m}^2$$
$$T_1 = 273 \text{ K},$$
$$T_2 = -10 + 273 = 263 \text{ K}$$
$$h_{fg} = 3.34 \times 10^5 \text{ J/kg}$$

$$V_{liq} = \frac{1}{\rho_l} = 1 \text{ cm}^3/\text{g} = 1 \times 10^{-3} \text{ m}^3/\text{kg}$$

$$V_{ice} = \frac{1}{\rho_i} = 1.091 \times 10^{-3} \text{ m}^3/\text{kg}$$

$$\therefore P_2 - 1.013 \times 10^5$$

$$= \frac{3.34 \times 10^5}{(1 - 1.091) \times 10^{-3}} \log \frac{263}{273} = 1.37 \times 10^8 \text{ N/m}^2$$

$$P_2 = 1.37 \times 10^8 + 1.013 \times 10^5$$
$$= 1.371 \times 10^8 \text{ N/m}^2$$

This pressure is high. If the temperature is less than –15°C, it is not possible to generate sufficient pressure to melt the ice, and conventional ice skating is not possible.

Problem 11.31: Show that for a perfect gas $\left(\frac{\partial U}{\partial T}\right)_p = C_P - pV\beta$, where β is the coefficient of volume expansion.

(b) Show that for a perfect gas $\left(\frac{\partial U}{\partial P}\right)_T = pV\kappa - \frac{C_P - C_v}{B\beta}$, where

B is the isothermal bulk modulus which is equal to $-V\left(\frac{\partial p}{\partial V}\right)_T$

Solution: (a) We know that, by definition

$$C_P = \left(\frac{\partial H}{\partial T}\right)_p = \left[\frac{\partial (U + pV)}{\partial T}\right]_p$$

$$= \left(\frac{\partial U}{\partial T}\right)_p + p\left(\frac{\partial V}{\partial T}\right)_p + V\left(\frac{\partial p}{\partial T}\right)_p$$

$$= \left(\frac{\partial U}{\partial T}\right)_p + p\left(\frac{\partial V}{\partial T}\right)_p$$

Hence $\left(\frac{\partial U}{\partial T}\right)_p = C_P - P\left(\frac{\partial V}{\partial T}\right)_p = C_P - pV\beta$

as $\beta = \frac{1}{V}\left(\frac{\partial V}{\partial T}\right)_p$

(b) From the first law we have

$$dQ = dU + pdV$$

Considering T and p as independent variables

$$dQ = \left(\frac{\partial U}{\partial T}\right)_p dT + \left(\frac{\partial U}{\partial p}\right)_T dp + p\left[\left(\frac{\partial V}{\partial T}\right)_p dT + \left(\frac{\partial V}{\partial p}\right)_T dp\right]$$

$$= \left[\left(\frac{\partial U}{\partial T}\right)_p + p\left(\frac{\partial V}{\partial T}\right)_p\right] dT + \left[\left(\frac{\partial U}{\partial p}\right)_T + p\left(\frac{\partial V}{\partial p}\right)_T\right] dp$$

But $\left(\frac{\partial U}{\partial T}\right)_p + p\left(\frac{\partial V}{\partial T}\right)_p = C_P$

so that $dQ = C_P\, dT + \left[\left(\frac{\partial U}{\partial p}\right)_T + p\left(\frac{\partial V}{\partial p}\right)_T\right] dp$

As this equation is true for any process, it will also be true for the case when V = constant, so that

$$(dQ)_V = C_v\,(dT)_V$$

$$= C_p\,(dT)_V + \left[\left(\frac{\partial U}{\partial p}\right)_T + p\left(\frac{\partial V}{\partial p}\right)_T\right] (dp)_V$$

hence $C_v = C_P + \left[\left(\frac{\partial U}{\partial p}\right)_T + p\left(\frac{\partial V}{\partial p}\right)_T\right]\left(\frac{\partial p}{\partial T}\right)_v$

but we can also write

$$\left(\frac{\partial p}{\partial T}\right)_v = \left(\frac{\partial T}{\partial V}\right)_p \left(\frac{\partial V}{\partial p}\right)_T = -1$$

so that
$$\left(\frac{\partial p}{\partial T}\right)_v = \frac{-1}{\left(\frac{\partial T}{\partial V}\right)_p \left(\frac{\partial V}{\partial p}\right)_T} \quad ...(1)$$

by definition,
$$\beta = \frac{1}{V}\left(\frac{\partial V}{\partial T}\right)_p$$

so that
$$\left(\frac{\partial V}{\partial T}\right)_p = V\,\beta$$

Also
$$\beta = -V\left(\frac{\partial p}{\partial V}\right)_T$$

so that
$$\left(\frac{\partial p}{\partial V}\right)_T = -\left(\frac{B}{V}\right)$$

Substituting these values in equation (1) we get,

$$\left(\frac{\partial p}{\partial V}\right)_T = \frac{-1}{\left[\frac{1}{V\beta}\right]\left[\frac{-V}{B}\right]} = B\,\beta$$

$\therefore$
$$C_v = C_P + \left[\left(\frac{\partial U}{\partial p}\right)_T - pV\kappa\right](B\,\beta)$$

so that
$$\left(\frac{\partial U}{\partial p}\right)_T = pVk - \frac{C_p - C_v}{B\,\beta}$$

Problem 11.32: (a) Prove that $\left(\frac{\partial C_P}{\partial p}\right)_T = T\left(\frac{\partial^2 V}{\partial T^2}\right)_p$

(b) Show that

(i) $C_v = T\left(\frac{\partial S}{\partial T}\right)_p$ and

(ii) $C_p = T\left(\frac{\partial S}{\partial T}\right)_p$

Solution: (a)

$$C_P = T\left(\frac{\partial S}{\partial T}\right)_p$$

$$\left(\frac{\partial C_P}{\partial p}\right)_T = T\,\frac{\partial^2 S}{\partial T \times \partial p}$$

By Maxwell's relation

$$\left(\frac{\partial S}{\partial p}\right)_T = -\left(\frac{\partial V}{\partial T}\right)_p$$

$$\frac{\partial^2 S}{\partial p \times \partial T} = -\left(\frac{\partial^2 V}{\partial T^2}\right)_p$$

$$\therefore \quad \left(\frac{\partial C_P}{\partial p}\right)_T = -T\left(\frac{\partial^2 V}{\partial T^2}\right)_p$$

(b) (i) The fundamental relation for dU is

$$dU = T\,dS - p dV$$

dU can also be expressed as

$$dU = \left(\frac{\partial U}{\partial S}\right)_v dS + \left(\frac{\partial U}{\partial V}\right)_s dV$$

Hence it can be concluded (by equating coefficients of dS) that

$$\left(\frac{\partial U}{\partial S}\right)_v = T$$

We also know that

$$C_v = \left(\frac{\partial U}{\partial T}\right)_v = \left(\frac{\partial U}{\partial S}\right)_v \left(\frac{\partial S}{\partial T}\right)_v = T\left(\frac{\partial S}{\partial T}\right)_v$$

(ii) Writing equations for dH we have

$$dH = T\,dS + V\,dp$$

$$\text{and } dH = \left(\frac{\partial H}{\partial S}\right)_p dS + \left(\frac{\partial H}{\partial p}\right)_s dp$$

$$\text{so that} \left(\frac{\partial H}{\partial S}\right)_p = T$$

$$\text{hence} \quad C_P = \left(\frac{\partial H}{\partial T}\right)_p$$

$$= \left(\frac{\partial H}{\partial S}\right)_p \left(\frac{\partial S}{\partial T}\right)_p$$

$$= T\left(\frac{\partial S}{\partial T}\right)_p$$

Problem 11.33: (a) Show that $\dfrac{C_p}{C_v} = \dfrac{\left(\frac{\partial p}{\partial V}\right)_S}{\left(\frac{\partial p}{\partial V}\right)_T}$

(b) Using Maxwell's relations show that when we compress the substances, which expand on heating, heat is produced.

Solution: Writing $T\,dS$ equations for C_P and C_V, we have

$$T\,dS = C_V\,dT + T\left(\frac{\partial p}{\partial T}\right)_v dV$$

$$T\,dS = C_p\,dT - T\left(\frac{\partial V}{\partial T}\right)_p dp$$

By setting $dS = 0$, we have from above equations

$$C_V\,(dT)_s = -T\left(\frac{\partial p}{\partial T}\right)_V (dV)_S$$

$$C_P\,(dT)_S = T\left(\frac{\partial V}{\partial T}\right)_p (dp)_S$$

Dividing, we get

$$\frac{C_P}{C_v} = -\frac{\left(\frac{\partial V}{\partial T}\right)_p (dp)_s}{\left(\frac{\partial p}{\partial T}\right)_v (dV)_S}$$

$$= -\frac{\left(\frac{\partial V}{\partial T}\right)_p}{\left(\frac{\partial p}{\partial T}\right)_v}\left(\frac{\partial p}{\partial V}\right)_S$$

Using the cyclic relation,

$$\left(\frac{\partial V}{\partial T}\right)_p \left(\frac{\partial T}{\partial p}\right)_V \left(\frac{\partial p}{\partial V}\right)_T = -1$$

We have $$\frac{\left(\frac{\partial V}{\partial T}\right)_p}{\left(\frac{\partial p}{\partial T}\right)_V} = -\left(\frac{\partial V}{\partial p}\right)_T$$

so that $$\frac{C_P}{C_v} = \frac{\left(\frac{\partial p}{\partial V}\right)_S}{\left(\frac{\partial p}{\partial V}\right)_T}$$

(b) From Maxwell's relaitons

$$\left(\frac{\partial S}{\partial p}\right)_T = -\left(\frac{\partial V}{\partial T}\right)_p$$

or $$T\left(\frac{\partial S}{\partial p}\right)_T = -T\left(\frac{\partial V}{\partial T}\right)_p$$

$$\text{or} \quad \left(\frac{\partial Q}{\partial p}\right)_T = -TV\alpha$$

where α is the coefficient of increase in volume at constant pressure.

where α is positive, i.e., the substance expands on heating, $\left(\frac{\partial Q}{\partial p}\right)_T$ is negative. Thus when we compress the substances, which expand on heating, heat is produced. Similarly, cooling effect will be produced when we compress a substance which contracts on heating.

> **Problem 11.34:** Calculate under what pressure the ice would freeze at 272 K, if the change in specific volume, when 1 kg of water freezes is 91×10^{-6} m^3, given latent heat of ice = 3.36×10^5 J kg^{-1} and one atmosphere = 10^5 Nm^{-2}.

Solution: From Maxwell relations

$$\left(\frac{\partial S}{\partial V}\right)_T = \left(\frac{\partial p}{\partial T}\right)_V$$

$$\text{or} \quad T\left(\frac{\partial p}{\partial V}\right)_T = T\left(\frac{\partial p}{\partial T}\right)_V$$

But $T \times dS = \partial Q$

$$\therefore \quad \left(\frac{\partial Q}{\partial V}\right)_T = T\left(\frac{\partial p}{\partial T}\right)_V \quad \text{(Clapeyron Equation)}$$

In this case $\partial Q = 3.36 \times 10^5$ J kg^{-1};

$\partial V = 91 \times 10^{-6}$ m^3

$T = 273$ K and $\partial T = 1$ K

Substituting,

$$\frac{3.36 \times 10^5}{91 \times 10^{-6}} = 273 \times \frac{\partial p}{1}$$

$$\therefore \quad \partial p = \frac{3.36 \times 10^5}{91 \times 10^{-6} \times 273} \text{ Nm}^{-2}$$

$$= \frac{3.36 \times 10^5}{91 \times 10^{-6} \times 273 \times 10^5} \text{ atm} = 135.2 \text{ atm}$$

Therefore the pressure under which ice would freeze

$$= 1 + 135.2 = 136.2 \text{ atmosphere}$$

Problem 11.35: Show the following for superheated vapours of a pure substance

(a) on a *T–S* diagram, lines of constant volume have a steeper slope than lines of constant pressure.

(b) isentropic processes can be described by the relation pV = constant, if the constant entropy lines on a ln *p–h* diagram are linear.

Solution: A superheated vapour behaves like a gas. Therefore pV = RT. From *TdS* relations

(a)
$$TdS = C_p\, dT - T\left(\frac{\partial V}{\partial T}\right)_p dp \quad \text{...(1)}$$

$$TdS = C_p\, dT + T\left(\frac{\partial p}{\partial T}\right)_V dV \quad \text{...(2)}$$

At constant pressure, from equation (1) and at constant volume from equation (2)

$$T = C_P\left(\frac{\partial T}{\partial S}\right)_p = C_V\left(\frac{\partial T}{\partial S}\right)_V$$

$$\therefore \quad \left(\frac{\partial T}{\partial S}\right)_p = \frac{T}{C_P}$$

and
$$\left(\frac{\partial T}{\partial S}\right)_V = \frac{T}{C_V}$$

Because $C_P > C_v$, the slope of a constant pressure line $\frac{T}{C_p}$ at any temperature will be smaller than that of a constant volume line at that temperature.

(b)
$$dh = TdS - Vdp$$

$$dh = -Vdp = RT\frac{dp}{p} \text{ (at constant entropy)}$$

$$h = RT \ln p \text{ i.e., } \ln p = \frac{h}{RT}$$

$$\frac{d}{dh}(\ln p) = \frac{1}{RT}$$

Left hand side is the slope of the constant entropy line on $\ln p$–h diagram. If the lines are linear, then $\frac{1}{RT}$ is constant. Because $pV = RT$, pV = constant.

Problem 11.37: Derive expressions relating the specific heat of the saturated liquid and of the saturated vapour to the specific heats at constant pressure and constant temperature.

Solution: We have $dQ = T\,dS = C\,dT$, or $C = T\left(\frac{dS}{dT}\right)$

Let $S = f(p, T)$;

then $$\frac{dS}{dT} = \left(\frac{\partial S}{\partial T}\right)_p + \left(\frac{\partial S}{\partial p}\right)_T \left(\frac{\partial p}{\partial T}\right) \text{ and}$$

$$C = T\left(\frac{dS}{dT}\right) = T\left(\frac{\partial S}{\partial T}\right)_p + T\left(\frac{\partial p}{\partial T}\right)\left(\frac{\partial S}{\partial P}\right)_T$$

But $$C_P = \left(\frac{\partial Q}{\partial T}\right)_p = T\left(\frac{\partial S}{\partial T}\right)_p;$$

also from the Maxwell relation,

$$\left(\frac{\partial S}{\partial p}\right)_T = -\left(\frac{\partial V}{\partial T}\right)_p$$

Thus $$C = C_P - T\left(\frac{\partial V}{\partial T}\right)_p \left(\frac{dp}{dT}\right)$$

If this equation is applied to the saturated liquid, we get

$$C_f = C_{pf} - T\left(\frac{\partial V_f}{\partial T}\right)_p \left(\frac{dp}{dT}\right)_f \text{ ; and}$$

for the saturated vapour,

$$C_g = C_{pg} - T\left(\frac{\partial V_g}{\partial T}\right)_p \left(\frac{\partial p}{\partial T}\right)_g$$

Problem 11.38: The free energy change (ΔG) accompanying a given process is –85.77 kJ/mole at 25°C and –83.68 kJ/mole at 35°C. Calculate the change in enthalpy (ΔH) for the process at 30°C.

Solution: $G = H - TS$

$$\Delta G = \Delta H - T\,\Delta S \text{ (for an isothermal process)}$$

Also $$\left(\frac{\partial G}{\partial T}\right)_p = -S$$

$$\therefore \quad -\Delta S = -(S_2 - S_1) = \left(\frac{\partial G_2}{\partial T}\right)_p - \left(\frac{\partial G_1}{\partial T}\right)_p$$

$$= \left(\frac{\partial \Delta G}{\partial T}\right)_p$$

$$\therefore \quad \Delta G = \Delta H - T\left(\frac{\partial \Delta G}{\partial T}\right)_p$$

$$\frac{\Delta G}{T^2} = \frac{\Delta H}{T^2} - \frac{1}{T}\left(\frac{\partial \Delta G}{\partial T}\right)_p$$

$$\left[\frac{\partial(\Delta G/T)}{\partial T}\right]_p - \frac{1}{T} \times \left(\frac{\partial \Delta G}{\partial T}\right)_p$$

$$= \frac{\Delta H}{T^2} - \frac{1}{T}\left(\frac{\partial \Delta G}{\partial T}\right)_p$$

$$\therefore \left[\frac{\partial(\Delta G / T)}{\partial T}\right]_p = -\frac{\Delta H}{T^2}$$

$$\Delta\left(\frac{\Delta G}{T}\right) = \left(-\frac{\Delta H}{T^2}\right) \Delta T$$

$$\therefore \frac{-85770}{298} - \frac{-83680}{308} = \frac{-\Delta H}{303^2} \times (308 - 298)$$

$$\therefore \quad \Delta H = -\ 148 \text{ kJ/mole}$$

Problem 11.39: Obtain the expressions for change in entropy of an ideal gas, starting from the $T\ dS$ relations.

Solution:

(a) $$T\ dS = dU + p\ dV$$

Dividing by T and using hte relation $pV = RT$

$$dS = \frac{dU}{T} + \frac{R}{V}\ dV$$

$$= C_V \frac{dT}{T} + R \times \frac{dV}{V}$$

Integrating

$$S_2 - S_1 = C_v\ ln \frac{T_2}{T_1} + R\ ln \frac{V_2}{V_1} \qquad \ldots(1)$$

(b) $$T\ dS = dH - V\ dp$$

$$\therefore \quad dS = \frac{dH}{T} - \frac{R}{p} \times dp = C_p \frac{dT}{T} - R \frac{dp}{p}$$

Integrating

$$S_2 - S_1 = C_p \ln \frac{T_2}{T_1} - R \ln \frac{P_2}{P_1} \qquad \ldots(2)$$

(c) Multiply equation (1) by C_p and equation (2) by C_v.

$$C_p\ (S_2 - S_1) = C_p \times C_v \ln \frac{T_2}{T_1} + C_p\ R \ln \frac{V_2}{V_1}$$

$$C_v (S_2 - S_1) = C_p \times C_v \ln \frac{T_2}{T_1} - C_v R \ln \frac{P_2}{P_1}$$

Subtracting

$$(C_p - C_v)(S_2 - S_1) = C_p R \ln \frac{V_2}{V_1} + C_v R \ln \frac{P_2}{P_1}$$

$$S_2 - S_1 = C_p \ln \frac{V_2}{V_1} + C_v \ln \frac{P_2}{P_1} \quad ...(3)$$

Problem 11.40: Show that pV^{γ} = constant for an isentropic process of a perfect gas.

Solution: Taking equation (3) of problem (11.39) above

$$S_2 - S_1 = C_p \ln \frac{V_2}{V_1} + C_v \ln \frac{P_2}{P_1} = 0$$

$$\frac{C_p}{C_v} \ln \frac{V_2}{V_1} + \ln \frac{P_2}{P_1} = 0$$

$$\ln \left(\frac{V_2}{V_1}\right)^{\gamma} + \ln \frac{P_2}{P_1} = 0$$

i.e.,
$$\ln \frac{p_2 V_2^{\gamma}}{p_1 V_1^{\gamma}} = 0$$

or
$$\frac{p_2 V_2^{\gamma}}{p_1 V_1^{\gamma}} = 1$$

$\therefore$
$$pV^{\gamma} = \text{Constant}$$

Problem 11.41: Express a general form of equation of state of a substance which has its internal energy as a function of only temperature.

Solution: For U to be function of temperature only, we have

$$\left(\frac{\partial U}{\partial V}\right)_T = 0 = T\left(\frac{\partial p}{\partial T}\right)_V - p$$

$$\text{or} \quad \frac{1}{p}\left(\frac{\partial p}{\partial T}\right)_V = \frac{1}{T}$$

Therefore, integrating with respect to T gives

$$\log_e p = \log_e T + \Psi(V)$$

This can be expressed as:

$$p = T\,\phi(V)$$

$$\text{or} \quad \frac{p}{T} = \phi(V)$$

Thus, for U to be function of temperature only, $\frac{p}{T}$ should be funciton of V only. If, in addition, pV is funciton of T only (Boyle's law), then

$$pV = \theta(T)$$

$$\text{or} \quad T\phi(V) = \frac{1}{V}\theta(T)$$

$$\text{or} \quad V\phi(V) = \frac{1}{T}\theta(T)$$

The only way it is satisfied is to have

$$\phi(V) = \frac{\text{constant}}{V}$$

$$\theta(T) = \text{constant}\,(T)$$

$$\text{Thus} \quad pV = \text{constant}\,(T) = RT$$

where R is a constant which varies from gas to gas.

CHAPTER 12

Vapour Power Cycles

Problem 12.1: A geothermal power plant utilizes steam produced by natural means underground. Steam wells are drilled to tap this steam supply which is available at 4.5 bar and 175°C. The steam leaves the turbine at 100 mm Hg absolute pressure. The turbine isentropic efficiency is 0.75. Calculate the efficiency of the plant. If the unit produces 12.5 MW, what is the steam flow rate?

Solution: Given data :

$p_1 = 4.5$ bar

$\eta_{isen} = 0.75$

$T_1 = 175°C = 448K$

$p_2 = 100$ mm Hg absolute

$= \frac{100}{760} \times 1.01325$

$= 0.1333$ bar

Turbine output

$= 12.5$ MW

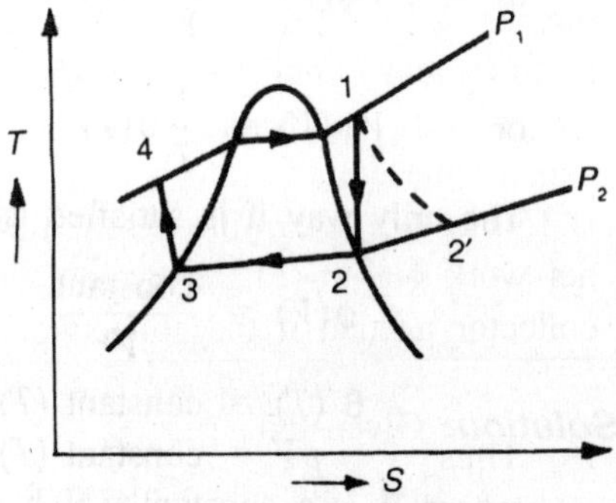

Fig. P. 12.1

From steam table at 4.5 bar and 175°C by interpolation,

$h_1 = 2802.7$ kJ/kg

$S_1 = 6.987$ kJ/kg.K

at 0.1333 bar

$S_f = 0.7172$ kJ/kg.K

$S_{fg} = 7.3412$ kJ/kg.K

$h_f = 213.67$ kJ/kg.K

$$h_{fg} = 2380.2 \text{ kJ/kg.K}$$

$$\therefore \quad S_1 = S_2 = 6.987 = 0.7172 + x \times 7.3412$$

$$x = 0.854$$

$$h_2 = h_f + xh_{fg}$$

$$= 213.67 + 0.854 \times 2380.2 = 2246.5 \text{ kJ/kg}$$

$$h_3 = 213.67 \text{ kJ/kg}$$

Now $$0.75 = \frac{h_1 - h_2'}{h_1 - h_3}$$

or, $$h_1 - h_2' = 417.15 \text{ kJ/kg}$$

$$\therefore \quad \eta_{plant} = \frac{h_1 - h_2'}{h_1 - h_3} = \frac{417.15}{2246.5 - 213.67} \quad 20.52\ \%$$

Let the steam flow rate = $\dot{m}$

$$\dot{m}\ (h_1 - h_2') = 12.5 \times 1000$$

$$\therefore \quad \dot{m} = \frac{12.5 \times 1000}{417.15} = 29.97 \text{ kg/s } \textbf{Ans}$$

Problem 12.2: A simple steam power cycle uses solar energy for the heat input. Water in the cycle enters the pump as a saturated liquid at 40°C, and is pumped to 2 bar. It then evaporates in the boiler at this pressure, and enters the turbine as saturated vapour. At the turbine exhaust the conditions are 40°C and 10% moisture. The flow rate is 150 kg/h. Determine (a) the turbine isentropic efficiency, (b) the net work output (c) the cycle efficiency, and (d) the area of solar collector needed if the collectors pick up 0.58 kW/m^2.

Solution: Given data :

flow rate = 150 kg/h

$$p_1 = 2 \text{ bar}$$

condenser temperature

$$T_3 = 40°\text{C}, \quad x_2' = 0.9$$

From steam table
at 2 bar,

$$h_1 = 2706.7 \text{ kJ/kg.K}$$

$$S_1 = 7.1271 \text{ kJ/kg.K}$$

at 40°C

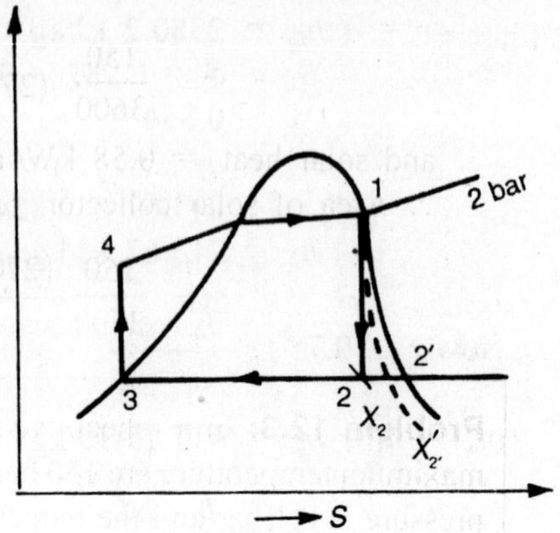

Fig. P. 12.2

$p_{sat} = 0.07384$ bar

$h_f = 167.57$ kJ/kg.K

$S_f = 0.5725$ kJ/kg.K

$S_g = 8.2570$ kJ/kg.K

$h_{fg} = 2406.7$ kJ/kg.K

$V_f = 0.001008$ m³/kg

$h_3 = 167.57$ kJ/kg

$h_2' = 167.57 + 0.9$

Now

$S_1 = S_2 \Rightarrow 7.1271$

$= 0.5725 + x_2$

$(8.2570 - 0.5725)$

$\therefore\ x_2 = 0.85296$

$\therefore\ h_2 = 167.57 + 0.85296 \times 2406.7$

$= 2220.398$ kJ/kg.K

(a) $$h_{isentropic} = \frac{h_1 - h_2'}{h_1 - h_2} = \frac{2706.7 - 2333.6}{2706.7 - 2220.398} = 0.7672$$

(b) pump input

$= v_{f3}\,(p_2 - p_1)$

$= 0.001008\,(2 - 0.07384) \times 100$

$= 0.1942$ kJ/kg

or, $h_4 - h_3 = 0.1942$

$\therefore\ h_4 = 167.764$ kJ/kg.K

$\therefore$ Net work out = turbine work – pump input

$= \dot{m}\,[(h_1 - h_2') - 0.1942]$

$= \frac{150}{3600}\,[12706.7 - 2333.6) - 0.1942]$

$= 15.54$ kW **Ans.**

(c) $$h_{cycle}\ \frac{(h_1 - h_2') - w_p}{h_1 - h_4}$$

$$= \frac{(2706.7 - 2333.6) - 0.1942}{2706.7 - 167.764} = 14.68\%$$

(d) heat needed to evaporate the water

$= \dot{m}\ (h_1 - h_4)$

$$= \frac{150}{3600}\,(2706.7 - 167.764)$$

and solar heat $= 0.58\ kW/m^2$

∴ Area of solar collector

$$= \frac{150}{3600}\,\frac{(2706.7 - 167.764)}{0.58}\ 182.4\ m^2$$

> **Problem 12.3:** In a reheat cycle, the initial steam pressure and the maximum temperature are 150 bar 550°C respectively. If the condenser pressure is 0.1 bar and the moisture at the condenser inlet is 5%, and assuming ideal process, determine (a) the reheat pressure, (b) the cycle efficiency, and (c) the steam rate.

Solution: Given data :

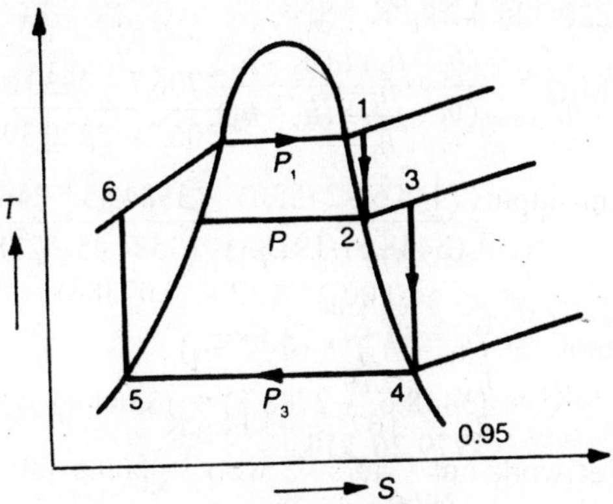

Fig. P. 12.3

$$p_1 = 150 \text{ bar},$$
$$T_1 = 550°C$$
$$p_3 = 0.1 \text{ bar},$$
$$x_4 = 0.95$$

From steam table

At 150 bar and 550°C

$$h_1 = 3448.6 \text{ kJ/kg}$$
$$S_1 = 6.5199 \text{ kJ/kg.K}$$
$$v_1 = 0.02293 \text{ m}^3\text{/kg}$$

At 0.1 bar and 5% moisture

$$h_f = 191.83 \text{ kJ/kg}$$
$$h_{fg} = 2392.8 \text{ kJ/kg}$$

S_f = 0 .6493 kJ/kg.K
S_g = 8.1502 kJ/kg.K
v_f = 0.00101 m³/kg
v_g = 14.674 m³/kg

∴ $h_4 = 191.83 + 0.95 \times 2392.8$
$= 2464.99$ kJ/kg.K

$S_4 = 0.6493 + 0.95\ (8.1502 - 0.6493\ )$
$= 7.7752$ kJ/kg.K

Now $S_1 = S_2 = 6.5199$ kJ/kg.K
and $S_3 = S_4 = 7.7752$ kJ/kg.K

From steam table,

at $S_3 = 7.7752$ kJ/kg.K and 550°C

(a) $p = 13.5$ bar **Ans**

At 13.5 bar,

$h_g = h_2 = 2788.8$ kJ/kg
$h_3 = 3584.45$ kJ/kg

(b)
$$\eta_{cycle} = \frac{(h_1 - h_2) + (h_3 - h_4)}{(h_1 - h_5) + (h_3 - h_2)}$$

$$= \frac{(3448.6 - 2788.8) + (3584.45 - 2469.99)}{(3448.6 - 191.83) + (3584.45 - 2788.8)}$$

$= 43.9\%$ **Ans**

(c) work output $= (h_1 - h_2) + (h_3 - h_4)$
$= (3448.6 - 2788.8) + (3584.45 - 2464.99)$
$= 1779.26$ kJ/kg

$$\therefore \text{ steam rate } = \frac{3600}{1779.26} = 2.03 \text{ kJ/kWh}$$ **Ans.**

Problem 12.4: In a nuclear power-plant heat is transferred in the reactor to liquid sodium. The liquid sodium is then pumped to a heat exchanger where heat is transferred to steam. The steam leaves this heat exchanger as saturated vapour at 55 bar, and is then superheated in an external gas-fired superheater to 650°C. The steam then enters the turbine, which has one extraction point at 4 bar where steam flows to an open feed water heater. The turbine efficiency is 75% and the condenser temperature is 40°C. Determine the heat transfer in the reactor and in the superheater to produce a power output of 80 MW.

Solution: Given data :

$p_1 = 55$ bar,
$p_2 = 4$ bar
$T_1 = 650°C$
$\eta_{isen} = 0.75$,
Condenser temperature
$= 40°C$

From steam table
at 55 bar and 650°C,
$h_1 = 3779.8$ kJ/kg.
$h_1' = 2789.3$ kJ/kg.

at 4 bar
$h_f = h_5 = 604.74$ kJ/kg.
$h_a = 2950$ kJ/kg.

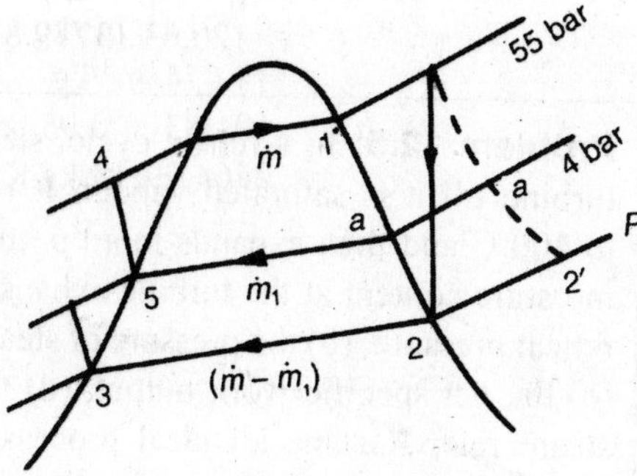

Fig. P. 12.4

$$\text{or,} \quad 0.75 = \frac{h_1 - h_a'}{h_1 - h_a} = \frac{3779.8 - h_a'}{3779.8 - 2950}$$

$$\therefore \quad h_a' = 3157.45 \text{ kJ/kg.}$$

At 40°C

$$h_2 = 2574.3 \text{ kJ/kg.}$$
$$h_3 = 167.57 \text{ kJ/kg.}$$

$$\eta_{isin} = 0.75 = \frac{h_1 - h_2'}{h_1 - h_2} = \frac{3779.8 - h_2'}{3779.8 - 2574.3}$$

$$\therefore \quad h_2' = 2875.675 \text{ kJ/kg}$$

Now, from heat balance in feed water heater

$$\dot{m}_1 = (h_a{}^1 - h_5) = (\dot{m} - \dot{m}_1)(h_5 - h_3)$$

$$\text{or,} \quad \dot{m}_1 = (\dot{m} - \dot{m}_1) \times 0.17$$

$$\Rightarrow \quad \dot{m}_1 = 0.855\, \dot{m}$$

power output = 80 MW

Now power output neglecting the pump input

$$= \dot{m}\,(h_1 - h_a') + (\dot{m} - \dot{m}_1)(h_a' - h_2')$$

or $\dot{m}\,(h_1 - h_a') + (\dot{m} - \dot{m}_1)(h_a' - h_2') = 80 \times 10^3$

or, $\dot{m} \times 622.35 + (\dot{m} - 0.855\, \dot{m}) \times 281.78 = 80,\,000$

$$\therefore \quad \dot{m} = 120.63 \text{kg/s}$$

$\therefore$ heat transfer in reactor

$$= \dot{m}\,(h_1' - h_5)$$
$$= 76.2\,(2789.3 - 604.74) = 263.51 \text{ MW}$$

Heat transfer in superheater

$$= \dot{m}\ (h_1 - h_1')$$
$$= 120.63\ (3779.8 - 2789.3) = 119.36 \text{ MW}$$

Problem 12.5: In a reheat cycle, steam at 500°C expands in a h.p. turbine till it is saturated vapour. It is reheated at constant pressure to 400°C and then expands in a l.p. turbine to 40°C. If the maximum moisture content at the turbine exhaust is limited to 15%, find (a) the reheat pressure, (b) the pressure of steam at the inlet to the h.p. turbine, (c) the net specific work output, (d) the cycle efficiency, and (e) the steam rate. Assume all ideal processes.
What would have been the quality, the work output, and the cycle efficiency without the reheating of steam? Assume that the other conditions remain the same.

Solution: Given data :

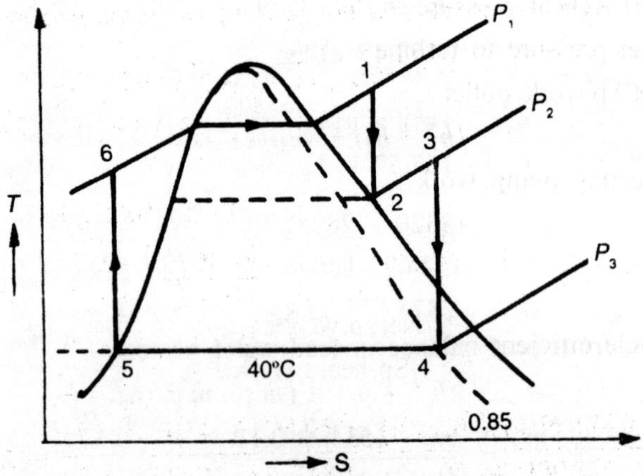

Fig. P. 12.5

$$T_1 = 500°\text{C}$$
$$T_3 = 400°\text{C}$$
$$x_4 = 0.85$$

From steam table

at 40°C and 0.85 quality

$$h_f = 167.57 \text{ kJ/kg}$$
$$h_{fg} = 2406.7 \text{ kJ.kg}$$
$$S_f = 0.5725 \text{ kJ/kg.K}$$
$$S_g = 8.2570 \text{ kJ/kg.K}$$
$$h_4 = 167.57 + 0.85 \times 2406.7$$

$= 2213.27$ kJ/kg

$S_4 = 0.5725 + 0.85\ (8.257 - 0.5725)$

$= 7.1043$ kJ/kg.K

$h_5 = 167.57$ kJ/kg

Now *at*

$S_3 = S_4 = 7.1043$ kJ/kg.K and $T_3 = 400°C$

$h_3 = 3246$ kJ/kg

$p_2 = 21$ bar **Ans**

at $p_2 = 21$ bar, $(S_2)_{sat} = 6.323$ kJ/kg.K

$h_2 = 2800.5$ kJ/kg.K

At $(S_2)_{sat} = 6.323$ kJ/kg.K and $T_1 = 500°C$

from superheated steam table by interpolation,

$h_1 = 3302.02$ kJ/kg

p_1 154.8 bar

∴ (a) Reheat pressure = P_2 = 21 bar

(b) Inlet pressure to turbine = 154.8 bar.

(c) Net sp.work outlet

$= (h_1 - h_2) + (h_3 - h_4)$,

neglecting pump work

$= (3320 - 2825) + (3250 - 2213.27)$

$= 1534.25$ kJ/kg **Ans**

(d) Cycle efficiency $= \dfrac{\text{Net sp. work output}}{\text{Sp. heat supplied}}$

$$= \frac{1534.25}{(h_1 - h_5) + (h_3 - h_2)}$$

$$= \frac{1534.25}{(3302.02 - 167.57) + (3246 - 2800.5)}$$

$= 42.86$ % **Ans**

(e) Steam rate $= \dfrac{3600}{\text{Sp. work output}}$

$= \dfrac{3600}{1534.25}$ kg/kWh = 2.35 kJ/kWh **Ans**

Problem 12.6: A regenerative cycle operates with steam supplied at 30 bar and 300°C, and condenser of 0.08 bar. The extraction points for two heaters (one closed and one open) are at 3.5 bar and 0.7 bar respectivelly. Calculate the thermal efficiency of the plant, neglecting pump work.

Solution: Given data :

p_1 = 30 bar
T_1 = 300°C
p_4 = 0.08 bar
p_2 = 3.5 bar
P_3 = 0.7 bar

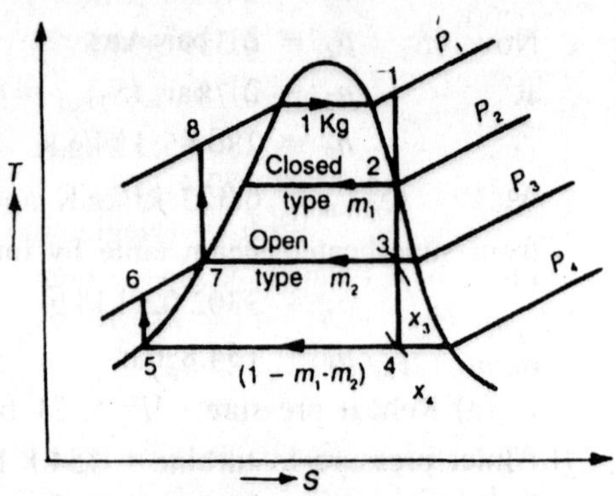

Fig. P. 12.6

From steam table

at 30 bar at 300°C

h_1 = 299305 kJ/kg
S_1 = 6.5390 kJ/kg.K

at 3.5 bar

S_f = 1.7275 kJ/kg.K
S_g = 6.9405 kJ/kg.K
h_f = 584.33 kJ/kg
h_{fg} = 2148.1 kJ/kg

$$S_1 = S_2 = 6.5390$$
$$= 1.7275 + x_2 (6.9405 - 1.7275)$$
$$\therefore \quad x_2 = 0923$$
$$\therefore \quad h_2 = 584.33 + x_2 \times 2148.1$$
$$= 2566.99 \text{ kJ/kg}$$
$$h_g = 584.33 \text{ kJ/kg}$$
$$= h_{10} \text{ [Neglecting pump work]}$$

at 0.7 bar

$$S_f = 1.1922 \text{ kJ/kg.K}$$
$$S_g = 7.4797 \text{ kJ/kg.K}$$
$$h_f = 375.61 \text{ kJ/kg}$$
$$h_{fg} = 2283.96 \text{ kJ/kg}$$
$$S_2 = S_3 = 6.5390$$
$$= 1.1922 + x_3 (7.4797 - 1.1922)$$
$$x_3 = 0.85$$
$$\therefore \quad h_3 = 375.61 + 0.85 \times 2283.96$$
$$= 2317.86 \text{ kJ/kg}$$

$h_7 = 375.61$ kJ/kg

$= h_8 =$ (Neglecting pump work)

at 0.08 bar

$S_f = 0.5928$ kJ/kg.K

$S_g = 8.2288$ kJ/kg.K

$h_f = 173.4$ kJ/kg

$h_{fg} = 2403.4$ kJ/kg

Now $S_3 - S_4 = 6.539 = 0.5928 + x_4\ (8.2288 - 0.5928)$

$\therefore \quad x_4 = 0.78$

$\therefore \quad h_4 = 173.4 + 0.78 \times 2403.4 = 2044.94$ kJ/kg

$h_5 = 173.4$ kJ/kg

$= h_6 =$ (Neglecting pump work)

From heat balance at closed type feed water heater

$m_1\ (h_2 - h_g) = (1 - m_1)\ (h_{10} - h_8)$

or $m_1\ (2666.99 - 584.33) = (1 - m_1)\ (584.33 - 375.61)$

$\therefore \quad m_1 = 0.091$ kg

Again from heat balance of heat at closed type feed water heater

$m_2h_3 + (1 - m_1 - m_2)\ h_5 + m_1h_g = 1 \times h_7$

or, $m_2 \times 2317.86 + (1 - 0.091 - m_2) \times 173.4 + 0.091 \times 584.33$
$= 375.61$

$\therefore \quad m_2 = 0.077$ kg

$\therefore\ \eta_{thermal}$ of Plant

$$= \frac{\text{Net work}}{\text{heat supplied}}$$

$$= \frac{(h_1 - h_2) + (1 - m_1)(h_2 - h_3) + (1 - m_1 - m_2)(h_3 - h_4)}{(h_1 - h_8)}$$

$$= \frac{(2993.5 - 2566.99) + (1 - 0.091)\ (2566.99 - 2317.86) + (1 - 0.091 - 0.077)(2317.86 - 2044.94)}{(2993.5 - 375.61)} = 33.62\ \%\ \textbf{Ans}$$

Problem 12.7: The net power output of the turbine in an ideal reheat-regenerative cycle is 100 MW, Steam enters the high-pressure (H.P) turbine at 90 bar, 550°C. After expansion to 7 bar, some of the steam goes to an open to an open heater and the balance is reheated to 400°C, after which it expands to 0.07 bar. (a) What is the steam flow rate to the H.P. turbine? (b) What is the total pump work? (c) Calculate the cycle efficiency. (d) If there is a 10°C rise in the temperature of the cooling water, what is the rate of flow of the cooling water in the condenser? (e) If the velocity of the steam flowing from the turbine to the condenser is limited to a maximum of 130 m/s, find the diameter of the connecting pipe.

Solution: Given data :

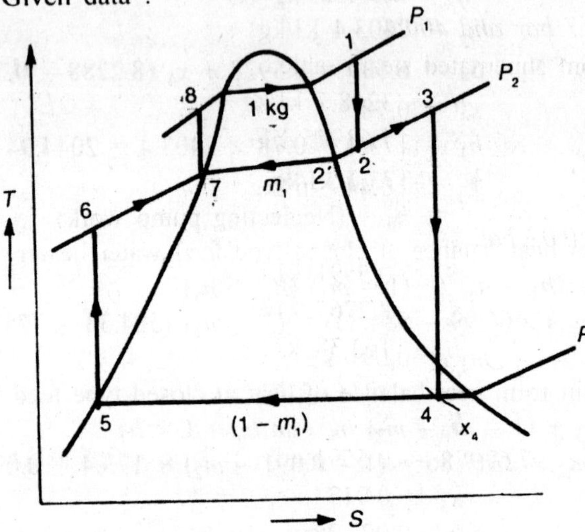

Fig. P. 12.7

$$p_1 = 90 \text{ bar}$$
$$p_2 = 7 \text{ bar}$$
$$p_3 = 0.07 \text{ bar}$$
$$T_3 = 400°\text{C}$$
$$T_1 = 550°\text{C}$$

Net power output = 100 MW

From steam table

at 90 bar and 550°C

$$h_1 = 3511.0 \text{ kJ/kg}$$
$$S_1 = 6.8142 \text{ kJ/kg.K}$$

at 7 bar

$$t_{sat} = 165°\text{C} = t_2$$
$$h_f = 697.22 \text{ kJ/kg}$$
$$h_{fg} = 2066.3 \text{ kJ/kg}$$
$$S_f = 1.9922 \text{ kJ/kg.K}$$
$$S_g = 6.708 \text{ kJ/kg.K}$$
$$v_f = 1.108 \times 10^{-3} \text{ m}^3\text{/kg}$$

since $S_g < S_1$, Hence will be in superheated state

$$h_7 = 697.22 \text{ kJ/kg}$$

$$V_{f_7} = 1.108 \times 10^{-3} \text{ m}^3\text{/kg}$$

by interpolation for points 2′, 2 and 3

$$t_2 = 185.88 = 186°C$$

Again at 7 bar and 186°C, from steam table by interpolation,

$$h_2 = 2812.28 \text{ kJ/kg}$$

at 7 bar and 400°C

from supernated steam table,

$$h_3 = 3268.7 \text{ kJ/kg}$$

$$S_3 = 7.63975 \text{ kJ/kg.K}$$

$$V_{f_3} = 0.449 \text{ m}^3\text{/kg}$$

at 0.07 bar.

$$S_f = 0.5591 \text{ kJ/kg.K}$$

$$S_g = 8.2767 \text{ kJ/kg.K}$$

$$h_f = 163.38 \text{ kJ/kg}$$

$$h_g = 2572.58 \text{ kJ/kg}$$

$$\therefore \quad S_3 = S_4$$

$$7.63975 = 0.5591 + x_4 (8.2767 - 0.5591)$$

$$\therefore \quad x_4 = 0.917$$

$$h_4 = 163.38 + 0.917 (2572.58 - 163.38)$$

$$= 2372.63 \text{ kJ/kg}$$

$$h_5 = 163.38 \text{ kJ/kg}$$

$$V_{f5} = 1.00.74 \times 10^{-3} \text{ m}^3\text{/kg}$$

From heat balance in open heater,

$$m_1 (h_2 - h_7) = (1 - m_1)(h_7 - h_6)$$

Now pump work in 6 to 5

$$= h_6 - h_5$$

$$= 0.0010074 (7 - 0.07) \times 100$$

$$\therefore \quad h_6 = h_5 + 0.0010074 \times (7 - 0.07) \times 100$$

$$= 164.12 \text{ kJ/kg}$$

$$\therefore m_1 (2812.28 - 697.22) = (1 - m_1)(697.22 - 164.12)$$

$$m_1 = 0.201 \text{ kg}$$

Again pump work to 8 to 7

$$= h_8 - h_7 = 1.108 \times 10^{-3} \times 100 (90 - 7)$$

$$h_8 = 706.42 \text{ kJ/kg}$$

(a) Net work output in H.P. turbine = work obtained in H.P.T – Work input in H.P.T

$= 1 \times (h_1 - h_2) - 0.001108 \times 100 \times 83 \times 1$
$= (3511 - 2812.28) - 0.001108 \times 8300$
$= 689.52$ kJ/kg

Network output in L.P. Turbine
$= (h_3 - h_4)(1 - 0.205) - 0.010074 \times (7 - 0.07) \times (1 - 0.201)$
$= 711.82$ kJ/kg

∴ total work output = work output of H.P. Turbine + work output in L.P. Turbine

= 1401.34 kJ/kg

∴ mass flow rate to H.P. Turbine

$$= \dot{m}_s = \frac{100 \times 1000}{1401.34} = 71.36 \text{ kg/s } \textbf{Ans}$$

(b) total pump work

$= \dot{m}_s \times [1 \times 0.001108 \times 100\,(70 - 7) + 0.0010074 \times 100\,(7 - 0.07) \times (1 - 0.2010)]$
$= 71.36 \times 7.535$ (kJ/kg) (kg/s)
$= 7.535 \times 71.36 = 537.7$ kW **Ans**

(c)
$$h_{cycle} = \frac{\text{Net work}}{\text{heat supplied}} = \frac{100 \times 10^3}{31.36 \times (h_1 - h_8)}$$
$= 49.96\%$ **Ans**

(d) Let $\dot{m}_w$ = rate of cooling water required

$\therefore \dot{m}_w\, Cp_w\, \Delta T = m_s\,(h_4 - h_5)(1 - m_1)$

or, $\dot{m}_w \times 4.187 \times 10$
$= 71.36 \times (2372.63 - 163.38)(1 - 0.201)$

∴ $\dot{m}_w = 3008.46$ kg/s **Ans**

(f) $V = 130$ m/s

$$\therefore \text{Volume flow rate} = A \times V = \frac{\pi}{4} d^2 \times V \frac{\text{m}^3}{\text{sec}}$$

$= \text{mass rate of flowing fluid} \times (1 - m_1) \times V_{f_3}$

$$\therefore \quad d = \sqrt{\frac{4}{\pi} \times \frac{71.36 \times (1 - 0.206) \times 0.449}{130}}$$

$= 0.50073$ m $= 50.07$ cm **Ans**

Problem 12.8: A mercury cycle is superposed on the steam cycle operating between the boiler outlet condition of 40 bar, 400°C and the condenser temperature of 40°C. The heat released by mercury condensing at 0.2 bar is used to impart the latent heat of vaporization to the water in the steam cycle. Mercury enters the mercury turbine as saturated vapour at 10 bar. Compute (a) kg of mercury circulated per kg of water, and (b) the efficiency of the combined cycle.

The property values of saturated mercury are given below.

p (bar)	t (°C)	h_f (kJ/kg)	h_g	s_f (kJ/kg.K)	s_g	v_f (m³/kg)	v_g
10	515.5	72.23	363.0	0.1478	0.5167	80.9×10^{-6}	0.0333
0.2	277.3	38.35	336.55	0.0967	0.6385	77.4×10^{-6}	1.163

Solution: Given data :

$p_1 = 40$ bar

$T_1 = 400$°C

$p_4 = 0.2$ bar

$p_3 = 10$ bar

condenser temparture

= 40°C

For Hg cycle

From question

$h_a = 363$ kJ/kg

$S_a = 0.5167$ kJ/kg.K

At 0.2 bar,

$s_f = 0.0947$

$s_g = 0.6385$ kJ/kg.K

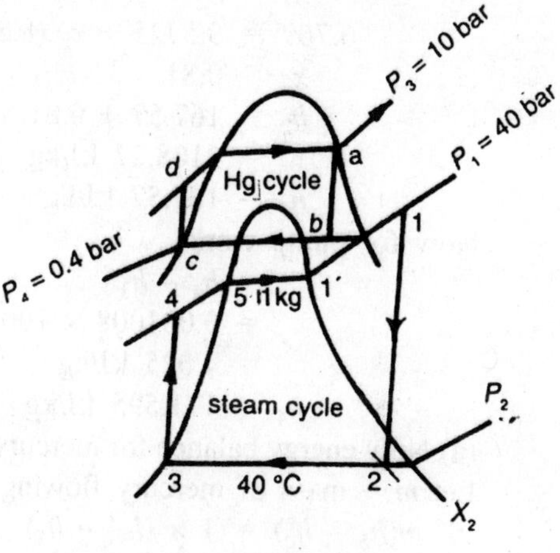

Fig. P. 12.8

Now $S_b = S_a$

$= 0.5167 = 0.0947 + x_b\,(0.6385 - 0.0967)$

$x_b = 0.7789$

At 0.2 bar,

$h_f = 38.35$ kJ/kg

$h_g = 336.55$ kJ/kg

$\therefore \quad h_b = 38.35 + x_b\,(336.55 - 38.35)$

$= 270.61$ kJ/kg

$h_c = h_f = 38.35$ kJ/kg

Now $h_d - h_c = v_c \times (p_{10} - p_{0.2}) \times 100$

$= 0.075852$ kJ/kg

$\therefore \quad h_d = 38.43$ kJ/kg

For steam cycle

From steam table,

at 40 bar and 400°C

$$h_1 = 3213.6 \text{ kJ/kg}$$
$$S_1 = 6.769 \text{ kJ/kg}$$
$$h_1' = 2554.4 \text{ kJ/kg}$$

at 40°C

$$h_f = 167.57 \text{ kJ/kg}$$
$$h_g = 2574.3 \text{ kJ/kg}$$
$$S_f = 0.5725 \text{ kJ/kg.K}$$
$$S_g = 8.2570 \text{ kJ/kg.K}$$
$$v_f = 1.008 \times 10^{-3} \text{ m}^3\text{/kg}$$
$$p_{sat} = 7.384 \text{ kPa}$$

Since ,

$$S_1 = S_2$$
$$6.769 = 0.5725 + x_2 (8.257 - 0.5725)$$
$$\therefore \quad x_2 = 0.81$$
$$\therefore \quad h_2 = 167.57 + 0.81 \times (2574.3 - 167.57)$$
$$= 2108.27 \text{ kJ/kg}$$
$$h_3 = 167.57 \text{ kJ/kg}$$

Now for pump work

$$= h_4 - h_3$$
$$= 0.001008 \times 100 (40 - 0.0738)$$
$$= 4.025 \text{ kJ/kg}$$
$$h_4 = 171.595 \text{ kJ/kg}$$

(a) Now energy balance for mercury condenser-steam boiler system

Let m = mass of mercury flowing in the circuit

$$\therefore \quad m(h_b - h_c) = 1 \times (h_1' - h_4)$$
$$\therefore \quad m = \frac{2554.4 - 171.595}{270.61 - 38.35}$$
$$= 10.26 \text{ kJ/kg of water } \textbf{Ans}$$

(b) $$Q_1 = \text{heat supplied} = m (h_a - h_d)$$
$$= 10.26 (363 - 38.43) = 3330.13 \text{ kJ}$$

Work obtained in turbine

$$= m (h_a - h_b) + 1.(h_1 - h_2)$$
$$= 10.26 (363 - 270.61) + (3213.6 - 2108.27)$$
$$= 2053.25 \text{ kJ}$$

Pump work $$= m (h_d - h_c) + 1 \times (h_4 - h_3)$$
$$= 10.26 \times 0.075852 + 4.025 = 4.803 \text{ kJ}$$

$\therefore$ Net work obtained = 2053.25 − 4.80 = 2048.45 kJ

$$\therefore \quad \eta_0 = \frac{W_{net}}{Q_1} = \frac{2048.45}{3330.13} = 61.55\% \textbf{ Ans}$$

> **Problem 12.9:** A 10, 000 kW steam turbine operates with steam at the inlet at 40 bar, 400°C and exhausts at 0.1 bar. Ten thousand kg/h of steam at 3 bar are to be extracted for process work. The turbine has 75% isentropic efficiency throughout. Find the boiler capacity required.

Solution: Given data :

p_1 = 40 bar
T_1 = 400°C
p_2 = 0.1 bar
$\dot{m}_1$ = 10, 000 kg/h
P_3 = 3 bar
Power = 10, 000 kW
η_{isen} = 0.75

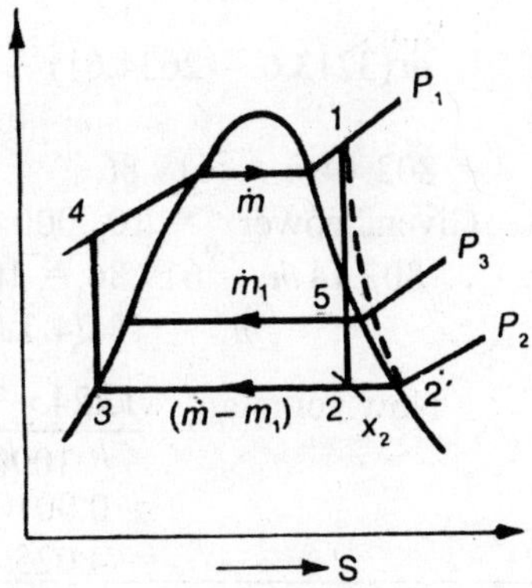

Fig. P. 12.9

From steam table
at *Inlet conditions*
at 40 bar bas and 400°C

h_1 = 3213.6 kJ/kg
s_1 = 6.7690 kJ/kg

at Exhaust condition at 0.1 bar

h_f = 191.83 kJ/kg
h_g = 2584.7 kJ/kg
s_f = 0.6493 kJ/kg.K
s_g = 8.1502 kJ/kg.K

$$s_1 = s_2 = 6.7690$$
$$= 0.6493 + x_2\,(8.1502 - 0.6493)$$
$$x_2 = 0.8159$$
$$\therefore \quad h_2 = 191.83 + 0.8159\,(2584.7 - 191.83)$$
$$= 2144.08 \text{ kJ/kg}$$
$$\therefore \quad \eta_{isen} = 0.75 = \frac{h_1 - h_2'}{h_1 - h_2}$$
$$\Rightarrow \quad h_1 - h_2' = 802.14 \text{ kJ/kg}$$
$$\therefore \quad h_2' = 2411.46 \text{ kJ/kg}$$

at 3 bar,

$$h_f = 561.47 \text{ kJ/kg}$$
$$h_{fg} = 2163.8 \text{ kJ/kg}$$
$$S_f = 1.6718 \text{ kJ/kg}$$
$$S_g = 6.9919 \text{ kJ/kg.K}$$

Since ,

$$s_1 = s_5 = 6.769$$
$$= 1.6718 + x_5 (6.9919 - 1.6718)$$
$$\therefore \quad h_3 = 561.47 + 0.958 \times 2163.8$$
$$= 2634.61 \text{ kJ/kg}$$

Now Net power available

$$= \dot{m}(h_1 - h_3) + (\dot{m} - \dot{m}_1)(h_3 - h_2')$$

$$= \dot{m}(3213.6 - 2634.61) + \left(\dot{m} - \frac{10{,}000}{3{,}600}\right) \times (2634.61 - 2411.46)$$

$$= 802.14\dot{m} - 619.86$$

Given, power $= 10,000$ kW

$$\therefore \; 802.14\dot{m} - 619.86 = 10,000$$
$$\therefore \quad \dot{m} = 13.24 \text{ kJ/s}$$
$$= \frac{13.24 \times 3600}{1000} = 47.664 \text{ ton/hr } \textbf{Ans}$$

Problem 12.10: A 50 MW steam plant built is 1935 operates with steam at the inlet at 60 bar, 450°C and exhausts at 0.1 bar, with 80% turbine efficiency. It is proposed to scrap the old boiler and put in a new boiler and a topping turbine of efficiency 85% operating with inlet steam at 180 bar, 500°C. The exhaust from the topping turbine at 60 bar is reheated to 450°C and admitted to the old turbine. The flow rate is just sufficient to produce the rated output from the old turbine. Find the improvement in efficiency with the new set up. What is the additional power developed ?

Solution: Given data :

For old turbine
From steam table
at 60 bar and 450°C

$$h_1 = 3301.8 \text{ kJ/kg}$$
$$s_1 = 6.7193 \text{ kJ/kg.K}$$

at 0.1 bar

$s_f = 0.6493$ kJ/kg.K

$s_g = 8.1502$ kJ/kg.K

$h_f = 191.83$ kJ/kg

$h_g = 2584.7$ kJ/kg

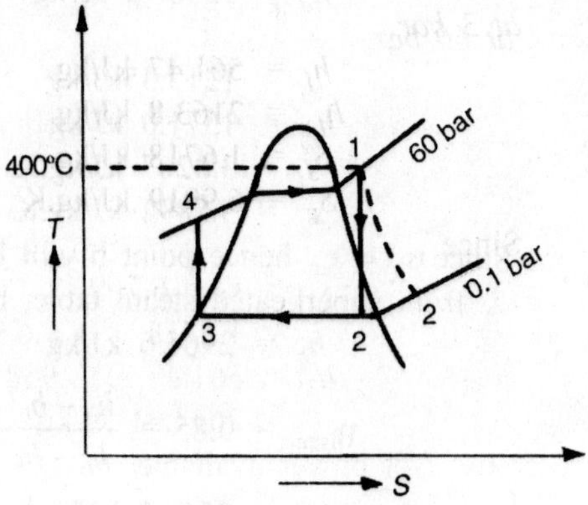

Fig. P. 12.10 (a)

Since,

$s_1 = s_2 = 6.7193$

$= 0.6493 + x_2$

$(8.1502 - 0.6493)$

$\therefore x_2 = 0.8092$

$\therefore h_2 = 191.83 + 0.8092$

$(2584.7 - 191.83)$

$= 2128.23$ kJ/kg

$$\eta_{isen} = 0.8 = \frac{h_1 - h_2'}{h_1 - h_2}$$

$\Rightarrow \quad h_1 - h_2' = 938.856$ kJ/kg

$\therefore \quad h_2' = 2362.944$ kJ/kg

$h_3 = 191.82$ kJ/kg

$= h_4$ (by neglecting pump work)

$\therefore$ Power available at old turbine

$= h_1 - h_2'$

$= 938.856$ kJ/kg

power given = 50 MW

$\therefore$ mass flow rate in old turbine

$$= \dot{m} = \frac{50{,}000}{938.856} = 53.26 \text{ kJ/kg}$$

η_{cycle} = cycle efficiency

$$= \frac{h_1 - h_1'}{h_1 - h_3}$$

$$= \frac{938.856}{3301.8 - 191.83}$$

$= 30.19\%$

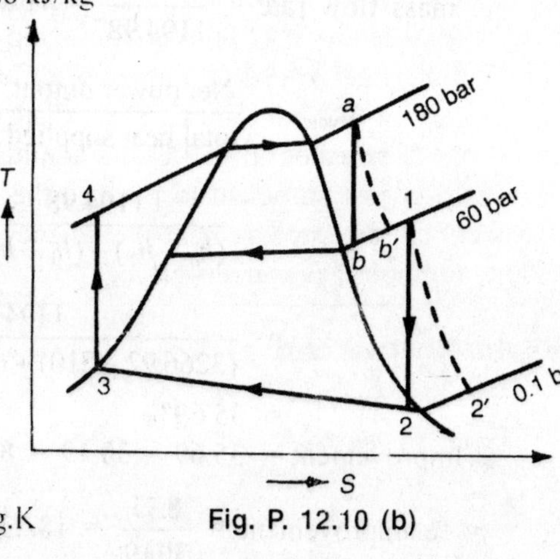

Fig. P. 12.10 (b)

For New set of turbine

From steam table

at 180 bar and 500°C

by interpolation

$s_a = 6.2176$ kJ/kg.K

$h_a = 3266.92$ kJ/kg

at 60 bar

$h_f = 1213.35$ kJ/kg

$h_{fg} = 1571.0$ kJ/kg

$s_f = 3.0267$ kJ/kg.K

$s_g = 5.8892$ kJ/kg.K

Since $s_g < s_a$, hence point b will be in superheated state,

∴ from superheated steam table, by interpolation

$h_b = 2965.6$ kJ/kg

$$\eta_{isen} = 0.85 = \frac{h_a - h_b'}{h_a - h_b}$$

$\Rightarrow \quad h_a - h_b' = 256.12$ kJ/kg

$\therefore \quad h_b' = 3010.8$ kJ/kg

∴ power availabe at topping turbine

$= h_a - h_b' = 256.12$ kJ/kg

∴ Net power output $= (h_a - h_b') + (h_1 - h_2')$

$= 938.856 + 256.12 = 1194.98$ kJ/kg

Again at 0.1 bar from steam table

$h_3 = 191.83$ kJ/kg

$V_f = 0.00101$ kJ/kg

and from old turbine

$h_2{}^1 = 2362.944$ kJ/kg

pump work $= h_4 - h_3 = 0.00101 \times (180 - 0.1) \times 100$

$\therefore \quad h_4 = 209.99$ kJ/kg $= 210$ kJ/kg

$$\therefore \text{ mass flow rate} = \frac{50{,}000}{1194.98} = 41.84 \text{ kJ/sec}$$

$$\eta_{cycle} = \frac{\text{Net power output}}{\text{total heat supplied}}$$

$$= \frac{1194.98}{(h_a - h_4) + (h_1 - h_b')}$$

$$= \frac{1194.98}{(3266.92 - 210) + (3301.8 - 3010.8)}$$

$= 35.69\%$

∴ Improvement $= 35.69 - 30.19 = 8.33$

$$\therefore \text{ \% improvement} = \frac{8.33}{30.19} = 18.23 \text{ \%}$$

Problem 12.11: A steam plant operates with an initial pressure at 20 bar and temperature 400°C, and exhausts to system at 2 bar. The condensate from the heating system is returned to the boiler at 65°C, and the heating system utilizes for its intended purpose 90% of the energy transferred from the steam it receives. The turbine efficiency is 70%. (a) What fraction of the energy supplied to the steam plant serves a useful purpose? (b) If two separate steam plants had been set up to produce the same useful energy, one to generate heating steam at 2 bar, and the other to generate power through a cycle working between 20 bar, 400°C and 0.07 bar, what fraction of the energy supplied would have served a useful purpose?

Solution: Given data :

$p_1 = 20$ bar
$T_1 = 400°C$
$p_2 = 2$ bar
$T_4 = 65°C$
$\eta_{turbine} = 70\%$

From steam table
at 20 bar and 400°C

$h_1 = 3247.6$ kJ/kg
$s_1 = 7.1271$ kJ/kg.K

at 2 bar

$t_{sat} = 120.33°C$
$h_f = 504.7$ kJ/kg
$h_g = 2706.7$ kJ/kg
$s_f = 1.5301$ kJ/kg.K
$s_g = 7.1271$ kJ/kg

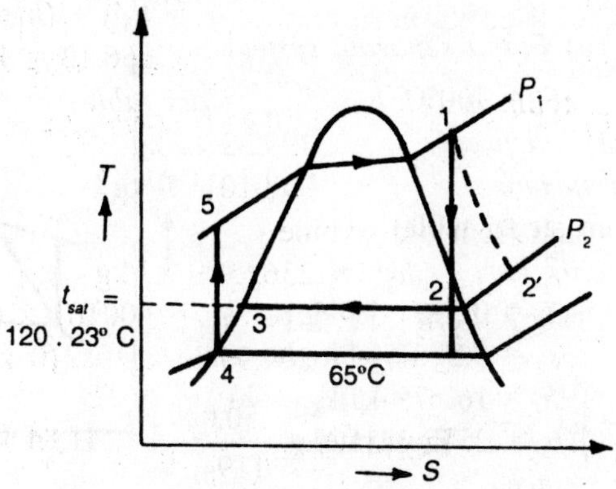

Fig. P. 12.11 (a)

Here

$$s_g = s_1$$

∴ point 2 will be in saturated state

$$\therefore \quad h_g = h_2 = 2706.7 \text{ kJ/kg}$$

given
$$\eta_T = 0.7 = \frac{h_1 - h_2'}{h_1 - h_2} \Rightarrow h_1 - h_2'$$
$$= 0.7\ (3247.6 - 2706.7) = 378.63 \text{ kJ/kg}$$
$$\therefore \quad h_2' = 2868.97 \text{ kJ/kg}$$

Now at 65°C,

$$p_{sat} = 25.03 \text{ kPa}$$
$$v_{f_4} = 0.00102 \text{ m}^3\text{/kg}$$

$h_f = 272.06$ kJ/kg

$\therefore \quad h_4 = 272.06$ kJ/kg

Theoretical heat supplied to the heating system

$= (h_2' - h_4) = (2868.97 - 272.06)$ kJ/kg

$\therefore$ Actual heat supplied to the heating system

$= 0.9 \times (2868.97 - 272.06)$

$= 2337.22$ kJ/kg

$\therefore$ Total heat supplied in the cycle

$= h_1 - h_4 = (3247.6 - 2868.97)$

$= 378.63$ kJ/kg

$\therefore$ fraction of energy utilises in useful purposes

$$= \frac{(h_1 - h_2') + 0.9 \times (h_2' - h_a)}{(h_1 - h_a)}$$

$$= \frac{378.63 + 2337.22}{2975.54} = 91.3 \text{ \% } \textbf{Ans}$$

(b) *For a separate system*

$T_1 = 400°C$

For 2nd plant

for turbine

from steam table

at 0.07 bar,

$s_f = 0.5591$ kJ/kg.K

$s_g = 8.2767$ kJ/kg.K

$h_f = 162.79$ kJ/kg $= h_3$

$h_g = 2572.14$ kJ/kg

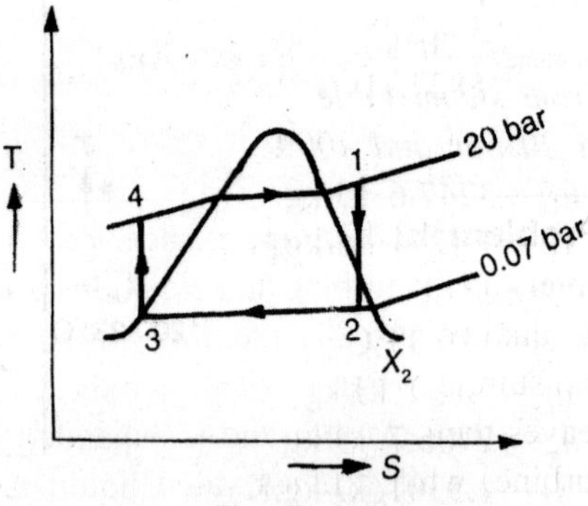

Fig. P. 12.11 (b)

since

$S_1 = S_2 = 7.1271$

$= 0.5591 + x_2 (8.2767 - 0.5591)$

$\therefore \quad x_2 = 0.85$

$\therefore \quad h_2 = 162.79 + 0.85 \times (2572.14 - 162.79)$

$= 2213.25$ kJ/kg

$\therefore$ Heat utilised in turbine

$= (h_1 - h_2) = 1034.35$ kJ/kg

For 1st plant

Total heat utilised = heat utilised in turbine + heat utilised in heating system

$h_1 = 3247.6$ kJ/kg

$h_2 = 2706.7$ kJ/kg

$h_{f3} = h_3 = 504.7$ kJ/kg

$\therefore$ total heat utilised

$= (h_1 - h_2) + (h_2 - h_3)$

$= (h_1 - h_3)$

$= (3247.6 - 504.7)$

$= 2742.9$ kJ/kg

Now total utilised in 1st and 2nd plant

$= 1034.35 + 2742.9$

$= 3777.25$ kJ/kg

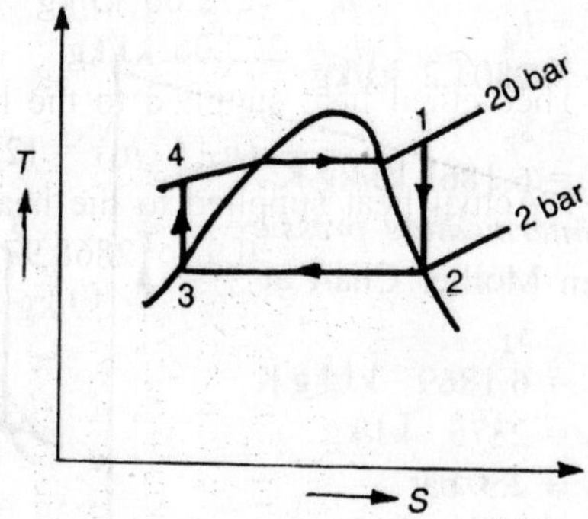

Fig. P. 12.11 (c)

total heat given to 1st and 2nd plant

$= (h_1 - h_3)_{\text{1st plant}} + (h_1 - h_3)_{\text{2nd plant}}$

$= (3247.6 - 504.7) + (3247.6 - 162.79)$

$= 5827.71$ kJ/kg

Hence, fraction of energy utilised

$$= \frac{3772.25}{5827.71} = 64.8\% \textbf{ Ans}$$

Problem 12.12: In a nuclear power plant saturated steam at 30 bar enters a H.P. turbine and expands isentropically to a pressure at which its quality is 0.841. At this pressure the steam is passed through a moisture separator which removes all the liquid. Saturated vapour leaves the separator and is expanded isentropically to 0.04 bar in I.P. turbine, while the saturated liquid leaving the separator is returned via a feed pump to the boiler. The condensate leaving the condenser at 0.04 bar is also returned to the boiler via a second feed pump. Calculate the cycle efficiency and turbine outlet quality taking into account the feed pump term. Recalculate the same quantities for a cycle with the same boiler and condenser pressure but without moisture separation.

Solution: Given data :

$p_1 = 30$ bar

$p_3 = 0.04$ bar

$x_2 = 0.841$

From steam table

at 30 bar

$h_1 = h_g$
$= 2804.2$ kJ/kg
$s_1 = s_g$
$= 6.1869$ kJ/kg.K

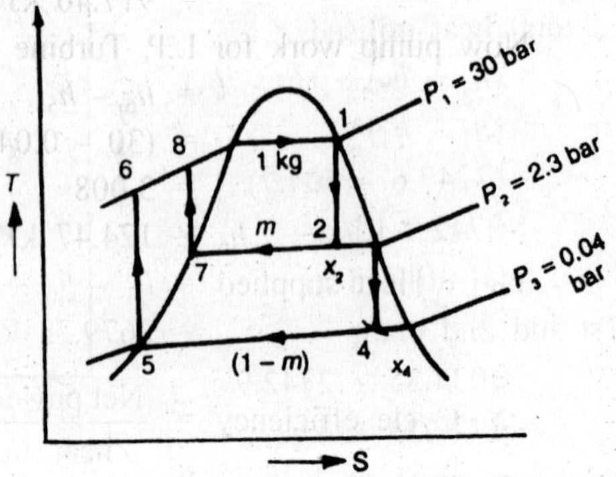

Fig. P. 12.12 (a)

At intermediate pressure
From Mollier Chart at
$s_1 = s_2$
$= 6.1869$ kJ/kg.K
$h_2 = 2375$ kJ/kg
$p_2 \approx 2.3$ bar
$h_3 = 2720$ kJ/kg
fro: steam table
at 2.3 bar
$h_f = h_7 = 520.37$ kJ/kg
$s_3 = 7.0899$ kJ/kg.K
$V_f = V_7 = 0.001065$ m³/kg
at 0.04 bar
from steam table

$$h_f = h_5 = 121.46 \text{ kJ/kg}$$
$$h_g = 2554.4 \text{ kJ/kg}$$
$$s_f = 0.4226 \text{ kJ/kg.K}$$
$$s_g = 8.4746 \text{ kJ/kg.K}$$
$$V_f = V_5 = 0.001004 \text{ m}^3\text{/kg}$$

Now since,

$$s_3 = s_4 = 7.0899$$
$$= 0.4226 + x_3 (8.4746 - 0.4226)$$
$$x_3 = 0.828 \textbf{ Ans}$$
$$\therefore \quad h_4 = 121.46 + x_3 (2554.4 - 121.46)$$
$$= 2135.93 \text{ kJ/kg}$$

since $x_2 = 0.841$
hence $m = 1 - 0.841 = 0.159$ kg

$\therefore$ Total turbine output $= 1 \times (h_1 - h_2) + 0.841 (h_3 - h_4)$
$= (2804.2 - 2375) + 0.841 (2720 - 2135.93)$
$= 920.4$ kJ/kg

total pump input
$= m \times V_7 (30 - 2.3) \times 100 + (1 - m)(30 - 0.04) \times V_5 \times 100$
$= 0.159 \times 0.001065 \times 100 (30 - 2.3) + 0.841 \times 0.001004 \times 100 \times (30 - 0.04)$
$= 2.9988$ kJ/kg

$\therefore$ Net power output $= 920.4 - 2.9988$

$= 917.40$ kJ/kg

Now pump work for L.P. Turbine

$= h_6 - h_5$

$= (30 - 0.04) \times 100 \times 0.001004$

$= 3.008$

$\therefore \quad h_6 = 124.47$ kJ/kg

$\therefore$ Heat supplied $= h_1 - h_6 = 2804.2 - 124.47$

$= 2679.73$ kJ/kg

$$\therefore \text{ Cycle efficiency} = \frac{\text{Net power output}}{\text{heat supplied}} = \frac{917.4}{2679.73}$$

$= 34.4$ % **Ans**

2nd part

At 30 bar

$h_1 = 2804.2$ kJ/kg

$s_1 = 6.1869$ kJ/kg

$h_2 = 2375$ kJ/kg

at 0.04 bar

$h_4 = h_f = 121.46$ kJ/kg

$h_g = 2554.4$ kJ/kg

$s_f = 0.4226$ kJ/kg.K

$s_g = 8.4746$ kJ/kg

$V_f = 0.001004$ m³/kg

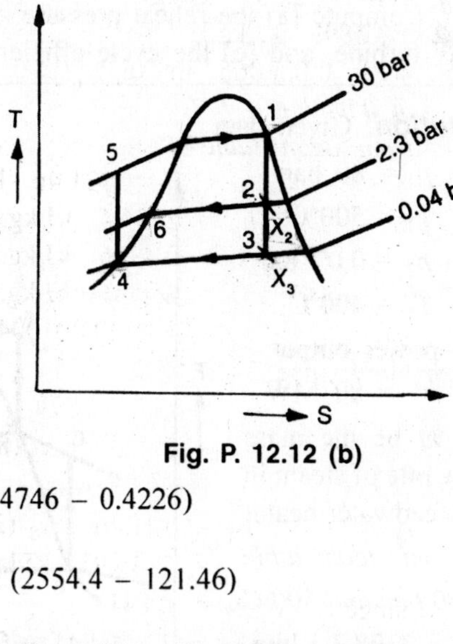

Fig. P. 12.12 (b)

since

$s_1 = s_3 = 6.1869$

$= 0.4226 + x_3 \times (8.4746 - 0.4226)$

$x_3 = 0.716$ **Ans**

$\therefore h_3 = 121.46 + 0.716 \times (2554.4 - 121.46)$

$= 1863.16$ kJ/kg

work output by turbine $= (h_1 - h_2) + (h_2 - h_3)$

$= (h_1 - h_3)$

$= 2804.2 - 1863.16$

$= 941.04$ kJ/kg

pump work input $= h_5 - h_4$

$= V_f \times (30 - 0.04) \times 100$

$= 0.001004 \times (30 - 0.04) \times 100$

$= 3.008$ kJ/kg

$\therefore \quad h_5 = h_4 + 3.008 = 124.47$ kJ/kg

$\therefore$ Net work $= 941.04 - 3.008 = 938.032$ kJ/kg

Heat supplied $= h_1 - h_5 = 2804.2 - 124.47$

$= 2679.73$ kJ/kg

$$\therefore \text{ Cycle efficiency} = \frac{\text{Net work}}{\text{heat supplied}} = \frac{938.032}{2679.73} = 35\% \textbf{ Ans}$$

Problem 12.13: The net power output of an ideal regenerative-reheat steam cycle is 80 MW. Steam enters the H.P. turbine at 80 bar, 500°C and expands till it becomes saturated vapour. Some of the steam then goes to an open to an open feed water heater and the balance is reheated to 400°C, after which it expands in the I.P. turbine to 0.07 bar. Compute (a) the reheat pressure, (b) the steam flow rate to the h.p. turbine, and (c) the cycle efficiency. Neglect pump work

Solution: Given data :

$p_1 = 80$ bar

$T_1 = 500°C$

$p_3 = 0.07$ bar

$T_3 = 400°C$

Net power output

$= 80$ MW

Let m be the mass flow rate of steam in the feedwater heater

$\therefore$ *From steam table at 80 bar and 500°C*

$h_1 = 3398.3$ kJ/kg

$s_1 = 6.724$ kJ/kg.K

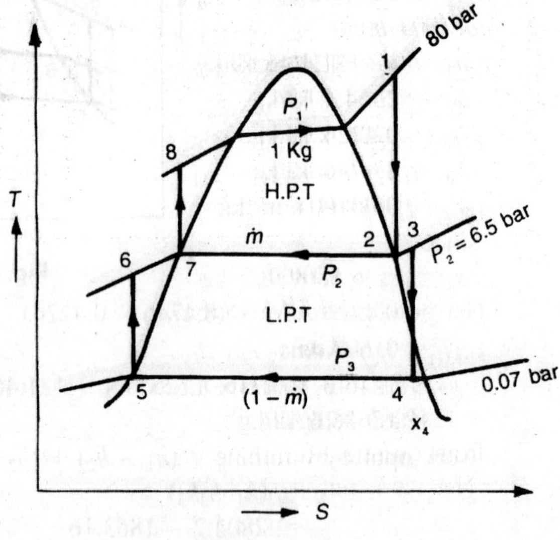

Fig. P. 12.13

from steam table

at

$s_1 = s_2$

$= 6.724$ kJ/kg.K

by interpolation

$p_2 = 6.5$ bar **Ans**

at 6.5 bar,

$$h_f = h_7 = 683.89 \text{ kJ/kg}$$
$$h_2 = h_g = 2760.15 \text{ kJ/kg}$$

at 6.5 bar and 400°C,

$$s_3 = 7.6738 \text{ kJ/kg.K}$$
$$h_3 = 3269.5 \text{ kJ/kg}$$

at 6.5 bar,

$$h_f = h_7 = 683.89 \text{ kJ/kg}$$
$$= h_8 = \text{(by neglecting pump work)}$$

at 0.07 bar

$$h_f = 162.596 \text{ kJ/kg}$$
$$h_g = 2572.14 \text{ kJ/kg}$$
$$s_f = 0.5564 \text{ kJ/kg.K}$$
$$s_g = 8.303196 \text{ kJ/kg.K}$$

since $s_3 = s_4 = 7.6738$

$$= 0.5564 + x_4 (8.303196 - 0.5564)$$

$\therefore \quad x_4 = 0.919$

$\therefore \quad h_4 = 162.596 + x_4(2572.14 - 162.596)$

$$= 2376.374 \text{ kJ/kg}$$

$h_5 = 162.596$ kJ/kg $= h_6 =$ [by neglecting the pump work]

From energy balance in open feed water heater

or m $(h_2 - h_7) = (1 - m)(h_7 - h_5)$

or, m (2760.15 – 683.89)

$$= (1 - m)(683.89 - 162.596)$$

$\therefore \quad m = 0.2$ kg

Net power output of turbine

$= (h_1 - h_2) + (1 - \dot{m})(h_3 - h_4)$

$= (3398.3 - 2760.15) + (1 - 0.2) \times (3269.5 - 2376.374)$

$= 1352.65$ kJ/kg

from question

Net work out = 80 MW

$\therefore \; \dot{m}_1$ = steam flow rate to the H.P. Turbine $= \dfrac{80 \times 1000}{1352.65}$

= 59.14 kg/s **Ans**

Heat supplied

$$= \dot{m}_1 (h_1 - h_8) + \dot{m}_1 (1 - \dot{m})(h_3 - h_2)$$
$$= 59.14 \, [(3398.3 - 683.89) + 0.8\,(3269.5 - 2760.15)]$$
$$= 184628.57 \text{ kW}$$

$$\therefore \text{ Cycle efficiency } = \frac{\text{Net work}}{\text{heat supplied}}$$

$$= \frac{1352.65 \times 59.14}{184628.57} = 43.3 \% \quad \textbf{Ans}$$

Problem 12.14: In a combined power process plant the boiler generates 21,000 kg/h of steam at a pressure of 17 bar, and temperature 230°C. A part of steam goes to a process heater which consumes 132.56 kW, steam leaving the process heater 0.957 dry at 17 bar being throttled to 3.5 bar. The remaining steam flows through a H.P. turbine which exhausts at a pressure of 3.5 bar. The exhaust steam mixes with the process steam before entering the L.P. turbine which develops 1337.5 kW. At the exhaust the pressure is 0.3 bar, and the steam is 0.912 dry. Draw a line diagram of the plant and determine (a) the steam quality at the exhaust from the H.P. turbine, (b) the power developed by the H.P. turbine, and (c) the isentropic efficiency of the H.P. turbine.

Solution: Given data :

$\dot{m}$ = 21, 000 kg/h

power consumed by process heater

= 132.56 kW

L.P. Turbine output

= 1337.5 kW

Let $\dot{m}_1$ be the mass flow rate to process heater and $\dot{m}_2$ be the mass flow rate to H.P. turbine

FOR H.P. TURBINE

From steam table

at 17 bar and 230°C at turbine inlet by interpolation,

h_1 = 2862.8 kJ/kg

s_1 = 6.54 kJ/kg.K

at 3.5 bar for turbine exhaust

h_f = 584.33 kJ/kg

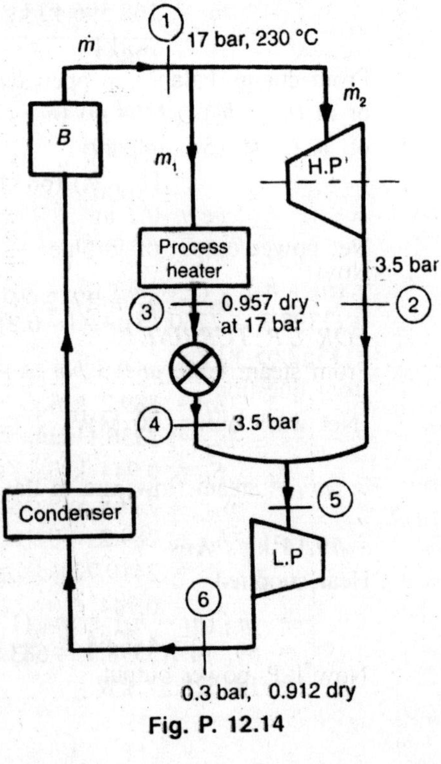

Fig. P. 12.14

$$h_g = 2732.4 \text{ kJ/kg}$$
$$s_f = 1.7275 \text{ kJ/kg.K}$$
$$s_g = 6.9405 \text{ kJ/kg.K}$$

since, $S_1 = S_2 = 6.54$

$$= 1.7275 + x_2 (6.9405 - 1.7275)$$

$\therefore$ $x_2 = 0.923$

$\therefore$ $h_2{}^1 = 584.33 + 0.923 (2732.4 - 584.33)$

$$= 2567.37 \text{ kJ/kg}$$

FOR PROCESS HEATER

$$h_1 = 2862.8 \text{ kJ/kg}$$

power consumed in process heater = 132.56 kW

$$= \dot{m}_1 (h_1 - h_3) \quad \ldots\text{(i)}$$

Now at 17 bar and 0.957 dry

from steam table,

$$h_f = 872.10 \text{ kJ/kg}$$
$$h_g = 2795.7 \text{ kJ/kg}$$

$\therefore$ $h_3 = 872.10 + 0.957 (2795.7 - 872.10)$

$$= 2712.98 \text{ kJ/kg}$$

$\therefore$ from (i),

$$\dot{m}_1 = \frac{132.56}{(2862.8 - 2712.98)} = 0.885 \text{ kJ/sec}$$

$$\dot{m} = \text{total steam flow rate}$$

$$= \frac{21{,}000}{3600} = 5.833 \text{ kg/sec}$$

$\therefore$ $\dot{m}_2 = \dot{m} - \dot{m}_1 = 4.949 \text{ kJ/kg}$

Now

$$h_3 = h_4 = 2712.98 \text{ kJ/kg}$$

FOR L.P. TURBINE

From steam table *at 0.3 bar & 0.912 dry*

$$h_f = 289.2 \text{ kJ/kg}$$
$$h_{fg} = 2336.1 \text{ kJ/kg}$$
$$S_f = 0.944 \text{ kJ/kg.K}$$
$$S_{fg} = 7.768 \text{ kJ/kg.K}$$

$\therefore$ $h_6 = 289.2 + 0.912 \times 2336.1$

$$= 2419.72 \text{ kJ/kg}$$

$$s_6 = 0.944 + 0.912 \times 7.768$$
$$= 8.0284 \text{ kJ/kg.K}$$

Now L.P. power output

$$= 1337.5 \text{ kW}$$

$$= \dot{m}(h_5 - h_6)$$

$$\therefore \quad h_5 = 2649.01 \text{ kJ/kg}$$

Now for mixing process

$$\dot{m}_1 h_4 + \dot{m}_2 h_2 = \dot{m} h_5$$

$$\text{or,} \quad h_2 = \frac{\dot{m}h_5 - \dot{m}_1 h_4}{\dot{m}_2}$$

$$= \frac{5.833 \times 2649.01 - 0.885 \times 2712.98}{4.949}$$

$$= 2637.04 \text{ kJ/kg}$$

$\therefore$ power developed by turbine

$$= \dot{m}_2 (h_1 - h_2)$$

$$= 4.949 \times (2862.8 - 2637.04)$$

$$= 1117.31 \text{ kJ/kg}$$

$$\therefore \eta_{isentropic} = \frac{h_1 - h_2}{h_1 - h_2'} = \frac{2862.8 - 2637.04}{2862.8 - 2567.37} = 76.42\ \%$$

Problem 12.15: In a cogeneration plant, the power load id 5.6 MW and the heating load is 1.163 MW. Steam is generated at 40 bar and 500°C and is expanded isentropically through a turbine to a condenser at 0.06 bar. The heating load is supplied by extracting steam from the turbine at 2 bar which condensed in the process heater to saturated liquid at 2 bar and then pumped back to the boiler. Compute (a) the steam generation capacity of the boiler in tonnes/h, (b) the heat input to the boiler in MW, and (c) the heat rejected to the condenser in MW.

Solution: Given data :

$$p_1 = 40 \text{ bar}$$

$$T_1 = 500°\text{C}$$

$$p_3 = 0.06 \text{ bar}$$

$$p_2 = 2 \text{ bar}$$

$$\text{power load} = 5.6 \text{ MW}$$

$$\text{heating load} = 1.163 \text{ MW}$$

Let m_1 and $\dot{m}_2$ be the total flow rate to urbine & flow rate in process heater respectively

from steam table

at 40 bar & 500°C

$h_1 = 3445.3$ kJ/kg

$s_1 = 7.0910$ kJ/kg.K

at 0.06 bar

by interpolation,

$h_f = 151.5$ kJ/kg

$h_g = 2567.4$ kJ/kg

$S_f = 0.5208$ kJ/kg.K

$S_g = 8.331$ kJ/kg.K

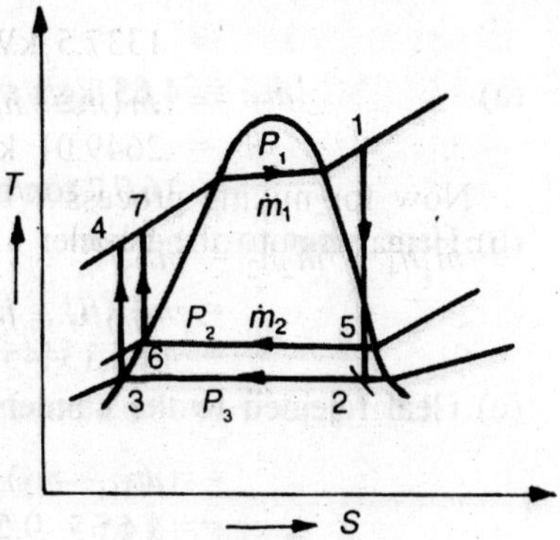

Fig. P. 12.15

Since,

$$S_1 = S_2$$
$$= 7.091$$
$$= 0.5208 + x_2 (8.331 - 0.5208)$$
$$x_2 = 0.841$$
$$\therefore \quad h_2 = 151.5 + 0.841 \times (2567.42 - 151.5)$$
$$= 2183.85 \text{ kJ/kg}$$

and $h_3 = 151.50$ kJ/kg

$= h_4$ (by neglucting pump work)

at 2 bar

$h_f = 504.7$ kJ/kg

$h_g = 2706.7$ kJ/kg

$s_f = 1.53$ kJ/kg.K

$s_g = 7.1271$ kJ/kg.K

since $S_1 = S_5 = 7.091$

$$= 1.5301 + x_5 (7.1271 - 1.5301)$$
$$\therefore \quad x_5 = 0.994$$
$$\therefore \quad h_5 = 504.7 + 0.994 (2706.7 - 504.7)$$
$$= 2692.50 \text{ kJ/kg}$$

and $h_6 = 504.7$ kJ/kg

$= h_7$ [by neglecting pump input]

$\therefore$ Heating load $= \dot{m}_2 (h_5 - h_6)$

$= 1.163 \times 10^3$ kW

$$\therefore \quad \dot{m}_2 = 0.5316 \text{ kg/s}$$

Now power load $= \dot{m}_1 (h_1 - h_5) + (\dot{m}_1 - \dot{m}_2)(h_5 - h_2)$

$= 5.6 \times 10^3$ kW

or, $\dot{m}_1 (3445.3 - 2692.5) + (\dot{m}_1 - 0.5316)(2692.5 - 2183.85)$

$= 5.6 \times 10^3$

or, $1265.13\,\dot{m}_1 = 5872.21449$

$$\therefore \quad \dot{m}_1 = 4.65 \text{ kg/s}$$

(a) $\dot{m}_1 = 4.65 \text{ kg/s} = \frac{4.65 \times 3600}{1000}$

$= 16.71$ ton/hr. **Ans.**

(b) Heat input to the boiler

$= \dot{m}_1 (h_1 - h_3)$

$= 4.65 (3445.3 - 151.5) = 15.32$ MW

(c) Heat rejected to the condenser

$= (\dot{m}_1 - \dot{m}_2)(h_2 - h_3)$

$= (4.65 - 0.5316)(2183.83 - 151.5)$

$= 8.4$ MW **Ans**

Problem 12.16: In a cogeneration plant, steam enters the h.p. stage of a two-stage turbine at 1 MPa, 200°Cand leaves it at 0.3 MPa. At this point some of the steam is bled off and passed through a heat exchanger which it leaves as saturated liquid at 0.3 MPa. The remaining steam expands in the I.p. stage of the turbine to 40 kPa. The turbine is required to produce a total power of 1 MW and the heat exchanger to provide a heating rate of 500 kW. Calculate the required mass flow rate of steam into the h.p. stage of the turbine. Assume (a) steady condition throughout the plant, (b) the velocity and gravity terms to be negligible, (c) both turbine stages are adiabatic with isentropic efficiencies of 0.80.

Solution: Given data :

$p_1 = 1$ MPa

$T_1 = 200$°C

$p_2 = 0.3$ MPa

$p_3 = 40$ kPa

total power produced by turbine = 1 MW

heating rate required in heat exchanger = 500 kW

Let $\dot{m}_1$ be total steam rate (kg/s) and $\dot{m}_2$ be mass rate of steam extracted (kg/s)

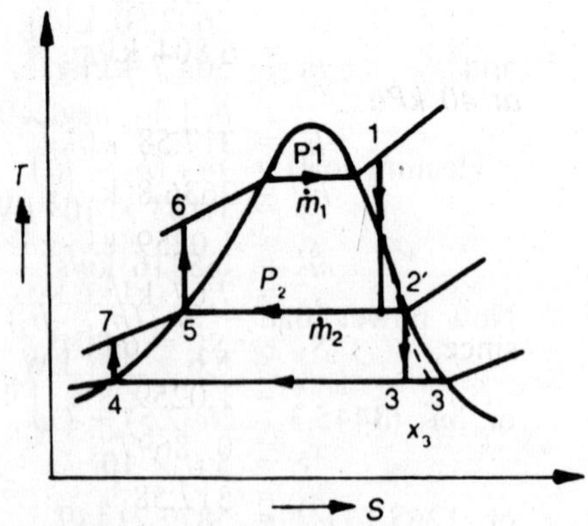

Fig. P. 12.16

∴ *From steam table*
at 1 MPa and 200°C

$$h_1 = 2827.9 \text{ kJ/kg}$$
$$s_1 = 6.6940 \text{ kJ/kg.K}$$

at 0.3 MPa,

$$h_f = 561.47 \text{ kJ/kg}$$
$$h_g = 2725.3 \text{ kJ/kg}$$
$$s_f = 1.6718 \text{ kJ/kg.K}$$
$$s_g = 6.9919 \text{ kJ/kg.K}$$

Since, $S_1 = S_2 = 6.694$
$= 1.6718 + x_2\,(6.9919 - 1.6718)$

∴ $x_2 = 0.944$

∴ $h_2 = 561.47 + 0.944\,(2725.3 - 561.47)$
$= 2604.14 \text{ kJ/kg}$

$$h_5 = 561.47 \text{ kJ/kg}$$

$$\eta_{isen} = 0.8 = \frac{h_1 - h_2'}{h_1 - h_2}$$

$$(h_1 - h_2') = 179 \text{ kJ/kg}$$

∴ $h_2' = 2648.89 \text{ kJ/kg}$

from inter polation,

$$S_2' = \left(\frac{h_2' - h_2}{h_g - h_2}\right)(S_g - S_2) + S_2$$

$$= \left(\frac{2648.89 - 2604.14}{2725.3 - 2604.14}\right)(6.9919 - 6.694) + 6.694$$

$$= 6.804 \text{ kJ/kg.K}$$

at 40 kPa

$$h_f = 317.58 \text{ kJ/kg}$$
$$h_g = 2636.8 \text{ kJ/kg}$$
$$s_f = 1.0259 \text{ kJ/kg.K}$$
$$s_g = 7.67 \text{ kJ/kg.K}$$

since, $S_3 = S_2' = 6.804$
$= 1.0259 + x_3\,(7.67 - 1.0259)$

∴ $x_3 = 0.8697$

∴ $h_3 = 317.58 + x_3\,(2636.8 - 317.58)$
$= 2334.52 \text{ kJ/kg}$

$$h_4 = 317.58 \text{ kJ/kg}$$

$$n_{isen} = 0.8 = \frac{h_2' - h_3'}{h_2' - h_3}$$

or, $(h_2' - h_3') = 251.496$ kJ/kg

$\therefore$ $h_3' = 2397.39$ kJ/kg

and $h_4 = 317.58$ kJ/kg

Now from heat balance in heat exchanger,
Neglecting pump work,

$$h_7 = h_4,$$
$$h_6 = h_5$$

$\therefore$ $\dot{m}_2 h_2' + (\dot{m}_1 - \dot{m}_2)\, h_7 = \dot{m}_1 h_5$

or, $\dot{m}_2\,(h_2' - h_7) = \dot{m}_1 (h_5 - h_7)$

or, $$\dot{m}_2 = \dot{m}_1 \left(\frac{h_5 - h_7}{h'_2 - h_7}\right)$$

or, $$\dot{m}_1 \left(\frac{561.47 - 317.58}{2648.89 - 317.58}\right) = 0.105\ \dot{m}_1 \quad \text{...(i)}$$

Now, total turbine power = 1000 kW

or, $\dot{m}_1\,(h_1 - h_2') + (\dot{m}_1 - \dot{m}_2)\,(h_2' - h_3') = 1000$

or, $\dot{m}_1\,(h_2 - h_3') - 0.895\ \dot{m}_1\,(h_2' - h_3') = 1000$

or, $\dot{m}_1$ [(2827.9 – 2397.39) – 0.105 (2648.89 – 2397.39)] = 1000

$\therefore$ $\dot{m}_1 = 2.475$ kg/sec **Ans**

Problem 12.17: In a reheat cycle steam at 550°C expands in an H.P. turbine till it is saturated vapour. It is reheated at constant to 400°C and then expands in a I.P. turbine to 40°C. If the moisture content at turbine exhaust is given to be 14.67%, find (a) the reheat pressure, (b) the pressure of steam at inlet to the H.P. turbine, (c) the net work ouput per kg, and (d) the cycle efficiency. Assume all process to be ideal.

Solution: Given data :

$$T_1 = 500°C$$
$$T_3 = 400°C$$
$$T_5 = 40°C$$
$$x_4 = 0.8533$$

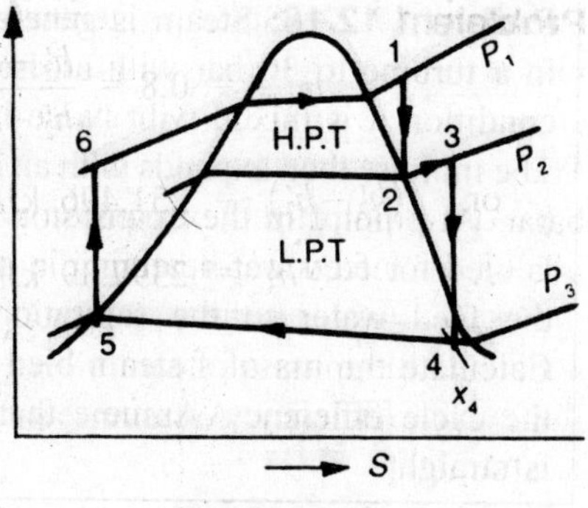

Fig. P. 12.17

From steam table

at 40°C

$x_4 = 0.8533,$

$P_{sat} = 0.074$ bar

$h_f = 167.57$ kJ/kg

$h_g = 2574.3$ kJ/kg

$s_f = 0.5725$ kJ/kg.K

$s_g = 8.2570$ kJ/kg.K

$V_5 = V_f = 0.001008$ m³/kg

$S_4 = 0.5725 + 0.8533 \times (8.257 - 0.5725)$

$= 7.1297$ kJ/kg

$h_4 = 167.57 + 0.8533\ (2574.3 - 167.57) = 2221.23$ kJ/kg.

$h_5 = 167.57$ kJ/kg

Now since, $S_3 = S_4 = 7.1297$ kJ/kg.K

From steam table by interpolation

(i) $P_2 \approx 20$ bar **Ans**

From steam table, at 20 bar

$h_2 = h_g = 2799.5$ kJ/kg

$S_2 = h_g = 6.3409$ kJ/kg.K

at 20 bar and 400°C,

$h_3 = 3247.6$ kJ/kg

(ii) From steam for $S_2 = 6.3409$ kJ/kg.K and 550°C by interpolation,

$p_3 \approx 200$ bar **Ans**

(iii) at 200 bar and 550°C,

$h_1 = 3393.5$ kJ/kg

Net work output $= (h_1 - h_2) + (h_3 - h_4)$

$= (3393.5 - 2799.5) + (3247.6 - 2221.23)$

$= 1620.37$ kJ/kg **Ans**

(iv) Net heat supplied $= (h_1 - h_6) + (h_3 - h_2)$

Now $h_6 - h_5$ = pump work $= v_5 \times (200 - 0.074) \times 100$

$\therefore h_6 = h_5 + v_5\ (200 - 0.074) \times 100$

$= 167.57 + (200 - 0.074) \times 100 \times 0.001003 = 187.72$ kJ/kg

$\therefore$ Net heat supplied $= (h_1 - h_6) + (h_3 - h_2)$

$= (3393.5 - 187.72) + (3247.6 - 2799.5)$

$= 3653.88$ kJ/kg

$$\therefore \text{Cycle efficiency} = \frac{\text{Net work output}}{\text{Net heat supplied}} = \frac{1620.37}{3653.88} = 44.3\%$$

Problem 12.18: Steam is generated at 70 bar, 500°C and expands in a turbine to 30 bar with an isentropic efficiency of 77%. At this condition it is mixed wiht twice its mass of steam at 30 bar, 400°C. The mixutre then expands with an isentropic efficiency of 80% to 0.06 bar. At a point in the exspansion where the pressure is 5 bar, steam is bled for feed water heating in a direct contact heater, which raises the feed water to the saturation temperature of the bled steam. Calculate the mass of steam bled per kg of high pressure steam and the cycle efficiency. Assume that the L.P. expansion condition line is straight.

Solution: Given data :

$$p_1 = 70 \text{ bar}$$
$$p_2 = 30 \text{ bar}$$
$$p_3 = 5 \text{ bar}$$
$$p_4 = 0.06 \text{ bar}$$
$$T_1 = 500°\text{C}$$

For H.P.T,

$$\eta_{isen} = 0.77$$

For L.P.T,

$$\eta_{isen} = 0.80$$

From steam table

at 70 bar & 500°C

$$h_1 = 3410.3 \text{ kJ/kg.K}$$
$$s_1 = 6.7975 \text{ kJ/kg.K}$$

at 30 bar

$$h_f = 100.42 \text{ kJ/kg}$$
$$h_g = 2804.2 \text{ kJ/kg}$$
$$s_f = 2.6457 \text{ kJ/kg.K}$$
$$s_g = 6.1869 \text{ kJ/kg.K}$$

since,

$$s_1 = s_2 = 6.7975$$
$$= 2.6457 + x_2 (6.1869 - 2.6457)$$

since

$$x_2 > 1$$

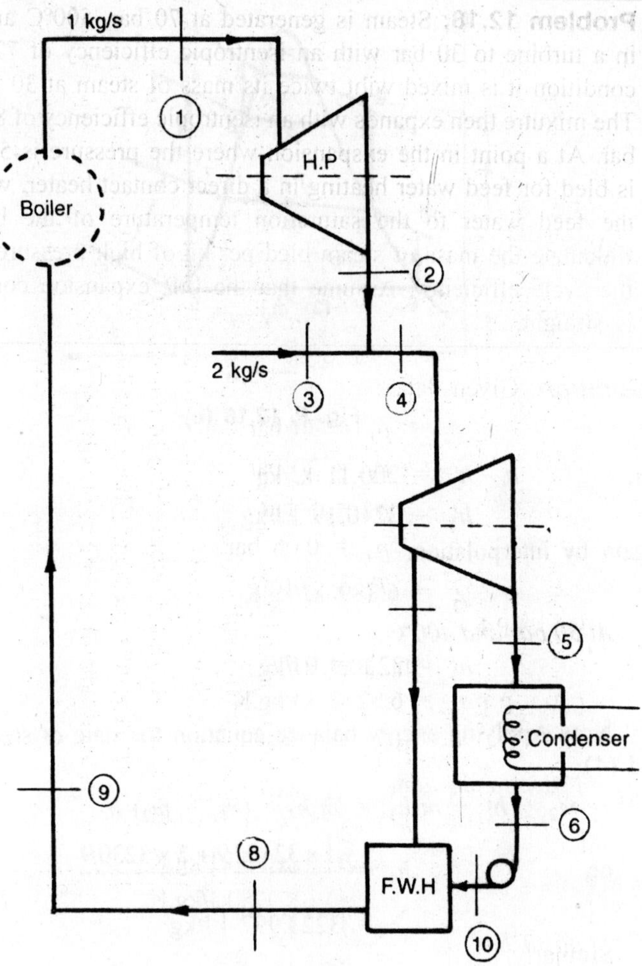

Fig. P. 12.18 (a)

Hence point 2 will be insepertuated state

∴ from super heated steam table at 30 bar

and $s_2 = 6.7975$ kJ/kg.K

by interpolation,

$$h_2 = 3150.42 \text{ kJ/kg}$$

$$\eta_{isen} = 0.77 = \frac{h_1 - h_2'}{h_1 - h_2}$$

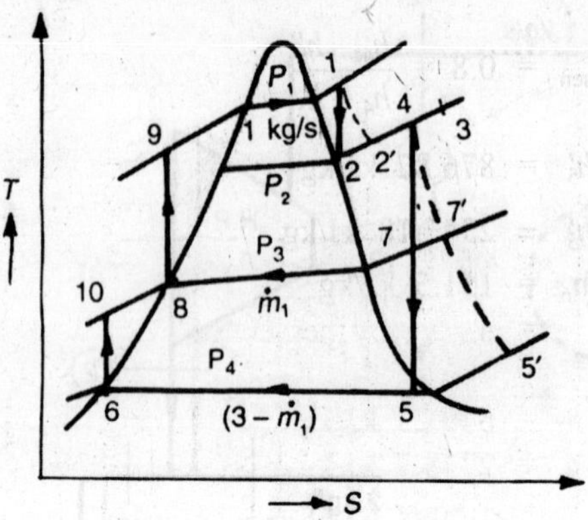

Fig. P. 12.18 (b)

or, $h_1 - h_2' = 200.11$ kJ/kg

∴ $h_2' = 3210.19$ kJ/kg

Again by interpolation,

$s_2' = 6.889$ kJ/kg.K

At 30 bar and 400°C

$h_3 = 3230.9$ kJ/kg

$s_3 = 6.9212$ kJ/kg.K

Now Applying energy balance equation for state of steam (2), (3) and (4)

$$\dot{m}_2 + h_2' + \dot{m}_3 h_3 = \dot{m}_4 h_4 = (\dot{m}_2 + \dot{m}_2)\, h_4$$

or, $$h_4 = \frac{1 \times 3210.19 + 2 \times 3230.9}{3}$$

$$= 3223.997 \text{ kJ/kg}$$

Similarly,

$$S_4 = \frac{1 \times 6.889 + 2 \times 6.9212}{3} = 6.9104 \text{ kJ/kg.K}$$

at 0.06 bar

$h_f = 151.50$ kJ/kg

$h_g = 2567.4$ kJ/kg

$s_f = 0.5208$ kJ/kg.K

$s_g = 8.331$ kJ/kg.K

Since $s_4 = s_5 = 6.9104 = 0.5208 + x_5\,(8.331 - 0.5208)$

$x_5 = 0.818$

∴ $h_5 = 151.5 + 0.818\,(2567.4 - 151.5) = 2127.97$ kJ/kg

$h_8 = 151.5 \times$ kJ/kg

since, $\eta_{isen} = 0.8 = \dfrac{h_4 - h_5'}{h_4 - h_5}$

or, $h_4 - h_5' = 876.82$ kJ/kg

$h_5' = 2347.18$ kJ/kg

$h_6 = 151.5$ kJ/kg

$= h_{10} =$ (by neglecting pubp work)

At 5 bar

$h_f = 640.23$ kJ/kg

$h_g = 2748.7$ kJ/kg

$s_f = 1.8607$ kJ/kg.K

$s_g = 6.8213$ kJ/kg.K

Hue $s_4 > s_g$

∴ point 7 will be in super heated region

∴ from steam table

∴ at $s_4 = s_7 = 6.9104$ kJ/kg.K

and 5 bar by interpolation,

$h_7 = 2812.68$ kJ/kg

$\eta_{isen} = 0.8 = \dfrac{h_4 - h_7'}{h_4 - h_7}$

or, $h_7' = h_4 - 0.8\,(h_4 - h_7) = 2894.92$ kJ/kg.

$h_8 = 640.23$ kJ/kg

pump work input $= h_g - h_8$

$= 0.0010093 \times (70 - 5) \times 100$

$= 7.1045$ kJ/kg

∴ $h_g = 647.33$ kJ/kg

From energy balance in feedwater heater

$m_1(h_7 - h_8) = (3 - m_1)\,(h_8 - h_6)$ or

$m_1\,(2894.92 - 640.23) = (3 - m_1)\,(640.23 - 151.5)$

∴ $m_1 = 0.5344$ kg

∴ Net work output

$= 1\left(h_1 - h_2'\right) + 3\,(h_4 - h_7') + (3 - 0.559)\,(h_7' - h_5')$

$= (3410.3 - 3210.19) + 3\,(3223.998 - 2894.92) + 2.441(2894.92 - 2347.18)$

$= 2524.38$ kJ/kg

Net heat supplied

$= 3\,(h_1 - h_g) + 2\,(h_4 - h_2')$

$= 3 \times (3410.3 - 647.33) + 2 \times (3223.997 - 3210.19)$
$= 8316.52$ kJ/kg

$$\therefore \text{ Cycle efficiency} = \frac{\text{Net work output}}{\text{Net heat supplied}} = 30.6\ \% \textbf{ Ans}$$

Problem 12.19: Geothermal energy from a natural geyser can be obtained as a continuous supply of steam 0.87 dry at 2 bar and at a flow rate of 2700 kg/h. This is utilized in turbine of 83% internal efficiency, which is supplied with 5500 kg/h of steam at 40 bar and 500°C. The mixing process is adiabatic and the mixture is expanded to a condenser pressure of 0.10 bar in a low pressure turbine of 78% internal efficiency. Determine the power output and the thermal efficiency of the plant.

Solution: Given data :

$p_1 = 40$ bar
$p_3 = 2$ bar
$T_1 = 500°C$
$x_3 = 0.87$ bar
$p_5 = 0.1$ bar
$\eta_{isen} = 0.83$
$\dot{m}_1$ = Steam flow rate to H.P. turbine

$$= \frac{5500}{3600} \text{ kg/s}$$

$\dot{m}_2$ = Steam flow rate from Natural geyser

$$= \frac{2700}{3600} \text{ kg/s}$$

From steam table
at 40 bar and 500°C
$h_1 = 3445.3$ kJ/kg
$s_1 = 7.0901$ kJ/kg.K
at 2 bar and 0.87 dry
$h_f = 504.7$ kJ/kg
$h_g = 2706.7$ kJ/kg

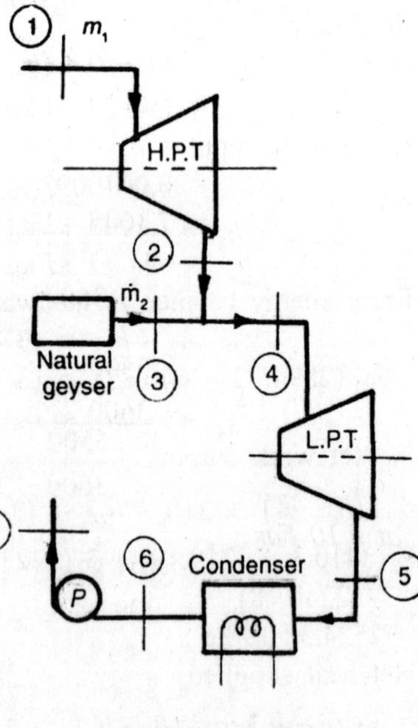

Fig. P. 12.19 (a)

$s_f = 1.5301$ kJ/kg.K
$s_g = 7.1271$ kJ/kg.K
$h_3 = 504.7 + 0.87$
$(2706.7 - 504.7)$
$= 2420.35$ kJ/kg
$s_3 = 1.5301 + 0.87$
$(7.1271 - 1.5301)$
$= 6.3995$ kJ/kg.K

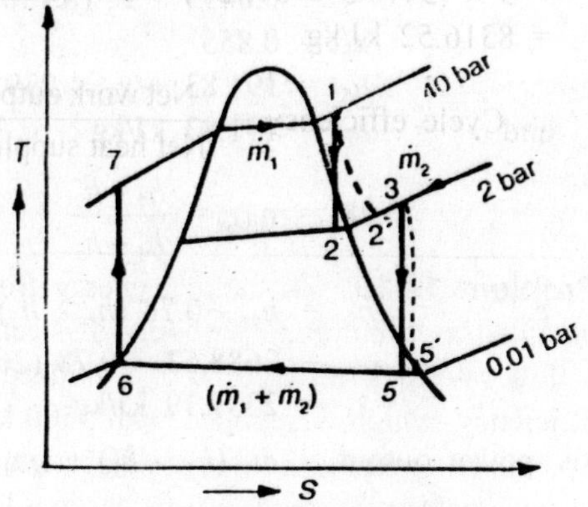

Fig. P. 12.19 (b)

at superheated exhaust from H.P. turbine at 2 bar

$s_1 = s_2 = 7.0901$
$= 1.5301 + x_2$
$(7.1271 - 1.5301)$
$\therefore x_2 = 0.9934$

$$\therefore \quad h_2 = 504.7 + 0.9934 \times (2706.7 - 504.7)$$
$$= 2692.14 \text{ kJ/kg}$$

$$\eta_{isen} = 0.83 = \frac{h_1 - h_2'}{h_1 - h_2}$$

$$\therefore \quad h_2' = h_1 - 0.83\,(h_1 - h_2) = 2820.18 \text{ kJ/kg}$$

by interpolation.

$$S_2' = 7.39 \text{ kJ/kg}$$

After mixing

$$h_4 = \frac{\frac{5500}{3600} \times h_2' + \frac{2700}{3600} h_3}{\frac{5500}{3600} + \frac{2700}{3600}} = 2688.53 \text{ kJ/kg}$$

$$s_4 = \frac{\frac{5500}{3600} \times s_2' + \frac{2700}{3600} s_3}{\frac{5500}{3600} + \frac{2700}{3600}} = 7.0639 \text{ kJ/kg}$$

at 0.10 bar

$$h_f = 191.83 \text{ kJ/kg.}$$
$$h_{fg} = 2392.8 \text{ kJ/kg.}$$
$$s_f = 0.6493 \text{ kJ/kg.K}$$
$$s_g = 8.1502 \text{ kJ/kg.K}$$

since.

$$s_4 = s_5 = 7.0639$$
$$= 0.6493 + x_5 (8.1502 - 0.6493)$$
$$\therefore \quad x_5 = 0.855$$
$$\therefore \quad h_5 = 191.83 + x_5 \times 2392.8 = 2238.098 \text{ kJ/kg}$$
$$\text{and} \quad h_6 = 191.83 \text{ kJ/kg}$$

$$0.78 = \eta_{isen} = \frac{h_4 - h_5'}{h_4 - h_5}$$

$$\therefore \quad h_5' = h_4 - 0.78 (h_4 - h_5)$$
$$= 2688.53 - 0.78 (2688.53 - 2238.098)$$
$$= 2337.19 \text{ kJ/kg}$$

$$\therefore \text{ power output} = \dot{m}_1 (h_1 - h_2') + (\dot{m}_1 + \dot{m}_2) (h_4 - h_5')$$

$$= \frac{5500}{3600} [(3445.3 – 2820.18) \times \frac{5500 + 2700}{3600} (2688.53 – 2337.19)]$$

= 1755.32 kW **Ans**

$$\text{Heat supplied} = \dot{m} (h_1 - h_6) = \frac{5500}{3600} (3445.3 - 191.83)$$

$$= 4970.58 \text{ kW}$$

$$\therefore \text{ Cycle efficiency} = \frac{\text{Power output}}{\text{Heat supplied}} = \frac{1755.32}{4970.58} = 35.3 \%$$

Problem 12.20: In a study for a space project it is thought that the condensation of a working fluid might be possible at - 40°C. A binary cycle is proposed, using Refrigerant-12 as the low temperature fluid, and water as the high temperature fluid. Steam is generated at 80 bar, 500°C and expands in a turbine of 81% isentropic efficiency to 0.06 bar, at which pressure it is condensed by the generation of dry saturated refrigerant vapour at 30°C from saturated liquid at –40°C. The isentropic efficiency of the R-12 turbine is 83%. Determine the mass ratio of R-12 to water and the efficiency of the cycle. Neglect all losses.

Solution: Given data :

$$p_1 = 80 \text{ bar}$$
$$T_1 = 500°\text{C}$$
$$p_2 = 0.06 \text{ bar}$$

$T_a = T_e = 30°C$

$T_b = T_c = -40°C$

η_{isen} of turbine

$= 0.81$

η_{isen} of R-12 cycle

$= 0.83$

From steam table

at 80 bar and 500°C

$h_1 = 3398.3$ kJ/kg.

$s_1 = 6.7240$ kJ/kg.

at 0.06 bar

$h_f = 150.21$ kJ/kg.

$h_{fg} = 2416.62$ kJ/kg.

$s_f = 0.5164$ kJ/kg.K.

$s_g = 8.3377$ kJ/kg.K.

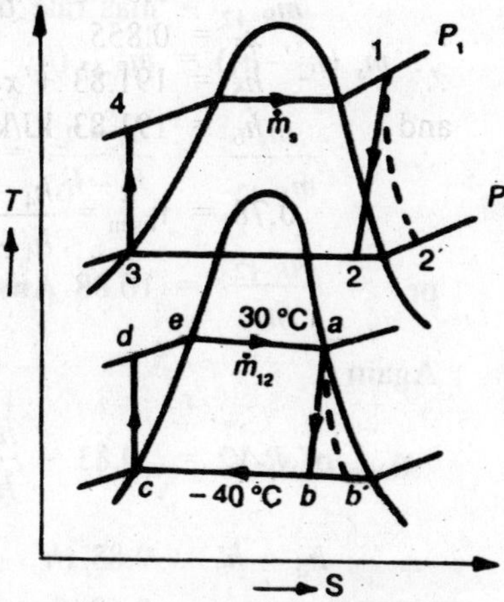

Fig. P. 12.20

since $s_1 = s_2 = 6.7240$

$= 0.5164 + x_2 (8.3377 - 0.5164)$

$x_2 = 0.7937$

$\therefore \quad h_2 = 150.21 + 0.7937 \times 2416.62 = 2068.3$ kJ/kg.

$$\eta_{isen} = 0.81 = \frac{h_1 - h_2'}{h_1 - h_2'}$$

or, $(h_1 - h_2') = 0.81 (3398.3 - 2068.23) = 1077.36$ kJ/kg.

$\therefore \quad h_2' = 2320.94$ kJ/kg

$h_3 = 150.21$ kJ/kg

For R-12 cycle at 30°C

$h_f = 64.539$ kJ/kg

$h_g = 199.475$ kJ/kg

$s_g = 0.6848$ kJ/kg

at –40°C

$h_f = 0.00$ kJ/kg

$h_g = 169.479$ kJ/kg

$s_f = 0.0$ kJ/kg

$s_g = 0.7269$ kJ/kg

$\therefore \quad h_a = 199.475$ kJ/kg

$h_e = 64.539$ kJ/kg

$h_c = 0.00$ kJ/kg

Now in steam condenser and R-12 boiler

Let $\dot{m}_s$ = mass rate of steam (kg/s)

$\dot{m}_{R-12}$ = mas rate of steam R-12 (kg/s)

$\therefore \dot{m}_s (h_2' - h_3) = \dot{m}_{R-12} (h_a - h_c)$

$$\therefore \frac{\dot{m}_s}{\dot{m}_{R-12}} = \frac{h_a - h_c}{h_2' - h_3} = \frac{199.475 - 0.009}{2320.94 - 150.21}$$

$$\text{or} \quad \frac{\dot{m}_{R-12}}{\dot{m}_s} = 10.88 \textbf{ Ans}$$

Again

$$\eta_{isen} \text{ of } R\text{-}12 = 0.83 = \frac{h_a - h_b'}{h_a - h_b}$$

$$\therefore \quad h_a - h_b' = 0.83 (h_a - h_b) \qquad \ldots\text{(i)}$$

since $s_a = s_b = 0.6848 = 0 + x_b \times 0.7269$

$\therefore \quad x_b = 0.942$

$\therefore \quad h_b = 169.479 \times .0942 = 159.66$ kJ/kg

$\therefore$ from (i)

$h_a - h_b' = 0.83 (199.475 - 159.66) = 33.04$ kJ/kg

$\therefore \quad h_b' = 166.43$ kJ/kg

$\therefore$ Net power output

$= \dot{m}_s (h_1 - h_2') + \dot{m}_{R-12} (h_a - h_b')$

$= \dot{m}_s [(3398.3 - 2320.94) + 10.88(199.475 - 166.43)]$

$= \dot{m}_s \times 1436.8896$ kW

Net heat supplied

$= \dot{m}_s (h_1 - h_3)$ (by neglecting pump work)

$= \dot{m}_s [(3398.3 - 150.21) = \dot{m}_s \times 3248.09$ kW

$$\therefore \text{cycle efficincy} = \frac{\text{Net power output}}{\text{Net heat supplied}} = \frac{\dot{m}_s \times 1436.8896}{\dot{m}_s \times 3248.09}$$

$= 44.24$ % **Ans**

Problem 12.21: Fig. P. 12.21 shows the arrangement of a steam plant in which steam is also required for industrial heating process. The steam leaves boiler *B* at 30 bar, 320ºC and expands in the H.P. turbine to 2 bar, the efficiency of the H.P. turbine being 75%. At this point one half of the steam passes to the process heater *P* and the remainder enters separator *S* which removes all the moisture. The dry steam enters the L.P. turbine at 2 bar and expands to the condenser pressure 0.07 bar, the efficiency of the L.P. turbine being 70%. The drainage from the separator mixes with the condensate from the process heater and the combined flow enters the hotwell H at 80°C. Trap are provided at the exits from P and S. A pump extract the condensate from condenser C and this enters the hotwell at 38°C. Neglecting the feed pump work and radiation loss, estimate the temperature of water leaving the hotwell which is at atmospheric pressure. Also calculate, as percentage of heat transferred in the boiler, (a) the heat transferred in the process heater, and (b) the work done in the turbines.

Solution: Given data :

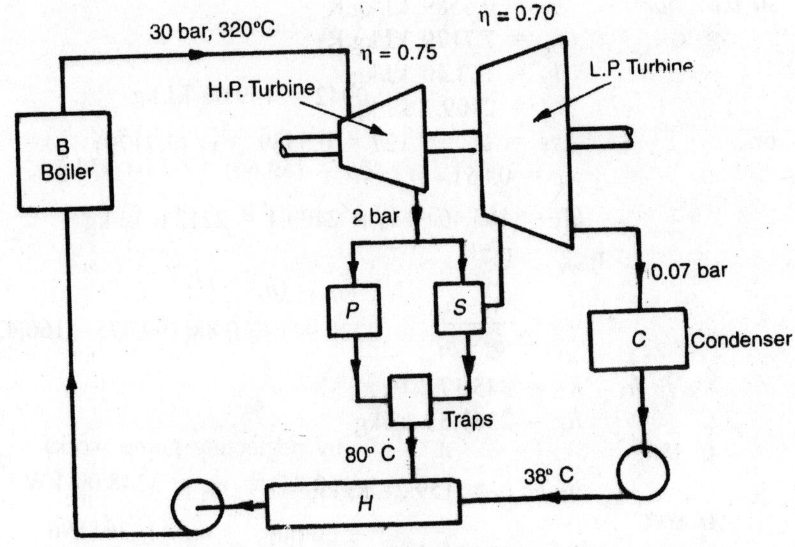

Fig. P. 12.21 (a)

From steam table

At 30 bar and 320°C, by interpolation

$$h_1 = 3042.22 \text{ kJ/kg}$$

$$s_1 = 6.6206 \text{ kJ/kg.K}$$

At 2 bar

$$t_{sat} = 120.2°\text{C}$$

$$h_f = 504.7 \text{ kJ/kg}$$

$$h_{fg} = 2201.9 \text{ kJ/kg}$$

$$s_f = 1.53 \text{ kJ/kg.K}$$

$$s_{fg} = 5.597 \text{ kJ/kg.K}$$

$$s_1 = s_2' = 6.6206$$

$$= 1.53 + x_2 (5.597)$$

$$x_2 = 0.9095$$

$$\therefore \quad h_2' = 5047 + x_2 \times 2201.9$$

$$= 2507.39 \text{ kJ/kg.K}$$

$$\eta_{isen} = 0.75 \frac{h_1 - h_2}{h_1 - h_2'}$$

$$\therefore \quad h_1 - h_2 = 401.12 \text{ kJ/kg}$$

$$\therefore \quad h_2 = 2641.1 \text{ kJ/kg}$$

$$h_3 = h_g = 2706.7 \text{ kJ/kg}$$

$$s_3 = 7.127 \text{ kJ/kg.K}$$

At 0.07 bar

$$s_f = 0.5589 \text{ kJ/kg.K}$$

$$s_{fg} = 7.7179 \text{ kJ/kg.K}$$

$$h_f = 163.40 \text{ kJ/kg}$$

$$h_{fg} = 2409.1 \text{ kJ/kg}$$

or,

$$s_3 = s_4' = 7.127 = 0.5589 + x_4 (7.7179)$$

$$\therefore \quad x_4 = 0.851$$

$$h_4' = 163.40 + x_4 \times 2409.1 = 2213.6 \text{ kJ/kg}$$

$$\eta_{isen} = 0.7$$

$$= \frac{h_3 - h_4}{h_3 - h_4'}$$

$$\therefore \quad h_3 - h_4 = 345.17 \text{ kJ/kg}$$

$$\therefore \quad h_4 = 2361.53 \text{ kJ/kg}$$

At 38°C

$$h_5 = h_f = 159.21 \text{ kJ/kg}$$

At 80°C

$$h_6 = 334.91 \text{ kJ/kg}$$

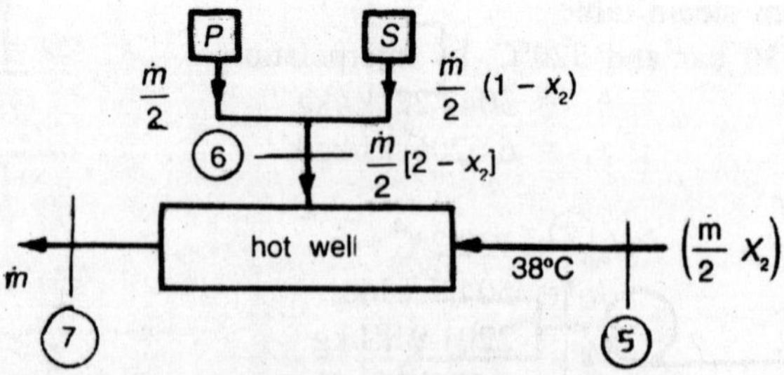

Fig. P. 12.21 (b)

Now applying energy balance in hot well

$$\frac{\dot{m}}{2} x_2 h_5 + \frac{\dot{m}}{2}(2 - x_2)\, h_6 = \dot{m}\, h_7$$

$$\therefore \quad \frac{0.9095}{2} \times 159.21 + \frac{(2 - 0.9095)}{2} \times 334.91 = h_7$$

$$\therefore \quad h_7 = 225.01 \text{ kJ/kg}$$

Heat supplied in boiler $= (h_1 - h_7)\ \dot{m}$

$$= \dot{m}\,(3042.22 - 225.01) \text{ kW}$$

$$= \dot{m}\, 2817.21 \text{ kW}$$

work done in turbine

$$= \dot{m}\ (h_1 - h_2) + \frac{\dot{m}}{2} x_2\ [h_3 - h_4]$$

$$= \dot{m}\ (3042.22 - 2641.1) + \frac{0.9095}{2}\ (2706.7 - 2361.53)]$$

$$= \dot{m}\ [558.09] \text{ kW}$$

heat transfer in process heater $= \dfrac{\dot{m}}{2}\ [h_2 - h_6]$

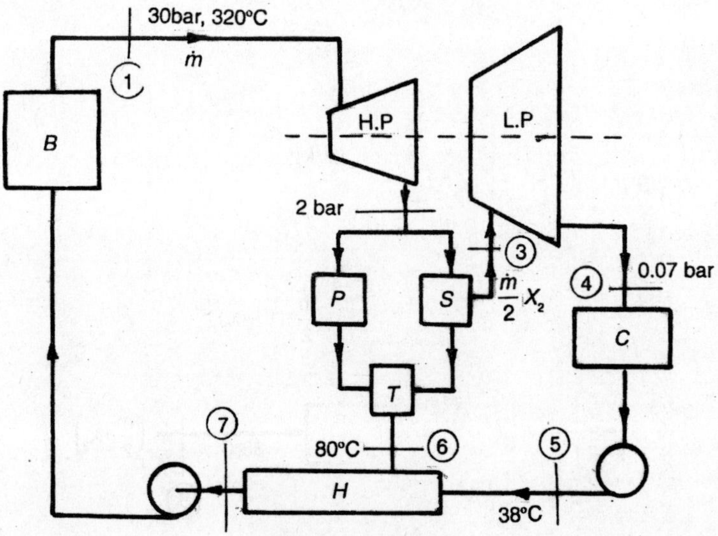

Fig. P. 12.21 (c)

$$= \dot{m}\left[\frac{2641.1 - 334.90}{2}\right]$$

$$= 1153.1 \ \dot{m}\,\text{kW}$$

$\therefore$ % of work done in turbine

$$= \frac{558.09}{2817.21} = 19.81\% \ \textbf{Ans}$$

$$\%\text{ of heat transfer in process heater} = \frac{1153.1}{2817.21} = 40.93\ \% \ \textbf{Ans}$$

Chapter 13

Gas Power Cycles

Problem 13.1 : In a stirling cycle the volume varies between 0.03 and 0.06 m^3, the maximum pressure is 0.2 MPa, and the temperature varies between 540° and 270°C. The working fluid is air (an ideal gas). (a) Find the efficiency and the work done per cycle for the simple cycle. (b) Find the efficiency and the work done per cycle for the cycle with an ideal regenerator, and compare with the cannot cycle having the same isothermal heat supply process and the same temperature range.

Solution : Given data

$$V_1 = V_4 = 0.03 \text{ m}^3$$
$$V_2 = V_3 = 0.06 \text{ m}^3$$
$$p_1 = 0.2 \text{ MPa}$$
$$T_3 = T_4 = 270°\text{C} = 270 + 273 = 543 \text{ K}$$
$$T_1 = T_2 = 540°\text{C} = 540 + 273 = 813 \text{ K}$$

(a) $$\eta = \frac{\text{Net work}}{\text{heat supplied}}$$

For process 1-2; Reversible isothermal process

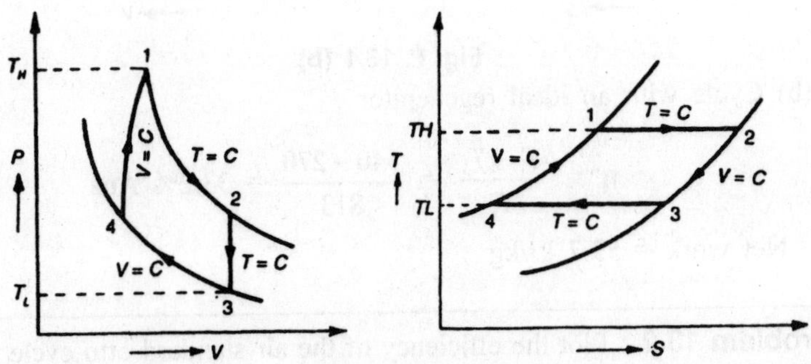

Fig. P. 13.1 (a)

$$w_{12} = Q_{1-2} = RT_1 \ln \frac{V_2}{V_1} \text{ kJ/kg}$$

For process 3-4; Reverible isothermal process

$$w_{12} = Q_{1\text{-}2} = RT_3 \ln \frac{V_4}{V_3} = -RT_3 \ln \frac{V_4}{V_3} = -RT_3 \ln \frac{V_2}{V_1} \text{ kJ/kg}$$

For proces 1-4; Reversible isochores proces

$$Q_{1-4} = C_v (T_1 - T_4) = C_v (T_1 - T_3) \text{ kJ/kg}$$

$$w_{1-4} = 0$$

$$\therefore \text{ Net work} = R \ln \frac{V_2}{V_1} (T_1 - T_3)$$

$$= 0.287 \ln \frac{0.06}{0.03} (540 - 270) = 53.7 \text{ kJ/kg } \textbf{Ans}$$

$$\therefore \quad \eta = \frac{R \ln \frac{V_2}{V_1} (T_1 - T_3)}{C_v (T_1 - T_3)} = \frac{0.287 \ln \frac{0.06}{0.02}}{0.718} = 27.7\% \textbf{ Ans}$$

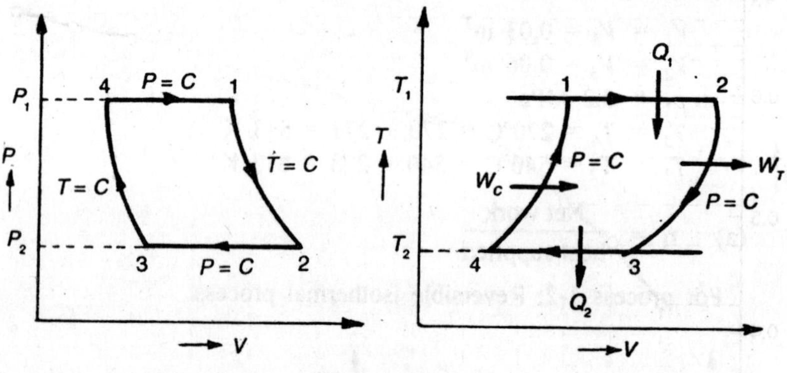

Fig. P. 13.1 (b)

(b) Cycle with an ideal regenerator

$$\eta = \frac{T_1 - T_3}{T_1} = \frac{540 - 270}{813} = 33.2\% \textbf{ Ans}$$

Net work = 53.7 kJ/kg

Problem 13.2 : Plot the efficiency of the air standard otto cycle as a function of the compression ratio for compression ratios from 4 to 16.

Solution : Given data

$$\eta_{otto} = 1 - \frac{1}{r^{\gamma-1}} \text{ where } r_k = \text{compression ratio}$$

$$= \frac{V_1}{V_2}$$

Hence, $r_k \rightarrow 4$ to 16

γ_K	η_{otto}	γ_K	η_{otto}
4	0.426	11	0.617
5	0.475	12	0.63
6	0.512	13	0.642
7	0.541	14	0.652
8	0.565	15	0.661
9	0.585	16	0.67
10	0.602		

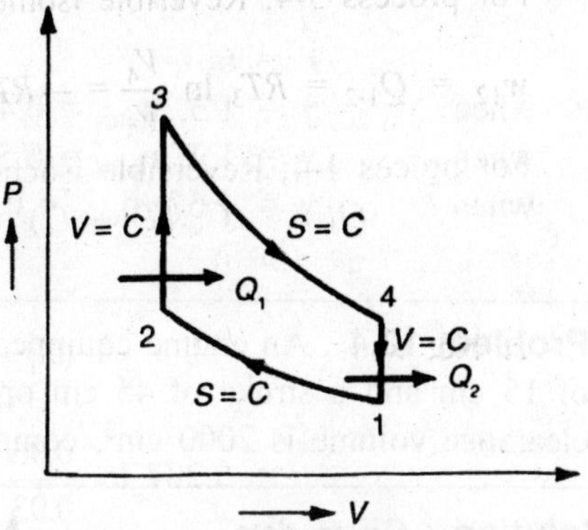

Fig. P. 13.2 (a)

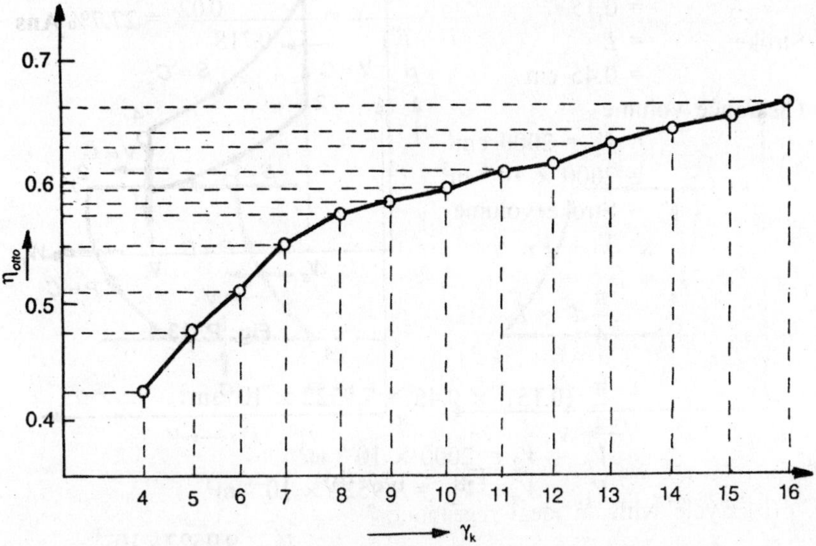

Fig. P. 13.2 (b)

Problem 13.3 : Find the air standard efficiencies for otto cycles with a compression ratio of 6 using ideal gases having specific heat ratio 1.3, 1.4 and 1.67. What are the advantages and disadvantages of using helium as the working fluid?

Solution : Given data

$$\eta = 1 - \left(\frac{1}{r}\right)^{\gamma - 1}$$

$r = 6$

when $\gamma = 1.3,\ \eta_{otto} = 0.416$

when $\gamma = 1.4,\ \eta_{otto} = 0.512$

when $\gamma = 1.67,\ \eta_{otto} = 0.699$

> **Problem 13.4** : An engine equipped with a cylinder having a bore of 15 cm and a stroke of 45 cm operates on an otto cycle. If the clearance volume is 2000 cm³, compute the air standard efficiency.

Solution : Given data

Cylinder bore $= d$

$= 0.15$

Stroke $= L$

$= 0.45$ cm

Clearance volume

$= V_c = 2000 \text{ cm}^3$

$= 2000 \times 10^{-6} \text{ m}^3$

V_s = Stroke volume

$= V_1 - V_2$

$= \frac{\pi}{4} d^2 \times L$

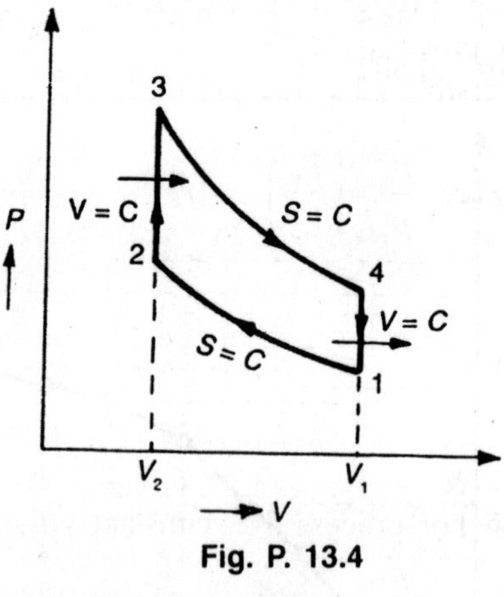

Fig. P. 13.4

$$= \frac{\pi}{4} (0.15)^2 \times 0.45 = 7.9522 \times 10^{-3} \text{ m}^3$$

$$V_c = V_2 = 2000 \times 10^{-6} \text{ m}^3$$

$$\therefore \quad V_1 = V_s + V_2 = 9.9522 \times 10^{-3} \text{ m}^3$$

$$\therefore \quad r_k = \text{compression ratio} = \frac{V_1}{V_2} = \frac{9.9522 \times 10^{-3}}{2000 \times 10^{-6}}$$

$$= 4.9761$$

$$\therefore \quad \eta_{otto} = 1 - \left(\frac{1}{r_k}\right)^{\gamma - 1} = 1 - \left(\frac{1}{4.9761}\right)^{1.4-1}$$

$$= 0.474 = 47.4\% \text{ **Ans**}$$

Problem 13.5 : In an air standard otto cycle the compression ratio is 7, and compression begins at 35°C, 0.1 MPa.The maximum temperature of the cycle is 1100°C. Find
(a) the temperature and pressure at the cardinal points of the cycle,
(b) the heat supplied per kg of air, (c) the work done per kg of air,
(d) the cycle efficiency, and (e) the m.e.p of the cycle.

Solution : Given data

Compression ratio,

$r_k = 7$

$T_1 = 35°C = 35 + 273 = 308$ K

$p_1 = 0.1$ MPa

$T_3 = 1100°C$

$= 1100 + 273 = 1373$ K

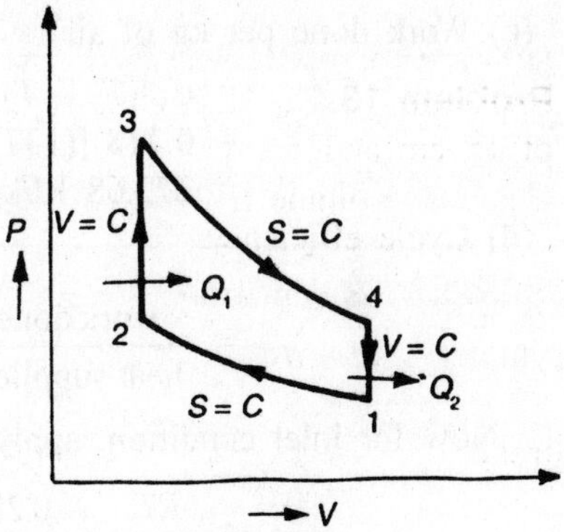

Fig. P. 13.5

For process 1-2,
reversible adiabaitc process

$$\therefore \quad \frac{p_2}{p_1} = \left(\frac{V_1}{V_2}\right)^\gamma = (r_k)^\gamma$$

$$\frac{T_2}{T_1} = \left(\frac{V_1}{V_2}\right)^{\gamma-1}$$

$$\therefore \quad P_2 = P_1\, r_k^\gamma = 0.1 \times (7)^{1.4} = 1.52 \text{ MPa } \textbf{Ans}$$

$$\& \quad T_2 = T_1(r_k)^\gamma = 308\ (7)^{1.4-1} = 670.8 \text{ K } \textbf{Ans}$$

For process 3-2, constant volume process

$$\frac{p_3}{p_2} = \frac{T_3}{T_2}$$

$$\Rightarrow \quad p_3 = p_2\left(\frac{T_3}{T_2}\right)$$

$$= 1.52\left(\frac{1373}{670.8}\right) = 3.12 \text{ MPa } \textbf{Ans}$$

For process 3–4, isentropic process

$$\frac{T_4}{T_3} = \left(\frac{V_2}{V_1}\right)^{\gamma-1}, \quad \frac{p_4}{p_3} = \left(\frac{V_2}{V_1}\right)^\gamma$$

$$\therefore \quad T_4 = \frac{T_3}{(r_k)^{\gamma-1}} = \frac{1373}{(7)^{0.4}} = 630.42 \text{ K } \textbf{Ans}$$

$$p_4 = \frac{p_3}{(r_k)^{\gamma}} = \frac{312}{(7)^{1.4}} = 0.20 \text{ MPa } \textbf{Ans}$$

(b) Heat supplied per kg of air =

$$= C_v (T_3 - T_2)$$
$$= 0.718 (1373 - 670.8) = 504.18 \text{ kJ/kg.}$$

(c) Work done per kg of air

$$= C_v (T_3 - T_2) - C_v (T_4 - T_1)$$
$$= 0.718 [(1373 - 670.8) - (630.42 - 308)$$
$$= 272.68 \text{ kJ/kg}$$

(d) Cycle efficiency

$$= \frac{\text{work done}}{\text{heat supplied}} = \frac{272.68}{504.18} = 54.08\%$$

Now for inlet condition, applying ideal gas eq.

$$v_1 = \frac{RT_1}{P_1} = \frac{0.287 \times 308}{0.1 \times 1000} = 0.884 \text{ m}^3\text{/kg}$$

given $\quad \dfrac{v_1}{v_2} = 7$

$$\Rightarrow \quad v_2 = \frac{v_1}{7} = 0.126 \text{ m}^3\text{/kg}$$

$$\therefore \quad v_1 - v_2 = 0.758 \text{ m}^3\text{/kg}$$

$$\therefore \quad mep = \frac{W_{net}}{(v_1 - v_2)} = \frac{272.68}{0.758} = 359.87 \text{ kW/m}^2$$

Problem 13.6 : An engine working on the otto cycle has an air standard cycle efficiency of 56% and rejects 544 kJ/kg of air. The pressure and temperature of air at the beginning of compression are 0.1 MPa and 60°C respectively. Compute (a) the compression ratio of the engine, (b) the work done per kg of air, (c) the pressure and temperature at the end of compression, and (d) the maximum pressure in the cycle.

Solution : Given data :

Standard cycle efficiency

$= \eta_{otto} = 56\%$

Heat rejected

$= Q_2 = 544$ kJ/kg

$p_1 = 0.1$ MPa

$T_1 = 60°C$

$= 60 + 273 = 333$ K

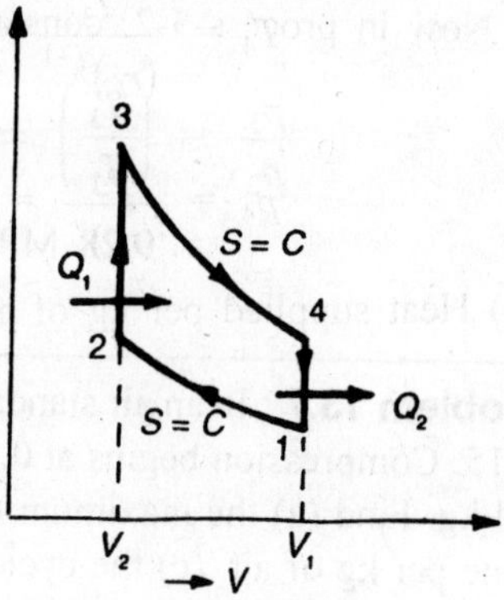

Fig. P. 13.6

(a) $\eta_{otto} = 1 - \dfrac{1}{r_k^{\gamma-1}}$

or, $$r_k = \left(\frac{1}{1-\eta_{otto}}\right)^{\frac{1}{\gamma-1}} = \left(\frac{1}{1-0.56}\right)^{\frac{1}{1.4-1}} = 7.79$$

(b) again $\eta_{otto} = 1 - \dfrac{Q_2}{Q_1}$

or, $$Q_1 = \frac{Q_2}{1-\eta_{otto}} = \frac{544}{1-0.56} = 1236.36 \text{ kJ/kg}$$

$\therefore$ work done per kg of air $= Q_1 - Q_2$

$= 1236.36 - 544 = 692.36$ kJ/kg

(c) For isentropic process 1-2

$$\frac{p_2}{p_1} = \left(\frac{V_1}{V_2}\right)^{\gamma} = (r_k)^{\gamma}$$

or, $$p_2 = P_1 (r_k)^{\gamma}$$
$$= 0.1 \times (7.79)^{1.4} = 1.77 \text{ MPa } \textbf{Ans}$$

$$\frac{T_2}{T_1} = \left(\frac{p_2}{p_1}\right)^{\frac{\gamma-1}{\gamma}} = (1.77)^{\frac{0.4}{1.4}}$$

$\therefore$ $$T_2 = T_1 (1.77)^{\frac{0.4}{1.4}} = 392.05 \text{ K } \textbf{Ans}$$

(d) $$Q_1 = C_v (T_3 - T_1) = 1236.36 \text{ kJ/kg}$$

$\therefore$ $$T_3 = \frac{1236.36}{0.718} + T_1 = 2054.95 \text{ K}$$

Now in process 3-2, constant volume process

$$\frac{p_3}{p_2} = \left(\frac{T_3}{T_2}\right) \Rightarrow P_3 = \left(\frac{2054.95}{392.05}\right) \times P_2$$

$$= 9.28 \text{ MPa } \textbf{Ans}$$

Problem 13.7 : In an air standard Diesel cycle, the compression ratio is 15. Compression begins at 0.1 MPa, 40°C. The heat added is 1.675 MJ/kg. Find (a) the maximum temperature of the cycle, (b) the work done per kg of air, (c) the cycle efficiency, (d) the temperature at the end of the isentropic expansion, (e) the cut-off ratio, (f) the maximum pressure of the cycle, and (g) m.e.p of the cycle.

Solution : Given data

r_k = Compression ratio

$$= 15 = \frac{v_1}{v_2}$$

$p_1 = 0.1$ MPa

$T_1 = 40°C$

$= 40 + 273 = 313$ K

Q_1 = heat added

$= 1.675$ MJ/kg

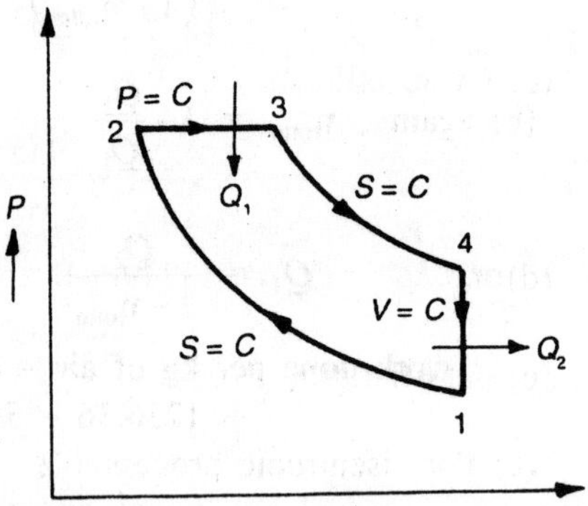

Fig. P. 13.7

For process 1-2. isentropic process

$$\frac{T_2}{T_1} = \left(\frac{v_1}{v_2}\right)^{\gamma-1} = r_k^{\gamma-1}$$

$\therefore$ $$T_2 = T_1 \; r_k^{\gamma-1} = 313\ (15)^{0.4} = 924.60 \text{ K}$$

(a) $$Q_1 = C_p\ (T_3 - T_2) = 1.675 \text{ MJ/kg}$$

$\therefore$ $$T_3 = \frac{1.675 \times 1000}{1.005} + T_2 = 2591.33 \text{ K } \textbf{Ans}$$

(b) $$Q_2 = \text{heat rejected} = C_v\ (T_4 - T_1)$$

For process 2-3, constant pressure process

$$\frac{T_3}{T_2} = \left(\frac{v_3}{v_2}\right) \Rightarrow v_3 = v_2\left(\frac{T_3}{T_2}\right)$$

For process 3-4, isentropic process

$$\frac{T_4}{T_3} = \left(\frac{v_3}{v_4}\right)^{\gamma-1} = \left(\frac{v_2\, T_3}{T_2\, v_1}\right)^{\gamma-1}$$

$$\therefore \quad T_4 = T_3 \times \left(\frac{T_3}{T_2} \times \frac{1}{r_k}\right)^{\gamma-1}$$

$$= 2591.33 \left(\frac{2591.33}{924.66 \times 15}\right)^{0.4} = 1324.65 \text{ K}$$

$$\therefore \text{Heat rejected} = C_v\,(T_4 - T_1)$$

$$= 0.718\,(1324.65 - 313) = 726.37 \text{ kJ/kg}$$

$\therefore$ Work done per kg.of air $= Q_1 - Q_2$

$$= 1675 - 726.37 = 948.63 \text{ kJ/kg}$$ **Ans**

(c) Cycle efficiency

$$= \frac{w_{net}}{Q_1} = \frac{948.63}{1675} = 56.63\%$$ **Ans**

(d) $T_4 = 1324.65$ K

(e) cut-off ratio $= r_c = \dfrac{v_3}{v_2} = \dfrac{v_2(T_3 / T_2)}{v_2}$

$$= \frac{T_3}{T_2} = \frac{2591.33}{924.66} = 2.8$$ **Ans**

(f) For process 1-2, isentropic process

$$\frac{p_2}{p_1} = \left(\frac{v_1}{v_2}\right)^{\gamma} = (r_k)^{\gamma}$$

$\therefore \quad p_2 = p_1 \times (15)^{1.4} = 4.43$ MPa **Ans**

(g) from ideal gas equation

$$p_1 v_1 = RT_1$$

$$\therefore \quad v_1 = \frac{RT_1}{p_1} = \frac{0.287 \times 313}{0.1 \times 1000} = 0.89831 \text{ m}^3\text{/kg}$$

$$\therefore \quad v_2 = 0.05989 \text{ m}^3\text{/kg}$$

$$\therefore \quad v_1 - v_2 = 0.8384 \text{ m}^3\text{/kg}$$

∴ mep of the cycle

$$= \frac{w_{net}}{v_1 - v_2} = \frac{948.63}{0.8384} = 1131.45 \text{ kN/m}^2$$

> **Problem 13.8** Two engines are to operate on otto and diesel cycles with the following data :
> Maximum temperature 1400 K, exhaust temperature 700 K. State of air at the beginning of compression 0.1 MPa, 300 K.
> Estimate the compression ratios, the maximum pressures, efficiencies, and rate of work outputs (for 1 kg/min of air) of the respective cycles.

Solution : Given data

T_3 = 1400 K
T_4 = 700 K
T_1 = 300 K
p_1 = 0.1 MPa
working fluid is air,
C_p = 1.005 kJ/kg.K
C_v = 0.718 kJ/kg.K
R = 0.287 kJ/kg.K

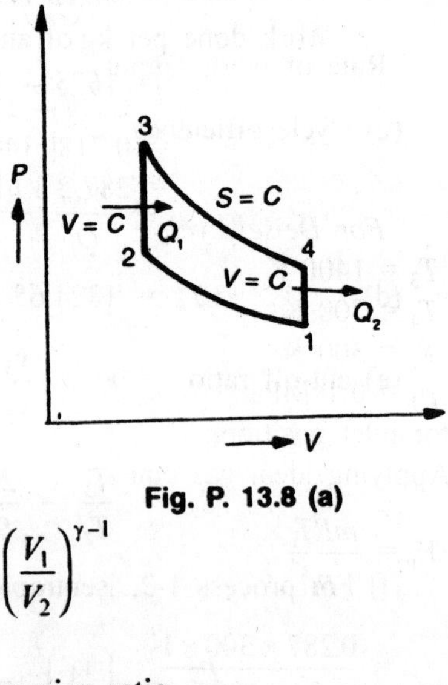

Fig. P. 13.8 (a)

FOR OTTO CYCLE

For process 3-4,
isentropic process

$$\frac{T_3}{T_4} = \left(\frac{V_4}{V_3}\right)^{\gamma-1} = \left(\frac{V_1}{V_2}\right)^{\gamma-1}$$

or, $$r_k = \frac{V_1}{V_2} = \text{compression ratio}$$

$$= \left(\frac{T_3}{T_4}\right)^{\frac{1}{\gamma-1}} = \left(\frac{1400}{700}\right)^{\frac{1}{1.4-1}} = 5.656 \textbf{ Ans}$$

For process 1-2,

$$\frac{T_2}{T_1} = \left(\frac{p_2}{p_1}\right)^{\frac{\gamma-1}{\gamma}} = \left(\frac{V_1}{V_2}\right)^{\gamma-1}$$

$$\therefore \quad T_2 = T_1 \times (5.656)^{0.4} = 600 \text{ K}$$

$$p_2 = p_1\left(\frac{V_1}{V_2}\right)^{\gamma} = 0.1 \times (5.656)^{1.4-1} = 1.1314 \text{ MPa}$$

For process 3-2, Constant volume process

$$\frac{p_3}{p_2} = \frac{T_3}{T_2}$$

$$\Rightarrow \quad p_3 = p_2 \times \frac{1400}{600}\ 2.64 \text{ MPa } \textbf{Ans}$$

$$\eta_{otto} = 1-\left(\frac{1}{r_k}\right)^{\gamma-1} = 1-\left(\frac{1}{5.656}\right)^{0.4} = 50\%$$

Rate of work output

$$= C_v(T_3 - T_4) - C_v(T_2 - T_1)$$
$$= 0.718[(1400 - 700) - (600 - 300)]$$
$$= 287.2 \text{ kJ/kg}$$

For Deisel cycle

$T_3 = 1400$ K
$T_4 = 700$ K
$T_1 = 300$ K
$p_1 = 0.1$ MPa

for inlet position,
Applying ideal gas eqn

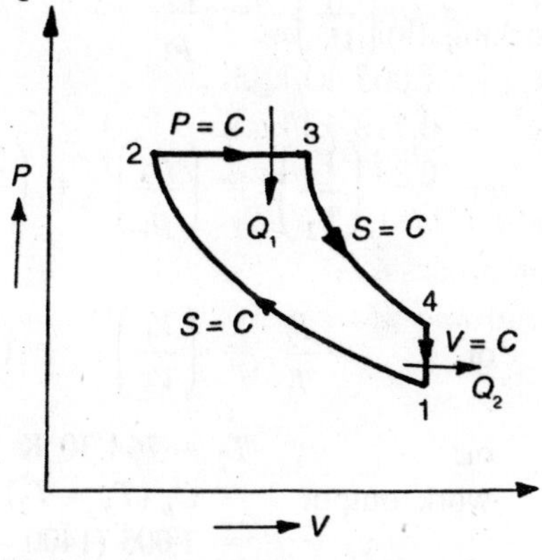

Fig. P. 13.8 (b)

$$V_1 = \frac{\dot{m}RT_1}{p_1}$$

$$= \frac{0.287 \times 300 \times 1}{0.1 \times 1000}$$

$= 0.861 \text{ m}^3/\text{min}$

For process 3-4, isentropic process

$$\frac{T_3}{T_4} = \left(\frac{V_4}{V_3}\right)^{\gamma-1}$$

$$\Rightarrow \quad V_3 = V_4\left(\frac{T_4}{T_3}\right)^{\frac{1}{\gamma-1}}$$

$$= 0.861\left(\frac{700}{1400}\right)^{\frac{1}{0.4}} = 0.152 \text{ m}^3/\text{min}$$

$$\frac{p_3}{p_4} = \left(\frac{V_4}{V_3}\right)^{\gamma} = \left(\frac{0.861}{0.152}\right)^{1.4} \quad 11.34$$

For process 4-1, constant volume process

$$\frac{p_4}{p_1} = \frac{T_4}{T_1}$$

$$\Rightarrow \quad p_4 = \left(\frac{T_4}{T_1}\right) p_1 = \left(\frac{700}{300}\right) \times 0.1 = 0.233 \text{ MPa}$$

$$\therefore \quad p_3 = 2.645 \text{ MPa} = p_2 \textbf{ Ans}$$

For process 1-2, isentropic process

$$\left(\frac{V_1}{V_2}\right)^{\gamma} = \frac{p_2}{p_1}$$

or $$\left(\frac{V_1}{V_2}\right) = \left(\frac{p_2}{p_1}\right)^{\frac{1}{\gamma}} = \left(\frac{2.645}{0.1}\right)^{\frac{1}{1.4}} = 10.36 \textbf{ Ans}$$

or $$\frac{T_2}{T_1} = \left(\frac{V_1}{V_2}\right)^{\gamma-1} = (10.36)^{0.4}$$

or, $T_2 = 764.30$ K

work output $= C_p (T_3 - T_2) - C_v (T_4 - T_1)$

$= 1.005 (1400 - 764.30) - 0.718 (700 - 300)$

$= 351.68$ kJ/kg **Ans**

$Q_1 = C_p (T_3 - T_2)$

$= 1.005 (1400 - 764.3)$

$= 638.88$ kJ/kg

$$\therefore \quad \eta = \frac{\text{Work output}}{Q_1}$$

$$= \frac{351.68}{638.88} = 55.04\% \textbf{ Ans}$$

Problem 13.9 : An air standard limited pressure cycle has a compression ratio of 15 and compression begins at 0.1 MPa, 40°C. The maximum pressure is limited to 6 MPa and the heat added is 1.675 MJ/kg. Compute (a) the heat supplied at constant volume per kg of air, (b) the heat supplied at constant pressure per kg of air, (c) the work done per kg of air, (c) the work done per kg of air, (d) the cycle efficiency, (e) the temperature at the end of the constant volume heating process, (f) the cut-off ratio, and (g) the m.e.p of the cycle.

Solution : Given data

r_k = compression ratio

$= \dfrac{V_1}{V_2} = 15$

$p_1 = 0.1$ MPa

$T_1 = 40°$ C $= 313$ K

$p_4 = p_3 = 6$ MPa

$Q_1 + Q_2 = 1.675$ MJ/kg

$= 1675$ kJ/kg

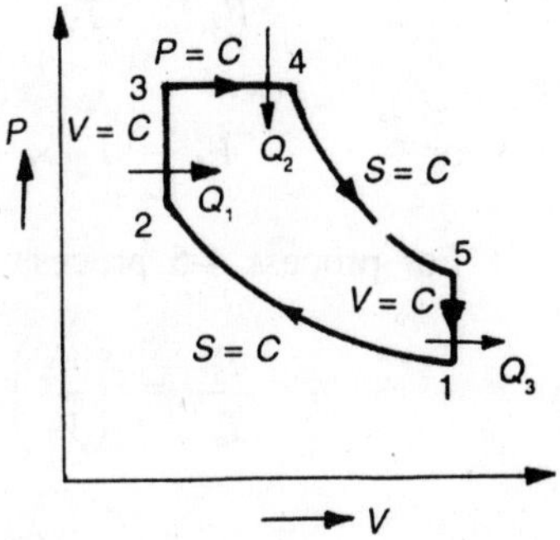

Fig. P. 13.9

For proces 1-2, isentropic process

$$\frac{T_2}{T_1} = \left(\frac{V_1}{V_2}\right)^{\gamma-1} = r_k^{\gamma} = (15)^{0.4}$$

$$\therefore \quad T_2 = 924.66 \text{ K}$$

$$\frac{p_2}{p_1} = \left(\frac{V_1}{V_2}\right)^{\gamma} = 15^{1.4}$$

$$\therefore \quad P_2 = 4.43 \text{ MPa}$$

For process 3-2, constant volume process

$$\frac{T_3}{T_2} = \frac{p_3}{p_2}$$

$$\Rightarrow \quad T_3 = T_2 \times \frac{p_3}{p_2} = 924.66 \times \frac{6}{4.43} = 1252.00 \text{ K}$$

∴ (a) Heat supplied at constant volume

$= Q_1 = C_v (T_3 - T_2)$

$= \cdot 0.718\ (1252 - 924.66) = 235.03$ kJ/kg **Ans**

(b) Q_2 = heat supplied at constant pressure

$= 1675 - 235.00$

$= 1440$ kJ/kg **Ans**

For process 3-4, constant pressure process

$$Q_2 = 1440 \text{ kJ/kg} = C_P (T_4 - T_3)$$

$$\therefore \quad T_4 = \frac{1440}{1.005} + T_3 = 2684.84 \text{ K}$$

$$\frac{V_4}{V_3} = \frac{T_4}{T_3}$$

or,
$$V_4 = V_3 \times \frac{T_4}{T_3} = V_3 \left(\frac{2684.84}{1252}\right)$$

For process 4-5 process, isentropic process

$$\frac{T_5}{T_4} = \left(\frac{V_4}{V_5}\right)^{\gamma-1}$$

$$\Rightarrow \quad T_5 = T_4 \left(\frac{V_4}{V_1}\right)^{\gamma-1} = T_4 \left(\frac{V_2}{V_1} \times \frac{2684.84}{1252}\right)^{\gamma-1}$$

$$= 2684.84 \left(\frac{2684.84}{15 \times 1252}\right)^{0.4} = 1233.13 \text{ K}$$

(c) Q_3 = heat rejected at constant volume

$= C_v (T_5 - T_1) = 0.718 (1233.13 - 313)$

$= 660.65$ kJ/kg

$\therefore$ Work done per kg of air $= Q_1 + Q_2 - Q_3$

$= 1675 - 660.65 = 1014.35$ kJ/kg **Ans**

(d) Cycle efficiency

$$= \frac{\text{Work done}}{\text{Heat supplied}} = \frac{1014.35}{1675} = 60.56 \%$$

(e) $T_3 = 1252$ K **Ans**

(f) Cut-off ratio $= \dfrac{V_4}{V_3} = \dfrac{V_3}{V_3}\left(\dfrac{2684.84}{1252}\right) = 2.14$ **Ans**

(g) At inlet point to compressor

applying ideal gas equation

$$V_1 = \frac{mRT_1}{p_1} = \frac{1 \times 0.287 \times 313}{0.1 \times 1000} = 0.898 \text{ m}^3/\text{kg}$$

$$V_2 = \frac{V_1}{15} = 0.0599 \text{ m}^3/\text{kg}$$

$$V_1 - V_2 = 0.898 - 0.0599 = 0.8381 \text{ m}^3/\text{kg}$$

$$\therefore \quad \text{m.e.p} = \frac{w_{net}}{V_1 - V_2} = \frac{1014.35}{0.8381} = 1.21 \text{ MP} \textbf{ Ans}$$

Problem 13.10 : Show that the air standard efficiency for a cycle comprising two constant pressure process and two isothermal processes (all reversible) is given by

$$\eta = \frac{(T_1 - T_2)\ln (r_p)^{(\gamma-1)/\gamma}}{T_1\left[1 + \ln (r_p)^{(\gamma-1)/\gamma} - T_2\right]}$$

where T_1 and T_2 are the maximum and minimum temperaturres of the cycle, and r_p is the pressure ratio.

Solution : Given data :

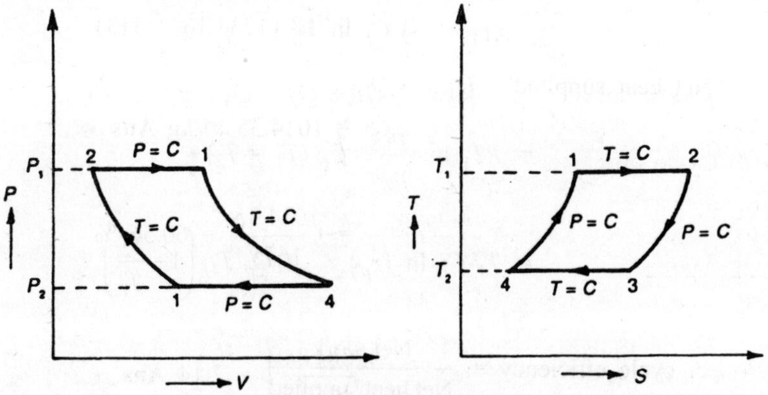

Fig. P. 13.10

$$r_p = \text{Pressure ratio} = \frac{p_1}{p_2}$$

For process 1-2, reversible isothermal process

$$Q_{1-2} = w_{1-2} = RT_1 \ln \frac{V_2}{V_1} = RT_1 \ln \frac{p_1}{p_2}$$

For Process2-3, reversible constant pressure process

$$Q_{2-3} = C_p (T_2 - T_1),\ w_{2-3} = p_2 (V_3 - V_2)$$
$$= R (T_2 - T_1)$$

For process 3-4, reversible isothermal process

Q_{3-4}, reverble constant pressure process

$$Q_{3-4} = w_{3-4} = RT_2 \ln \frac{V_4}{V_3} = RT_2 \ln \frac{p_2}{p_1}$$

For process 4-1, reversible canstant pressure process

$$Q_{4-1} = C_p (T_1 - T_2)$$
$$w_{4-1} = p_2 (V_1 - V_2) = \text{R} (T_1 - T_2)$$

$$\text{Net work} = T_1 R \ln \frac{p_1}{p_2} - T_2 R \ln \frac{p_1}{p_2}$$

$$= (T_1 - T_2) C_p \frac{(\gamma - 1)}{\gamma} \ln \frac{p_1}{p_2} \quad \left[\because R = Cp\frac{\gamma - 1}{\gamma}\right]$$

$$= (T_1 - T_2) C_p \ln (r_p)^{\frac{\gamma-1}{\gamma}}$$

$$\text{Net heat supplied} = Q_{1-2} + Q_{2-3}$$

$$= RT_1 \ln \frac{p_1}{p_2} + C_p (T_1 - T_2)$$

$$= T_1 C_p \ln (r_p)^{\frac{\gamma-1}{\gamma}} + C_p T_1 \left(1 - \frac{T_2}{T_1}\right)$$

$$\therefore \text{cycle efficiency} = \frac{\text{Net work}}{\text{Net heat supplied}}$$

$$= \frac{(T_1 - T_2)\, C_p \ln(r_p)^{\frac{\gamma-1}{\gamma}}}{C_p T_1\left[1 + \ln(r_p)^{\frac{\gamma-1}{\gamma}} - \frac{T_2}{T_1}\right]} = \frac{(T_1 - T_2) \ln(r_p)^{\frac{\gamma-1}{\gamma}}}{T_1\left[1 + \ln(r_p)^{\frac{\gamma-1}{\gamma}} - \frac{T_2}{T_1}\right]} \quad \textbf{Ans}$$

Problem 13.11: Obtain an expression for the specific work done by an engine working on the otto cycle in terms of the maximum and minimum temperatures of the cycle, the compression ratio r_k and constants of the working fluid (assumed to be an ideal gas). Hence show that the compression ratio for maximum specific work output is given by

$$r_k = \left(\frac{T_{min}}{T_{max}}\right)^{1/2(1-\gamma)}$$

Solution : Given data :

r_k = compression ratio

$$= \frac{V_1}{V_2} = \frac{V_4}{V_3}$$

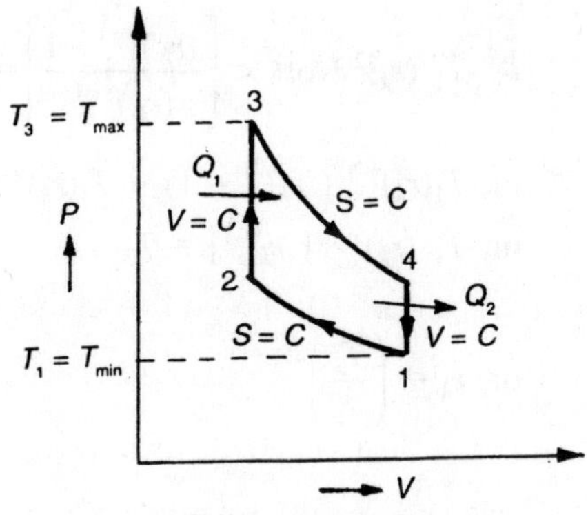

Fig. P. 13.11

Q_1 = heat added

$= C_v\,(T_3 - T_2)$ kJ/kg

Q_2 = heat rejected

$= C_v\,(T_4 - T_1)$ kJ/kg

$\therefore\ w$ = Net work

$= Q_1 - Q_2$

$= C_v\,[(T_3 - T_2) - (T_4 - T_1)]$

For process 1-2

$$\frac{T_2}{T_1} = \left(\frac{V_1}{V_2}\right)^{\gamma-1} = r_k^{\gamma-1},\ T_2 = T_1\, r_k^{\gamma-1}$$

For process 3-4

$$\frac{T_4}{T_3} = \left(\frac{V_3}{V_4}\right)^{\gamma-1} = \left(\frac{1}{r_k}\right)^{\gamma-1},\ T_4 = \frac{T_3}{(r_k)^{\gamma-1}}$$

$$\therefore \quad w = C_v \left[T_3 - T_1 \, r_k^{\gamma-1} \frac{T_3}{(r_k)^{\gamma-1}} + T_1 \right] \text{ kJ/kg}$$

$$= C_v \left[T_3 \left\{ 1 - \frac{1}{(r_k)^{\gamma-1}} \right\} + T_1 \left\{ 1 - (r_k)^{\gamma-1} \right\} \right]$$

$$= C_v \left[1 - \frac{1}{(r_k)^{\gamma-1}} \right] \left[T_3 - (r_k)^{\gamma-1} T_1 \right] \text{ kJ/kg}$$

For maximum work

$$\frac{dw}{d(r_k)} = 0 = C_v[0 - (\gamma - 1)(r_k)^{\gamma-2} T_1]\left[1 - \frac{1}{(r_k)^{\gamma-1}}\right]$$

$$+ C_v [T_3 - T_1 (r_k)^{\gamma-1}] [(\gamma - 1)(r_k)^{-\gamma}]$$

$$\Rightarrow (\gamma - 1)(r_k)^{\gamma-2} T_1 \left[1 - \frac{1}{(r_k)^{\gamma-1}}\right] = [T_3 - T_1 (r_k)^{\gamma-1}] [(\gamma - 1)(r_k)^{-\gamma}]$$

$$Þ \quad T_1 (r_k)^{2(\gamma-1)} \times \left[\frac{(r_k)^{\gamma-1} - 1}{(r_k)^{\gamma-1}}\right] = T_3 - T_1 (r_k)^{\gamma-1}$$

or, $T_1 (r_k)^{\gamma-1} (r_k^{\gamma-1} - 1) + T_1 (r_k)^{\gamma-1} = T_3$

or, $T_1 (r_k)^{\gamma-1} [r_k^{\gamma-1}] = T_3$

$$\text{or, } r_k = \left(\frac{T_3}{T_1}\right)^{\frac{1}{2(\gamma-1)}}$$

$$\text{or, } T_k = \left(\frac{T_{max}}{T_{min}}\right)^{\frac{1}{2(\gamma-1)}} \quad \textbf{Ans}$$

Problem 13.12 : A dual combustion cycle operates with a volumetric compression ratio $r_k = 12$, and with a cut-off ratio 1.615. The maximum pressure is given by $p_{max} = 54\, p_1$, where p_1 is the pressure before compression. Assuming indices compression and expansion of 1.35, show that the m.e.p of the cycle

$p_m = 10\ p_1$
Hence evaluate(a) temperatures at cardinal points with $T_1 = 335$K, and (b) cycle efficiency.

Solution : Given data

Compression ratio $= r_k$
$= 12$

Cut of ratio
$= r_c = 1.615$

$p_{max} = p_3 = p_4 = 54\ p_1$

n = indices of compression and expansion = 1.35

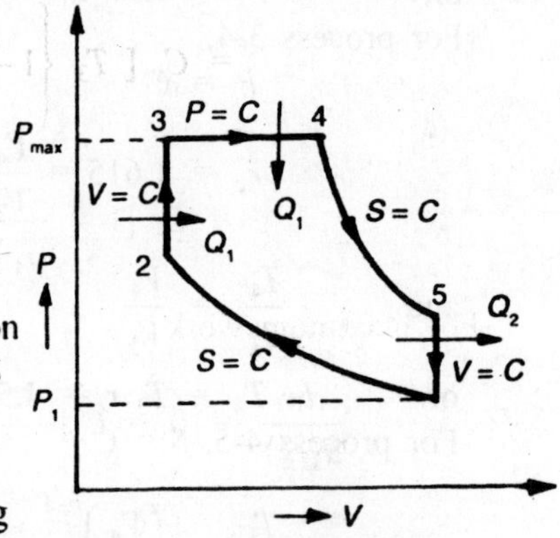

Fig. P. 13.12

Heat supplied
$= Q_1$
$= C_v (T_3 - T_2) + C_p (T_4 - T_3)$. kJ/kg

Heat rejected $= Q_2 = C_v (T_5 - T_1)$

$\therefore$ Work done per kg of air $= w = Q_1 - Q_2$

$= C_v (T_3 - T_2) + Cp (T_4 - T_3) - C_v (T_5 - T_1)$

For process 1-2,

$$S = C$$

$$\frac{V_1}{V_2} = 12$$

or,
$$V_2 = \frac{V_1}{12}$$

$$\frac{T_2}{T_1} = \left(\frac{V_1}{V_2}\right)^{n-1} = r_k^{n-1} = (12)^{0.35}$$

or
$$T_2 = 2.39\ T_1$$

$$\frac{p_2}{p_1} = \left(\frac{V_1}{V_2}\right)^n = 12^{1.35}$$

or,
$$p_2 = 28.63\ p_1$$

For process 2-3,

$$V = C$$

$$\frac{T_3}{T_2} = \frac{p_3}{p_2} \Rightarrow T_3 = T_2 \left(\frac{p_{\max}}{28.63 p_1}\right)$$

$$= 2.39\ T_1 \left(\frac{54\ p_1}{28.63\ p_1}\right)$$

or, $$T_3 = 4.51\ T_1$$

For process 3-4,

$$P = C$$

$$r_c = 1.615 = \frac{V_4}{V_3}$$

and $$\frac{T_4}{T_3} = \frac{V_4}{V_3}$$

or $$T_4 = T_3\ r_c = 4.51 \times 1.615\ T_1 = 7.284\ T_1$$

For process 4-5, $S = C$

$$\frac{T_5}{T_4} = \left(\frac{V_4}{V_5}\right)^{n-1} = \left(\frac{V_3\, r_c}{V_1}\right)^{n-1}$$

or, $$T_5 = T_4 \times \left(\frac{r_c \times V_1}{12 \times V_1}\right)^{1.35-1}$$

or, $$T_5 = 7.284\ T_1 \left(\frac{1.615}{12}\right)^{0.35}$$

or, $$T_5 = 3.61\ T_1$$

$$\therefore \quad w = C_v\ \left[(T_3 - T_2) + \frac{C_p}{C_v}\ (T_4 - T_3) - (T_5 - T_1)\right]$$

$$= C_v \left[\ (4.51\ T_1 - 2.39\ T_1) + \frac{C_p}{C_v}\ (7.284\ T_1 - 4.51\ T_1) - (3.61\ T_1 - T_1)\right]$$

[∵ for air standard cycle

$$\left[\frac{C_p}{C_v} = \gamma,\ C_p - C_v = R,\quad (\gamma - 1)\ C_v = R\right.$$

$$\left.\text{Considering } \gamma = n,\ C_v = \frac{R}{n-1}\right]$$

$$\therefore \quad w = \frac{R}{0.35}\ [3.2549]\ T_1$$

$$= 9.29971\ RT_1\ \text{kJ/kg}$$

$$V_1 - V_2 = V_1 - \frac{V_1}{12} = \frac{11}{12}\ V_1 = \frac{11}{12}\ \frac{RT_1}{P_1}\ \text{m}^3/\text{kg}$$

$$\therefore \qquad mep = \frac{w}{V_1 - V_2} = \frac{9.29971RT_1}{11RT_1} \times 12\ p_1$$

$$= 10.1\ P_1 \approx 10\ p_1 \quad \textbf{Ans}$$

(a) When $T_1 = 335$ k

$$T_2 = 2.39\ T_1 = 2.4\ T_1 = 804\ \text{K}$$
$$p_2 = 28.63\ p_1$$
$$T_3 = 4.51\ T_1 = 1507.5\ \text{K}$$
$$T_4 = 7.284\ T_1 = 7.2\ T_1 = 2412\ \text{K}$$
$$T_5 = 3.61\ T_1 = 1206\ \text{K}$$

(b) Cycle efficiency

$$= \frac{Q_1 - Q_2}{Q_1} = 1 - \frac{Q_2}{Q_1}$$

$$= 1 - \frac{C_v(T_5 - T_1)}{C_V(T_3 - T_2) + C_p(T_4 - T_3)}$$

$$\text{Cycle Efficiency} = 1 - \frac{(T_5 - T_1)}{(T_3 - T_2) + n(T_4 - T_3)}$$

$$= 1 - \frac{(1206 - 335)}{(1507.5 - 804) + 1.35(2412 - 1507.5)} = 0.5574\ \textbf{Ans}$$

Problem 13.13 : In an air standard Brayton cycle the compression ratio is 7 and the maximum temperature of the cycle is 800°C. The compression begins at 0.1 MPa, 35°C. Compare the maximum specific volume and the maximum pressure with the otto cycle of problem 13.6. Find (a) the heat supplied per kg of air, (b) the net work done per kg of air, (c) the cycle efficiency, and (d) the temperature at the end of the expansion process.

Solution : Given data

$$r_k = \text{compression ratio} = 7$$
$$T_3 = 800°\text{C} = 1073\ \text{K}$$

$$p_1 = 0.1 \text{ MPa}$$

$$T_1 = 35°\text{C}$$

$$= 35 + 273$$

$$= 308 \text{ K}$$

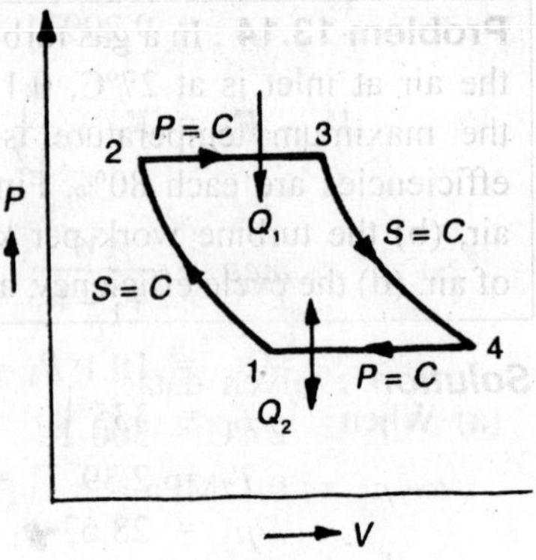

Fig. P. 13.13

For process 1-2

$$\frac{T_2}{T_1} = \left(\frac{V_1}{V_2}\right)^{\gamma-1}$$

$$= (7)^{0.4}$$

$$\therefore \quad T_2 = 670.8 \text{ K}$$

For process 3-4

$$\frac{T_4}{T_3} = \left(\frac{p_4}{p_3}\right)^{\frac{\gamma-1}{\gamma}} = \left(\frac{p_1}{p_2}\right)^{\frac{\gamma-1}{\gamma}}$$

$$= \left[\left(\frac{V_2}{V_1}\right)^{\gamma}\right]^{\frac{\gamma-1}{\gamma}} = \left(\frac{V_2}{V_1}\right)^{\gamma-1}$$

$$\Rightarrow T_4 \frac{T_3}{\left(\frac{V_1}{V_2}\right)^{\gamma-1}} = \frac{1073}{(7)^{0.4}} = 492.67 \text{ K}$$

(a) Heat supplied per kg of air

$$= Q_1 = C_p (T_3 - T_2)$$

$$= 1.005 (1073 - 670.8) = 404.21 \text{ kJ/kg}$$

(b) Heat rejected

$$= Q_2 = C_p (T_4 - T_1)$$

$$= 1.005 (492.67 - 308) = 185.6 \text{ kJ/kg}$$

$\therefore$ Net work done per kg of air $= Q_1 - Q_2$

$= 404.21 - 185.6 = 218.62$ kJ/kg **Ans**

(c) Cycle efficiency $= \dfrac{\text{Net work}}{\text{heat supplied}} = \dfrac{218.62}{404.21} = 54.1$ % **Ans**

(d) $T_4 = 492.62$ **Ans**

Problem 13.14 : In a gas turbine plant working on the Brayton cycle the air at inlet is at 27°C, 0.1 MPa. The pressure ratio is 6.25 and the maximum temperature is 800°C. The turbine and compressor efficiencies are each 80%. Find (a) the compressor work per kg of air, (b) the turbine work per kg of air, (c) the heat supplied per kg of air, (d) the cycle efficiency, and (e) the turbine exhaust temperature.

Solution : Given data

$T_1 = 27°C = 300\text{ K}$

$p_1 = 0.1\text{ MPa}$

$\frac{P_2}{P_1} = 6.25$

$T_3 = 800°C = 1073\text{ K}$

$\eta_c = \eta_T$

= isentropic efficiency

= 80%

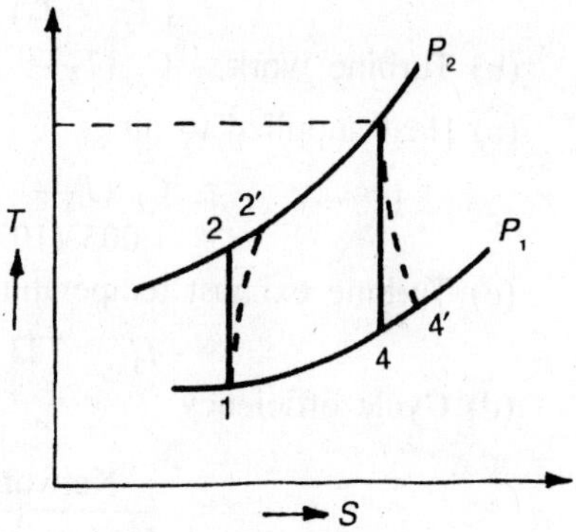

Fig. P. 13.14

For process 1-2

$$\frac{T_2}{T_1} = \left(\frac{p_2}{p_1}\right)^{\frac{\gamma-1}{\gamma}}$$

$$\Rightarrow \quad T_2 = T_1 \times (6.25)^{\frac{0.4}{1.4}} = 506.7\text{ K}$$

$$\eta_c = 0.8 = \frac{T_2 - T_1}{T_2' - T_1}$$

$$\Rightarrow (T_2' - T_1) = \frac{T_3 - T_4}{\eta_c} = \frac{(506.7 - 300)}{0.8} = 258.4\text{ K}$$

or, $T_2' = 558.4\text{ K}$

For process 3-4

$$\frac{T_4}{T_3} = \left(\frac{p_1}{p_2}\right)^{\frac{\gamma-1}{\gamma}}$$

$$\Rightarrow \quad T_4 = \frac{T_3}{(6.25)^{\frac{0.4}{1.4}}} = 635.3\text{ K}$$

Now $\eta_T = 0.8 = \dfrac{T_3 - T_4'}{T_3 - T_4}$

$\Rightarrow \quad T_3 - T_4' = 0.8\,(T_3 - T_4)$

$= 0.8\,(1073 - 635.3) = 350.2 \text{ K}$

or, $T_4' = 722.84 \text{ K}$

(a) Compression work

$= (T_2' - T_1)'\, C_p = 1.005\,(258.4) = 259.7 \text{ kJ/kg}$

(b) Turbine work $= C_p\,(T_3 - T_4') = 1.005\,(350.2) = 351.95 \text{ kJ/kg}$

(c) Heat supplied to air

$= C_p\,(T_3 - T_2')$

$= 1.005\,(1073 - 558.4) = 517.17 \text{ kJ/kg}$

(e) Turbine exhaust temperature

$= T_4' = 723 \text{ K}$

(d) Cycle efficiency

$$= \frac{\text{Network}}{\text{heat supplied}} = \frac{351.95 - 259.7}{517.17} = 17.84\ \%$$

Problem 13.15 : Solve problem 13.14 if a regenerator of 75% effectiveness is added to the plant.

Solution : From the previous figure :

$T_2' = 588.4 \text{ K}$

$T_4' = 722.84 \text{ K}$

Regenerature efficiency

$= 75\%$

Let T_5 be the temperature of air after the regeneration

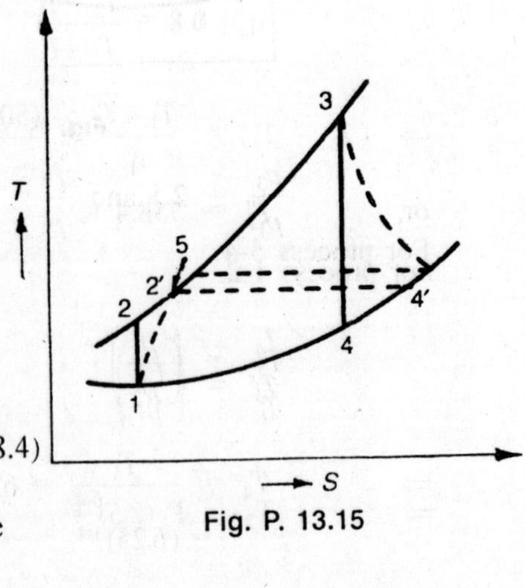

Fig. P. 13.15

or $0.75 = \dfrac{T_5 - T_2'}{T_4' - T_2'}$

or $T_5 - T_2'$

$= 0.75\,(722.84 - 588.4)$

or $T_5 = 689.23 \text{ K}$

(a), (b), (c) will be same

but (c) Heat supplied

$$= C_p (T_3 - T_5)$$
$$= 1.005 (1073 - 689.23)$$
$$= 385.69 \text{ kJ/kg} \textbf{ Ans}$$

(d) Cycle efficiency

$$= \frac{\text{Network}}{\text{heat supplied}} = \frac{351.95 - 259.7}{385.69} = 23.92\% \textbf{ Ans}$$

> **Problem 13.16** : Solve problem 13.15 if the compression is divided into two stages, each of pressure ratio 2.5 and efficiency 80%, with intercooling to 27°C.

Solution : Given data

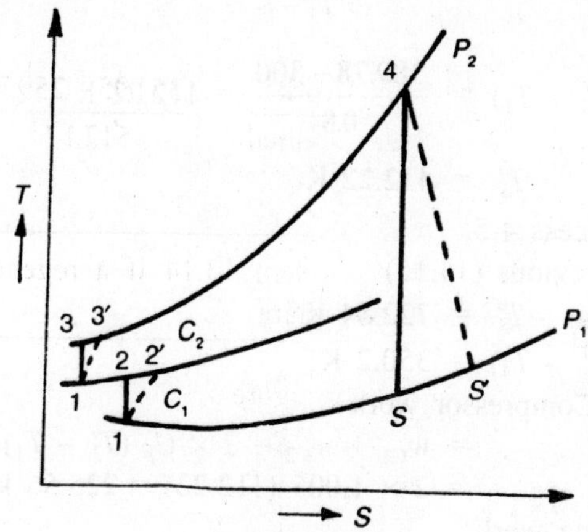

Fig. P. 13.16

$$\frac{p_3}{p_1} = 2.5 \text{ and } \frac{p_2}{p_3} = 2.5$$

For process 1-2

$$\frac{T_2}{T_1} = \left(\frac{p_3}{p_1}\right)^{\frac{\gamma-1}{\gamma}}$$

$$\Rightarrow \quad T_2 = T_1 (2.5)^{\frac{0.4}{1.4}} = 300 \times (2.5)^{\frac{0.4}{1.4}} = 389.78 \text{ K}$$

$$\eta_c = 0.8 = \frac{T_2 - T_1}{T_2' - T_1}$$

or, $$(T_2' - T_1) = \frac{389.78 - 300}{0.8} \quad 112.22 \text{ K}$$

$$T_2' = 412.22 \text{ K}$$

For process 1-3

$$\frac{T_3}{T_1} = \left(\frac{p_2}{p_3}\right)^{\frac{\gamma-1}{\gamma}}$$

$$\Rightarrow \quad T_3 = T_1 \ (2.5)^{\frac{0.4}{1.4}} = 300 \ (2.5)^{\frac{0.4}{1.4}} = 389.78 \text{ K}$$

$$n_c = 0.8 = \frac{T_3 - T_1}{T_3' - T_1}$$

$$\Rightarrow \ (T_3' - T_1) = \frac{389.78 - 300}{0.8} = 112.22 \text{ K}$$

$$T_3' = 412.22 \text{ K}$$

For process 4-5,
from previous (13.15)

$$T_5' = 722.84 \text{ K}$$

$$(T_5' - T_4) = 350.2 \text{ K}$$

∴ (a) Compressor work

$$= w_{C1} + w_{C2} = 2 \times C_p \ (T_2' - T_1)$$

$$= 2 \times 1.005 \ (112.22) = 225.56 \text{ kJ/kg}$$

(b) Turbine work

$$= C_p \ (T_5' - T_4)$$

$$= 1.005 \ (350.2) = 351.95 \text{ kJ/kg}$$

(c) Heat supplied $= C_p \ (T_4 - T_3')$

$$= 1.005 \ (1073 - 412.22) = 664.1 \text{ kJ/kg}$$

(d) Cycle efficiency

$$= \frac{\text{Net work}}{\text{heat supplied}}$$

$$= \frac{351.95 - 225.56}{664.1} = 19.03 \ \%$$ **Ans**

Problem 13.17 : Solve problem 13.16 if a regenerator of 75% effectiveness is added to the plant.

Solution : Given data :

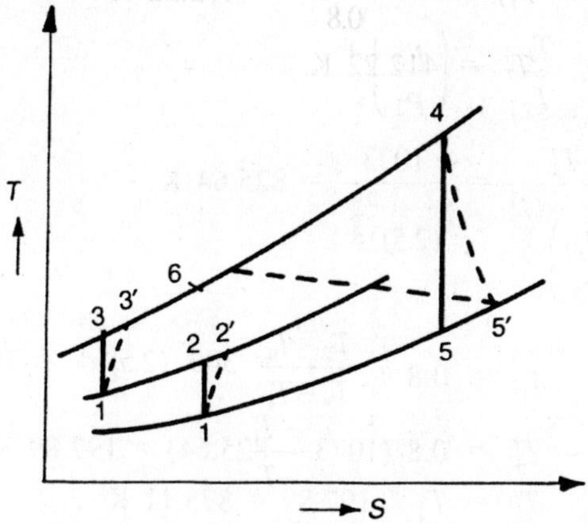

Fig. P. 13.17

$$\text{Regenerator effectiveness} = 0.75 = \frac{T_6 - T_3'}{T_5' - T_3'}$$

or,
$$T_6 = 0.75\,(T_5' - T_3') + T_3'$$
$$= 0.75\ (722.84 - 412.22) + 412.22 = 645.19\ \text{K}$$

∴ (a), (b), (e) will be same as

(c) Heat supplied $= C_p\ (T_4 - T_6)$
$$= 1.005(1073 - 645.19) = 427.82\ \text{kJ/kg}$$

(d) Cycle efficiency $= \dfrac{\text{Net work}}{\text{heat supplied}}$

$$= \frac{351.95 - 225.56}{427.82} = 29.54\%\ \textbf{Ans}$$

Problem 13.18 : Solve problem 13.14 if a reheat cycle is used. The turbine expansion is divided into two stages, each of 2.5 and efficiency 80%, with regeat to 800°C.

Solution : Given data

$$\frac{p_2}{p_3} = \frac{p_3}{p_1} = 2.5,\ T_5 = T_1 = 1073\ \text{K}$$

from previous problem (13.14)

$$(T_2' - T_1) = 258.4 \text{ K}$$

$$T_2' = 558.4 \text{ K}$$

For process 3-4

$$\frac{T_4}{T_3} = \left(\frac{p_3}{p_2}\right)^{\frac{\gamma-1}{\gamma}}$$

$$\Rightarrow T_4 = \frac{T_3}{\left(\frac{p_2}{p_3}\right)^{\frac{\gamma-1}{\gamma}}} = \frac{1073}{(2.5)^{\frac{0.4}{1.4}}} = 825.64 \text{ K}$$

$$\eta_T = 0.8 = \frac{T_3 - T_4'}{T_3 - T_4}$$

or, $T_3 - T_4' = 0.8\,(1073 - 825.64) = 197.89$ K

or, $T_4' = T_3 - 197.89 = 875.11$ K

For process 5-6

$$\frac{T_6}{T_5} = \left(\frac{p_1}{p_3}\right)^{\frac{\gamma-1}{\gamma}}$$

$$\Rightarrow T_6 = \left(\frac{T_5}{(p_3 / p_1)^{\frac{\gamma-1}{\gamma}}}\right) = \frac{1073}{(2.5)^{\frac{0.4}{1.4}}} = 825.64 \text{ K}$$

$$(T_5 - T_6') = 0.8\,(1073 - 825.64) = 197.89 \text{ K}$$

$$T_6' = 875.11 \text{ K}$$

∴ (a) Comprassor work

$$= C_p\,(T_2' - T_1) = 1.005\,(258.4) = 259.7 \text{ kJ/kg}$$

(b) Turbine work $= 2 \times C_p\,(T_3 - T_4')$

$$= 2 \times 1.005 \times 197.89 = 397.76 \text{ kJ/kg}$$

(c) Heat supplied $= C_p\,(T_3 - T_2')$

$$= 1.005 \times (1073 - 558.4) = 517.17 \text{ kJ/kg}$$

(d) Cycle efficiency $= \dfrac{\text{Net work}}{\text{heat supplied}}$

$$= \frac{397.76 - 259.7}{517.17} = 26.7\% \textbf{ Ans}$$

(e) Turbine exhaust temperature

$$= T_6' = 875.11 \text{ K } \textbf{Ans}$$

Problem 13.19 : Solve problem13.18 if a regenerator of 75% effectiveness is added to the plant.

Solution : Given data

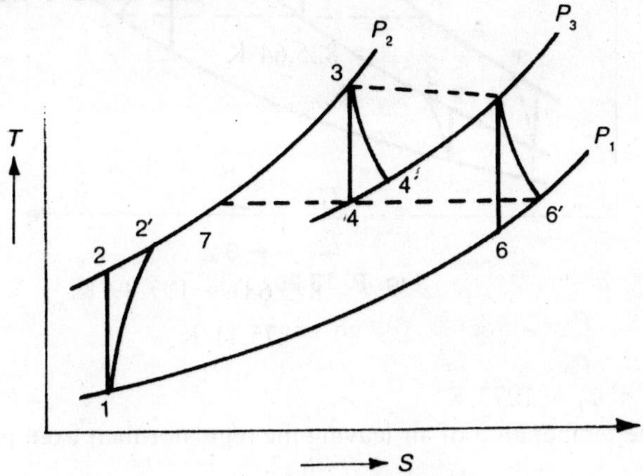

Fig. P. 13.19

Let T_7 be temperature of air leaving the regenerator

If regenerator is introduced then

$$\Rightarrow 0.75 = \frac{T_7 - T_2'}{T_6' - T_2'}$$

or $(T_7 - T_2') = (T_6' - T_2') \times 0.75$

$\therefore$ $T_7 = (T_6' - T_2') \times 0.75 + T_2'$

$= (875.11 - 558.4) \times 0.75 + 558.4 = 795.93$ K

$\therefore$ (a), (b) & (c) will be same as in (13.18)

(c) heat supplied $= C_p\,(T_3 - T_7)$

$= 1.005\,(1073 - 795.93) = 278.45$ kJ/kg

(d) Cycle efficiency $= \dfrac{\text{Net work}}{\text{heat supplied}} = \dfrac{397.76 - 259.7}{278.45} = 49.58\%$ **Ans**

Problem 13.20 : Solve problem 13.19 if the staged compression of Problem 13.16 is used in the plant.

Solution : *Given data*

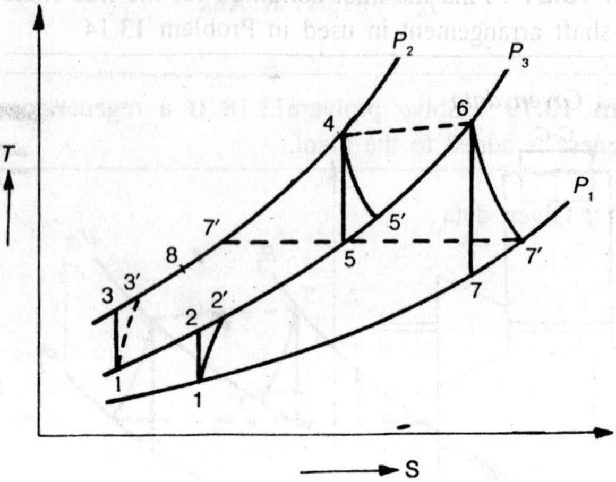

Fig. P. 13.20

$$\frac{p_3}{p_1} = \frac{p_2}{p_3} = 2.5$$

$$T_4 = T_6 = 1073 \text{ K}$$

Let T_8 be the temperature of air leaving the regulator then from problem 13.19.

$$(T_3' - T_1) = 112.22 \text{ K}$$

$$T_3' = 412.22 \text{ K}$$

from problem 13.18, $(T_5' - T_4) = 197.89$ K

$$(T_6 - T_7') = 197.89 \text{ K}$$

$$T_7' = 875.11 \text{ K}$$

Now using regulator, $0.75 = \dfrac{T_8 - T_3'}{T_7' - T_3'}$

or, $T_8 = 0.75\,(T_7' - T_3') + T_3'$

$= 0.75\,(875.11 - 412.22) + 412.22 = 759.39$ K

∴ (a) Compressor work $= 2\,C_p\,(T_2' - T_1) = 2 \times 1.005\,(112.22)$

$= 225.56$ kJ/kg. **Ans**

(b) Turbine work $= 2 \times C_p\,(T_5' - T_4)$

$= 2 \times 1.005\,(197.89) = 397.76$ kJ/kg. **Ans**

(c) Heat supplied $= C_p\,(T_4 - T_8)$

$= 1.005\,(1073 - 759.39) = 315.18$ kJ/kg

(d) Cycle efficiency $= \dfrac{\text{Net work}}{\text{heat supplied}} = \dfrac{172.1989}{315.18} = 54.64\%$ **Ans**

(e) Turbine exhaust temperature $= T_7' = 875.11$ K. **Ans**

Problem 13.21 : Find the inlet condition for the free-shaft turbine if a two-shaft arrangement in used in Problem 13.14

***Solution** : Given data*

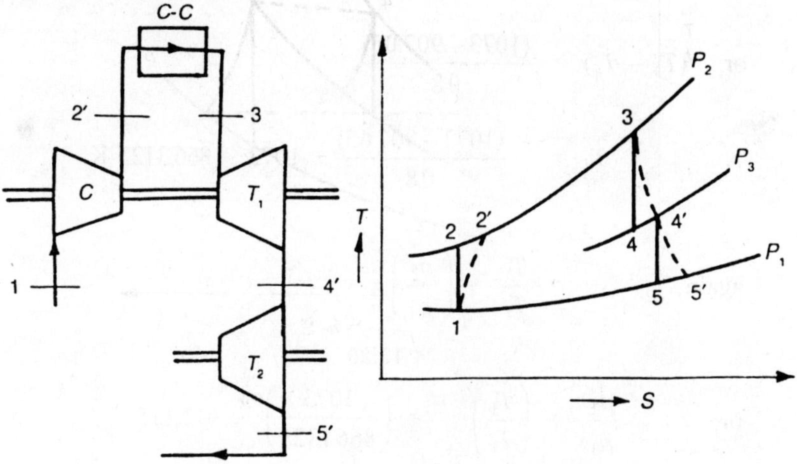

Fig. P. 13.21

$$p_1 = 0.1 \text{ MPa} \qquad T_1 = 300 \text{ K}$$

$$\eta_c = \eta_T = 80\% \qquad T_3 = 1073 \text{ K}$$

$$\frac{p_2}{p_1} = 6.25$$

For free shaft turbine $\quad W_c = W_T$

From process 1-2

$$\frac{T_2}{T_1} = \left(\frac{p_2}{p_1}\right)^{\frac{\gamma-1}{\gamma}}$$

$$\Rightarrow \quad T_2 = T_1 (6.25)^{\frac{0.4}{1.4}} = 506.7 \text{ K}$$

$$\eta_c = 0.8 = \frac{T_2 - T_1}{T_2' - T_1}$$

or, $\quad (T_2' - T_1) = 0.8\ (506.7 - 300) = 165.35 \text{ K}$

$\therefore \quad T_2' = 465.35 \text{ K}$

since $\quad w_c = w_T$

or, $\quad C_p\ (T_2' - T_1) = C_p\ (T_3 - T_4')$

or, $\quad 165.35 = 1073 - T_4'$

or, $T_4' = 1073 - 165.35 = 907.65$ K

Now since 3–4 process in isentropic,

$$\therefore \quad \eta_T = 0.8 = \frac{T_3 - T_4'}{T_3 - T_4}$$

$$\text{or,} \quad (T_3 - T_4) = \frac{(1073 - 907.05)}{0.8}$$

$$\therefore \quad T_4 = -\frac{(1073 - 907.65)}{0.8} + 1073 = 866.3125 \text{ K}$$

$$\text{again,} \quad = \frac{T_3}{T_4} = \left(\frac{p_2}{p_3}\right)^{\frac{\gamma-1}{\gamma}}$$

$$\text{or} \quad \frac{p_2}{p_3} = \left(\frac{T_3}{T_4}\right)^{\frac{\gamma}{\gamma-1}} = \left(\frac{1073}{866.1325}\right)^{\frac{1.4}{0.4}} = 2.115$$

$$\text{or,} \quad p_3 = \frac{p_2}{2.115} = \frac{6.25 p_1}{2.115} = 2.956\ p_1$$

$$p_3 = 2.956\ P_1$$

$\therefore$ Inlet condition for free shaft turbine are

$$p_3 = 2.950\ p_1$$

$$T_4' = 866.31 \text{ K } \textbf{Ans}$$

Problem 13.22 : Obtain an expression for the specific work output of a gas turbine unit in terms of pressure ratio, isentropic efficiencies of the compressor and turbine, and the maximum and minimum temperatures, T_3 and T_1. Hence show that the pressure ratio r_p for maximum power is given by

$$r_p = \left(\eta_T \eta_c \frac{T_3}{T_1}\right)^{\gamma/2(\gamma-1)}$$

If $T_3 = 1073$ K, $T_1 = 300$ K, $n_c = 0.8$ $n_T = 0.8$ and $\gamma = 1.4$ compute the optimum value of pressure ratio, the maximum net work output per kg of air, and corresponding cycle efficiency.

Solution : Given data

$$\frac{p_2}{p_1} = r_p$$

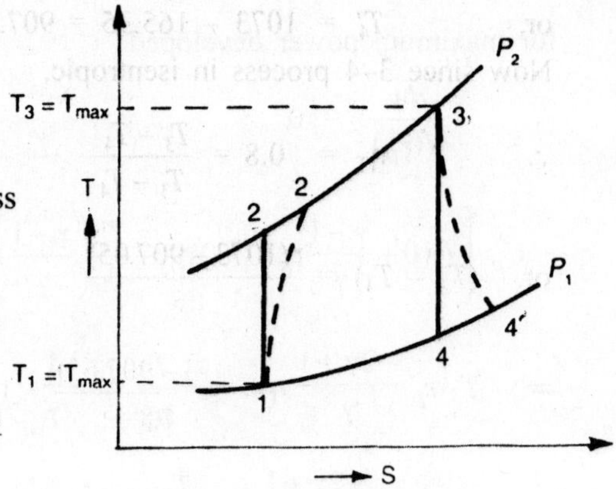

Fig. P. 13.22

Now for 1-2 process

$$\frac{T_2}{T_1} = \left(\frac{p_2}{p_1}\right)^{\frac{\gamma-1}{\gamma}}$$

$$\text{or,}\quad T_2 = T_1\, r_p^{\frac{\gamma-1}{\gamma}}$$

$$\eta_c = \frac{T_2 - T_1}{T_2' - T_1}$$

$$\text{or,}\quad (T_2' - T_1) = \frac{T_2 - T_1}{\eta_c} = \left(r_p^{\frac{\gamma-1}{\gamma}} - 1\right)\frac{T_1}{\eta_c}$$

For process 3-4

$$\frac{T_3}{T_4} = \left(\frac{p_2}{p_1}\right)^{\frac{\gamma-1}{\gamma}} \quad \text{or,}\quad T_4 = \frac{T_3}{r_p^{\frac{\gamma-1}{\gamma}}}$$

$$\eta_T = \frac{T_3 - T_4'}{T_3 - T_4}$$

$$\text{or,}\quad (T_3 - T_4') = (T_3 - T_4)\,\eta_T$$

$$\text{or,}\quad (T_3 - T_4') = T_3\left(1 - \frac{1}{r_p^{\frac{\gamma-1}{\gamma}}}\right)\eta_T$$

$\therefore$ Sp work output

$$= w_T - w_C = C_p\,[(T_3 - T_4') - (T'_2 - T_1]$$

$$w = C_p\left[T_3\left(1 - \frac{1}{r_p^{\frac{\gamma-1}{\gamma}}}\right)\eta_T - (r_p\frac{\gamma-1}{\gamma})\frac{T_1}{\eta_c}\right]$$

for maximum power developed

$$\frac{dw}{d(r_p)} = 0$$

$$= C_p\left[T_3(0+\frac{\gamma-1}{\gamma}r_p^{\frac{-2\gamma+1}{\gamma}})\eta_T - \frac{T_1}{\eta_c}\left(\frac{\gamma-1}{\gamma}\right)r_p^{\frac{-1}{\gamma}}\right]$$

$$\Rightarrow \quad T_3\, r_p\, \frac{-2\gamma+1}{\gamma}\eta_T = +\frac{T_1}{\eta_c}\, r_p - \frac{1}{\gamma}$$

$$\Rightarrow \quad r_p^{-\frac{1}{\gamma}} - \frac{-2\gamma+1}{\gamma} = \frac{T_3}{T_1}\eta_c\,\eta_T$$

$$\Rightarrow \quad r_p^{2\left(\frac{\gamma-1}{\gamma}\right)} = \frac{T_3}{T_1}\eta_c\,\eta_T$$

or $$(r_p)_{optim} = \left(\frac{T_{max}}{T_{min}}\eta_c\,\eta_T\right)^{\frac{\gamma}{2(\gamma-1)}}$$

If
$$T_3 = 1073 \text{ K}$$
$$T_1 = 300 \text{ K}$$
$$n_c = n_T = 0.8$$
$$\gamma = 1.4$$

$$(r_p)_{optim} = \left(\frac{1073}{300}\times 0.8\times 0.8\right)^{\frac{1.4}{2\times 0.4}} = 4.26 \textbf{ Ans}$$

$$T_2' = T_1 + \left(r_p^{\frac{1.4-1}{1.4}} - 1\right)\frac{T_1}{\eta_c} = 492.6 \text{ K}$$

Max^m Net work output

$$= C_p\left[T_3\left(1-\frac{1}{r_p^{\frac{\gamma-1}{\gamma}}}\right)\frac{T_1}{n_c}\, n_T\left(r_p\frac{\gamma-1}{\gamma}-1\right)\right]$$

$$= 1.005\left[1073\left(1-\frac{1}{4.26^{\frac{1.4-1}{1.4}}}\right)\times 0.8-\frac{300}{0.8}\left(4.26^{\frac{1.4-1}{1.4}}-1\right)\right]$$

$= 99.17$ kJ/kg. **Ans**

Heat supplied $= C_p\ (T_2 - T_3')$

$= 1.005\ (1073 - 492.6) = 583.3$ kJ/kg

$$\therefore \text{ cycle efficincy } = \frac{\text{Max}^{\text{m}} \text{ Net work output}}{\text{heat supplied}}$$

$$= \frac{99.17}{583.3} = 17.1\% \textbf{ Ans}$$

Problem 13.23 : An ideal air cycle consists of isentropic compression, constant volume heat transfer, isothermal expansion to the original pressure, and constant pressure heat transfer to the original temperature. Deduce and expression for the efficiency in term of volumetric compression ratio r_k, and isothermal expansion ratio, r_c. In such a cycle, the pressure and temperature at the start of compression are 1 bar and 40°C, the compression ratio is 8, and the maximum pressure is 100 bar. Determine the cycle efficiency and the m.e.p.

Solution : Given data

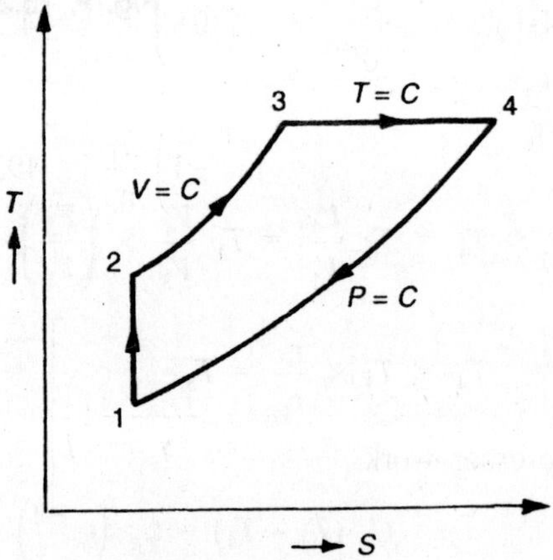

Fig. P. 13.23 (a)

$$r_k = \text{Compression ratio (volumetric)}$$

$$= \frac{V_1}{V_2} = \left(\frac{p_2}{p_1}\right)^{\frac{1}{\gamma}}$$

$$r_e = \text{isothermal expansion ratio} = \frac{V_4}{V_3}$$

Let p_1 and T_1, be the initial pressure and temperature of air before compresion.

given

$$p_1 = p_4, \; V_3 = V_2$$

For process 1-2

$$\frac{T_2}{T_1} = \left(\frac{V_1}{V_2}\right)^{\gamma-1}$$

$$= r_k^{\gamma-1} \text{ or,}$$

$$T_2 = T_1 \, r_k^{\gamma-1}$$

For process 3-4,

$$T = \text{Constant}$$

$$p_3V_3 = p_4V_4$$

or, $p_3V_3 = p_1V_4$

or, $\frac{p_3}{p_1} = \frac{V_4}{V_3} = r_e$

$\therefore \quad p_3 = p_1 \, r_e$

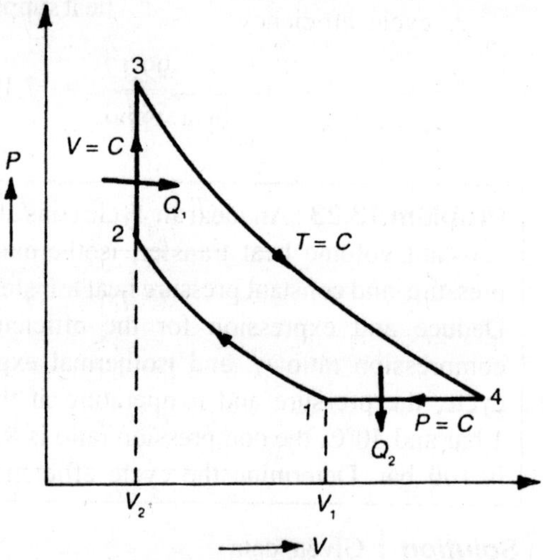

Fig. P. 13.23 (b)

For process 4-1

$$\frac{V_4}{T_4} = \frac{V_1}{T_1}$$

$$\Rightarrow \quad T_4 = T_1 \frac{V_4}{V_1} = T_1 \frac{V_4}{V_3} \times \left(\frac{V_3}{V_1}\right)$$

$$\Rightarrow \quad T_4 = T_1 \times \frac{r_e}{r_k} = T_3$$

$\therefore$ Compressor work

$$= C_p (T_2 - T_1) = C_p \left(r_k^{\gamma-1}\right) T_1$$

and turbine work

$$p_3 V_3 \ln \frac{V_4}{V_3} = RT_4 \ln r_e = RT_1 \frac{r_e}{r_k} \ln r_e$$

$$Q_1 = \text{heat supplied} = C_v (T_3 - T_2) = C_v (T_4 - T_2)$$

$$= C_v \left[T_1 \frac{r_e}{r_k} - T_1 r_k^{\gamma-1} \right]$$

$$= C_v T_1 \left[\frac{r_e}{r_k} - r_k^{\gamma-1} \right]$$

$\therefore$ cycle efficiency

$$= \frac{w_{net}}{\text{heat supplied}}$$

$$= \frac{RT_1 \frac{r_e}{r_k} \ln r_e - C_p T_1 \left(r_k^{\gamma-1} - 1\right)}{C_v T_1 \left[\frac{r_e}{r_k} - r_k^{\gamma-1} \right]}$$

$$= \frac{(\gamma - 1) \frac{r_e}{r_k} \ln r_e - \gamma \left(r_k^{\gamma-1} - 1\right)}{\frac{r_e}{r_k} - r_k^{\gamma-1}} \qquad \left[\because C_v = \frac{R}{\gamma - 1} \right]$$

$$\text{cycle efficiency} = \frac{(\gamma - 1) r_e \ln r_e - \gamma r_k \left(r_k^{\gamma-1} - 1\right)}{r_e - r_k^{\gamma}}$$

$$Q_2 = \text{heat rejected} = C_p (T_4 - T_1)$$

$$= C_p \left(T_1 \frac{r_e}{r_k} - T_1 \right) = \frac{C_p T_1}{r_k} (r_e - r_k)$$

when,

$$p_1 = 1 \text{ bar}$$

$$T_1 = 40^\circ\text{C} = 273 + 40 = 313 \text{ K}$$

$$r_k = 8$$

$$P_3 = 100 \text{ bar}$$

$$r_e = \frac{V_4}{V_3} = \frac{p_3}{p_4} = \frac{p_3}{p_1} = 100$$

Problem 13.24 : A gas turbine for use as an automotive engine is shown in Fig. P.13.24. In the first turbine, the gas expands to just a low enough pressure p_5, for the turbine to drive the compressor. The gas is then expanded through a second turbine connected to the drive wheels. Consider air as the working fluid, and assume that all process are ideal. Determine (a) pressure p_5, (b) the net work per kg and mass flow rate, (c) temperature T_3 and cycle thermal efficiency, and (d) the *T-s* diagram for the cycle.

Solution : Given data

$$\frac{p_2}{p_1} = 4$$

$$p_1 = 1 \text{ atm}$$

$$T_1 = 25°C = 25 + 273 = 298 \text{ K}$$

$$T_4 = 920°C = 920 + 273 = 1193 \text{ K}$$

$$P_7 = 1 \text{ atm}$$

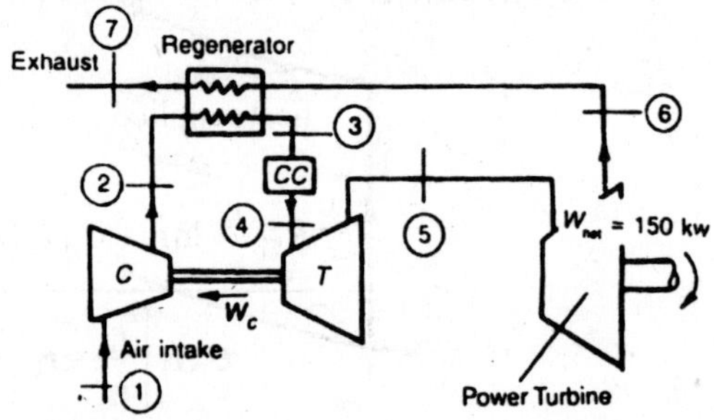

p_1 = 1 atm, t_1 = 25°C, $\frac{p_2}{p_1}$ = 4.0, t_4 = 920°C, p_7 = 1 atm

Fig. P. 13.24

For process 1-2, $S = C$

$$\frac{T_2}{T_1} = \left(\frac{p_2}{p_1}\right)^{\frac{\gamma-1}{\gamma}} = (4)^{\frac{0.4}{1.4}}$$

$$\therefore T_2 = 443 \text{ K}$$

Now since $w_c = w_{T1}$

$$\Rightarrow C_p (T_2 - T_1) = C_p (T_4 - T_5)$$

$$\Rightarrow 443 - 298 = 1193 - T_5$$

$$\Rightarrow T_5 = 1048 \text{ K}$$

Now. For process 4 –5

$$\frac{T_4}{T_5} = \left(\frac{p_2}{p_5}\right)^{\frac{\gamma-1}{\gamma}}$$

or

$$\frac{p_2}{p_5} = \left(\frac{T_4}{T_5}\right)^{\frac{\gamma}{\gamma-1}} = \left(\frac{1193}{1048}\right)^{\frac{1.4}{0.4}}$$

(a) $\Rightarrow \quad p_5 = \dfrac{p_2}{1.15739} = \dfrac{4}{1.15739} = 2.54$ atm **Ans**

For process 5-6

$$\frac{T_6}{T_5} = \left(\frac{p_1}{p_5}\right)^{\frac{\gamma-1}{\gamma}} \quad \text{or,}\ T_6 = T_5\left(\frac{1}{2.54}\right)^{\frac{0.4}{1.4}} = 802.83 \text{ K}$$

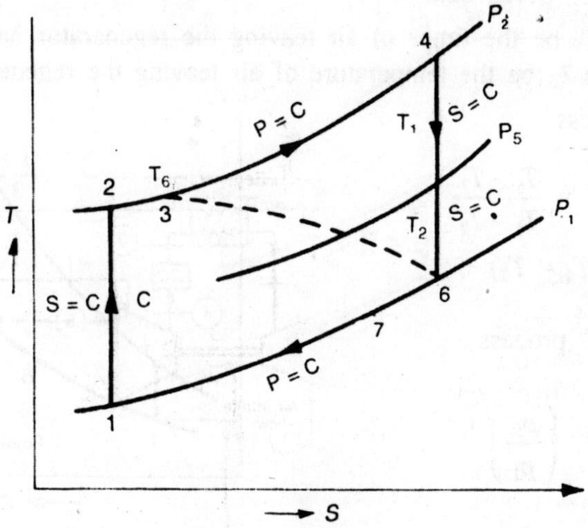

Fig. P. 13.24 (a)

(b) ∴ Net work per kg

$$= C_p\,(T_5 - T_6)$$
$$= 1.005\,(1048 - 802.83) = 246.4 \text{ kJ/kg.}$$

Mass flow rate $= \dfrac{150}{246.4} = 0.61$ kg/sec **Ans**

(c) $\quad T_3 = T_6 = 802.83$ K = for ideal regenerator

Cycle thermal efficinecy $= \dfrac{\text{Net work}}{\text{heat supplied}}$

∴ heat supplied $= mC_p\,(T_4 - T_3)$

$$= 0.61 \times 1.005\,(T_4 - T_6)$$
$$= 0.61 \times 1.005\,(1193 - 802.83) = 239.2 \text{ kW}$$

∴ Cycle efficiency$= \dfrac{150}{239.2}$ 62.71% **Ans**

Problem 13.25 : Develop expressions for work output per kg and the efficiency of an ideal Brayton cycle with regeneration, assuming maximum possible regeneration, assuming maximum possible regeneration. For fixed maximum and minimum temperature, how do the efficiency and work outputs vary with the pressure ratio? what is the optimum pressure ratio ?

Solution : Given data

Let T_5 be the timpr of air leaving the regenerator having 100% effectivess T_6 be the temperature of air leaving the regenerator.

Effectiviness

$$= 1 = \frac{T_5 - T_2}{T_4 - T_2}$$

$\therefore$ $T_5 = (T_4 - T_2) +$
$T_2 = T_4$

$\therefore$ For 1-2 process

$$T_2 = T_1 \left(\frac{p_2}{p_1}\right)^{\frac{\gamma-1}{\gamma}}$$

$$= T_1 \, r_p^{\frac{\gamma-1}{\gamma}}$$

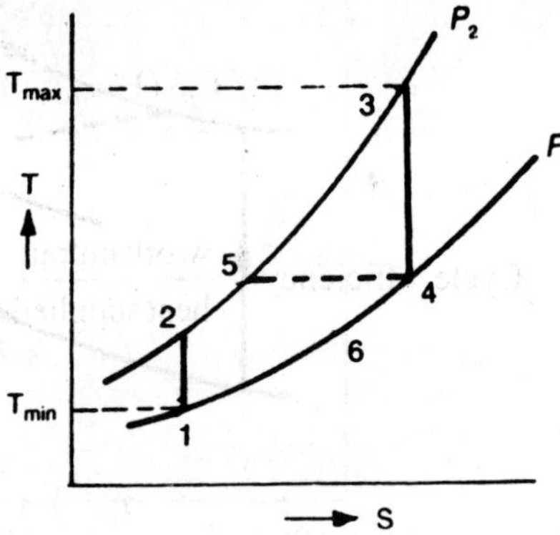

Fig. P. 13.25 (a)

where r_p = pressure ratio similarly for 3–4 process.

$$\frac{T_4}{T_3} = \left(\frac{p_1}{p_2}\right)^{\frac{\gamma-1}{\gamma}}$$

or, $T_4 = \dfrac{T_3}{r_p^{\frac{\gamma-1}{\gamma}}}$

$\therefore$ Compressor work = w_c

$= C_p (T_2 - T_1)$

$$= C_p \, T_1 \left(r_p^{\frac{\gamma-1}{\gamma}} = 1\right)$$

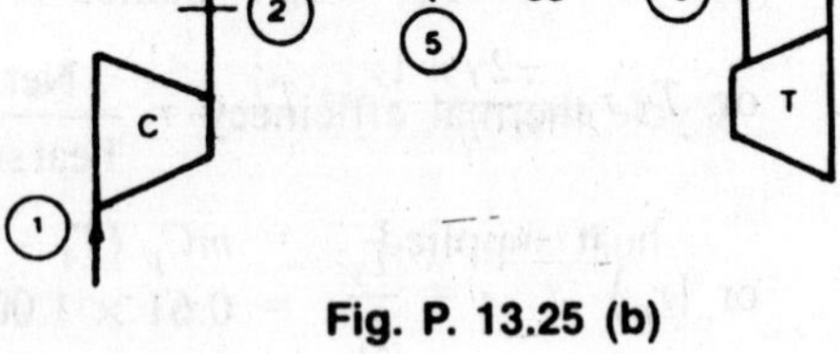

Fig. P. 13.25 (b)

Turbine work $= w_T = C_p (T_3 - T_4) = C_p \, T_3 \left(1 - \dfrac{1}{r_p^{\frac{\gamma-1}{\gamma}}}\right)$

Heat supplied $= Q_1 = C_p\,(T_3 - T_5) = C_p\,(T_3 - T_4)$

$$= C_p\left(T_3 - \frac{T_3}{r_p^{\frac{\gamma-1}{\gamma}}}\right) = C_p\,T_3\left(1 - \frac{1}{r_p^{\frac{\gamma-1}{\gamma}}}\right)$$

$\therefore$ Work output $= w_T - w_C = w$

$$= C_P\left(1 - \frac{1}{r_p^{\frac{\gamma-1}{\gamma}}}\right)T_3 - C_P\,T_1\,(-1 + r_p^{\frac{\gamma-1}{\gamma}})$$

$$= C_p\,(1 - r_p^{\frac{\gamma-1}{\gamma}})\left(T_3 - T_1 \times r_p^{\frac{\gamma-1}{\gamma}}\right)$$

$$\text{Cycle efficiency} = \frac{\text{work output}}{\text{heat supplied}} = 1 - \frac{T_1\left(-1 + r_p^{\frac{\gamma-1}{\gamma}}\right)}{T_3\left(1 - \frac{1}{r_p^{\frac{\gamma-1}{\gamma}}}\right)}$$

$$\eta = 1 - \frac{T_1}{T_3}\,r_p^{\frac{\gamma-1}{\gamma}}$$

Now for maximum work output,

$$\frac{d_w}{d_{rp}} = 0$$

$$= C_p\,T_3\left[0 + \left(\frac{\gamma-1}{\gamma}\right)r_p^{\frac{-2\gamma+1}{\gamma}}\right] - C_p\,T_1\left[\left(\frac{\gamma-1}{\gamma}\right)r_p^{\frac{-1}{\gamma}} - 0\right]$$

$$\text{or, } T_3\,r_p^{\frac{-2\gamma+1}{\gamma}} = T_1\,r_p^{\frac{-1}{\gamma}}$$

$$\text{or } (r_p)^{\frac{2(\gamma-1)}{\gamma}} = \frac{T_3}{T_1}$$

$$\therefore (r_p)_{opt} = \left(\frac{T_3}{T_1}\right)^{\frac{\gamma}{2(\gamma-1)}} \quad \textbf{Ans}$$

Problem 13.26 : Show that the mean effective pressure, p_m, for the otto cycle is given by

$$p_m = \frac{\left(P_3 - P_1 r_k^{\gamma}\right)\left(1 - \frac{1}{r_k^{\gamma-1}}\right)}{(\gamma-1)(r_k-1)}$$

where $P_3 = P_{max}$, $P_1 = P_{min}$ and r_k is the compression ratio.

Solution : Given data

$$r_k = \text{Compression ratio} = \frac{V_1}{V_2} = \frac{V_4}{V_3}$$

$$p_{\max} = p_3$$

$$p_{\min} = p_1$$

Net work done in cycle is equal to the area under curve.

$\therefore W_{net}$ = (Area under the curve 3 – 4) – (Area under the curve 1–2)

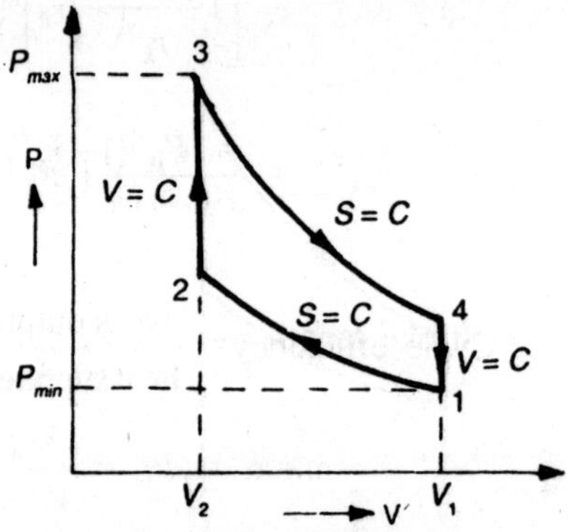

Fig. P. 13.26

$$= \frac{p_3 V_3 - p_4 V_4}{\gamma - 1} - \frac{p_2 V_2 - p_1 V_1}{\gamma - 1}$$

$$= \frac{p_3 V_3}{\gamma - 1}\left[1 - \frac{p_4}{p_3}\frac{V_4}{V_3}\right] - \frac{p_1 V_1}{\gamma - 1}\left[\frac{p_2 V_2}{p_1 V_1} - 1\right]$$

Now For 1-2 process

$$\frac{p_2}{p_1} = \left(\frac{V_1}{V_2}\right)^{\gamma} = (r_k)^{\gamma}$$

$$\therefore \quad \frac{p_2}{p_1} \times \frac{V_2}{V_1} = r_k^{\ \gamma-1}$$

and $$V_2 = \frac{V_1}{r_k} = V_3$$

For process 3-4

$$\frac{p_4}{p_3} = \left(\frac{V_3}{V_4}\right)^{\gamma} = \frac{1}{r_k^{\ \gamma}}$$

$$\therefore \quad \frac{p_4}{p_3}\frac{V_4}{V_3} = \frac{1}{r_k^{\gamma-1}}$$

$$\therefore \quad W_{net} = \frac{p_3}{(\gamma-1)}\frac{V_1}{r_k}\left[1-\frac{1}{r_k^{\gamma-1}}\right] - \frac{p_1 V_1}{(\gamma-1)}\left[r_k^{\gamma-1}-1\right]$$

$$= \left[1-\frac{1}{r_k^{\gamma-1}}\right]V_1\left[\frac{p_3}{r_k} - p_1 r_k^{\gamma-1}\right]\times\frac{1}{(\gamma-1)}$$

$$= \frac{V_1}{(\gamma-1)r_k}[\,P_3 - P_1\,r_k^{\gamma}\,]\left[1-\frac{1}{r_k^{\gamma-1}}\right]$$

$$\text{Stroke length} = V_1 - V_2 = V_1 - \frac{V_1}{r_k} = \frac{V_1}{r_k}[r_k - 1]$$

$$\therefore \quad \text{m.e.p} = P_m = \frac{w_{net}}{V_1 - V_2}$$

$$= \frac{\left(p_3 - p_1 r_k^{\gamma}\right)\left(1-\dfrac{1}{r_k^{\gamma-1}}\right)}{(\gamma-1)(r_k-1)} \quad \textbf{Ans}$$

Problem 13.27 : A gas turbine plant operates on the Brayton cycle using as optimum pressure ratio maximum net work output and a regenerator of 100% effectiveness.

Derive expressions for net work output per kg of air and corresponding efficiency of the cycle in tems of the maximum and the minimum temperatures.

If the maximum and minimum temperatures are 800°C and 30°C respectively, compute the optimum value of pressure ratio, the maximum net work output per kg and the corresponding cycle efficiency.

Solution : Given data :

Let T_5 be the temperature of air leaving the regenerator of 100% effectiveness

$\therefore \quad T_4 = T_5$

$r_p = p_2/p_1$

$\therefore$ For 1-2 process

$$T_2 = T_1 \left(\frac{p_2}{p_1}\right)^{\frac{\gamma-1}{\gamma}}$$

$$= T_1 \, r_p^{\frac{\gamma-1}{\gamma}}$$

For process 3-4 process

$$T_4 = \frac{T_3}{r_p^{\frac{\gamma-1}{\gamma}}}$$

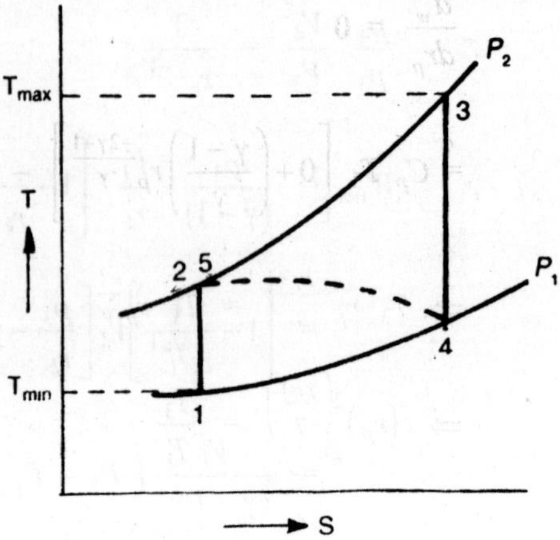

Fig. P. 13.27

$\therefore \quad w_c$ = Compressor work

$$= C_p (T_2 - T_1) = C_p T_1 \left(r_p^{(\gamma-1)/\gamma} - 1\right)$$

w_T = Turbine work

$$= C_p (T_3 - T_4) = C_p \left(1 - \frac{1}{r_p^{\frac{\gamma-1}{\gamma}}}\right)$$

Q_1 = Heat supplied

$$= C_p (T_3 - T_5) = C_p (T_3 - T_4)$$

$$= C_p T_3 \left(1 - \frac{1}{r_p^{\frac{\gamma-1}{\gamma}}}\right)$$

$$\therefore \quad \eta = \frac{w_{net}}{Q_1} = 1 - \frac{T_1}{T_3} r_p^{\frac{\gamma-1}{\gamma}}$$

$$w = w_T - w_C$$

$$= C_P T_3 \left[1 - \frac{1}{r_p^{\frac{\gamma-1}{\gamma}}}\right] - C_p T_1 \left[r_p^{\frac{\gamma-1}{\gamma}} - 1\right]$$

For maximum work output,

$$\frac{d_w}{dr_p} = 0$$

$$= C_p\ T_3\left[0+\left(\frac{\gamma-1}{\gamma}\right)r_p^{\frac{-2\gamma+1}{\gamma}}\right] - C_p\ T_1\left[\left(\frac{\gamma-1}{\gamma}\right)r_p^{\frac{-1}{\gamma}}-0\right]$$

$$\Rightarrow T_3\ r_p^{\frac{-2\gamma+1}{\gamma}} = T_1\ r_p^{\frac{-1}{\gamma}}$$

$$\Rightarrow (r_p)^{2\left(\frac{\gamma-1}{\gamma}\right)} = \frac{T_3}{T_1}$$

$$\Rightarrow \qquad r_p = \left(\frac{T_3}{T_1}\right)^{\frac{\gamma}{2(\gamma-1)}}$$

$$\therefore \qquad W_{max} = C_p\ T_3\left[1-\frac{1}{\left(\frac{T_3}{T_1}\right)^{\frac{1}{2}}}\right] - C_p\ T_1\left[\left(\frac{T_3}{T_1}\right)^{\frac{1}{2}}-1\right]$$

$$= C_p\ T_3\left[1-\frac{\sqrt{T_1}}{\sqrt{T_3}}\right] - C_p\ T_1\left[\frac{\sqrt{T_3}}{\sqrt{T_1}}-1\right]$$

$$= C_p\ \sqrt{T_3}\left(\sqrt{T_3}-\sqrt{T_1}\right) - C_p\ \sqrt{T_1}\left(\sqrt{T_3}-\sqrt{T_1}\right)$$

$$= C_p\ \left(\sqrt{T_3}-\sqrt{T_1}\right)^2 \quad \textbf{Ans}$$

$$\eta_{max} = 1 - \frac{T_1}{T_3}\left(\frac{T_3}{T_1}\right)^{\frac{1}{2}} = 1 - \left(\frac{\sqrt{T_1}}{\sqrt{T_3}}\right) \quad \textbf{Ans}$$

When $T_1 = 30°C = 30 + 273 = 303$ K

$T_3 = 800°C = 800 + 273 = 1073$ K

$$\therefore \qquad (w_{net})_{max} = C_p\ \left(\sqrt{T_3}-\sqrt{T_1}\right)^2$$

$$= 1.005\left[\sqrt{1073}-\sqrt{303}\right] = 236.79 \text{ kJ.kg.}$$

$$\eta_{max} = 1-\sqrt{\frac{T_1}{T_3}} = 1-\sqrt{\frac{300}{1073}} = 0.469 \textbf{ Ans}$$

$$(r_p)_{opt} = \left(\frac{T_3}{T_1}\right)^{\frac{\gamma}{2(\gamma-1)}} = \left(\frac{1073}{303}\right)^{\frac{1.4}{2\times0.4}} = 9.14 \textbf{ Ans}$$

Problem 13.28 : A stationary gas turbine power plant operates on the Brayton cycle and delivers 20 MW to an electric generator. The maximum temperature is 1200 K and the minimum 290 K. The minimum pressure is 95 kPa and the maximum 380 kPa. Determine: (a) the power output of the turbine, (b) the fraction of the output of the turbine used to drive the compressor, (c) the mass flow rate of air to the compressor, and (d) the volume flow rate to the compressor.

Solution : Given data

Power developed

$= 20$ MW

$T_3 = 1200$ K

$T_1 = 290$ K

$p_1 = 95$ kPa

$p_2 = 380$ kPa

Let $\dot{m}$ = mass rate of flow of air

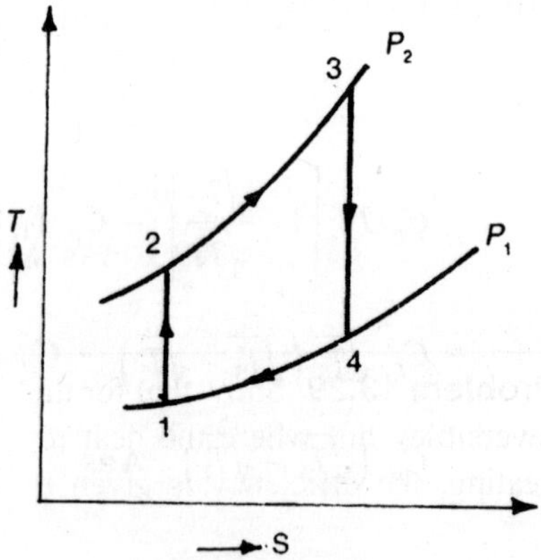

Fig. P. 13.28

For process 1-2

$$\frac{T_2}{T_1} = \left(\frac{p_2}{p_1}\right)^{\frac{\gamma-1}{\gamma}} = \left(\frac{300}{95}\right)^{\frac{0.4}{1.4}}$$

$$T_2 = 431.11 \text{ K}$$

For process 3-4

$$\frac{T_4}{T_3} = \left(\frac{p_1}{p_2}\right)^{\frac{\gamma-1}{\gamma}}$$

$$\Rightarrow \quad T_4 = \frac{T_3}{\left(\frac{p_2}{p_1}\right)^{\frac{0.4}{1.4}}} = \frac{1200}{\left(\frac{380}{95}\right)^{\frac{0.4}{1.4}}} = 807.22 \text{ K}$$

$$w_c = \text{Compressor work} = \dot{m}\, c_p\, (T_2 - T_1)$$

$$= \dot{m} \times 1.005\ (431.11 - 290) = 141.82\ \dot{m}\,\text{kW}$$

$$w_T = \text{Turbine work} = \dot{m}\, c_p\, (T_3 - T_4)$$

$$= m \times 1.005\ (1200 - 807.22)$$

$$= 394.74\ \dot{m}\,\text{kW}$$

$$\therefore \quad w_{net} = w_T - w_C = 252.92 \text{ mkW}$$

Now $$\dot{m} = \frac{20{,}000}{252.92} = 79.1 \text{ kg/sec } \textbf{Ans}$$

(a) Power output of the turbine = 394.74 × 79.1 = 31.22 MW **Ans**

(b) Fraction $$= \frac{141.82}{394.74} = 0.359 \textbf{ Ans}$$

(c) $$\dot{m} = 79.1 \text{ kg/sec}$$

(d) Applying ideal gas eq·

$$p_1 v_1 = RT_1\, \dot{m}$$

or, $$v_1 = \frac{\dot{m}\, RT_1}{p_1} = \frac{79.1 \times 0.287 \times 290}{95}$$

$$= 69.3 \text{ m}^3\text{/s } \textbf{Ans}$$

Problem 13.29 :Show that for the cycle with all the process occuring reversibley but where the heat rejected is not used for regenerative heating, the efficiency is given by

$$\eta = 1 - \frac{\left(\frac{T_1}{T_2} - 1\right) + (\gamma - 1)\ln r}{\left(\frac{T_1}{T_2} - 1\right) + (\gamma - 1)\frac{T_1}{T_2}\ln r}$$

where r is the compression ratio and T_1/T_2 the maximum to minimum temperature ratio.

Solution : Given data

Here without regeneration a point of heat is transfused to constant volume process and constant temperature process T_1 and same for the heat rejection

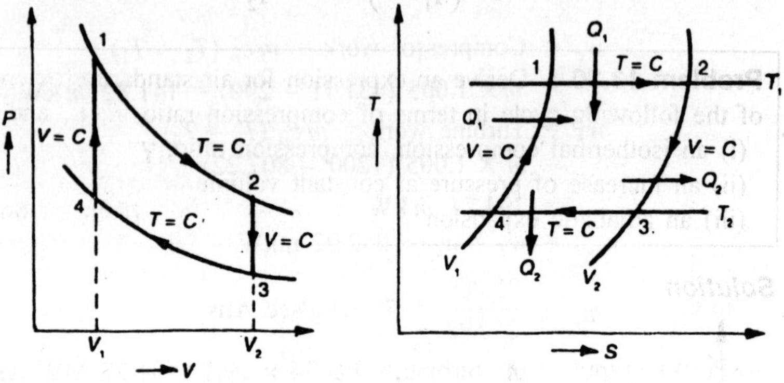

Fig. P. 13.29

$$\therefore \quad Q_{1-2} = w_{1-2} = RT_1 \ln \frac{V_2}{V_1} = RT_1 \ln r$$

$$w_{2-3} = 0,\ Q_{2-3} = -C_v (T_2 - T_1) = C_v (T_1 - T_2)$$

$$Q_{3-4} = w_{3-4} = -RT_2 \ln \frac{V_1}{V_2}$$

$$= RT_2 \ln \frac{V_2}{V_1} = RT_2 \ln r$$

$$Q_{4-1} = C_v (T_1 - T_2),\ W_{4-1} = 0$$

$\therefore$ Heat added $= R\,T_1 \ln r + C_v (T_1 - T_2)$

and heat rejected $= R\,T_2 \ln r + C_v (T_1 - T_2)$

$$\therefore \text{Thermal efficiency} = 1 - \frac{\text{heat rejected}}{\text{heat added}}$$

$$= 1 - \frac{RT_2 \ln r + C_v (T_1 - T_2)}{RT_1 \ln r + C_v (T_1 - T_2)}$$

$$= 1 - \frac{R \ln r + C_v \left(\frac{T_1}{T_2} - 1\right)}{\frac{T_1}{T_2} R \ln r + C_v \left(\frac{T_1}{T_2} - 1\right)}$$

$$= 1 - \frac{\left(\frac{T_1}{T_2} - 1\right) + C_v(\gamma - 1)\ln r}{\left(\frac{T_2}{T_1} - 1\right) + (\gamma - 1)\frac{T_1}{T_2}\ln r}$$

Problem 13.30 : Derive an expression for air standard efficiency of the following cycle in terms of compression ratio, R, C_v and γ

(i) an isothermal compression, compression ratio, γ

(ii) an increase of pressure at constant volume

(iii) an adiabatic expansion (ESE – 1997)

Solution :

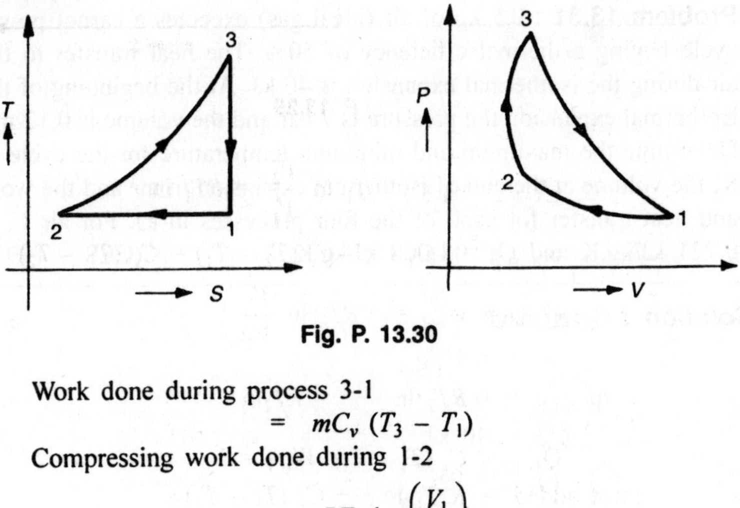

Fig. P. 13.30

Work done during process 3-1

$$= mC_v\,(T_3 - T_1)$$

Compressing work done during 1-2

$$= mRT_1 \ln\left(\frac{V_1}{V_2}\right)$$

Net work output $= mC_v\,T_1\left(\frac{T_3}{T_1} - 1\right) - mRT_1 \ln\left(\frac{V_1}{V_2}\right)$

Work output $= mC_v\,T_1\left(\frac{T_3}{T_1} - 1\right)$

Efficiency, $$\eta = \frac{mC_vT_1\left(\frac{T_3}{T_1} - 1\right) - mRT_1 \ln\left(\frac{V_1}{V_2}\right)}{mC_vT_1\left(\frac{T_3}{T_1} - 1\right)}$$

$$= 1 - \frac{R}{C_v} \frac{\ln r}{\left(\frac{T_3}{T_1} - 1\right)}$$

$$\frac{T_3}{T_1} = \left(\frac{V_1}{V_3}\right)^{\gamma-1} = (r)^{\gamma-1}$$

$$\therefore \quad \eta = 1 - \frac{R}{C_v} \frac{\ln r}{\left(r^{\gamma-1} - 1\right)}$$

Problem 13.31 : 0.5 kg of air (ideal gas) executes a carnot power cycle haging a thermal efficiency of 50%. The heat transfer to the air during the isothermal expansion is 40 kJ. At the beginning of the isothermal expansion the pressure is 7 bar and the volume is 0.12 m^3. Determine the maximum and minimum temperature for the cycle in K, the volume at the end of isothermal expansion in m^3 and the work and heat transfer for each of the four processes in kJ. For air C_v = 0.721 kJ/kg.K and C_p = 1.008 kJ/kg.K. (CES – 1993)

Solution : Given data

$$m = 0.5 \text{ kg}$$

$$\eta_{\text{thermal}} = 0.5$$

$$Q_{in} = 40 \text{ kJ}$$

$$p_1 = 7 \text{ bar}$$

$$V_1 = 0.12 \text{ m}^3$$

$$C_v = 0.721 \text{ kJ/kg.K}$$

$$C_p = 1.008 \text{ kJ/kg.K}$$

To find: (i) T_{max}, T_{min}, V_2

(ii) Work and Heat transfer for all processes

(i) $\quad p_1 V_1 = m RT_1$

$$R = C_p - C_v = (1.008 - 0.721)$$

$$= 0.287 \text{ kJ/kg.K}$$

$$T_{max} = T_1 \text{ or } T_2 = \frac{7 \times 1.03 \times 10^5 \times 10^{-3} \times 0.12}{0.5 \times 0.287}$$

$$= 602.9 \text{ K}$$

$$\eta_{thermal} = 0.5 = \frac{T_{max} - T_{min}}{T_{max}}$$

$$T_{min} = T_3 \text{ or } T_4 = 602.9 - 0.5 \times 602.9$$
$$= 301.45 \text{ K}$$

For the process 1-2

$$Q_{in} = p_1 V_1 \ln \frac{V_2}{V_1}$$

$$40 = 7 \times 1.03 \times 10^5 \times 10^{-3} \times 0.12 \ln \frac{V_2}{0.12}$$

$$V_2 = 0.191 \text{ m}^3$$

Work and Heat transfer for the four process.

Process 1-2

$$Q_{in} = + 40 \text{ kJ}$$
$$\Delta Q = \Delta U + \Delta W$$

As the temperature is constant

$$T_1 = T_2;\ \Delta U = 0$$
$$\Delta W = \Delta Q = + 40 \text{ kJ}$$

Process 2-3

$$\Delta Q = 0 \text{ (Adiabatic process)}$$
$$\Delta Q = \Delta U + \Delta W$$
$$\Delta U = mC_v \Delta T = mC_v (T_3 - T_2)$$
$$= 0.12 \times 0.721 (301.45 - 602.9)$$
$$= - 26.08 \text{ kJ}$$
$$\Delta W = - \Delta U = + 26.08 \text{ kJ}$$

Process 3-4

$$\eta_{thermal} = 0.5 = \frac{Q_{in} - Q_{rejected}}{Q_{in}}$$

$Q_{in} = 40$ kJ Substituting this value in the above equation we get,

$$Q_{rejected} = 20 \text{ kJ}$$
$$\Delta Q = \Delta U + \Delta W$$

$\Delta U = 0$ as this is an isothermal process.

$$- 20 = 0 + \Delta W$$
$$\therefore \quad \Delta W = - 20 \text{ kJ}$$

(Negative sign is used for ΔQ as the heat is rejected by the system)

Process 4-1

$\Delta U = 0$ as this is an isothermal process.

$$\Delta Q = \Delta U + \Delta W$$
$$\Delta U = mC_v\ (T_1 - T_4)$$
$$= 0.12 \times 0.721\ (602.9 - 301.45)$$
$$= 26.08\ \text{kJ}$$
$$\Delta W = -\Delta U = -26.08\ \text{kJ}$$

(ii) Heat and work transfer

Process	*Heat transfer*	*Work transfer*
1-2	+ 40 kJ	+ 40 kJ
2-3	0	+ 26.08 kJ
3-4	– 20 kJ	– 20 kJ
4-1	0	–26.08 kJ

$+\Delta Q \rightarrow$ Heat supplied to the system, $-\Delta Q \rightarrow$ Heat rejected by the system.

$+\Delta W \rightarrow$ Work done by the system, $-\Delta W \rightarrow$ Work done on the system.

Problem 13.32 : One kg of air initially at 20°C and 1 kg/cm^2 is to be reversibly compressed to a final state of 10 kg/cm^2 and 100°C. This may be achieved by any of the following three process:
(i) Heating at constant volume followed by cooling at constant pressure.
(ii) Isothermal compression followed by heating at constant pressure.
(ii) Adiabatic compression followed by cooling at constant volume.
Assuming air to behave like an ideal gas with C_v = 0.1729 kcal/kg.K and C_p = 0.2421 kcal/kg.K, determine the work required, heat transferred and changes in the internal energy and enthalpy of the air for each of the process. Sketch the path followed by each of the process on a single diagram. (CES – 1983)

Solution : Given data

Process 1-a-2 represents constant volume heat addition and constant pressure cooling.

Process 1-b-2 represents isothermal compression and constant pressure heat addition.

Process 1-c-2 represents adiabatic compression and constant volume cooling.

$$p_1 = 1 \text{ kg/cm}^2$$
$$T_1 = 20 + 273 = 293 \text{ K}$$
$$p_2 = 10 \text{ kg/cm}^2$$
$$T_2 = 100 + 273 = 373 \text{ K}$$
$$V_1 = \frac{mRT_1}{P_1}$$

$$= \frac{1(\text{kg})(0.2421 - 0.1729)\frac{\text{kcal}}{\text{kg.K}} \times 293(\text{K})}{1 \times 10^4\left(\frac{\text{kg}}{\text{m}^2}\right)} \times 427\left(\frac{\text{kg.m}}{\text{kcal}}\right) = 0.866 \text{ m}^3$$

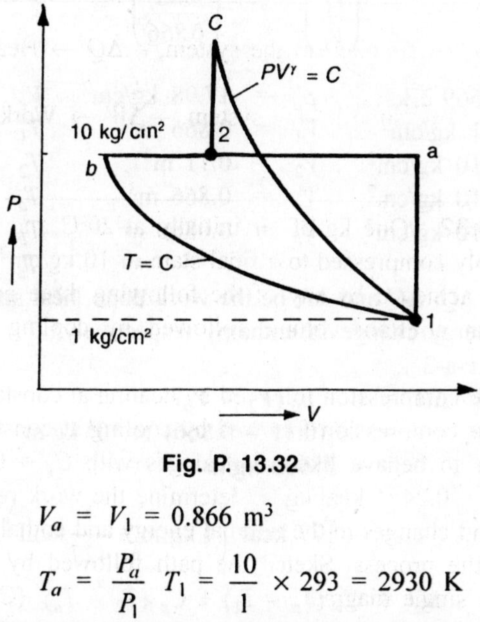

Fig. P. 13.32

$$V_a = V_1 = 0.866 \text{ m}^3$$
$$T_a = \frac{P_a}{P_1} T_1 = \frac{10}{1} \times 293 = 2930 \text{ K}$$
$$V_2 = \frac{T_2}{T_a} V_a = \frac{373}{2930} \times 0.866 = 0.11 \text{ m}^3$$
$$T_b = T_1 = 293 \text{ K}$$
$$V_b = \left(\frac{P_1}{P_b}\right) V_1 = \frac{1}{10} (0.0866) \text{ m}^3$$
$$P_1 V_1^{\gamma} = P_C V_C^{\gamma}$$

$$P_C = P_1 \left(\frac{V_1}{V_c}\right)^{\gamma}$$

$$= 1\left(\frac{0.866}{0.11}\right)^{\frac{0.2421}{0.1729}} = 17.98 \text{ kg/cm}^2$$

$$\frac{T_1}{T_c} = \left(\frac{P_1}{P_c}\right)^{\frac{\gamma-1}{\gamma}} = \left(\frac{V_c}{V_1}\right)^{\gamma-1}$$

$$T_c = \frac{T_1}{\left(\frac{V_c}{V_1}\right)^{\gamma-1}} = \frac{293}{\left(\frac{0.11}{0.866}\right)^{0.4}}$$

T_c = 669.2 k	p_c = 17.98 kg/cm^2	V_c = 0.11 m^3
p_1 = 1 kg/cm^2	V_1 = 0.866 m^3	T_1 = 293 K
p_2 = 10 kg/cm^2	V_2 = 0.11 m^3	T_2 = 373 K
p_a = 10 kg/cm^2	V_a = 0.866 m^3	T_a = 2930 K
p_b = 10 kg/cm^2	V_b = 0.0866 m^3	T_b = 293 K
		T_c = 669.2 K

Let ΔW, ΔQ, ΔU and ΔH be the work done, heat supplied, change in internal energy change in enthalpy respectively.

Process 1-a-2

$$\begin{aligned}
\Delta W &= 0 + p_a (V_2 - V_a) \\
&= 10\ (0.11 - 0.866) \times 10^4 \text{ kg.m} \\
&= -\ 75600 \text{ kg.m} \\
&= -\ 177 \text{ kcal} \quad \left(1 \text{ kg.m} = \frac{1}{427} \text{kcal}\right) \\
\Delta Q &= C_v (T_a - T_1) + C_p (T_2 - T_a) \\
&= 0.1729\ (2930 - 293) + 0.2421\ (373 - 2930) \\
&= 455.94 - 619.04 = -\ 163.11 \text{ kcal} \\
\Delta U &= C_v (T_a - T_1) + C_v (T_2 - T_a) \\
&= C_v (T_2 - T_1) \\
&= 0.1729\ (373 - 293) = +\ 13.83 \text{ kcal} \\
\Delta H &= C_p (T_a - T_1) + C_p (T_2 - T_a) \\
&= C_p (T_2 - T_1) \\
&= 0.2421\ (373 - 293) = +\ 1937 \text{ kcal}
\end{aligned}$$

Process 1-b-2

$$\Delta W = P_1 V_1 \ln\left(\frac{V_b}{V_1}\right) + P_b (V_2 - V_b)$$

$$= 1 \times 10^4 \times 0.866 \ln\left(\frac{1}{10}\right) + 10 \times 10^4 (0.11 - 0.0866)$$

$$= -19940 + 2340 = -17600 \text{ kg.m} = -41.22 \text{ kcal}$$

$$\Delta Q = P_1 V_1 \ln\left(\frac{V_b}{V_1}\right) + C_p (T_2 - T_b)$$

$$= \frac{10^4 \times 0.866 \ln\left(\frac{1}{10}\right)}{427} + 0.2421 (373 - 293)$$

$$= -46.70 + 19.37 \text{ kcal} = -27.33 \text{ kcal}$$

$$\Delta U = 0 + C_v (T_2 - T_b)$$

$$= 0.1729 (373 - 293) = 13.83 \text{ kcal}$$

$$\Delta H = 0 + C_p (T_2 - T_b)$$

$$= 0.2421 (373 - 293) = 19.37 \text{ kcal}$$

Process 1-c-2

$$\Delta W = \frac{P_1 V_1 - P_1 V_2}{\gamma - 1} + 0 = C_v (T_1 - T_c) + 0$$

$$= 0.1729 (293 - 669.2) = -65.04 \text{ kcal}$$

$$\Delta Q = 0 + C_v (T_2 - T_c)$$

$$= 0.1729 (373 - 669.2) = -51.21 \text{ kcal}$$

$$\Delta E = C_v (T_c - T_1) + C_v (T_2 - T_c)$$

$$= C_v (T_2 - T_1)$$

$$= 0.1729 (373 - 293) = 13.83 \text{ kcal}$$

$$\Delta H = C_p (T_c - T_1) + C_p (T_2 - T_c)$$

$$= C_p (T_2 - T_1) = 0.2421 (373 - 293)$$

$$= 19.37 \text{ kcal}$$

Problem 13.33 : An ideal diesel engine operates with air as the working substance. The temperature and pressure of the air at the beginning of the compression process are 25°C and 1.005 bar. The compression ratio is 18 and cut-off occurs at 6.5% of the expansion stroke.

(i) Draw the P-V and T-S diagrams for the cycle indicating clearly each of the process.
(ii) Determine the pressure and temperature at the end of each process.
(iii) Determine the air standard efficiency of the cycle assuming γ = 1.4 for air.
(iv) The work done per cycle (assume C_p = 1.005 kJ/kg.K)
(v) The mean effective pressure. (CES – 1985)

Solution : given data

1-2 Compression process - (Adiabatic)
2-3 Heat addition at constant pressure process
3-4 Expanstion process (Adiabatic)
4-1 Heat rejection at constant volume process
Diesel cycle is represented in *P-V* and *T-S* diagram as shown above

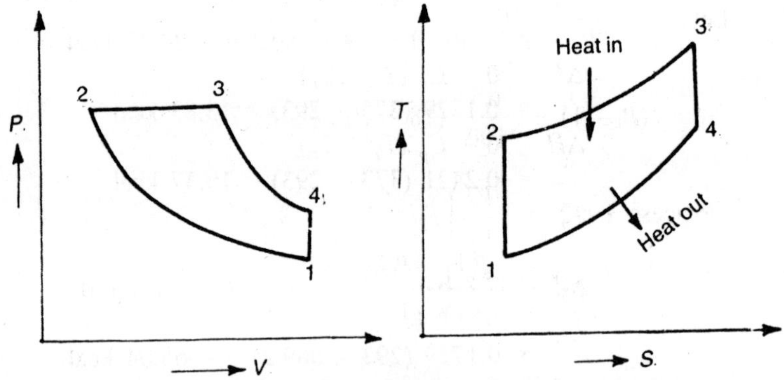

Fig. P. 13.33

$$\frac{T_1}{T_2} = \left(\frac{P_1}{P_2}\right)^{\frac{\gamma-1}{\gamma}} = \left(\frac{V_2}{V_1}\right)^{\gamma-1}$$

$$\frac{V_2}{V_1} = \frac{1}{18}, \gamma = 1.4$$

$$T_2 = \frac{T_1}{\left(\frac{V_2}{V_1}\right)^{\gamma-1}} = \frac{298}{\left(\frac{1}{18}\right)^{0.4}} = 946.9 \text{ K}$$

$$\left(\frac{P_1}{P_2}\right)^{1/\gamma} = \frac{V_2}{V_1}; p_2 = \left(\frac{V_1}{V_2}\right)^{\gamma} p_1 = (18)^{1.4} \times 1.005$$

$$T_1 = 298 \text{ K}$$

$$p_1 = 57.48 \text{ bar} = 1.005 \text{ bar}$$
$$T_2 = 946.9 \text{ K}$$
$$p_2 = 57.48 \text{ bar}$$
$$p_3 = p_2 = 57.48 \text{ bar}$$
$$T_3 = \left(\frac{V_3}{V_2}\right) T_2$$

$$(V_3 - V_2) = \frac{6.5}{100}(V_4 - V_3)$$

Dividing by V_2, we get

$$\left(\frac{V_3}{V_2} - 1\right) = \frac{6.5}{100}\left(\frac{V_4}{V_2} - \frac{V_3}{V_2}\right)$$

Let $\quad \dfrac{V_3}{V_2} = P$

$$(P - 1) = \frac{6.5}{100}(18 - P) \text{ solving } P = 2.03$$

$$T_3 = \left(\frac{V_3}{V_2}\right) T_2 = 2.03 \times 946.9 = 1922.2 \text{ K}$$

$$p_3 = 57.8 \text{ bar}$$
$$T_3 = 1922.2 \text{ K}$$

$$\frac{T_3}{T_4} = \left(\frac{p_3}{p_4}\right)^{\frac{\gamma-1}{\gamma}} = \left(\frac{V_4}{V_3}\right)^{\gamma-1}$$

$$T_4 = \frac{T_3}{\left(\dfrac{V_4}{V_3}\right)^{\gamma-1}} = \frac{1922.2}{\left(\dfrac{V_4}{V_3}\right)^{0.4}}$$

$$(V_3 - V_2) = \frac{6.5}{100}(V_4 - V_3)$$

Divide by V_3

$$\left(1 - \frac{V_2}{V_3}\right) = \frac{6.5}{100}\left(\frac{V_4}{V_3} - 1\right)$$

$$\left(1-\frac{1}{2.03}\right) = \frac{6.5}{100}\left(\frac{V_4}{V_3}-1\right)$$

$$7.81 = \left(\frac{V_4}{V_3}-1\right)$$

$$\frac{V_4}{V_3} = 8.81$$

$$T_4 = \frac{1922.2}{\left(\frac{V_4}{V_3}\right)^{0.4}} = \frac{1922.2}{(8.81)^{0.4}} = 805 \text{ K}$$

$$\frac{T_3}{T_4} = \left(\frac{p_3}{p_4}\right)^{\frac{\gamma-1}{\gamma}} = \left(\frac{V_4}{V_3}\right)^{\gamma-1}$$

$$p_4 = \frac{p_3}{\left(\frac{T_3}{T_4}\right)^{\frac{\gamma}{\gamma-1}}} = \frac{57.48}{\left(\frac{1922.2}{805}\right)^{\frac{1.4}{0.4}}} = 2.732 \text{ bar}$$

$$p_4 = 2.732 \text{ bar}$$
$$T_4 = 805 \text{ K}$$

Air standard efficiency

$$= \frac{\text{Work done}}{\text{Heat supplied}}$$

$$= \frac{\text{Heat input} - \text{Heat output}}{\text{Heat input}}$$

$$= \frac{C_p(T_3-T_2)-C_v(T_4-T_1)}{C_p(T_3-T_2)}$$

$$= 1-\frac{1}{\gamma}\frac{(T_4-T_1)}{(T_3-T_2)}$$

$$= 1-\frac{1}{1.4}\frac{(8.5-298)}{(1922.2-946.9)}$$

$$= 0.6287 = 62.87\ \%$$

Work done/cycle

$$= C_p\ (T_3 - T_1) - C_v\ (T_4 - T_1)$$

$$= 1\ (1922.2 - 946.9) - \frac{1}{1.4}\ (805 - 298)$$

$$= 975.3 - 362.1$$

$$= 613.2 \text{ kJ/kg of air}$$

Mean effective pressure

$$= \frac{\text{Work done}}{\text{Swept volume}} = \frac{613.2 \text{ kJ / kg}}{(V_1 - V_2) \text{ m}^3 \text{ / kg}}$$

$$V_1 = \frac{RT_1}{P_1}$$

$$= \frac{0.286\left(\frac{\text{kJ}}{\text{kg.K}}\right) \times 298(\text{K})}{1.005\left(\frac{\text{kg}}{\text{cm}^2}\right) \times 10^4 \left(\frac{\text{cm}^2}{\text{m}^2}\right)} \times \frac{427}{4.18}\left(\frac{\text{kgm}}{\text{kJ}}\right)$$

$$= \frac{0.286 \times 298 \times 427 \text{ m}^3}{1.005 \times 4.18 \times 10^4 \text{ kg}} = 0.866\ \frac{\text{m}^3}{\text{kg}}$$

Mean effective pressure

$$= \frac{613 \text{kJ / kg}}{\left(0.866 - \frac{0.866}{18}\right)\frac{\text{m}^3}{\text{kg}}} \times \frac{427}{4.18}\left(\frac{\text{kgm}}{\text{kJ}}\right) 10 \times 10^{-3} \text{ kPa}$$

$$= 765.6 \text{ kPa}$$

Problem 13.34 : A reciprocating engine works on the otto cycle where the temperatures at the beginning and end of the compression storke are found to be 40°C and 360°C respectively. Assuming the ratio of the specific heats at constant pressure and constant volume to be 1.4 determine the air standard efficiency and the compression ratio. (CES – 1984)

Solution : Given data :

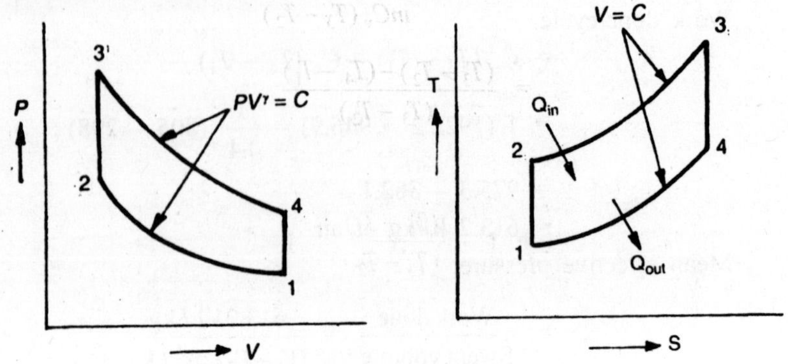

Fig. P. 13.34

$$T_1 = 40 + 273 = 313 \text{ K}$$
$$T_2 = 360 + 273 = 633 \text{ K}$$

For the Process 1-2

$$pV^\gamma = C$$

$$P_1 V^\gamma_1 = P_2 V_2^\gamma \text{ and also } = \frac{p_1 V_1}{T_1} = \frac{p_2 V_2}{T_2}$$

$$\frac{p_1}{P_2} = \left(\frac{V_2}{V_1}\right)^\gamma \qquad \frac{P_1}{P_2} = \left(\frac{V_2}{V_1}\right)\left(\frac{T_1}{T_2}\right)$$

Hence $\left(\frac{V_2}{V_1}\right)^\gamma = \left(\frac{V_2}{V_1}\right)\left(\frac{T_1}{T_2}\right)$

$$\frac{T_1}{T_2} = \left(\frac{V_2}{V_1}\right)^{\gamma-1}$$

$$\left(\frac{313}{633}\right) = \left(\frac{V_2}{V_1}\right)^{\gamma-1}$$

$$\frac{V_1}{V_2} = \text{Compression ratio} = 5.816$$

$$\text{Air standard } \eta = \frac{\text{Heat added} - \text{Heat rejected}}{\text{Heat added}}$$

$$= \frac{mC_v(T_3 - T_2) - mC_v(T_4 - T_1)}{mC_v(T_3 - T_2)}$$

$$= \frac{(T_3 - T_2) - (T_4 - T_1)}{(T_3 - T_2)}$$

$$= 1 - \frac{(T_4 - T_1)}{(T_3 - T_2)} = 1 - \frac{T_1\left(\frac{T_4}{T_1} - 1\right)}{T_2\left(\frac{T_3}{T_2} - 1\right)}$$

$$\frac{T_1}{T_2} = \left(\frac{V_2}{V_1}\right)^{\gamma-1} = \left(\frac{V_3}{V_4}\right)^{\gamma-1} = \frac{T_4}{T_3}$$

as $V_2 = V_3$ and $V_4 = V_1$

Compression ratio

$$= r = \frac{V_1}{V_2} = \frac{V_4}{V_3}$$

$$\eta = 1 - \frac{1}{r^{\gamma-1}} = 1 - \frac{1}{(5.816)^{0.4}}$$

$$= 0.5055 = 50.55\%$$

Chapter 14

Refrigeration Cycles

Problem 14.1 : Develop an expression to calculate ideal COP of a vapour absorption refrigeration system. Describe its components. A vapour absorption system has a source temperature of 110°C and refrigeration temperature of –5°C and amsbient temperature of 35°C. Calculate its COP. IES – 2001

Solution : Given

$T_1 = 110°C$

$T_R = -5°C$

$T_2 = 35°C$

$\therefore$ *COP*

$$= \frac{T_R}{T_2 - T_R} \times \frac{T_1 - T_2}{T_1}$$

$$= \frac{(-5+273)}{(35+5)} \times \frac{110-35}{(110+273)}$$

= 1.312 **Ans**

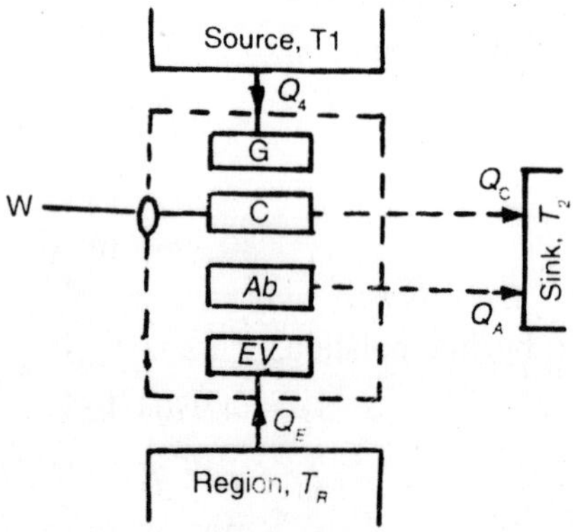

Fig. P. 14.1

Problem 14.2 : A four cylinder single acting ammonia compressor with cylinder dimensions as 7.5 × 10 cm operates at 600 r.p.m. condenser and evaporator pressures are 12 and 2 bar respectively. The vapour from the evaporator to suction of compressor is dry and saturated and there is no undercooling in the condenser. Compression

takes place according to law $PV^{1.2}$ = constant. If clearance is 2% of the stroke, calculate :
(i) Refrigerating capacity in tons of refrigeration.
(ii) Power required to drive the compressor in kW
(iii) Heat rejected to the condenser in kJ/min
Take value of γ for NH_3 as 1.31. (IES – 2000)

Solution : Given :

dia. of cylinder = d = 7.5 cm

stroke length = L = 10 cm

N = 600 rpm

p_1 = evaporate pressure = 2 bar

p_2 = condenser pressure = 12 bar

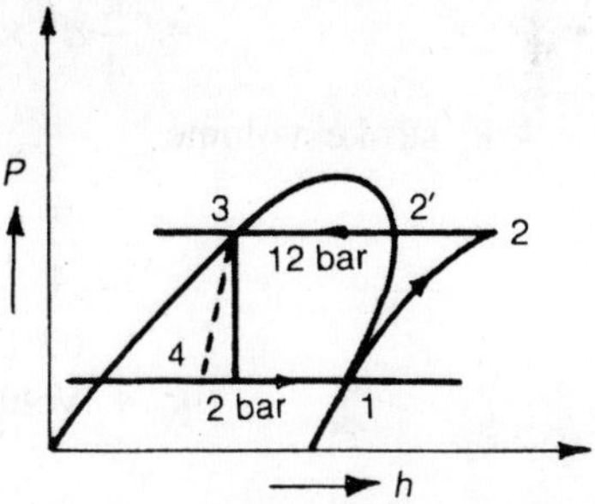

Fig. P. 14.2

Compression law,

$$pV^{1.2} = C$$

$$v_c = 2\%\ v_s$$

Now from Ammonia chart

at 2 bar

by interpolation,

$$T_1 = -18.882°C$$

$$h_1 = 1420.51 \text{ kJ/kg}$$

$$s_1 = 5.6035 \text{ kJ/kg.K} = s_2$$

$$v_1 = 0.5961 \text{ m}^3\text{/kg}$$

At 12 bar

by interpolation,

$$S_2' = 4.9703 \text{ kJ/kg.K}$$

$$t_{sat} = t_3 = t_2' = 30.95°C$$

$$h_3 = 327.48 \text{ kJ/kg} = h_4$$

Now $$R = \frac{\bar{R}}{17} = \frac{8.314}{17} = 0.489 \text{ kJ/kg.K}$$

$$C_p = \gamma C_v = 1.31\ C_v$$

$$\therefore \quad C_v = \frac{R}{\gamma - 1} = \frac{0.4891}{0.31} = 1.57761 \text{ kJ/kg.K}$$

$$C_p = \gamma C_v = 2.0982 \text{ kJ/kg.K}$$

$\therefore$ from property relation

$$S_2 - S_2' = C_p \ln \frac{t_2}{t_2'}$$

or $t_2 = 138.0°C$

∴ from superheated table

at 12 bar and 138°C

∴ $h_2 = 1753.93$ kJ/kg

$$= \frac{\pi}{4} d^2 \times L\ N \times n$$

V_s stroke volume

$$= \frac{\pi}{4} (7.5)^2 \times 10 \times 4 \times \frac{600}{60} \times 10^{-6} \text{ m}^3\text{/sec}$$

$$= 0.01767 \text{ m}^3\text{/sec}$$

η_v = Volumetric efficiency

$$= 1 + k - k \left(\frac{p_2}{p_1}\right)^{1/n}, \ k = \frac{V_c}{V_s}$$

$$= 1 + 0.02 - 0.02 \left(\frac{12}{6}\right)^{\frac{1}{1.2}} = 0.9844$$

∴ Actual volume flow rate

$$= \eta_v \times V_s = 0.017395 \text{ m}^3\text{/sec}$$

∴ mass of flow rate

$$= \dot{m} = \frac{\text{Volume flow rate}}{v_1}$$

$$= \frac{0.017394}{0.5961} = 0.0292 \text{ kg/sec}$$

(i) Refrigerating capacity

$$= \dot{m} = (h_1 - h_4)$$

$$= 0.0292\ (1420.51 - 327.48) = 31.916 \text{ kW}$$

∴ tonne of refrigerating

$$= \frac{31.916}{3.517} = 9.08 \textbf{ Ans}$$

(ii) Power required to drive the compressor

$$= \dot{m}\ (h_2 - h_1)$$

$= 0.0292\ (1753.93 - 1420.51)$
$= 9.74$ kW **Ans**

(iii) heat rejected in condenser

$= 60 \times 0.0292\ (h_2 - h_3)$
$= 60 \times 0.0292\ (1753.93 - 327.98)$
$= 2499.14$ kJ/min **Ans**

Problem 14.3 : A food storage locker requires a refrigeration capacity of 50 kW. It works between a condenser temperature of 35°C an evaporator temperature of –10°C. The refrigerant is ammonia. It is subcooled by 5°C before entering the expansion valve by the dry saturated vapour leaving the evaporator. Assuming a single cylinder, single-acting compressor operating at 1000 rpm with stroke equal to 1.2 times the bore, determine (a) the power required, and (b) the cylinder dimensions.

Saturation	Temperature °C	–10	35
Pressure	bar	2.9157	13.522
Enthalpy kJ/kg	Liquid	154.056	366.072
	Vapour	1450.22	1488.57
Specific volume m^3/kg	Liquid	—	1.7023
	Vapour	0.417477	0.095629
Specific heat kJ/kg.K	Liquid	—	4.556
	Vapour	2.492	2.903

(IES – 1997)

Solution : Given

Refrigerator capacity

$= 50$ kW

$N = 1000$ rpm

$L =$ stroke $= 1.2 \times$ bore (d)

at –10°C

$h_1 = 1450.22$ kJ/kg

$s_1 = 5.755$ kJ/kg.K $= s_2$

$v_1 = 0.417477$ m^3/kg

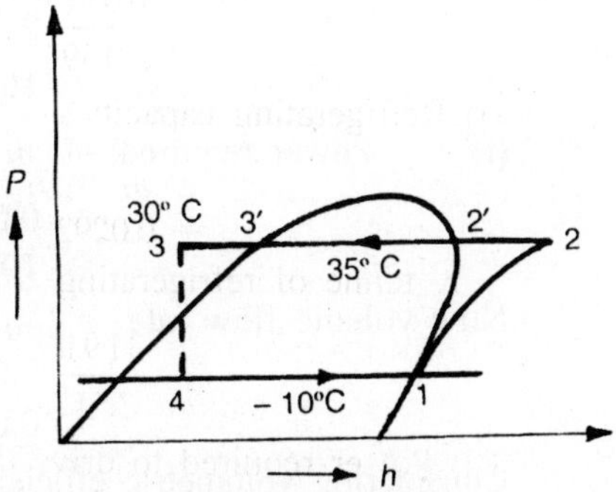

Fig. P. 14.3

at 35°C

$$h_2' = 1488.57 \text{ kJ/kg}$$
$$s_2' = 5.2086 \text{ kJ/kg.K}$$
$$Cp_v = 2.903 \text{ kJ/kg.K}$$

$\therefore$ from property relation at constant pressure

$$s_2 - s_2' = Cp_v \ln \frac{T_2}{T_2'}$$

or $5.755 - 5.2086 = 2.903 \ln \dfrac{t_2}{(35+273)}$

$\therefore$ $t_2 = 371.79$ K

Now $h_3' =$ at 35°C $= 366.072$ kJ/kg

and $Cp_w = 4.556$ kJ/kg.K

$$\therefore \quad h_3 - h_3' = Cp_w (T_3 - T_3')$$
$$= 4.556 (30 - 35) = -22.78$$

Assuming Cp_w constant upto 3 to 31

$\therefore$ $h_3 = 343.292$ kJ/kg $= h_4$

$\therefore$ Refrigerating load $= (h_1 - h_4)$

$= (1450.22 - 343.292) = 1106.93$ kJ/kg

$\therefore$ mass of refrigerating flow

$$= \frac{\text{Refrigerating capacity}}{(h_1 - h_4)}$$

$$= \frac{50}{1106.93} = 0.0452 \text{ kg/sec}$$

$$h_2 - h_2' = Cp_v (T_2 - T_2')$$
$$= 2.903 [371.79 - (35 + 273)] = 183.183$$

$\therefore$ $h_2 = 1673.75$ kJ/kg

(i) Power required $= \dot{m} (h_2 - h_1)$

$= 0.0452 (1673.75 - 1450.22)$

$= 10.104$ kW **Ans**

Now volume flow rate $= \dot{m} . v_1$

$= 0.0452 \times 0.417477$

$= 0.0189$ m³/sec

Considering volumetric efficiency 100 × volume displacement

$= 0.0189$ m³/sec

$$\text{or} \quad \frac{\pi}{4} d^2 LN = 0.0189$$

$$\therefore \quad d^2 L = 1.442 \times 10^{-3}$$

$$d^2 \times 1.2d = 1.442 \times 10^{-3}$$

$$\therefore \quad d = 0.1063 \text{ m} = 10.63 \text{ cm}$$

$$\therefore \quad L = 0.1276 \text{ m} = 12.76 \text{ cm} \quad \textbf{Ans}$$

Problem 14.4 : A Freon-12 refrigerator producing a cooling effect of 20 kJ/s operates on a simple cycle with pressure limits of 1.509 bar and 9.607 bar. The vapour leaves the evaporator dry saturated and there no undercooling. Determine the power required by the machine. If the compressor operates at 3000 rpm and has a clearance volume of 3% of stroke volume determine the piston displacement of the compressor. For compressor assume that the expansion follows the law $pv^{1.13}$ = constant. (IES – 1993)

Given :

Temp.	p_s	v_g	Enthalpy	kJ/kg	Entropy	kJ/kg K	Specific heat kJ/kg.K
°C	bar	m³/kg	h_f	h_g	s_f	s_g	s_g
–20	1.509	0.1088	17.8	178.61	0.073	0.7082	
40	9.607		74.53	203.05	0.2716	0.682	0.747

Solution : Given :

Cooling effect = 20 kJ/s

N = 3000 rpm

v_c = 3% V_s

Compression law,

$pV^{1.3}$ = const.

at 1.509 of bar

h_1 = 178.61 kJ/kg

s_1 = 0.7082 kJ/kg.K = s_2

v_1 = 0.1088 m³/kg

at 9.607 bar

h_2' = 203.05 kJ/kg

s_2' = 0.682 kJ/kg.K

h_3 = 74.53 kJ/kg

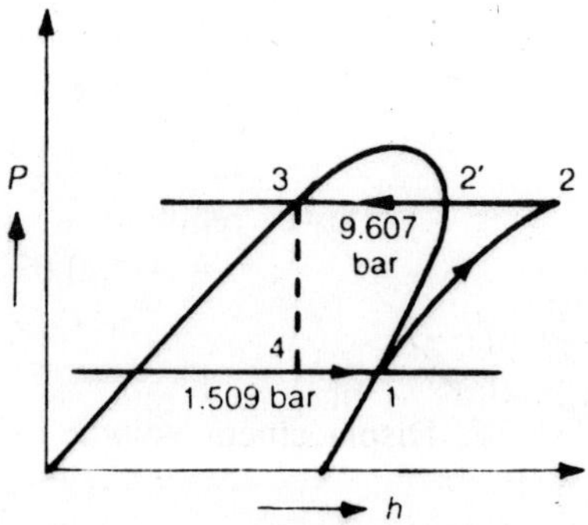

Fig. P. 14.4

From property relation at constant pressure

$$s_2 - s_2' = C_p \ln \frac{t_2}{t_2'}$$

$$\Rightarrow \quad 0.7082 - 0.682 = 0.747 \ln \frac{t_2}{(40+273)}$$

$$t_2 = 324.173 \text{ K}$$

$$\therefore \quad h_2 - h_2' = C_p (t_2 - t_2')$$

$$= 0.747 [324.173 - (40 + 273)]$$

$$\therefore \quad h_2 = 211.4 \text{ kJ/kg}$$

∴ Cooling effect

$$= (h_1 - h_4)$$

$$= (178.61 - 74.53) = 104.08 \text{ kJ/kg}$$

∴ mass flow rate

$$= \dot{m} = \frac{\text{Cooling effect}}{h_1 - h_4}$$

$$= \frac{20}{104.08} = 0.1922 \text{ kg/sec}$$

∴ Volume flow rate

$$= \dot{m} v_1 = 0.1922 \times 0.1088$$

$$= 0.02091 \text{ m}^3/\text{sec}$$

∴ Power required

$$= \dot{m} (h_2 - h_1)$$

$$= 0.1922 (211.4 - 178.61) = 6.30 \text{ kW } \textbf{Ans}$$

$$\eta_v = \text{volumetric efficiency}$$

$$= 1 + K - K \left(\frac{P_2}{P_1}\right)^{1/n}, \quad K = \frac{V_c}{V_s}$$

$$= 1 + 0.03 - 0.03 \left(\frac{9.607}{1.509}\right)^{\frac{1}{1.13}} = 87.564\%$$

$$\therefore \text{Displacement volume} = \frac{\text{Volume flow rate}}{\eta_v}$$

$$= \frac{0.02091}{0.87564} = 0.0238797 \text{ m}^3/\text{sec}$$

or $\frac{\pi}{4} \times d^2 \times L \times N = 0.0238797$

suppose $\frac{d}{L} = 1$

then $d = 0.08472\ m = 8.472$ cm

$L = 8.472$ cm **Ans**

Problem 14.5 : A vapor compression heat pump is driven by a power cycle having a thermal efficiency of 25%. For the heat pump condenser pressure of 12 bars. The isentropic efficiency of the compressor is 80%. Saturated liquid enters the expansin valve at 12 bars. For the power cycle, 80% of the heat rejected by it is transferred to the heated space which has a total heating requirement of 500 kJ/min. Determine the power input to the heat pump compressor. The following data for refrigerant 12 may be used. (IES – 1995)

Pressure, bar	Temperature °C	Enthalpy, Liquid	kJ/kg Vapor	Entropy, Liquid	kJ/kg.K Vapor
2.0	–12.53	24.57	182.07	0.0992	0.7035
12.0	49.31	84.21	206.24	0.3015	0.6799

Vapor specific heat at constant pressure = 0.7 kJ/kg.K

Solution : Given

$\eta_{isen} = 80\%$

$Q'_c = 500$ kJ/min

From table

$h_1 = 182.07$ kJ/kg

$h_3 = h_4 = 84.21$ kJ/kg

$h''_2 = 206.24$ kJ/kg

$s_1 = 0.7035$ kJ/kg.K $= s_2$

$s''_2 = 0.6799$ kJ/kg.K

At constant pressure

$$s_2 - s''_2 = C_p \ln \frac{t_2}{t''_2}$$

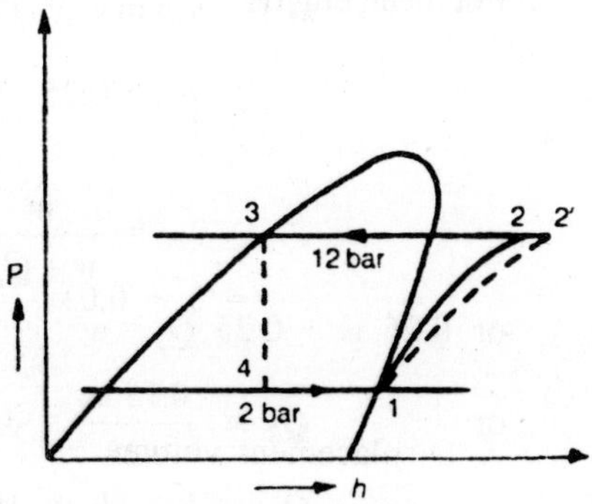

Fig. P. 14.5 (a)

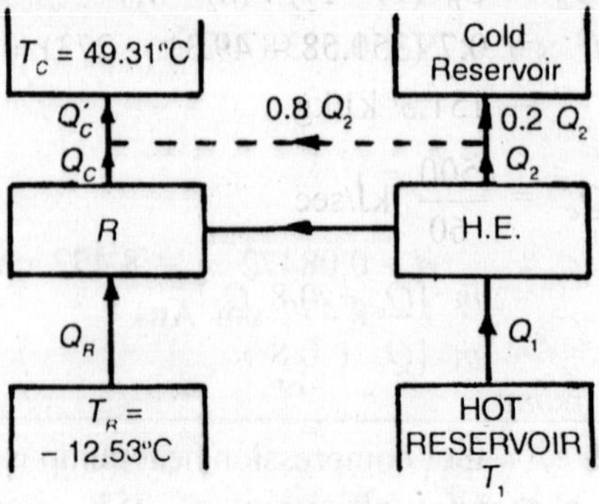

Fig. P. 14.5 (b)

or $\quad 0.7035 - 0.6799 = 0.7 \ln \dfrac{t_2}{(49.31+273)}$

or $\quad t_2 = 333.36 \text{ K}$

$$\eta_{isen} = 0.8 = \frac{T_2 - T_1}{T_2' - T_1}$$

or $\quad T_2' - T_1 = \dfrac{T_2 - T_1}{0.8}$

$$= \frac{333.36 - 260.47}{0.8}$$

$$= 91.115$$

$\therefore \quad T_2' = 351.58 \text{ K}$

For heat engine

$$\eta_{th} = 0.25 = \frac{w}{Q_1}$$

$$= \frac{w}{w + Q_2}$$

or $\quad 0.25\, w + 0.25\, Q_2 = w$

or $\quad Q_2 = \dfrac{0.75\, w}{0.25} = 3w$

$$Q_R = h_1 - h_4 = 182.07 - 84.21$$

$$= 97.86 \text{ kJ/kg}$$

$$Q_c = C_p\ (T_2' - T_2'') + (h_2'' - h_3)$$
$$= 0.7\ (351.58 - 49.31 - 273) + (206.24 - 84.21)$$
$$= 151.3 \text{ kJ/kg}$$

Now $$Q_c' = \frac{500}{60} \text{ kJ/sec}$$
$$= \dot{m}\ [Q_c + 0.8\ Q_2]$$
$$= \dot{m}\ [Q_c + 0.8 \times 3w] \quad \text{...(i)}$$

or *For heat pump*

$$w + Q_R = Q_c$$

or $$w = Q_c - Q_R$$
$$= 151.3 - 97.86$$
$$= 53.44 \text{ kJ/kg}$$

$\therefore$ From (i) $$\dot{m} = \frac{500}{60\left[Q_c + 2.4w\right]}$$
$$= \frac{500}{60\left[151.3 + 2.4 \times 53.44\right]}$$
$$= 0.02981 \text{ kg/sec}$$

$\therefore$ $$w = 53.44 \times 0.02981 \text{ kW}$$
$$= 1.593 \text{ kW } \textbf{Ans}$$

Problem 14.6 : In a residential heat system the refrigerant flow rate is 412.2 kg/hr and the electric energy supplied to the compressor is 25,480 kJ/hr. The properties of the refrigerant at the inlet (state 1) and at exit (state 2) of the compressor are as given below :

State	t (°C)	p (bars)	h (kJ/kg)	s (kJ/kg.K)
1	−1	2.6	226.35	0.8636
2	93	15.0	286.40	0.9222

Determine the amount of available energy destroyed in the compressor in kJ/hr. Al lheat transfer, if it occurs, takes place at T_0. Assume T_0 = 273 K, p_0 = 1 atm. (IES – 1994)

Solution : Given :

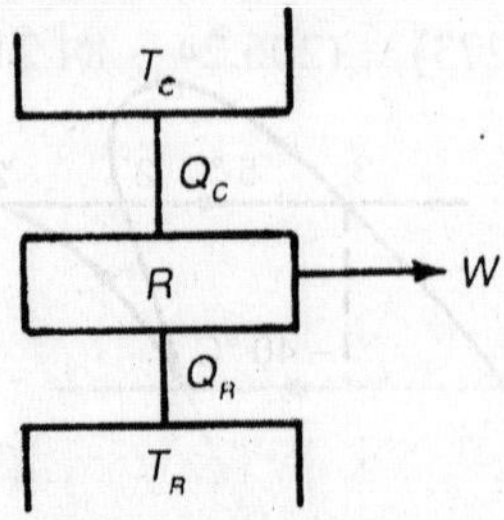

Fig. P. 14.6 (a)

$\dot{m}$ = refrigerant flow rate
= 412.2 kg/hr

w = 25,480 kJ/hr

Available energy of the compressor

$= \dot{m}\ [(h_2 - h_1) - T_0\ (s_2 - s_1)]$
$= 412.2\ [(286.4 - 226.35)$
$- 273\ (0.9222 - 0.8636)]$
$= 18{,}158.32$ kJ/hr

$W_{actual} = 25{,}480$ kJ/hr

∴ Available energy lost
$= W_{actual}$ − Available energy of the compressor
$= 25{,}480 - 18{,}158.32 = 7321.68$ kJ/hr **Ans**

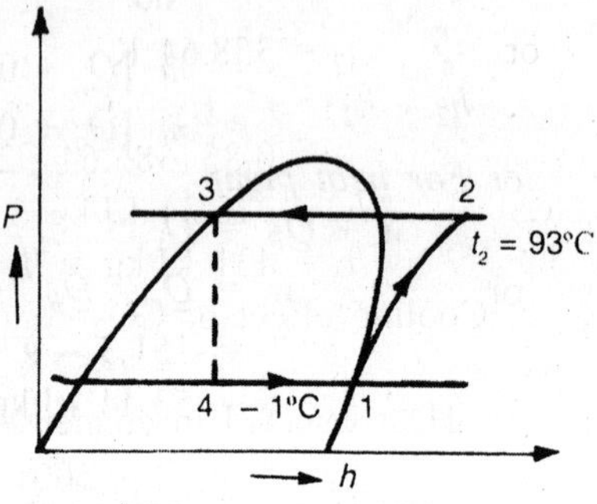

Fig. P. 14.6 (b)

Problem 14.7 : A cascade refrigeration system of 100 tons (350 kW) capacity uses amoña and carbon dioxide. The evaporating and condensing temperatures of CO_2 are—40°C and 5°C, respectively. The evaporating temperature of NH_3 is—7°C. The power supplied to the ammonia compressor is 96.5 kW. In the CO_2 circuit, the liquid leaving the condenser is saturated, the vapor leaving the evaporator is dry and saturated, and compression is isentropic. Calculate the mass flow rate of CO_2 and the C.O.P. of the system. The following data for CO_2 may be used : (IES – 1994)

Temperature, °C	Pressure, bars	Enthalpy, kJ/kg Liquid	Enthalpy, kJ/kg Vapor	Entropy, kJ/kgK Liquid	Entropy, kJ/kgK Vapor
–40	10.05	332.7	652.8	3.8531	5.2262
5	39.70	431.0	649.8	4.2231	5.0097

Specific heat of superheated vapor = C_p = 0.85 kJ/kg K

Solution : Given

Refrigerating capacity = 350 kW

w_{input} in ammonia system = 96.5 kW

For CO_2 *system* $h_1 = 652.8$ kJ/kg

$h_2' = 649.8$ kJ/kg

$s_1 = s_2 = 5.2262$ kJ/kg.K
$s_2' = 5.0097$ kJ/kg.K
$t_2' = 5°C = 278$ K

$\therefore \quad s_2 - s_2' = C_p \ln \frac{t_2}{t_2'}$

or $\quad t_2 = 358.64$ K

$\therefore \quad h_2 - h_2' = C_p (t_2 - t_2')$
$= 0.85 (385.64 - 278)$

$\therefore \quad h_2 = 718.34$ kJ/kg
$h_3 = 431$ kJ/kg $= h_4$

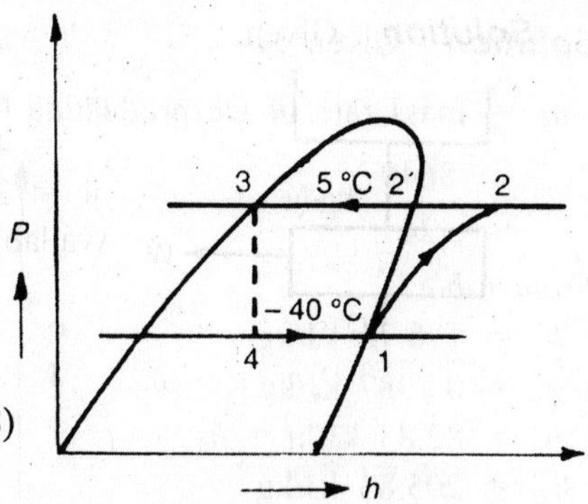

Fig. P. 14.7

$\therefore$ Cooling effect of CO_2 system $= h_1 - h_4$
$= (652.8 - 431) = 221.8$ kJ/kg

Heat rejected in condenser
$= h_2 - h_3 = 718.34 - 431 = 287.34$ kJ/kg

Let $\dot{m}_c$ = mass flow rate CO_2 and $\dot{m}_N$ be mass flow rate of NH_3.

Now for ammonia system

Cooling effect of NH_3 system
$= 287.34$ kJ/kg

Problem 14.8 : A vapour compression refrigeration system using refrigerant 12 is employed to produce 8640 kg of ice per day. The condensing and evaporating temperatures of the refrigerant are 48°C and –20°C, respectively. Saturated liquid leaves the condenser and saturated vapour leaves the evaporator. Compression is isentropic water at 35°C is used to form ice. The temperature of ice should be –8°C. Heat flows into the brine tank from the surroundings may be taken to be 10% of the total heat removed from water to form ice at the specified temperature. Determine the power required to drive the compressor, in kW. Take specific heat of ice = 2.26 kJ/kg. K, latent heat of ice = 334.72 kJ/kg and specific heat of water = 4.187 kJ/kg. K. The following data for refrigerant 12 may be used :

Temperature, °C	Pressure, bars	Enthalpy, kJ/kg Liquid	Enthalpy, kJ/kg Vapor	Entropy, kJ/kg.K Liquid	Entropy, kJ/kg.K Vapor
48	11.64	82.83	205.83	0.2973	0.6802
–20	1.51	17.82	178.74	0.0731	0.7087

Specific heat of vapour = 0.82 kJ/kg.K (IES – 1993)

Solution : Given :

$\dot{m}$ = mass rate of ice production

$= \dfrac{8640}{24}$ kg/hr

From table

h_1 = 178.74 kJ/kg

s_1 = 0.7087 kJ/kg.K = s_2

h_3 = 82.83 kJ/hr = h_4

h_2' = 205.83 kJ/kg

s_2' = 0.6802 kJ/kg.K

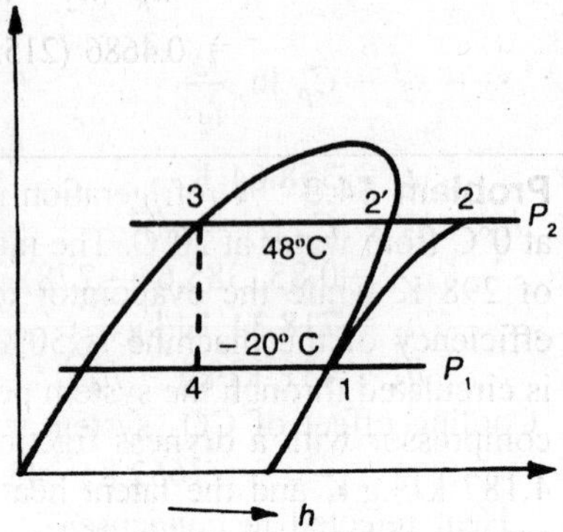

Fig. P. 14.8

For constant pressure system

$$s_2 - s_2' = C_p \ln \frac{t_2}{t_2'}$$

$$\Rightarrow \quad 0.7087 - 0.6802 = 0.82 \times \ln \frac{t_2}{(48+273)}$$

$$\therefore \quad t_2 = 332.35 \text{ k}$$

$$\therefore \quad h_2 = h_2' + C_p (t_2 - t_2')$$

$$= 205.83 + 0.82\ (332.35 - 321)$$

$$= 215.14 \text{ kJ/kg}$$

Now heat taken out from the evaporator

$$= \dot{m}_R\ (h_1 - h_4) + 0.1Q_2$$

$$= \dot{m}_R\ (178.74 - 82.83) + 0.1Q_2$$

$$= 0.1Q_2 + 95.91\ \dot{m}_R \quad \text{...(i)}$$

[where $\dot{m}_R$ = mass flow rate of refrigerant]

∴ heat rejected to produce 8640 kg of ice per day

$$Q_2 = \dot{m}\ [Cp_w\ (35 - 0) + L_{ice} + Cp_{ice}\ (8)]$$

$$= \frac{8640}{24}\ [4.187 \times 35 + 334.72 + 2.26 \times 8]$$

$$= 49.9345 \text{ kJ/sec} \quad \text{...(ii)}$$

Equating (i) and (ii) we get

$$\dot{m}_R = \frac{0.9 \times 49.9345}{95.91} = 0.4686 \text{ kg/s}$$

∴ Power required to drive the compressor

$$= \dot{m}_R \ (h_2 - h_1)$$

$$= 0.4686 \ (215.14 - 178.74) = 17.06 \text{ kW } \textbf{Ans}$$

Problem 14.9 : A refrigeration machine is required to produce ice at 0°C from water at 20°C. The machine has a condenser temperature of 298 K while the evaporator temperature is 268 K. The relative efficiency of the machine is 50% and 6 kg of Freon-12 refrigerant is circulated through the system per minute. The refrigerant enters the compressor with a dryness fraction of 0.6. Specific heat of water is 4.187 kJ/kg.K and the latent heat of ice is 335 kJ/kg. Calculate the amount of ice produced in 24 hours. The table of properties of Freon-12 is given below : (IES – 1992)

Temperature K	Liquid heat kJ/kg	Latent heat kJ/kg	Entropy of liquid kJ/kg.K
298	59.7	138.0	0.2232
268	31.4	154.0	0.1251

Solution : Given :

Relative efficiency of m/c = 50%

$\dot{m}$ = mass flow rate of refrigerant

= 6 kg/min

From table

$$h_3 = h_4 = 59.7 \text{ kJ/kg}$$

$$h_2' = 197.7 \text{ kJ/kg}$$

$$h_1 = h_{f_{268K}} + x \ h_{fg_{268\,K}}$$

$$= 31.4 + 0.6 \times 154$$

$$= 123.8 \text{ kJ/kg}$$

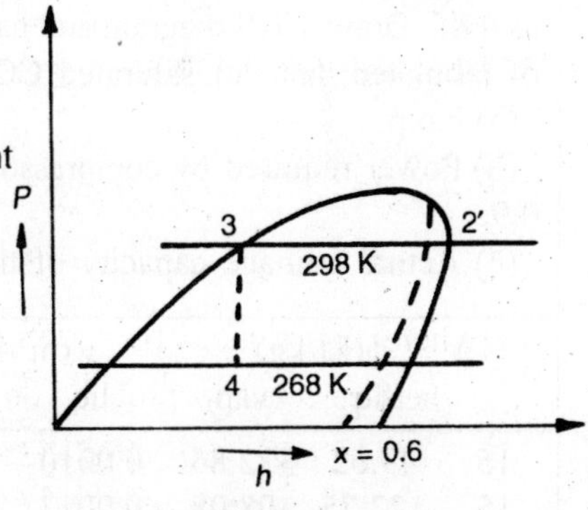

Fig. P. 14.9

∴ Refrigerating effect $= h_1 - h_4$

$$= 123.8 - 59.7 = 64.1 \text{ kJ/kg}$$

or Refrigerating effect $= \dot{m} \times 64.1$

$$= \frac{6}{60} \times 64.1 = 6.41 \text{ kW} \quad \ldots(i)$$

Now heat rejected to produce $\dot{m}_1 \ \frac{kg}{sec}$ of water into ice at 0°C

$$= \dot{m}_1 [Cp_w (20 - 0) + L]$$
$$= \dot{m}_1 \ [4.187 \times 20 + 335]$$
$$= 418.74 \ \dot{m}_1 \ kW \qquad ...(ii)$$

or Equating (i) and (ii)

$$\dot{m}_1 = \frac{6.41}{418.74}$$
$$= 0.01531 \text{ kg/sec} \times 3600 \text{ sec/hr} \times 24 \text{ hr/day}$$
$$= 1322.6 \text{ kg/day } \textbf{Ans}$$

Problem 14.10 : The compressor of a refrigeration plant having stroke volume of 500 c.c. runs at 500 rpm. and works with CO_2 gas. The evaporator and condenser temperatures are –15°C and 25°C respectively. The liquid in the condenser (using another Fr-12 refrigerating system) is undercooled to 15°C, before passing on to the expansion valve. Assume isentropic compression with an initial quality of CO_2 as $x = 1$ and volumetric efficiency of the compressor as 0.85. Draw a T-S diagram and calculate the following with the help of tabulated data for saturated CO_2 —

(a) c.o.p.

(b) Power required by compressor assuming motor efficiency to be 0.9.

(c) Actual tonnage capacity of the compressor

t (C)	h(kJ/kg)		v (m³/kg)		s (kJ/kg C)		p (bar)
	liq	vap	liq	vap	liq	vap	
15	49.62	322.86	0.0010	0.0166	0.1976	1.2567	22.88
15	127.75	308.08	0.0013	0.0063	0.4697	1.0959	50.92
25	164.17	283.63	0.0015	0.0042	0.5903	0.9912	64.32

(Specific heat of vapour CO_2 = 2.4 kJ/kg.C) (IES – 1990)

Solution : Given

$$\eta_v = \text{volumetric efficiency} = 0.85$$
$$v_s = 500 \text{ cm}^3 = 500 \times 10^{-6} \text{ m}^3$$
$$N = 500 \text{ rpm}$$

∴ Volume flow rate

$$= V_s \times \eta_v \times \frac{500}{60}$$

$$= 0.003542 \text{ m}^3/\text{sec}$$

Now from table

$h_1 = 322.86$ kJ/kg

$v_1 =$ sp. volume of vapour

$= 0.0166$ m³/kg

$s_1 = 1.2567$ kJ/kg.K $= s_2$

$s_2' = 0.9912$ kJ/kg.K

$h_2' = 283.63$ kJ/kg

$h_3 = 127.75$ kJ/kg $= h_4$

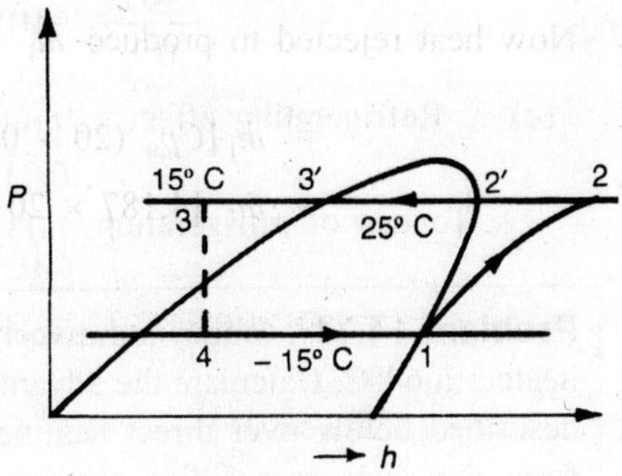

Fig. P. 14.10

$$\text{Now } s_2 - s_2' = C_p \ln \frac{t_2}{t_2'}$$

$$\text{or} \quad 1.2567 - 0.9912 = 2.4 \times \ln \frac{t_2}{(25+273)}$$

$$t_2 = 332.86 \text{ K}$$

$$\therefore \quad h_2 = h_2' + C_p (t_2 - t_2')$$

$$= 367.29 \text{ kJ/kg}$$

Now $\dot{m}$ = mass flow rate

$$= \frac{\text{Volume flow rate}}{v_1}$$

$$= \frac{0.003542}{0.0166} = 0.2134 \text{ kg/sec}$$

(a) $$\text{C.O.P.} = \frac{h_1 - h_4}{h_2 - h_1}$$

$$= \frac{322.86 - 127.75}{367.29 - 322.86}$$

$$= 4.3914 \textbf{ Ans}$$

(b) Compressor work $= \dot{m}\,(h_2 - h_1)$

$$= 0.2134\,(367.29 - 322.86) = 9.4814 \text{ kW}$$

∴ Power required by compressor

$$= \frac{9.4814}{0.9} = 10.535 \text{ kW } \textbf{Ans}$$

(c) $\therefore$ Refrigerating effect $= \dot{m}\ (h_1 - h_4)$

$$= 0.2134\,(322.86 - 127.75) = 41.743 \text{ kW}$$

$\therefore$ Tonnes of refrigeration $= 11.87$ **Ans**

Problem 14.11 : Water is drawn from a river at 7°C and has to be heated to 64°C. Calculate the advantage of using the heat pump plant described below over direct heating of the water. The water drawn from the river passes first through the Freon condenser and then through the steam condenser. Assume that a heat source above 450°C is available.

Freon-12 Heat Pump compressor is driven by steam turbine. Its condenser temperature 60°C, and evaporator temperature —5°C. Freon-12 enters compressor as saturated vapour and enters throttle value as saturated liquid.

Steam pressure and temperature leaving boiler 30 kg/cm² and 450°C respectively. Steam condenser temperature 76°C.

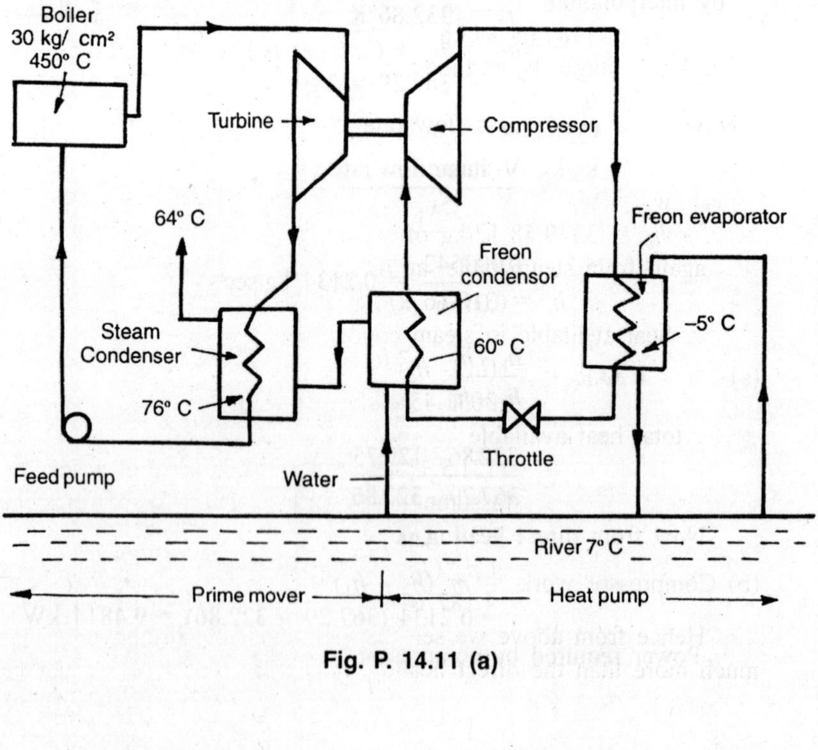

Fig. P. 14.11 (a)

Enthalpy values of Freon-12 : At entry to compressor 187 kJ/kg, at discharge of compressor 221.5 kJ/kg and at exit of condenser 90 kJ/kg. (4.18 kJ = 1 kcal).

Solution : Given :

h_1 = 187 kJ/kg

h_2 = 221.5 kJ/kg

h_3 = 90 kJ/kg

∴ work input in compressor

= $h_2 - h_1$

= 221.5 – 187

= 34.5 kJ/kg of refrig.

∴ heat gained in from condensor

= $(h_2 - h_3)$

= (221.5 – 90)

= 131.5 kJ/kg of refri.

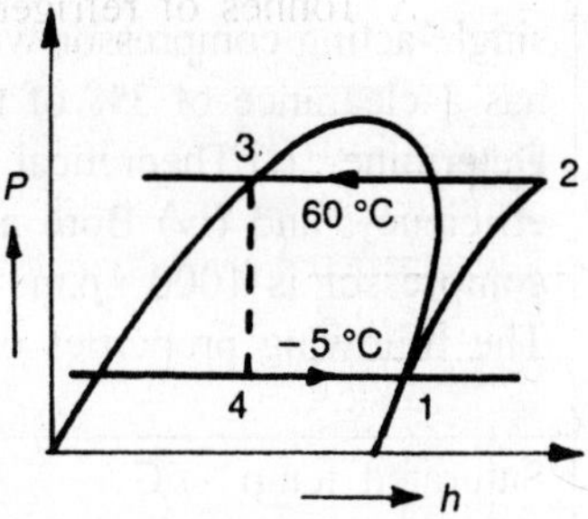

Fig. P. 14.11 (b)

For turbine

Now from steam table at 30 kg/cm^2 and 450°C by interpolation

h_a = 3384.08 kJ/kg

∴ Work output by turbine

= $h_a - h_b$

= $(3384.08 - h_b)$ kg/kg of steam

∴ $w_c = w_T$

∴ h_b = 3349.58 kJ/kg of steam

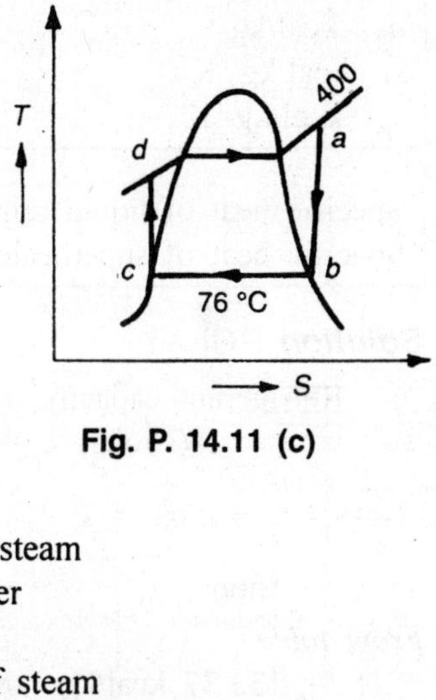

Fig. P. 14.11 (c)

again from steam table, at 76°C,

h_c = 318.16 kJ/kg of steam

∴ heat available in steam condenser

= $(h_b - h_c)$

= 3031.45 kJ/kg of steam

∴total heat available

= 131.5 + 3031.45 = 3162.95 kJ/kg of water

[assume 1 kg of steam and 1 kg of refrigerant]

Now from direct heating, heat available

= $Cp_w (T_2 - T_1)$

= 4.187 (64 – 7) = 238.66 kJ/kg of water

Hence from above we see that heat available in above system is much more than the direct heating. **Ans**

Problem 14.12 : A vapour compression refrigeration machine, with Freon-12 as refrigerant, has a capacity of 20 tons of refrigeration operating between —28°C and 26°C. The refrigerant is subcooled by 4°C before entering the expansion value and the vapour is superheated by 5°C before leaving the evaporator. The machine has a six-cylinder single-acting compressor with stroke equal to 1.25 times the bore. It has a clearance of 3% of the stroke volume.

Determine : (i) Theoretical power required, (ii) COP, (iii) Volumetric efficiency, and (iv) Bore and stroke of cylinder. The speed of the compressor is 1000 r.p.m.

The following properties of Freon-12 may be used :

Saturated temp. °C	28	26
Saturated pressure kgf/cm^2	1.1149	6.8175
v_g m^3/kg	0.1792	0.027
h_f kcal/kg	93.98	106.01
h_g kcal/kg	133.77	139.7
s_f kcal/kg-°K	0.977	1.0207
s_g kcal/kg-°K	1.139	1.1334

Specific heat of liquid refrigerant = 0.23 kcal/kg.°K

Specific heat of superheated vapour = 0.147 kcal/kg.°K (IES-1987)

Solution : Given

Refrigerant capacity = 20 tons

(stroke) $L = 1.2\, d$ (bore)

$$\frac{v_c}{v_s} = k = 0.03$$

$N = 1000$ rpm

From table

$h_1' = 133.77$ kcal/kg

$s_1' = 1.139$ kcal/kg.K

$h_2' = 139.7$ kcal/kg

$s_2' = 1.1334$ kcal/kg.K

$h_3' = 106.01$ kcal/kg

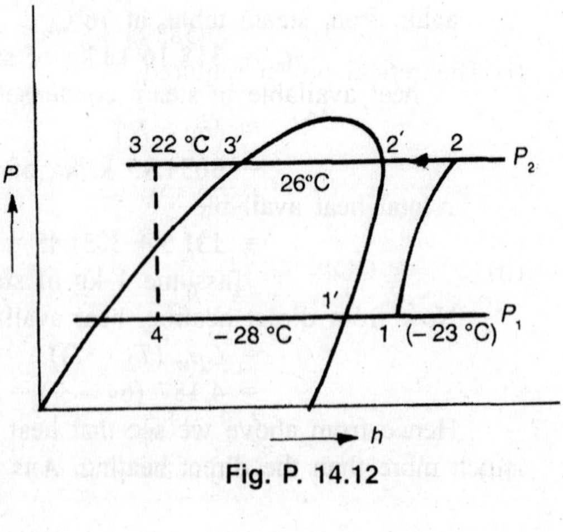

Fig. P. 14.12

Now

$$h_1 - h_1' = Cp_v\,(t_1 - t_1')$$

$$= 0.147 \times (5)$$

$$= 0.735 \text{ kcal/kg}$$

$\therefore \quad h_1 = 134.51$ kcal/kg

and at constant pressure

$$s_1 - s_1' = Cp_v \ln \frac{t_1}{t_1'}$$

$$= 0.147 \ln \frac{(-23+273)}{(-28+273)}$$

$$\therefore \quad s_1 = 1.142 \text{ kcal/kg.K} = s_2$$

Again at constant pressure

$$s_2 - s_2' = Cp_v \ln \frac{t_2}{t_2'}$$

$$\therefore \quad t_2 = 316.95 \text{ k}$$

$$\therefore \quad h_2 = h_2' + Cp_v (T_2 - T_2')$$

$$= 139.7 + 0.147 (316.95 - 299)$$

$$= 142.34 \text{ kcal/kg}$$

Similarly $h_3 = h_3' + Cp_w (T_3 - T_3')$

$$= 106.01 + 0.23 \times (-4)$$

$$= 105.09 \text{ kcal/kg} = h_4$$

$\therefore$ refrigerating effect

$$= h_1' - h_4$$

$$= 133.77 - 105.09 = 28.68 \text{ kcal/kg}$$

$\therefore$ ton of refrigeration

$$= 20 \times 3.517 \text{ kJ/sec}$$

$\therefore$ mass flow rate $= \dot{m} = \dfrac{20 \times 3.517}{4.2 \times 28.68}$ kg/sec

$$= 0.58395 \text{ kg/sec}$$

(i) Theoretical power required

$$= \dot{m} (h_2 - h_1)$$

$$= 0.58395 (142.34 - 134.51)$$

$$= 4.572 \text{ kW } \textbf{Ans}$$

(ii) $$\text{COP} = \frac{h_1' - h_4}{h_2 - h_1}$$

$$= \frac{(133.77 - 105.09)}{(142.34 - 134.51)} = 3.663 \textbf{ Ans}$$

(iii) η_v = volumetric efficiency

$$= 1 + 0.03 - 0.03 \left(\frac{6.8175}{1.1149}\right)^{\frac{1}{\gamma}}$$

$$\gamma = \frac{c_p}{c_v}$$

= Assuming reversible adiabatic process

M = molecular wt of CCl_2F_2 = 120.91 = 121

$$\therefore \quad R = \frac{\overline{R}}{M} = 0.06871 \text{ kJ/kg.K}$$

$$\therefore \quad C_v = -\frac{0.06871}{4.2} + 0.147 = 0.131 \text{ kcal/kg.K}$$

$$\therefore \quad V = 1.13$$

$$\therefore \quad \eta_v = 0.88$$ **Ans**

(iv) From table $v_1' = 0.1492 \text{ m}^3\text{/kg}$

At constant pressure process

$$\frac{v_1}{T_1} = \frac{v_1'}{T_1'}$$

$$\text{or,} \quad v_1 = \frac{0.1492}{250} \times 245 = 0.1462 \text{ m}^3\text{/kg}$$

$$\therefore \quad \text{volume flow rate} = \dot{m}\, v_1 = 0.58395 \times 0.1462 = 0.0854 \text{ m}^3\text{/sec}$$

$$\text{or} \quad \text{displacement volume} = \frac{\text{volume flow rate}}{\eta_v}$$

$$= \frac{0.0854}{0.88} = 0.09702 \text{ m}^3\text{/sec}$$

$$\text{or} \quad \frac{\pi}{4} d^2 \times LN \times 6 = 0.09702 \text{ m}^3\text{/sec}$$

$$\text{or} \quad \frac{\pi}{4} \times 1.2d^3 \times \frac{1000}{60} \times 6 = 0.09702$$

$$\therefore \quad d = 0.10097 \text{ m} = 10.1 \text{ cm}$$

$$L = 0.12621 \text{ m} = 12.62 \text{ cm}$$ **Ans**

Problem 14.13 : An ammonia refrigerator operates on the sample vapor compression cycle. The pressure of liquid ammonia before throttling is 12.3 kgf/cm^2. Ammonia gas leaves the evaporator at –9.5°C. The power input to the compressor is 2.12 kW and the mass flow rate of ammonia is 27 kg/h. Determine the dryness fraction of the gas after throttling, the heat absorbed per hour in the evaporator and the coefficient of performance of the refrigerator. Sketch the cycle on the T-s and the p-h diagrams. Properties of ammonia are given below :

Temperature, °C	Pressure, kgf/cm^2	Saturation liquid	enthalpy, kcal/kg gas
–13.8	2.53	27.5	341
+31.1	12.30	77.5	351

Specific heat of liquid at 12.3 kgf/cm^2 may be taken to be 1.14 kcal/kg°C and the specific heat of superheated vapor at 2.53 kgf/cm^2 may be taken to be 0.6 kcal/kg°C. (IES – 1985)

Solution : Given

power input to the compressor

$= 2.12$ kW

$\dot{m}$ = mass flow rate

$= 27$ kg/h

From table

$h_1' = 341$ kcal/kg

$h_3' = 77.5$ kcal/kg

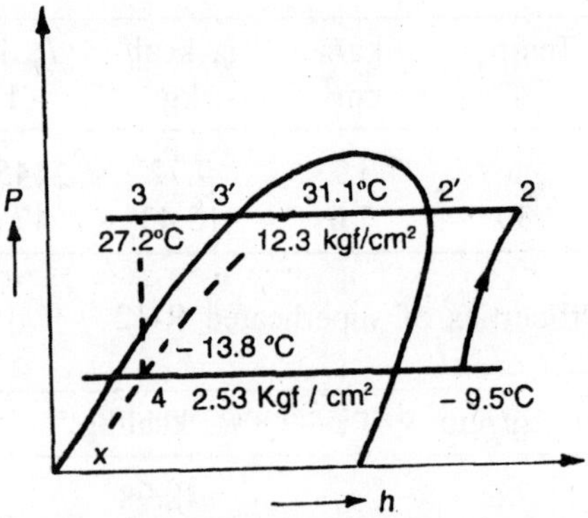

Fig. P. 14.13

$h_3 - h_3' = C_p\,(T_3 - T_3')$

$= 1.14\,(27.2 - 31.1)$

$\therefore \quad h_3 = 73.05$ kcal/kg

Now $h_3 = h_4$

$= 73.05$ kcal/kg

or, $\quad 27.5 + x\,(341 - 27.5) = 73.05$

$\therefore \quad x = 0.15$ **Ans**

$h_1 - h_1' = Cp\,(T_1 - T_1')$

$\therefore \quad h_1 = 341 + 0.6\,(-9.5 + 13.8)$

$= 343.58$ kcal/kg

$\therefore$ heat absorbed in evaporator

$= (h_1' - h_4)\,\dot{m}$

$$= (341 - 73.05) \times 27$$
$$= 7234.65 \text{ kcal/hr } \textbf{Ans}$$

$$\therefore \quad \text{COP} = \frac{\text{Refrigerating effect}}{\text{Work input}}$$

$$= \frac{7234.65 \times 4.2}{3600 \times 2.12} = 3.98 \textbf{ Ans}$$

Problem 14.14 : Refrigerant-12 is employed as the working medium in a heat pump. The evaporator temperature (outdoor temperature) is 5°C and the condenser temperature (the indoor temperature) is 30°C. The efficiency of the compressor is 85%. The heating requirement is 60,000 kcal/h. Determine the co-efficient of performance of the heat pump. If direct electric resistance heating is employed for this purpose what would be the relative gain of the heat pump. Assume that the drive for the heat pump compressor is an electric motor of 70% efficiency. (IES – 1984)

Properties of R-12 are given below :

Temp.p, °C	kgf/cm^2	i_f, kcal/kg	I_g, kcal/kg	s_f, kcal/kg. K	s_g kcal/kg. K
5	3.7	9.72	45.3	0.0379	0.1658
30	7.6	15.43	47.675	0.0573	0.1637

Properties of superheated R-12

p, kgf/cm^2	i_g, kcal/kg	s_g, kcal/kg. K
7.6	48.98	0.1658

Solution : Given

$$\eta_c = 0.85$$

$$\text{Heating requirement} = Q_c = 60{,}000 \text{ kcal/hr}$$

$$\eta_{motor} = 90\%$$

From table

$$h_1 = 45.3 \text{ kcal/kg}$$
$$s_1 = 0.1658 \text{ kcal/kg.K} = s_2$$
$$h_2 = 48.98 \text{ kcal/kg}$$

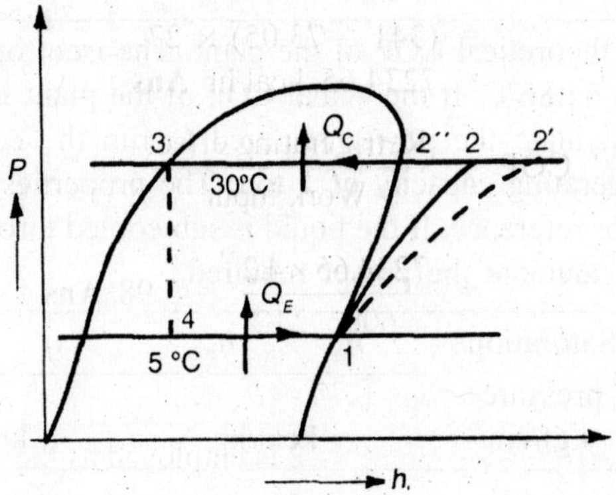

Fig. P. 14.14

$$h_3 = 15.43 \text{ kcal/kg} = h_4$$

$$\eta_c = 0.85 = \frac{h_2 - h_1}{h'_2 - h_1}$$

or $$h'_2 - h_1 = \frac{h_2 - h_1}{0.85}$$

$$= \frac{48.98 - 45.3}{0.85}$$

$$= 4.3294$$

$\therefore$ $$h'_2 = 49.63 \text{ kcal/kg}$$

heating requirement

$$= 60{,}000 \text{ kcal/hr} = \dot{m} \ (h'_2 - h_3)$$

$\therefore$ $$\dot{m} = \frac{60{,}000}{49.63 - 15.43} = 1754.39 \text{ kg/hr}$$

$\therefore$ COP of heat pump $$= \frac{Q_c}{\dot{m}(h'_2 - h_1)}$$

$$= \frac{60{,}000}{1754.39(49.63 - 45.3)} = 7.9 \textbf{ Ans}$$

Probiem 14.15 : A commercial refrigerator operates with R-12 between 1.2603 kgf/cm^2 and 13.9272 kgf/cm^2. The vapour is dry and saturated at the compressor inlet. Assuming isentropic compression determine the theoretical COP of the plant. The isentropic discharge temperature is 64.86°C. If the actual COP of the plant is 80% of the theoretical, calculate the power required to run the compressor to obtain a refrigerating capacity of 1 ton. The properties of R-12 are given below for reference. If the liquid is sub-cooled through 10° after condensation calculate the power required.

Saturation temp. °C	Saturation pressure kgf/cm²	h_f (kcal/kg)	h_g (kcal/kg)	s_f (kcal/kg°C)	g (kcal/kg°C)
–25	1.2603	3.19	42.17	0.0132	0.1703
55	13.9272	22.00	50.11	0.0779	0.1636

Properties of superheated R-12

Temp. °C	Pressure kgf/cm²	h kcal/kg	s kcal/gk°C
64.86	13.9272	52.684	0.1703

Assume specific heat of liquid to be 0.252 kcal/kg°C

Solution : Given :

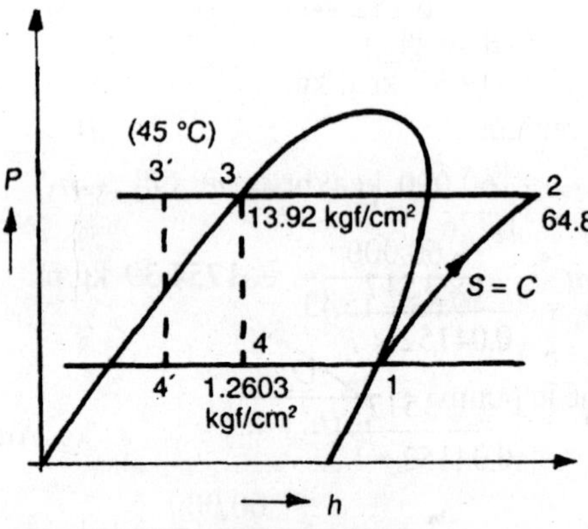

Fig. P. 14.15

Actual COP = 80% theoretical COP

From table

$$h_1 = 42.17 \text{ kcal/kg}$$
$$h_2 = 52.684 \text{ kcal/kg}$$
$$h_3 = 22.00 \text{ kcal/kg} = h_4$$

$$\therefore (COP)_{theoretical} = \frac{h_1 - h_4}{h_2 - h_1}$$

$$= \frac{42.17 - 22}{52.68 - 42.17} = 1.92$$

$$\therefore \text{Actual } (COP) = 0.80 \times (COP)_{theoretical}$$

$$= 0.8 \times 1.92 = 1.5353$$

Refrigerating effect

$$= 1 \text{ ton} = 3.517 \text{ kW} = (h_1 - h_4) \times \dot{m}$$

$$\therefore \quad \dot{m} = \frac{3.517}{(42.17 - 22) \times 4.2} = 0.04152 \text{ kg/sec}$$

$$\therefore \quad (COP)_{actual} = \frac{\text{Refrigerating effect}}{\text{Power required}}$$

$$\therefore \text{ Power required} = \frac{1 \times 3.517 \text{ kW}}{0.04152 \times 1.5353}$$

$$= 55.18 \text{ kJ/kg } \textbf{Ans}$$

2nd care

$$h_3' - h_3 = cp\,(t_3' - t_3)$$

$$\therefore \quad h_3' = 22 + 0.252\,(45 - 55)$$

$$= 19.48 \text{ kcal/kg} = h_4'$$

$$\therefore \quad (COP)_{theo} = \frac{h_1 - h_4'}{h_2 - h_1} = \frac{42.17 - 19.48}{52.684 - 42.17} = 2.16$$

$$\therefore \quad (COP)_{actual} = 1.726$$

$$\therefore \text{ Now } 1.726 = \frac{1 \times 3.517}{0.04152 \times P}$$

$$\therefore \quad P = \frac{3.517}{0.04152 \times 1.726} = 49.06 \text{ kJ/kg } \textbf{Ans}$$

Problem 14.16 : Consider the process indicated in the figure for producing liquid Freon–12. Freon-12 at 2.75 MPa, 100°C enters a heat exchanger and is cooled by the saturated vapour being with drawn from the insulated liquid receiver. The high pressure gas is then throttled across a valve to the liquid receiver pressure. The liquid receiver contains liquid and vapour in equillibrium at –20°C and

vapour leaving the heat exchanger at (2) as at 80°C. Neglecting all pressure losses except across the expansion valve, determine,
(a) the fraction of high pressure gas which is liquid.
(b) the pressure and pressure if (superheated) or quality at the inlet and outlet of the valve.

Solution : Given

points	*P* (MPa)	*T*°C	*H* (kJ/kg)
1	2.75	100°C	225.2
2	0.15	80°C	241.371
3	0.15	–20°C	17.8
3′			178.610
4	2.75		166.9
5	0.15		166.98

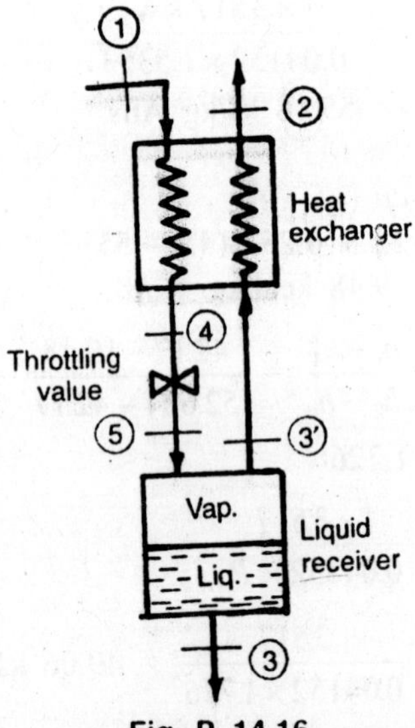

Fig. P. 14.16

From Table of Properties of Freon 12
$p_3 = p_5 = p_2 = 0.1509$ MPa say 0.15 MPa.
∴ enthalpies at 2 and 3 are filled in the above table.

$$\text{Heat in} = \text{Heat out}$$

$$h_1 = mh_3 + (1 - m)\, h_2$$

where mass in is 1 kg and liquid out is *m* kg.

$$225.2 = m \times 17.8 + (1 - m)(241.371)$$
$$= m \times 17.8 + 241.371 - 241.371\ m$$

$$\therefore \quad m = \frac{241.371 - 225.2}{241.371 - 17.8} = 0.0723 \text{ kg}$$

$$(1 - m)\ h_3' + q = (1 - m)\ h_2$$
$$q = (1 - m)(h_2 - h_3')$$
$$= 0.9277 \times (241.371 - 178.610) = 58.22 \text{ kJ}$$

$$\therefore \quad h_4 = h_1 - q$$
$$= 225.2 - 58.22 = 166.98 \text{ kJ/kg}$$

condition at (4) is 2.75 MPa

Quality of vapour (from *p–h* chart) is 0.47

$h_5 = h_4$ pressure at (5) is 0.15 MPa

$\therefore$ Quality of vapour at 5 is 0.93.

Chapter 15

Psychrometrics

Problem 15.1 : An air-water vapour mixture at .01 MPa, 30°C, 80% RH has a volume of 50 m³. Calculate the specific humidity, dew point, wbt, mass of dry air, and mass of water vapour.
If the mixture is cooled at constant pressure to 5°C, calculate the amount of water vapour condensed.

Solution : Given data :

$P = 0.1$ MPa
$= 1$ bar
$T_1 = 30°C$
$= 303$ k
$\phi = 80\%$
V = Volume
$= 50\ m^3$

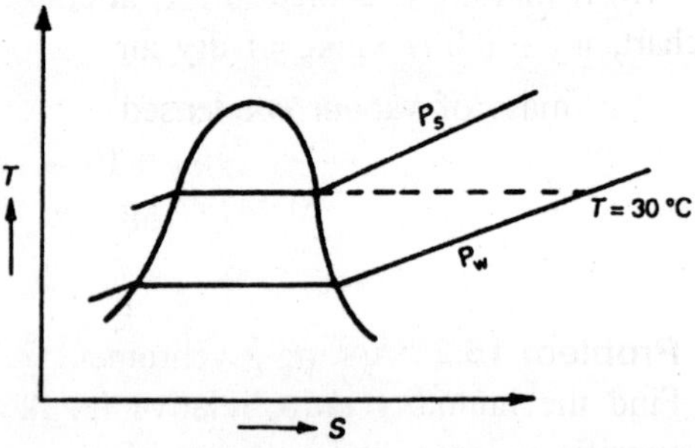

Fig. P. 15.1

From steam table
at 30°C
⇒ $P_{sat} = 4.246$ kPa
$= 0.04246$ bar

$$\because \qquad \phi = \frac{p_{w\ at\ 30°C}}{p_{sat\ at\ 30°C}}$$

$$p_{w\ at\ 30°C} = 0.8 \times 0.04246 = 0.033968 \text{ bar}$$

$$\therefore \text{sp. humidity} = w = .0622 \frac{p_w}{p - p_w}$$

$$= \frac{0.622 \times 0.033968}{1 - 0.033908}$$

$$= 0.02187 \text{ kg/kg of d.a}$$

Hence from steam table,

$$\text{at } p_w = 3.3968 \text{ kPa}$$
$$t_{sat} = 26.06°C$$

$\therefore$ dew point tempr = 26.06°C

Now from ideal gas eqn,

$$pv = m\ RT$$

$$\Rightarrow \quad m_w = \frac{p_w\ v}{R_w\ T} = \frac{3.3968 \times 50}{\frac{8.314}{18} \times 303} = 1.214 \text{ kg}$$

$$p_a = p - p_w$$
$$= 100 - 3.3968 = 96.6032 \text{ kPa}$$

$$\therefore \quad m_a = \frac{p_a\ V}{R_a\ T} = \frac{96.6032 \times 50}{0.287 \times 303} = 55.544 \text{ kg}$$

From psychrometric chart, at 80% *RH* and 30°C dbt

wbt = ≈ 28°C Ans

(b) If mixture is cooled to 5°C at constant pressure then at 5°C from chart, w_2 = 0.004 kg/kg of dry air

$\therefore$ mass of vapour condensed

$$= w_1 - w_2 = (0.02187 - 0.004) \times m_a$$
$$= 0.99257 \text{ kg } \textbf{Ans}$$

Problem 15.2 : A sling psychrometer reads 40°C dbt and 36°C wbt. Find the humidity ratio, relative humidity, dew point temperature, specific volume, and enthalpy of air.

Solution : Given data :

dbt = 40°C

wbt = 36°C

From steam table,

at 40°C,

p_{sat} = 7.384 kPa,

h_{g_1} = 2574.3 kJ/kg,

v_1 = 19.52 × 101 m³/kg

at 36°C

p_{sat} = 5.9792 kPa,

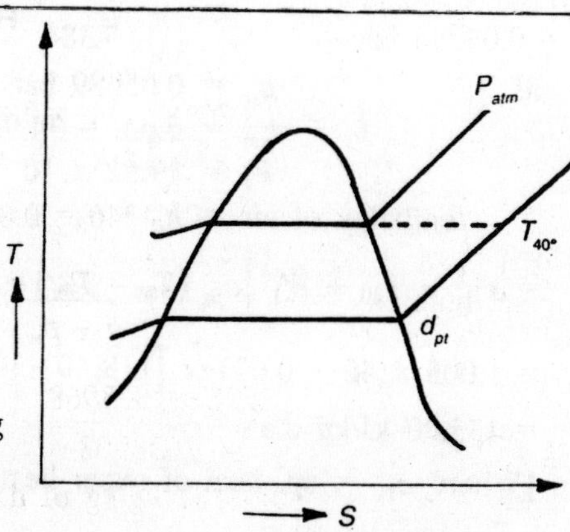

Fig. P. 15.2

$$h_{fg_2} = 2416.22 \text{ kJ/kg}$$

$$h_{f_2} = 150.86 \text{ kJ/kg}$$

Now from adiabatic saturation eqn.

$$\text{humidituy ratio} = w_1 = \frac{c_{P_a}(T_2 - T_1) + w_2 \times h_{fg2}}{h_{g_1} - h_{f_2}}$$

Where $$w_2 = 0.622 \frac{p_{w\,36°C}}{p - p_{w\,36°C}}$$

$$= 0.622 \times \frac{5.9792}{101.325 - 5.9792}$$

$$= 0.039 \text{ kg/kg of d.a}$$

$\therefore$ humidity ratio $= w_1$

$$= \frac{1.005(36 - 40) + 0.039 \times 2416.22}{2574.3 - 150.86}$$

$$= 0.037 \text{ kg/kg of d.a } \textbf{Ans}$$

$$w_1 = 0.037 = 0.622 \times \frac{p_w}{p - p_w}$$

$\therefore$ $$p_w = 0.05689 \text{ bar}$$

$$\therefore \text{Relative humidity} = \frac{p_w}{p_{\text{sat at }40°C}}$$

$$= \frac{0.05689 \times 100}{7.384} = 77.04\% \textbf{ Ans}$$

at $$p_w = 0.05689 \text{ bar}$$

$$t_{sat} = \text{d.p.t.} = 34.92°C \textbf{ Ans}$$

$$V_1 = 19.52 \times 10^{-3} \text{ m}^3/\text{kg}$$

$$\text{enthalpy of air} = h_a + h_w$$

$$= c_{P_a} \times T_{40} + W_1 \left[c_{P_s}(T_{40} - T_{dpt}) + h_{fg_{40}}\right] + c_{P_w} \times \text{dpt}$$

$$= 1.005 \times 40 + 0.037 \times \left[1.18(40 - 34.92) + 2406.7 + 4.18 \times 34.92\right]$$

$$= 135.00 \text{ kJ/kg d.a}$$

[Where c_{P_s} = sp. heat of super heated steam]

Problem 15.3 : Calculate the amount of heat removed per kg of dry air if the initial condition of air is 35°C, 70% *RH*, and the final condition is 25°C, 60% *RH*.

Solution : Given data :

Initial condition,

$$T_1 = 35°C = 35 + 273 = 308 \text{ K}$$

$$\phi_1 = 70\%$$

final condition,

$$T_2 = 25°C = 25 + 273 = 298 \text{ K}$$

$$\phi_2 = 60\%$$

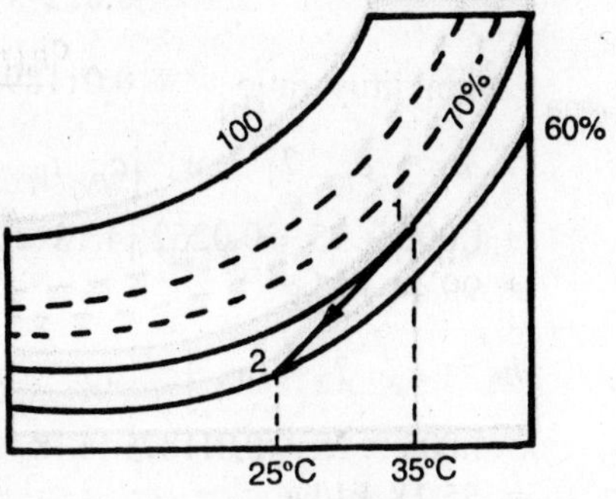

Fig. P. 15.3

$$\phi_1 = 70\% = \frac{p_{w_{at\,35}}}{p_{sat_{at\,35}}}$$

from steam table

at 35°C, $p_{sat} = 5.628$ kPa

$$h_{fg_1} = 2418.6 \text{ kJ/kg}$$

$$\therefore \quad p_{w_1} = 0.7 \times 5.628 = 3.9396 \text{ kPa}$$

at this pressure

$$t_{sat} = \text{dpt}_1 = 28.67°C$$

$$w_1 = 0.622 \frac{p_{w_1}}{p - p_{w_1}}$$

$$= 0.622 \times \frac{3.9396}{101.325 - 3.9396}$$

$$= 0.0252 \text{ kg/kg d.a}$$

$$\phi_2 = 0.60 = \frac{p_{w_{at\,25°C}}}{p_{sat_{at\,25°C}}}$$

From steam table

at 25°C $\Rightarrow p_{sat} = 3.169$ kPa

$$h_{fg_2} = 2442.3 \text{ kJ/kg}$$

$$\therefore \quad p_{w_2} = 0.6 \times 3.169 = 1.9014 \text{ kPa}$$

at this pressure t_{sat} = dpt_2 = 16.62°C

$$w_2 = 0.622 \frac{p_{w_2}}{p - p_{w_2}}$$

$$= 0.622 \times \frac{1.9014}{101.325 - 1.9014}$$

$$= 0.011895 \text{ kg/kg d.a}$$

$$\therefore h_1 = c_{p_a} T_1 + w_1 \left[c_{p_w} t_{dp_1} + h_{fg_1} + c_{p_s}\left(t_{35} - t_{dp_1}\right)\right]$$

= 1.005 × 35 + 0.0252 [4.18 × 28.67 + 2418.6 + 1.81 (35 – 28.67)]
= 99.44 kJ/kg

$$h_2 = c_{p_a} T_2 + w_2 \left[c_{p_w} t_{dp_2} + h_{fg_2} + c_{p_s}\left(t_{25} - t_{dp_2}\right)\right]$$

= 1.005 × 25 + 0.011895 [4.18 × 16.62 + 2442.3 + 1.88 (25 – 16.62)]
= 55.18 kJ/kg

∴ heat removed per kg of dry air

$= h_1 - h_2$

= 99.44 – 55.11 = 44.26 kJ/kg **Ans**

Problem 15.4 : Two streams of air 25°C, 50% *RH* and 25°C, 60% *RH* are mixed adiabatically to obtain 0.3 kg/s of dry air at 30°C. Calculate the amounts of air drawn from both the streams and the humidity ratio of the mixed air.

Solution : Given data

T_1 = 25°C
= 25 + 273
= 298 K
ϕ_1 = 50%
T_2 = 25°C
= 25 + 273
= 298 K
ϕ_2 = 60%
G_3 = 0.3 kg/s
T_3 = 30°C = 30 + 273
= 303 K

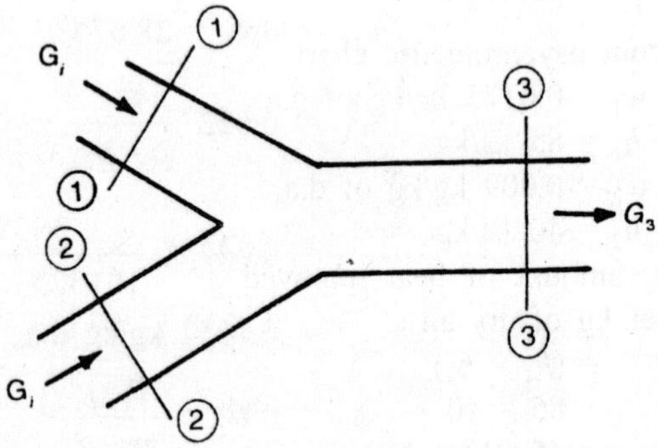

Fig. P. 15.4 (a)

$$G_1 + G_2 = G_3 \qquad \text{...(a)}$$

$$G_1 w_1 + G_2 w_2 = G_3 w_3 \qquad \text{...(b)}$$

$$G_1 h_1 + G_2 h_2 = G_3 h_3 \qquad \text{...(c)}$$

from solving, we get,

$$\frac{G_1}{G_2} = \frac{h_3 - h_2}{h_1 - h_3} = \frac{w_3 - w_2}{w_1 - w_3}$$

Now from psychrometric chart at 25°C and 50%

w_1 = 0.010 kg/kg of d.a.

h_1 = 50 kJ/kg

at 25°C and 60%

w_2 = 0.0117 kg/kg of d.a.

h_2 = 56 kJ/kg

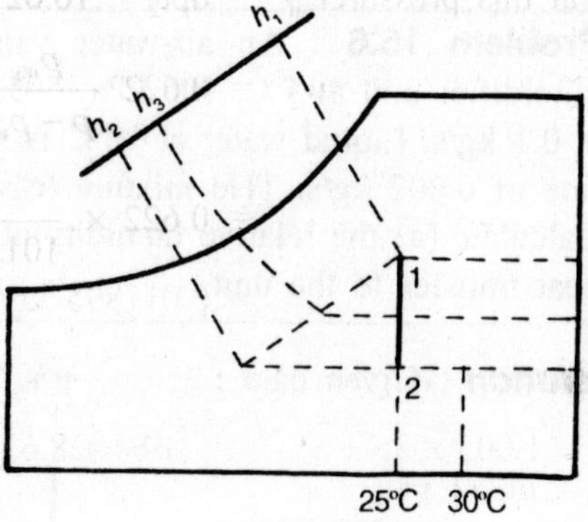

Fig. P. 15.4 (b)

Problem 15.5 : Air at 40°C dbt and 27°C wbt is to be cooled and dehumidified by passing it over a refrigerant–filled coil to give an final condition of 15°C and 90% *RH*. FInd the amounts of heat and moisture removed per kg of dry air.

Solution : Given data :

Inlet Conditions

t'_{db} = 40°C

t'_{wb} = 27°C

Outlet Conditions,

t^{2}_{db} = 15°C

ϕ_2 = 90% *RH*

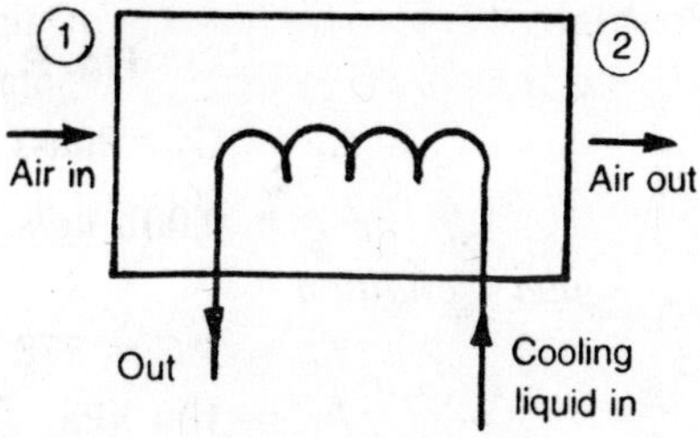

Fig. P. 15.5 (a)

from psychrometric chart

w_1 = 0.0175 kg/kg of d.a.

h_1 = 85 kJ/kg

w_2 = 0.009 kg/kg of d.a.

h_2 = 40 kJ/kg

∴ amount of heat removed per kg of dry air

= $(h_1 - h_2)$

= 85 – 40

= 45 kJ/kg **Ans**

amount of moisture removed per kg of dry air

= $w_1 - w_2$

= 0.0175 – 0.009

= 0.0085 kg/kg of d.a.

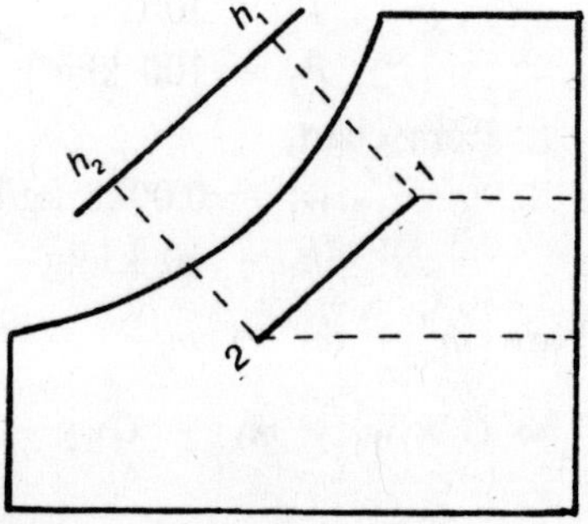

Fig. P. 15.5 (b)

Problem 15.6 : An air-water vapour mixture enters a heater-humidifier unit at 5°C, 100 kPa, 50% *RH*. The flow rate of dry air is 0.1 kg/s. Liquid water at 10°C is sprayed into the mixture at the rate of 0.002 kg/s. THe mixture leaves the unit at 30°C, 100 kPa. Calculate (a) the relative humidity at the outlet, and (b) the rate of heat transfer to the unit.

Solution : Given data :

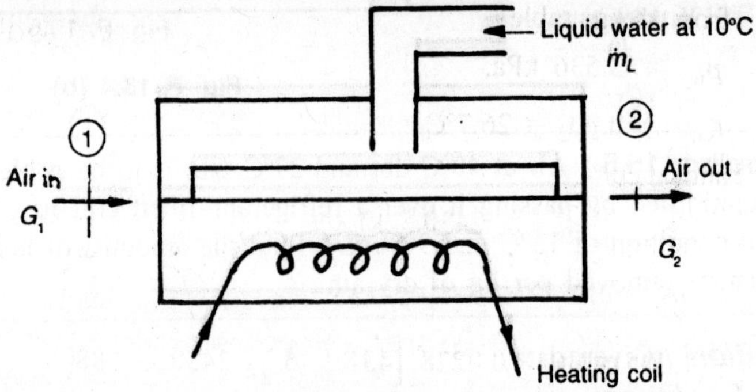

Fig. P. 15.6 (a)

$$G_1 = G_2 = \text{mass of dry air} = 0.1 \text{ kg/s}$$

$$\dot{m}_L = 0.002 \text{ kg/s}$$

Inlet Condition

$$T_1 = 5°C = 278 \text{ K}$$

$$P_1 = 100 \text{ kPa}$$

$$\phi_1 = 50\%$$

outlet Condition

$$T_2 = 30°C$$

$$P_2 = 100 \text{ kPa}$$

$\therefore$ from chart,

$$w_1 = 0.0028 \text{ kg/kg of d.a.}$$

$$h_1 = 10 \text{ kJ/kg}$$

$$\text{Now } \dot{m}_{w_1} + \dot{m}_L = \dot{m}_{w_2}$$

$$\Rightarrow G \times w_1 + \dot{m}_L = Gw_2$$

$$\Rightarrow \quad w_2 = w_1 + \frac{\dot{m}_L}{G_w}$$

$$= 0.0028 + \frac{0.002}{0.1}$$

$$= 0.0228 \text{ kg/kg of air}$$

$$w_2 = 0.0228$$

$$= \frac{p_{w_2}}{p - p_{w_2}} \times 0.622$$

$$p_{w_2} = 9.536 \text{ kPa}$$

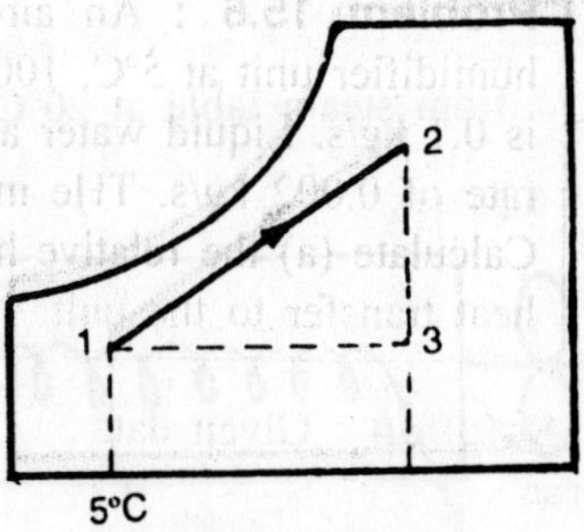

Fig. P. 15.6 (b)

from steam table

at $p_{w_2} = 3.536$ kPa,

$$t_{sat_2} = \text{d.p.t}_2 = 26.7°\text{C}$$

and at 30°C,

$$h_{fg_2} = 2430.5 \text{ kJ/kg}$$

$$\therefore\ h_2 = c_{p_a}\ T_2 + w_2 \left[c_{p_w}\ t_{dp_2} + h_{fg_2}\ c_{p_s}(T_{30}) - t_{dp_2}\right]$$

$$= 1.005 \times 30 + 0.0228\ \left[4.18 \times 26.7 + 2430.5 + 1.88(30 - 26.7)\right]$$

$$= 88.25 \text{ kJ/kg}$$

$$\therefore \text{ Heat removed} = h_2 - h_1$$

$$= 88.25 - 10 = 78.25 \text{ kJ/kg } \textbf{Ans}$$

Problem 15.7 : A laboratory has a volume of 470 m^3, and is to be maintained at 20°C, 52.5% *RH*. THe air in the room is to be completely changed once every hour and is drawn from the atmosphere at 1.05 bar, 32°C, 86%. *RH*, by a fan absorbing 0.45 kW. This air passes through a cooler which reduces its temperature and causes condensation, the condensate being drained off at 8°C. The resulting saturated air is heated to room condition. The total pressure is constant threoughout. Determine (a) the temperature of the air leaving the cooler, (b) the rate of condensation, (c) the heat transfer in the cooler, and (d) the heat transfer in the heater.

Solution : Given :

Power absorbed by fan = 0.45 kw

volume of room = U = 470 m^3

Now at Room Condition (3)

$$T_3 = 20°C,$$
$$\phi = 0.525$$

from steam table at 20°C

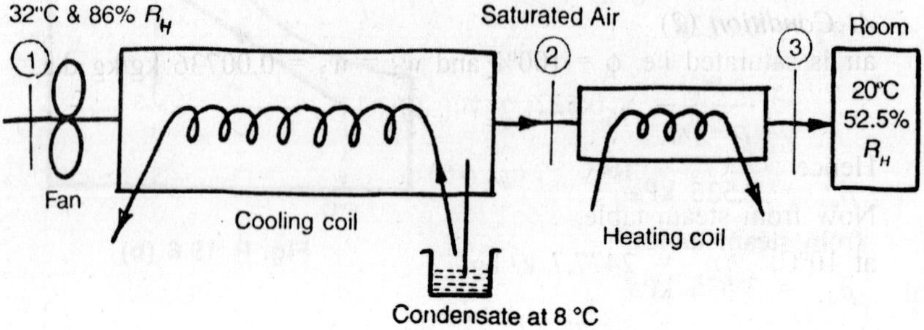

Fig. P. 15.7 (a)

$$p_{sat} = 2.339 \text{ kPa}$$
$$h_{fg_3} = 2454.1 \text{ kJ/kg}$$

$$\phi = \frac{p_{w_3}}{p_{sat}}$$

$$\therefore p_{w_3} = 0.525 \times 2.339 = 1.22798 \text{ kPa}$$

$$\therefore \; w_3 = 0.622 \times \frac{p_{w_3}}{p - p_{w_3}}$$

$$= 0.622 \times \frac{1.22798}{105 - 1.22798}$$

$$= 0.00736 \text{ kg/kg of d.a.}$$

$$p_a = p - p_w = 105 - 1.22978$$
$$= 103.77203 \text{ kPa}$$

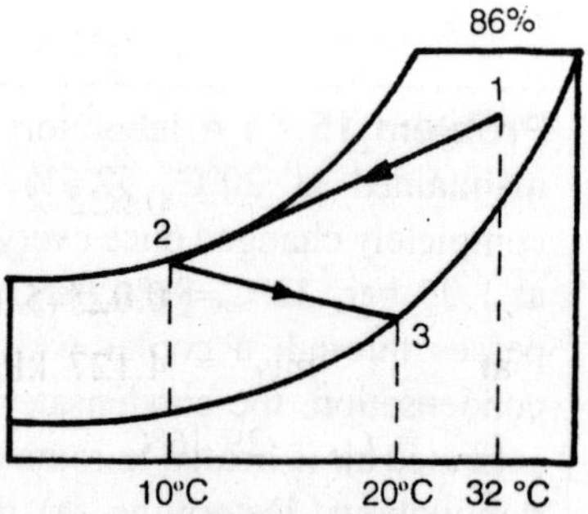

Fig. P. 15.7 (b)

$\therefore$ from ideal gas eqn

$$m_a = \frac{p_a \times V}{R\, T_3}$$

$$= \frac{103.77203 \times 470}{0.287 \times 293} = 580 \text{ kg/hr}$$

from steam table

$$\text{at } p_{w_3} = 1.22798 \text{ kPa},$$

$$t_{sat} = t_{dp_3} = 9.7385 \approx 10°C$$

$\therefore\ h_3 = c_{p_a}\ T_3 + w_3\ [4.18\ t_{dp_3}\ + 1.88\ (20 - 9.7385) +\ h_{fg_3}\ ]$

$= 1.005 \times 20 + 0.00736\ [4.18 \times 9.75.385 + 1.88\ (20 - 9.7385) + 2454.1]$

$= 38.6$ kJ/kg

At Condition (2)

air is saturated i.e. $\phi = 100\%$ and $w_2 = w_3 = 0.00736$ kg/kg d.a.

$\therefore \qquad p_{sat} = p_{w_2} = 1.22798$ kPa

Hence $\qquad T_2 = 10°C = t_{dp_2}$ **Ans**

Now from steam table,

at 10°C, $\quad h_{fg_2} = 2477.7$ kJ/kg

$$\therefore \qquad h_2 = c_{p_a} T_2 + w_2 \left[4.18\, t_{dp_2} + h_{fg_2}\right]$$

$= 1.005 \times 10 + 0.00736\ [4.18 \times 10 + 2477.7]$

$= 28.59$ kJ/kg

At Condition 1

32°C and 86% *RH*

from steam table,

$$p_{sat} = 4.7988 \text{ kPa}$$

$$h_{fg_1} = 2425.74 \text{ kJ/kg}$$

$$\therefore \qquad pw_1 = \phi \times p_{sat} = 4.7988 \times 0.86 = 4.127 \text{ kPa}$$

$$\therefore \qquad w_1 = 0.622 \times \frac{pw_1}{p - pw_1}$$

$$= 0.622 \times \frac{4.127}{105 - 4.127}$$

$$= 0.02545 \text{ kg/kg d.a.}$$

at $\qquad pw_1 = 4.127$ kPa

$t_{dp_1} = t_{sat} = 29.46°C$

$$\therefore\ h_1 = c_{pa'}\, T_1 + w_1 \left[c_{pw} \times T_1 + 1.88\left(T_1 - t_{dp_1}\right) + h_{fg_1}\right]$$

$= 1.005 \times 32 + 0.02545\ [4.18 \times 32 + 1.88\ (32 - 29.46) + 2425.74]$

$= 97.42$ kJ/kg

at 8°C,

from steam table

$$h_f = 33.6 \text{ kJ/kg}$$

(b) Rate of condensation

$$= (w_1 - w_2)\ \dot{m}_a$$

$$= \frac{580}{hr} \times (0.2545 - 0.00730]$$

$$(\dot{m}_w) = 10.49 \text{ kg/hr}$$

(c) Applying S.F.E.E. in between (1) and (2) neglecting ΔP.E. and ΔK.E.

$$\Rightarrow \quad \dot{m}_a [h_1] + \frac{dQ}{d\tau} = \dot{m}_a [h_2] + \dot{m}_w \times h_f + \frac{dw}{dt}$$

$$\Rightarrow \quad \frac{dQ}{d\tau} = \dot{m}_a [h_2 - h_1] + \dot{m}_w\ h_f - 0.45$$

$$= \frac{580}{3600} [28.59 - 97.42] + \frac{10.49}{3600} \times 33.6 - 0.45 = -11.44 \text{ kW}$$

= $-v_e$ sign for heat removed in cooler

(d) Heat transfer in heater

$$= \dot{m}_a\ (h_3 - h_2)$$

$$= \frac{580}{3600} (38.6 - 28.59) = 1.62 \text{ kw } \textbf{Ans}$$

Problem 15.8 : In an air conditioning system, air is to be cooled and dehumidified by means of a cooling coil. The data are as follows:
Initial condition of the air at inlet to the cooling coil: *dbt* = 25°C, partial pressure of water vapour = 0.019 bar, absolute total pressure = 1.02 bar

Final condition of air at exit of the cooling coil:
dbt = 15°C, *RH* = 90%, absolute total pressure = 1.02 bar.

Other data are as follows:
Characteristic gas constant for air = 278 J/kg.K
Characteristic gas constant for water vapour = 461.5 J/kg.K
Saturation pressure for water at 15°C = 0.017 bar
Enthalpy of dry air = 1.005 *t* kJ/kg
Enthalpy of water vapour = (2500 + 1.88 *t*) kJ/kg where *t* is in °C

Determine (a) the moisture removed from air per kg of dry air, (b) the heat removed by the cooling coil per kg of dry air.

Solution : Given data

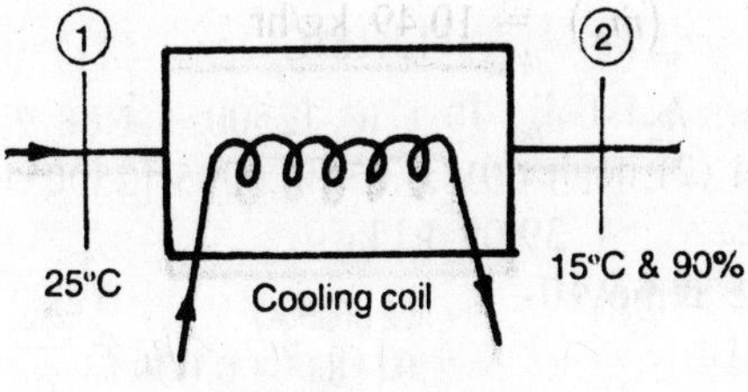

Fig. P. 15.8

$$p = \text{total pressure} = 102 \text{ kPa}$$

At Condition 1

$$p_{w1} = 0.019 \text{ bar}$$

$$w_1 = \text{sp. humidity} = 0.622 \times \frac{p_{w1}}{p - p_{w1}}$$

$$= 0.622 \times \frac{0.019}{1.02 - 0.019} = 0.01181 \text{ kg/kg of d.a}$$

from steam table

at 25°C, $p_{sat} = 3.169$ kPa

$$h_{fg_1} = 2442.3 \text{ kJ/kg}$$

at $p_{w1} = 0.019$ bar $= 1.9$ kPa,

$$t_{sat} = t_{dp_1} = 16.61°C$$

$$h_1 = Cp_a\, T_1 + w_1\, [2500 + 1.88\, t]$$

$$= 1.005 \times 25 + 0.01181\, [2500 + 1.88 \times 25]$$

$$= 55.21 \text{ kJ/kg}$$

at Condition 2

at 15°C

from steam table,

$$p_{sat} = 1.7051 \text{ kPa}$$

$$h_{fg_2} = 2465.9 \text{ kJ/kg}$$

$$\therefore \quad p_{w2} = p_{sat} - \phi = 1.7051 \times 0.9 = 1.5346 \text{ kPa}$$

$$\therefore \quad w_2 = 0.622 \times \frac{p_{w2}}{p - p_{w2}}$$

$$= 0.622 \times \frac{1.5346}{102 - 1.5346}$$
$$= 0.0095 \text{ kg/kg d.a.}$$

From steam table

at
$$p_{w2} = 1.5346 \text{ kPa},$$
$$t_{sat} = t_{dp_2} = 13.18°\text{C}$$
$$h_2 = c_{pa}\, T_2 + w_2\, [2500 + 1.88\, T_2]$$
$$= 1.005 \times 15 + .0095\, [2500 + 1.88 \times 15]$$
$$= 39.09 \text{ kJ/kg}$$

(a) moisture removed $= w_1 - w_2$

$= .01181 - 0.0095$

$= 2.31 \times 10^{-3}$ kg/kg of d.a. **Ans**

(b) heat removed by the cooling coil per kg of d.a.

$= h_1 - h_2 = 55.21 - 34.04$

$= 16.12$ kJ/kg of d.a. **Ans**

Problem 15.9 : Air at 30°C, 80% *RH* is cooled by sprayin gin water at 12°C. This causes saturation, followed by condensation, the mixing being assumed to take place adiabatically and the condensate being drained off at 16.7°C. The resulting saturated mixture is then heated to produce the required conditions of 60% *RH* at 25°C. The total pressure is constant at 101 kPa. Determine the mass of water supplied to the sprays to provide 10 m³/h of conditioned air. What is the heater power required?

Solution : Given data :

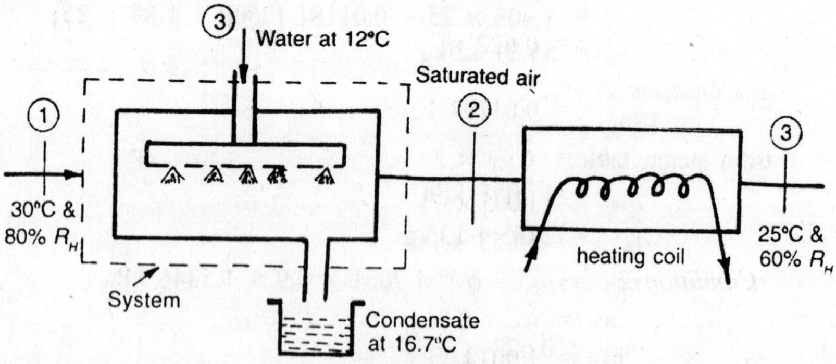

Fig. P. 15.9 (a)

$$p = \text{total pressure} = 101 \text{ kPa}$$
$$V_5 = 10 \text{ m}^3/\text{h}$$

Condition 1

at 30°C and 80% *RH*

from steam table,

$$p_{sat} = 4.246 \text{ kPa}$$
$$\therefore \quad p_{w1} = \phi \times p_{sat}$$
$$= 0.8 \times 4.246$$
$$= 3.3968 \text{ kPa}$$

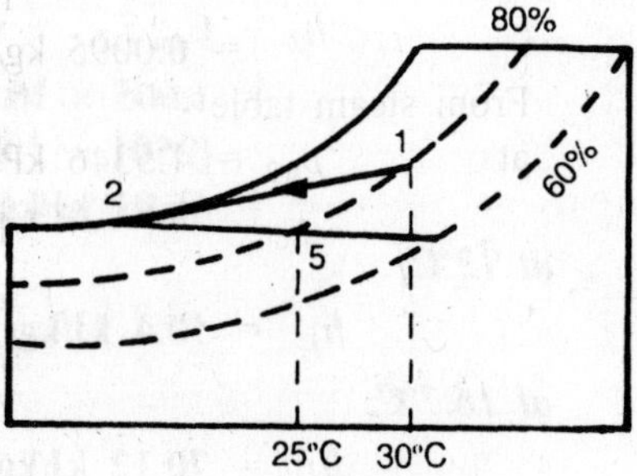

Fig. P. 15.9 (b)

$$\therefore \quad w_1 = 0.622 \frac{p_{w1}}{p - p_{w1}}$$
$$= 0.622 \times \frac{3.3968}{101 - 3.3968}$$
$$= 0.02165 \text{ kg/kg of d.a.}$$
$$h_1 = C_{Pa} T_1 + w_1 [2501 + 1.88 \times T_1^0]$$
$$= 1.005 \times 30 + .02165 [2501 + 1.88 \times 30]$$
$$= 85.51 \text{ kJ/kg}$$

Condition 5,

at 25°C and 60% *RH*

from steam table,

$$p_{sat} = 3.169 \text{ kPa},$$
$$Vg_5 = 43.36 \times 10^{-3} \text{ m}^3/\text{kg}$$
$$\therefore \quad p_{w5} = \phi \times p_{sat}$$
$$= 0.6 \times 3.169 = 1.9014 \text{ kPa}$$
$$\therefore \quad w_5 = 0.622 \times \frac{p_{w5}}{p - p_{w5}}$$
$$= 0.622 \times \frac{1.9014}{101 - 1.9014}$$
$$= 0.01193 \text{ kg/kg of d.a.}$$
$$\therefore \quad h_5 = C_{Pa} \times T_5 + w_5 [2501 + 1.88 \times T_5]$$
$$= 1.005 \times 25 + 0.01193 [2501 + 1.88 \times 25]$$
$$= 55.53 \text{ kJ/kg}$$

Condition 2,

$$w_2 = w_5 = 0.01193 \text{ kg/kg of d.a.}$$
$$p_{w2} = 1.9014 \text{ kPa}$$

from steam table at p_{w2},

$$t_{sat} = t_{dp_2} = 16.62°C$$

$$\therefore \quad h_2 = C_{pw} T_{dp_2} + w_2 \, [2801 + 1.88 \times T_{dp_2}]$$

$$= 1.005 \times 16.62 + 0.01193 \, [2501 + 1.88 \times 16.62]$$

$$= 46.91 \text{ kJ/kg}$$

at 12°C,

$$h_{f_3} = 50.4 \text{ kJ/kg}$$

at 16.7°C,

$$h_{f_4} = 70.12 \text{ kJ/kg}$$

From ideal gas equation at condition (5)

$$\dot{m}_a = \frac{P_a v_5}{RT_5} = \frac{99.1 \times 10}{0.287 \times 298} = 11.5847 \text{ kg/hr}$$

$$\therefore \quad p_a = p - p_w = 101 - 1.9014 = 99.1 \text{ kPa}$$

$$\therefore \text{ mass of condensate } = \dot{m}_a \, (w_1 - w_2)$$

$$\left(\dot{m}_{w_4}\right) = 11.587 \, (0.02165 - 0.01193)$$

$$= 0.11263 \text{ kg/hr}$$

Now applying S.F.E.E. in spraying water system, neglecting $\frac{dw}{d\tau}$, ΔP.E. and ΔK.E.

$$\dot{m}_a h_1 + \dot{m}_3 h_{f_3} + \frac{dQ}{d\tau} = \dot{m}_a h_2 + \dot{m}w_4 h_{f_4} + \dot{m}_3 h_{f_4}$$

$$\Rightarrow \quad \frac{dQ}{d\tau} = \dot{m}_a \, (h_2 - h_1) + \dot{m}w_4 h_{f_4} - \dot{m}_4 \, h_{f_4}$$

For adiabatic mixing

$$\frac{dQ}{d\tau} = 0$$

$$\Rightarrow \quad \dot{m}_3 \left[h_{f_3} - h_{f_4}\right] = \dot{m}_a \, [h_2 - h_1] + \dot{m}_{w4} \, hf_4$$

$$\Rightarrow \quad \dot{m}_3 = \frac{\dot{m}_a [h_2 - h_1] + \dot{m}_{w4} \, h_{f_4}}{h_{f_3} - h_{f_4}}$$

$$= \frac{11.587 \times [46.91 - 85.51] + 0.11263 \times 70.12}{50.4 - 70.12}$$

= 22.28 kg/hr **Ans**

Heater power required

$$= \dot{m}_a \ (h_5 - h_2)$$

$$= \frac{11.587}{3600} \ (55.53 - 46.91) = 27.74 \text{ W } \textbf{Ans}$$

Problem 15.10 : An air-conditioned room requires 30 m^3/min of air at 1.013 bar, 20°C, 52.5% *RH*. The steady flow conditioner takes in air at 1.013 bar, 77% *RH*, which it cools to adjust the moisture content and reheats to room temperature. Find the temperature to which the air is cooled and the thermal loading on both the cooler and heater. Assume that a fan before the cooler absorbs 0.5 kW, and that the condensate is discharged at the temperature to which the air is cooled.

Solution : Given

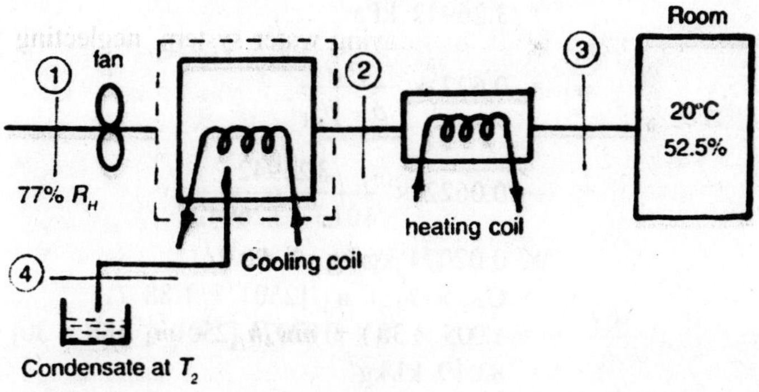

Fig. P. 15.10

$$V_3 = 30 \text{ m}^3\text{/min}$$

$$\text{fan power} = 0.5 \text{ kW}$$

Assume atmosphere air is at 30°C

At Condition 3 or Room Condition

at 20°C & 52.5%

$$p_{sat} = 2.339 \text{ kPa}$$

$$p_{w1} = \phi \times p_{sat} = 1.22798 \text{ kPa}$$

$$w_3 = 0.622 \frac{p_{w1}}{101.3 - p_{w1}} = 0.622 \times \frac{1.22798}{101.3 - 1.22798}$$

$$= 0.007633 \text{ kg/kg of d.a.}$$

$$h_1 = C_{P_a} T_3 + w_3 [2501 + 1.88 T_3]$$

$$= 1.005 \times 20 + 0.007633 [2501 + 1.88 \times 20]$$

$$= 39.48 \text{ kJ/kg}$$

For saturated Condition at 2,

$$w_2 = w_3 = 0.007633 \text{ kg/kg of d.a.}$$

$$\therefore \quad p_{w2} = 1.22798 \text{ kPa}$$

From steam table,

$$\text{at } p_{w2} = 1.22798 \text{ kPa}$$

$$t_{sat} = T_2 = 9.74 \approx 10^\circ\text{C Ans}$$

$$\therefore \quad h_2 = C_{Pa} \times T_2 + w_2 [2501 + 1.88 T_2]$$

$$= 1.005 \times 10 + 0.007633 [2501 + 1.88 \times 10]$$

$$= 29.28 \text{ kJ/kg}$$

at Condition 1

at 30°C and 77% *RH*

$$p_{sat} = 4.246 \text{ kPa}$$

$$\therefore \quad p_{w1} = 4.246 \times 0.077$$

$$= 3.26942 \text{ kPa}$$

$$\therefore \quad w_1 = 0.622 \times \frac{p_{w1}}{p - p_{w1}}$$

$$= 0.0622 \times \frac{3.26942}{101.3 - 3.26942}$$

$$= 0.02074 \text{ kg/kg of d.a.}$$

$$h_1 = C_{Pa} \times T_1 + w_1 [2501 + 1.88 T_1]$$

$$= 1.005 \times 30 + 0.02074 [2501 + 1.88 \times 30]$$

$$= 83.19 \text{ kJ/kg}$$

Now at 10°C,

$$h_{f_4} = 42.01 \text{ kJ/kg}$$

At room condition

Applying ideal gas eqn

$$p_a V = \dot{m}_a R T_3,$$

$$p_a = p - p_w = 101.3 - 1.22798$$

$$= 100.07 \text{ kPa}$$

$$\Rightarrow \quad \dot{m}_a = \frac{100.07 \times 30}{0.287 \times 293} = 35.70 \text{ kg/min}$$

$$\therefore \quad \dot{m}_{w_4} = \text{rate of condensation}$$

$$= \dot{m}_a (w_1 - w_2)$$

$$= 35.07\ (0.02074 - 0.007633)$$

$$= 0.4679 \text{ kg/min}$$

Applying S.F.E.E. eqn between eqn (1) and (2) as a system, neglecting, ΔP.E. and ΔK.E.

$$\dot{m}_a (h_1) \times \frac{dQ}{d\tau} = \dot{m}_a (h_2) + \dot{m}_{w_4}\, h_{f_4} + \frac{dw}{d\tau}$$

$$\Rightarrow \quad \frac{dw}{d\tau} = \dot{m}_a (h_2 - h_1) + \dot{m}_{w_4}\, h_{f_4} - 0.5$$

$$= \frac{35.70}{60}(29.28 - 83.19) + \frac{0.4679}{3600} \times$$

$$42.01 - 0.5 = -32.57 \text{ kW}$$

($-v_e$ sign shows heat is taken out from the system) **Ans**

$$\text{heating load} = \dot{m}_a (h_3 - h_2)$$

$$= (39.48 - 29.28) \times \frac{35.7}{60}$$

$$= 6.07 \text{ kW } \textbf{Ans}$$

Problem 15.11 : An industrial process requires an atmosphere having a *RH* of 88.4% at 22°C, and invovles a flow rate of 2000 m³/h. The external conditions are 44.4% RH, 15°C. The air intake is heated and then humidified by water spray at 20°C. Determine the mass flow rate of spray water and the power required for heating, if the pressure throughout is 1 bar.

Solution : Given data,

$$\text{air flow rate} = V_a = 2000 \text{ m}^3\text{/hr},$$

$$p = 1 \text{ bar}$$

at Condition 1, 44.4% *RH* and 15°C

from steam table

at 15°C, $p_{sat} = 1.7051$ kPa

$$\therefore \quad p_{w1} = \phi_1 \times p_{sat} = 0.444 \times 1.7051 = 0.7571 \text{ kPa}$$

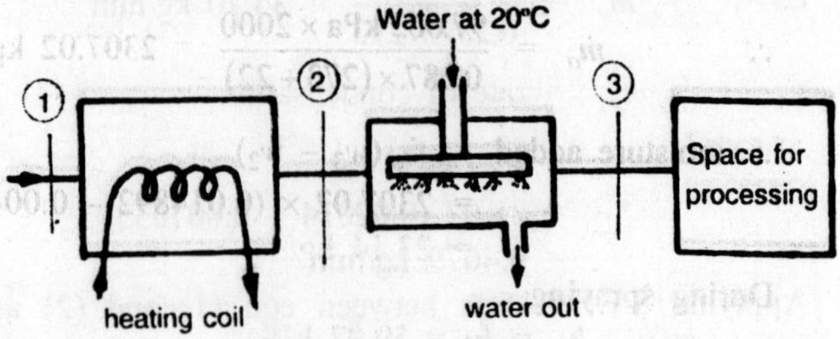

Fig. P. 15.11

$$w_1 = 0.622 \times \frac{p_{w1}}{100 - p_{w1}}$$

$$= 0.622 \times \frac{0.7571}{100 - 0.7571}$$

$$= 4.745 \times 10^{-3} \text{ kg/kg of d.a.}$$

$$h_1 = c_{Pa} \times T_1 + w_1 [2501 + 1.88 \times T_1]$$

$$1.005 \times 15 + 4.745 \times 10^{-3} [2501 + 1.88 \times 15]$$

$$= 27.08 \text{ kJ/kg}$$

at Condition 2

$$w_1 = w_2 = 4.745 \times 10^{-3} \text{ kg/kg of d.a.}$$

$$\therefore \quad p_{w1} = p_{w2} = 0.7571 \text{ kPa}$$

at Condition 3

at 22°C and 88.4% *RH*

from steam table, at 22°C,

$$p_{sat} = 2.645 \text{ kPa}$$

$$\therefore \quad p_{w3} = p_{sat} \times \phi_3 = 2.33818 \text{ kPa}$$

$$w_3 = 0.0622 \frac{p_{w3}}{p - p_{w3}}$$

$$= 0.622 \times \frac{2.33818}{(100 - 2.33818)}$$

$$= 0.014892 \text{ kg/kg d.a.}$$

$$h_3 = c_{Pa} T_3 + w_3 [2501 + 1.88 T_3]$$

$$= 1.005 \times 22 + 0.014792 [2501 + 1.88 \times 22]$$

$$= 59.87 \text{ kJ/kg}$$

From ideal gas equations

$$p_a V_a = \dot{m}_a R T_3$$

or $$p_a = p - p_{w_3} = 100 - 2.33818 = 97.662 \text{ kPa}$$

$$\therefore \quad \dot{m}_a = \frac{97.662 \text{ kPa} \times 2000}{0.287 \times (273 + 22)} = 2307.02 \text{ kg/hr}$$

$$\therefore \text{ moisture added} = \dot{m}_a (w_3 - w_2)$$
$$= 2307.02 \times (0.014892 - 0.004745)$$
$$= 23.14 \text{ kg/hr}$$

During spraying,

$$h_3 = h_2 = 59.97 \text{ kJ/kg}$$

Hence power require for heating

$$= \dot{m}_a (h_2 - h_1)$$

$$= \frac{2307.02}{3600} (59.97 - 27.08)$$

$$= 21.08 \text{ kJ/kg } \textbf{Ans}$$

Problem 15.12 : Cooling water enters a cooling tower at a rate of 1000 kg/h and 70°C. Water is pumped from the base of the tower at 24°C and some make-up water is added afterwards. Air enters the tower at 15°C, 50% *RH*, 1.013 bar, and is drawn from the tower saturated at 34°C, 1 bar. Calculate the flow rate of the dry air in kg/h and the make–up water required per hour.

Solution : Given data

mass flow rate of water = $\dot{m}_w$ = 1000 kg/h

Assumption

(1) Make up water is added at 24°C

Condition 1

at 15°C and 50% *RH* and p = 1.013 bar

from steam table

$$\text{at } 15°\text{C}, \; p_{sat} = 1.7051 \text{ kPa}$$

$$p_{w1} = p_{sat} \times \phi_1$$
$$= 1.7051 \times 0.5 = 0.85255 \text{ kPa}$$

$$w_1 = 0.622 \frac{p_{w1}}{101.3 - p_{w1}}$$

$$= 0.622 \times \frac{0.85255}{101.3 - 0.85255}$$

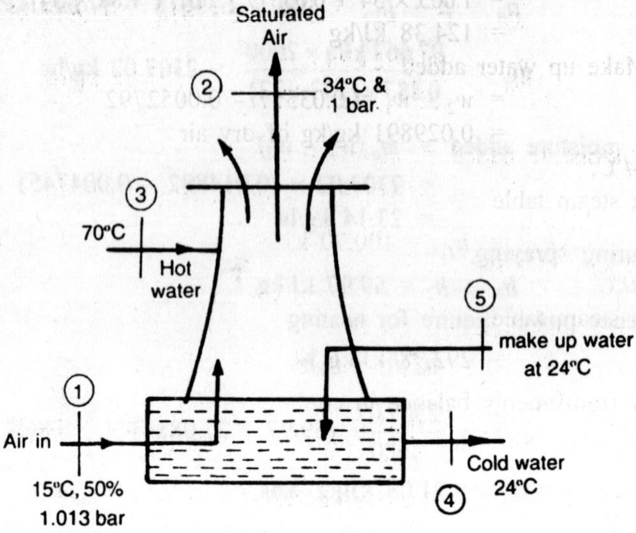

Fig. P. 15.12

$$= 0.0052792 \text{ kg/kg of d.a.}$$

$$\therefore \quad h_1 = c_{p_a} T_1 + w_1 [2501 + 1.88 \times T_1]$$

$$= 1.005 \times 15 + 0.0052792 [2501 + 1.88 \times 15]$$

$$= 28.43 \text{ kJ/kg}$$

Condition 2,
for saturated air,

$$\phi = 100\% = 1$$

Now at 34°C from steam table,

$$p_{sat} = 5.3516 \text{ kPa}$$

$$\therefore \quad p_{w2} = 5.3516 \text{ kPa}$$

$$\therefore \quad w_2 = 0.622 \times \frac{p_{w2}}{100 - p_{w2}}$$

$$= 0.0622 \times \frac{5.3516}{100 - 5.3516}$$

$$= 0.03517 \text{ kg/kg of d.a.}$$

$$\therefore \quad h_2 = c_{p_a} T_2 + w_2 [2501 + 1.88 \times T_2]$$

$$= 1.005 \times 34 + 0.03517\,[2501 + 1.88 \times 34]$$
$$= 124.38 \text{ kJ/kg}$$

∴ Make up water added

$$= w_2 - w_1 = 0.03517 - 0.0052792$$
$$= 0.029891 \text{ kg/kg of dry air}$$

at 24°C,
from steam table

$$h_{f_4} = h_{f_5} = 100.70 \text{ kJ/kg}$$

at 70°C,
from steam table,

$$h_{f_3} = 292.98 \text{ kJ/kg}$$

Now from energy balance in cooling tower

$$\dot{m}_w\, h_{f_3} + \dot{m}_a\, h_1 + (w_2 - w_1)\ \dot{m}_a\, h_{f5} = \dot{m}_w h_{f_4} + \dot{m}_a h_2$$

$$\Rightarrow 1000 \times 292.98 + \dot{m}_a \times 28.43 + 0.029891 \times \dot{m}_a \times 100.7$$
$$= 1000 \times 100.7 + \dot{m}_a \times 124.28$$

$$\therefore \quad \dot{m}_a = \text{flow rate of dry air}$$
$$= 2071.1 \text{ kg/hr}$$ **Ans**

Now make up water required

$$= 0.029891 \text{ kg/kg of d.a.}$$
$$= 0.029891 \times 2071.1 \text{ kg/hr}$$
$$= 61.9 \text{ kg/hr}$$ **Ans**

Problem 15.13 : A grain dryer consists of a vertical cylindrical hopper through which hot air is blown. The air enters the base at 1.38 bar, 65°C, 50% *RH*. At the top, saturated air is discharged into the atmosphere at 1.035 bar, 60°C.
Estimate the moisture picked up by 1 kg of dry air, and the total enthalpy change between the entering tand leaving streams expressed per unit mass of dry air.

Solution : *For Condition 1*

at 65°C, 50% *RH*

$$p_{sat} = 25.03 \text{ kPa}$$
$$p_{w1} = \phi \times p_{sat}$$
$$= 25.03 \times 0.5 = 12.52 \text{ kPa}$$

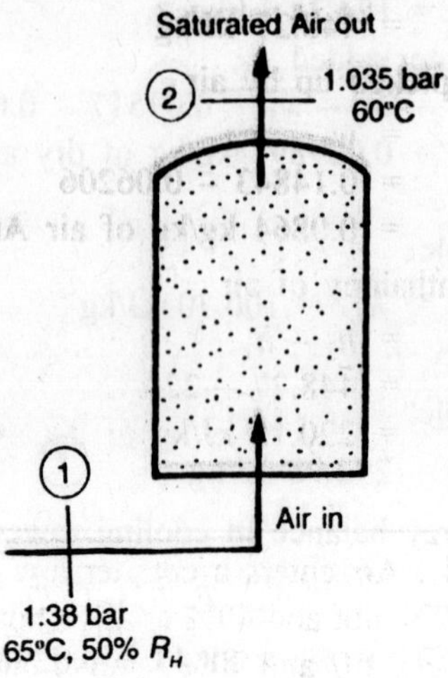

Fig. P. 5.13

$$w_1 = 0.622 \times \frac{p_{w1}}{p - p_{w1}}$$

$$= 0.622 \times \frac{12.52}{138 - 12.52}$$

$$= 0.06206 \text{ kg/kg of d.a.}$$

$$h_1 = c_{Pa}\, T_1 + w_1\, [2501 + 1.88 \times T_1]$$

$$= 1.005 \times 65 + 0.06206\, [2501 + 1.88 \times 65]$$

$$= 228.12 \text{ kJ/kg}$$

from steam table,

at 60°C, $p_{sat} = 19.94$ kPa

since $\phi = 1$

$\therefore$ $p_{w2} = p_{sat} = 19.94$ kPa

$$\therefore \quad w_2 = 0.622 \times \frac{p_{w2}}{103.5 - p_{w2}}$$

$$= 0.622 \times \frac{19.94}{103.5 - 19.94}$$

$$= 0.14843 \text{ kg/kg of d.a.}$$

$$h_2 = c_{Pa}\, T_2 + w_2\, [2501 + 1.88 \times T_2]$$

$= 1.005 \times 60 + 0.14843 [2500 + 1.88 \times 60]$
$= 448.27$ kJ/kg

$\therefore$ Moisture picked up by air

$= w_2 - w_1$
$= 0.14843 - 0.06206$
$= 0.0864$ kg/kg of air **Ans**

Change in enthalopy of air

$= h_2 - h_1$
$= 448.27 - 228.12$
$= 220.15$ kJ/kg of d.a. **Ans**

> **Problem 15.14** : Air enters a counterflow cooling tower at a rate of 100 m³/s at 30°C dbt and 40% relative humidity. Air leaves at the top of the tower at 32°C and 90% relative humidity. Water enters the tower at 35°C and the water flow rate is 1.2 times the mass flow rate of air. Make-up water is supplied at 20°C. What are the range and approach of the tower? At what rate is heat absorbed from the load by the stream of water on its way back to the top of the tower? What percentage of the water flow rate must be supplied as make-up water to replace the water evaporated into the air stream?

Solution : Given data

(water flow rate) $\dot{m}_w = 1.2 \times \dot{m}_a$ (air flow rate)

$\dot{V}_a$ = flow rate of air = 100 m³/s

let cold water leaves the tower at T°C

From Psychrometry chart

at 30°C and 40% $RH \rightarrow$

$h_1 = 58.5$ kJ/kg of d.a

$t_{wb_1} = 20°C$

$w_1 = 0.0105$ kg/kg d.a.,

$v_1 = 0.884$ m³/kg

at 32°C and 90% RH

$h_2 = 105$ kJ/kg

$t_{wb_2} = 30.5°C$

$w_2 = 0.0278$ kg/kg of d.a.

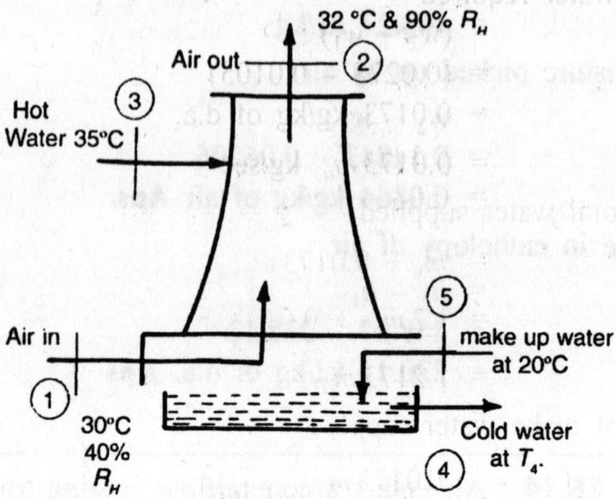

Fig. P. 15.14

at 20°C from steam table,

$$h_{f5} = 83.96 \text{ kJ/kg}$$

at 35°C from steam table,

$$h_{f3} = 146.68 \text{ kJ/kg}$$

From energy balance eqn. of tower, we get,

$$\dot{m}_w (h_{f3} - h_{f4}) = \dot{m}_a [h_2 - h_1] - \dot{m}_a (w_2 - w_1) h_{f5}$$

$$\Rightarrow C_{pw} (t_3 - t_4) = \frac{\dot{m}_a}{\dot{m}_w} [(h_2 - h_1) - (w_2 - w_1) h_{f5}]$$

$$= \frac{1}{1.2} [(105 - 58.5) - (0.0278 - 0.0105)\, 83.96]$$

$$= 37.54 \text{ kJ/kg}$$

$$\Rightarrow t_3 - t_4 = \frac{37.54}{4.18}$$

$$= 8.980°\text{C}$$

$$\Rightarrow t_4 = t_3 - 8.980$$

$$= 26.02°\text{C}$$

$$\therefore \text{Approach} = t_4 - t_{wb_1}$$

$$= 26.02 - 20 = 6.02°\text{C} \textbf{ Ans}$$

$$\text{Range} = t_3 - t_4$$

$$= 35 - 26.02 = 8.98°\text{C} \textbf{ Ans}$$

Make water required

$$= (w_2 - w_1)$$
$$= (0.0278 - 0.0105)$$
$$= 0.0173 \text{ kg/kg of d.a.}$$
$$= 0.0173\, \dot{m}_a \text{ kg/sec}$$

Now total water supplied

$$= \dot{m}_w + 0.0173\, \dot{m}_a$$
$$= 1.2\, \dot{m}_a + 0.0173\, \dot{m}_a$$
$$= 1.2173\, \dot{m}_a$$

∴ % of make water supply

$$= \frac{0.0173\, \dot{m}_a}{1.2173\, \dot{m}_a} \times 100$$
$$= 1.42\% \textbf{ Ans}$$

$$\therefore \quad v_1 = 0.884 \text{ m}^3\text{/kg}$$
$$v_a = 100 \text{ m}^3\text{/sec}$$
$$\therefore \quad \dot{m}_a = \frac{v_a}{v_1} = \frac{100}{0.884} \text{ kg/sec}$$
$$= 113.12 \text{ kg/sec}$$

∴ heat absorbed

$$= \dot{m}_w \ (h_3 - h_4)$$
$$= 1.2 \times \dot{m}_a \ C_{pw} \ (t_3 - t_4)$$
$$= 1.2 \times 113.12 \times 4.18 \ (35 - 26.02)$$
$$= 5095.34 \text{ kW } \textbf{Ans}$$

Problem 15.15 : Air enters a chamber at 5°C dry bulb temperature (DBT) and 2.5°C wet bulb temperature at the rate of 100 m³/min and at the pressure of 1 bar. While passing through the chamber, the air absorbs 50 kW heat and picks up 35 kg/hr of saturated steam at 110°C. Show the process on a psychrometric chart and find the dry and wet bulb temperatures of the leaving air. At 110°C, enthalpy of saturated steam is 2691.3 kJ/kg. (IES – 2001)

Solution : Given

$$Q = \text{heat added in air} = 50 \text{ kW}$$
$$(\dot{m}_2 - \dot{m}_1) = 35 \text{ kg/hr}$$

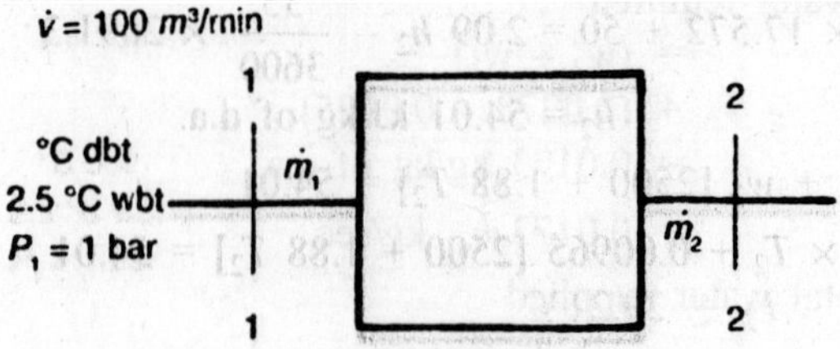

Fig. P. 15.15

where $\dot{m}_1$ = mass of water vapour of entering air

$\dot{m}_2$ = mass of water vapour of leaving air

Let $\dot{m}_a$ = mass of dry air

h_{110} = 2691.3 kJ/kg

Now $\dot{m}_2 - \dot{m}_1 = \dot{m}_a (w_2 - w_1) = 35$ kg/hr

where w_2 = sp. humidity of entering air

w_1 = sp. humidity of leaving air

Now applying S.F.E.E. between (1) and (2), neglecting ΔK.E. and ΔP.E.

$$\dot{m}_a [h_1] + \frac{dQ}{d\tau} = \dot{m}_a (h_2) + \frac{dw}{d\tau} - (\dot{m}_2 - \dot{m}_1) h_{110} \quad \text{...(i)}$$

From ideal gas equation

$$\dot{m}_a = \frac{p_1 v_1}{RT_1} = \frac{100 \times 100}{60 \times 0.287 \times 278} = 2.09 \text{ kg/sec}$$

From psychrometric chart at 5°C dbt and 2.5 wbt.

$$w_1 = 0.005 \text{ kJ/kg of d.a.}$$

$$\therefore \quad h_1 = C_{p_a} T_1 + w_1 (2500 + 1.88\ T_1)$$

$$= 1.005 \times 5 + 0.005\ [2500 + 1.88 \times 5]$$

$$= 17.572 \text{ kJ/kg of d.a.}$$

again

$$\dot{m}_a\ (w_2 - w_1) = \frac{35}{3600} \text{ kg/sec}$$

$$\therefore \quad w_2 = 0.00965 \text{ kg/kg of d.a.}$$

from equation (1) putting $\frac{dw}{d\tau} = 0$

$$\Rightarrow 2.09 \times 17.572 + 50 = 2.09\, h_2 - \frac{35}{3600} \times 2691.3$$

$\therefore$ $h_2 = 54.01$ kJ/kg of d.a.

or $C_{p_a} T_2 + w_2\,[2500 + 1.88\, T_2] = 54.01$

or $1.005 \times T_2 + 0.00965\,[2500 + 1.88\, T_2] = 54.01$

$\therefore$ $T_2 = 29.21°C$

= dry bulb temperature of leaving air Ans

Now from chart at 29.21°C and $w_2 = 0.00965$

T w.b.t. of leaving air

= 19.5°C **Ans**

Problem 15.16 : 100 m³ per minute outdoor air at 43.3°C d.b.t. and 37% relative humidity is mixed with 200 m³ per minute of air at 38.2°C dbt and 24.5°C wet bulb temperature. The mixed air is dehumidified first by a cooling coil having by-pass factor of 0.32 and apparatus dew point of 15°C and then by a chemical dehumidifier. Air leaves the chemical dehumidifier at 30°C dry bulb temperature. Air is then passed over a cooling coil whose surface temperature is 15°C and by-pass factor 0.26. Calculate

(i) Capacity of two cooling coils in tons of refrigeration

(ii) Capacity of coil dehumidifier in kg/min.

(iii) Capacity of chemical dehumidifier in kg/min.

(iv) Total humidifying capacity of the system in kg/min. (IES – 2000)

Solution : Given

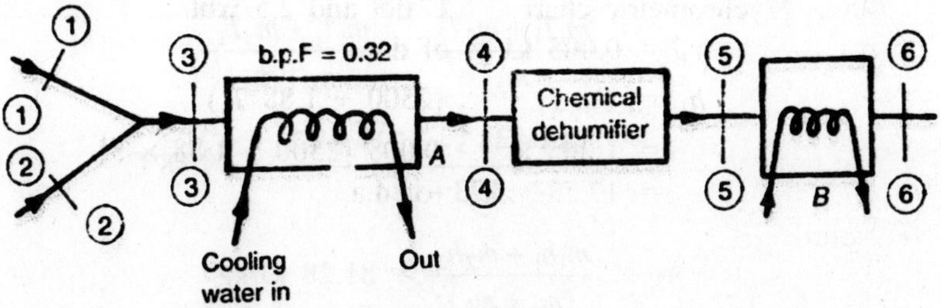

Fig. P. 5.16

$\dot{v}_1 = 100$ m³/min

$\dot{v}_2 = 200$ m³/min

$T_1 = 43.3°C$ dbt

$T_2 = 38.2°C$ dbt

$$\phi_1 = 37\%$$
$$T_2' = 24.5 \text{ wbt}$$
$$(\text{D.P.F.})_A = 0.32$$
$$(\text{A.D.P.})_A = 15°\text{C}$$
$$T_5 = 30°\text{C}$$
$$(\text{A.D.P.})_B = 15°\text{C}$$
$$(\text{D.P.F.})_B = 0.26$$

Let atmospheric pressure = p_a = 100 kPa

$$\therefore \quad \dot{m}_1 = \frac{p_a V_1}{RT_1}$$

$$= \frac{100 \times 100}{60 \times 0.287 \times (43.3 + 273)} = 1.84 \text{ kg/sec}$$

Similarly $$\dot{m}_2 = \frac{100 \times 200}{0.287 \times (38.2 + 273) \times 60} = 3.73 \text{ kg/sec}$$

From psychrometric chart
for stream of air at section 1-1
at 43.3°C dbt and 37% ϕ,

$$w_1 = 0.01967 \text{ kg/kg of d.a.}$$
$$h_1 = 94 \text{ kJ/kg of d.a.}$$

for stream 2,
at 38.2°C and 24.5 wbt

$$w_2 = 0.0143 \text{ kg/kg of d.a.}$$
$$h_2 = 75 \text{ kJ/kg of d.a.}$$

∴ *For stream 3,*

$$T_3 = \frac{m_1 T_1 + m_2 T_2}{m_3} = \frac{m_1 T_1 + m_2 T_2}{m_1 + m_2}$$

$$= \frac{1.84 \times 43.3 + 3.73 \times 38.2}{3.73 + 1.84} = 39.88°\text{C}$$

$$h_3 = \frac{m_1 h_1 + m_2 h_2}{(m_1 + m_2)} = 81.28 \text{ kJ/kg}$$

$$w_3 = \frac{w_1 T_1 + w_2 T_2}{T_3}$$

$$= 0.035 \text{ kg/kg of d.a.}$$

$$\dot{m}_3 = 5.57 \text{ kg/sec}$$

For cooling coil A

$$\text{D.P.F.} = \frac{T_4 - T_{ADP}}{T_3 - T_{ADP}} = \frac{h_4 - h_{ADP}}{h_3 - h_{ADP}}$$

from chart,

at 15°C ADP, h_{ADP} = 43 kJ/kg

$$\text{or} \qquad 0.32 = \frac{T_4 - 15}{39.88 - 15} = \frac{h_4 - 43}{81.28 - 43}$$

$\therefore \quad T_4 = 22.96°C,$

$h_4 = 55.25$ kJ/kg

or $\quad h_4 = C_{Pa}T_4 + w_4\,[2500 + 1.88 \times T_4]$

$\therefore \quad w_4 = 0.0126516$ kg/kg of d.a.

Now between (4) and (5)

$\dot{m}\ C_P\,(T_5 - T_4) = \dot{m}\ (h_5 - h_4)$

$\therefore \quad h_5 = 1.005\,(30 - 22.96) + h_4$

$= 62.33$ kJ/kg

or, $\quad h_5 = C_{Pa}T_5 + w_5\,(2500 + 1.88\,T_5)$

$\therefore \quad w_5 = 0.012588$ kg/kg of air

For cooling coil B

$$\text{D.P.F.} = 0.26 = \frac{T_6 - T_{ADP}}{T_S - T_{ADP}}$$

[$\because$ surface temperature = Apparatus dew point temperature]

$$= \frac{h_6 - h_{ADP}}{h_5 - h_{ADP}}$$

h_{ADP} at 15 = 43 kJ/kg

or $\quad T_6 = 18.9°C$

$h_6 = 48.0258$ kJ/kg

$\therefore \quad w_6 = 0.0114498$ kg/kg of d.a.

$\therefore$ (i) Capacity of two cooling coil

$= \dot{m}_3\,[(h_3 - h_4) + (h_5 - h_6)]$

$= 5.57\,[(81.28 - 55.25) + (62.33 - 48.03)]$

$= 224.6381$ kJ/sec

$= 63.87$ ton of refrigeration Ans

(ii) Capacity of coil dehumidifier

$= \dot{m}_3\,[(w_3 - w_4) + (w_5 - w_6)]$

$= 5.57\ [(0.03.51 - 0.0126516) + (0.012588 - 0.0114498)]$
$= 0.01314$ kg/sec
$= 7.883$ kg/min **Ans**

(iii) Capacity of chemical dehumidifier

$$= \dot{m}_3\ [w_4 - w_5]$$
$$= 5.57\ [0.0126516 - 0.012588]$$
$$= 0.000354252 \text{ kg/sec}$$
$$= 0.0213 \text{ kg/min } \textbf{Ans}$$

(iv) total dehumidifying capacity

$$= 7.883 + 0.0213 = 7.904 \text{ kg/min } \textbf{Ans}$$

Problem 15.17 : 300 m³/min of air at 10°C dry bulb temperature and 90% relative humidity is to be heated and humidified to 35°C dry bulb temperature and 22.5°C wet bulb temperature. The required conditions are achieved by heating, humidifying and then again by heating. The relative humidity of the air coming out of the humidifier is 90%. Find

(i) the heating capacity of the first heater and the by-pass factore if the surface temperatur e of the coil is 40°C

(ii) the capacity of the humidifier in kg/hour

(iii) the heating capacity of the second heater and the coil surface temperature if the bypass factor is 0.5, and

(iv) the humidifying efficiency of air washer (humidifier).

(IES – 1997)

Solution : Given

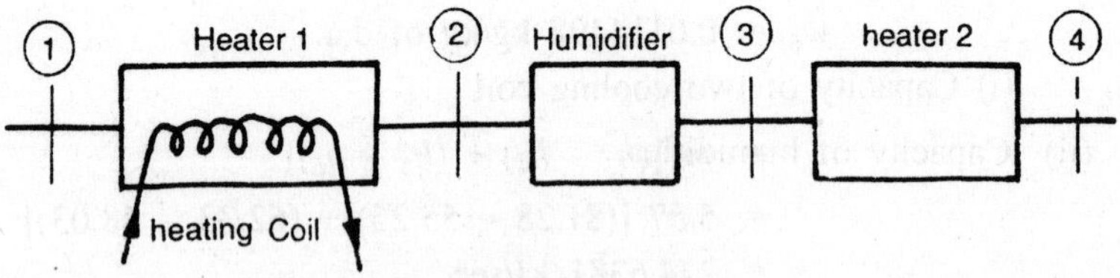

Fig. P. 15.17

$$\dot{V}_0 = 300 \text{ m}^3/\text{min}$$
$$T_1 = 10°\text{C}$$
$$\phi_1 = 90\%$$

$T_4 = 35°C$
$T_4' = 22.5°C$
$\phi_3 = 90\%$

Let atmosphere pressure = 100 kPa

∴ From ideal gas equation

$$\dot{m} = \frac{p_a \dot{V}}{RT_1}$$

$$= \frac{100 \times 300}{60 \times 0.287 \times 283} = 6.16 \text{ kg/sec}$$

at $T_1 = 10°C$, $Q_1 = 90\%$ from psychrometric chart

$h_1 = 29$ kJ/kg

$w_1 = 0.006$ kg/kg of d.a. $= w_2$

at 22.5°C wbt and 35°C dbt

$h_4 = 67$ kJ/kg

$w_4 = 0.0011$ kg/kg of d.a.

Now $w_3 = w_4 = 0.011$ kg/kg of d.a. [since sensible heating]

and $\phi_3 = 90\%$,

$h_3 = 43$ kJ/kg

$T_3 = 15.5°C = T_2$ [sensible humidification]

∴ at $T_2 = 15.5°C$, $w_2 = 0.006$ kg/kg of d.a.

∴ $h_2 = C_{p_a} T_2 + w_2 [2500 + 1.88 T_2]$

$= 30.75$ kJ/kg

(i) heating capacity $= \dot{m} C_P (T_2 - T_1)$

$= 6.16 \times 1.005 (15.5 - 10) = 34.15$ kW

$$\therefore \text{ B.P.F.} = \frac{T_s - T_2}{T_s - T_1}$$

$$= \frac{40 - 15.5}{40 - 10} = 0.82 \textbf{ Ans}$$

(ii) Capacity of humidifier $= \dot{m} (w_3 - w_2)$

$= 6.16 (0.11 - 0.006)$

$= 0.0308$ kg/sec

$= 110.88$ kg/hr **Ans**

(iii) Heat capacity of 2nd heater

$= \dot{m} C_P (T_4 - T_3)$

$= 1.005 \times 6.16 (22.5 - 15.5)$

$$= 43.34 \text{ kW } \textbf{Ans}$$

$$\text{B.P.F.} = 0.5 = \frac{t_s - t_4}{t_s - t_3}$$

$$\text{or} \quad 0.5\,[t_s - 15.5] = [t_s - 22.5]$$

$$t_s = 29.5°\text{C } \textbf{Ans}$$

Problem 15.18 : The properties of air in a room are temperature 35°C and relative humidity 70 percent. Determine (i) the partial pressure of air, (ii) the humidity ratio, (iii) the saturation ratio, (iv) dew point temperature, (v) density of the air–vapour mixture, (vi) the enthalpy of mixture per kg of dry air. Assume room pressure to be 1.0132 bar and the specific heat for the super-heated vapour as 1.863 kJ/kg.K. Solve the problem with the help of steam tables.

(IES – 1996)

Solution : Given :

$$\text{d.B.T.} = T = 35°\text{C}$$

$$\phi = 70\%$$

$$p_a = \text{atmospheric pressure}$$

$$= 101.32 \text{ kPa} = 1.0132 \text{ bar}$$

from steam at 35°C

$$p_{sat} = 0.05628 \text{ bar}$$

$$\therefore \quad \phi = 70\%$$

$$\therefore \quad p_v = \text{partial pressure of water vapour}$$

$$= 0.7 \times 0.05628 = 0.039396 \text{ bar}$$

(a) ∴ partial pressure of air

$$= p_a - p_v$$

$$= 1.0132 - 0.039396 = 0.9738 \text{ bar } \textbf{Ans}$$

$$\text{(b) humidity ratio} = 0.622 \; \frac{p_v}{p_a - p_v}$$

$$= 0.622 \times \frac{0.039396}{101.32 - 0.039396}$$

$$= 0.02516 \text{ kg/kg of d.a. Ans}$$

(c) Saturation humidity ratio

$$= w_s = 0.622 \times \frac{P_{sat}}{p_a - p_{sat}}$$

$$= \frac{0.622 \times 0.05628}{1.0132 - 0.05628}$$

$$= 0.04080 \text{ kg/kg of d.a.}$$

$\therefore$ Saturation ratio $= \dfrac{w}{w_s} = 0.61589$ **Ans**

(d) from steam table

at $p_v = 0.973804$ bar

t_{sat} = dew point temperature = 28.98°C **Ans**

(v) ρ_a = density of air

$$= \frac{\text{partial pressure of air}}{\text{R T}}$$

$$= \frac{0.973804 \times 100}{0.287 \times (273 + 35)} = 1.102 \text{ kg/m}^3$$

ρ_v = density of water vapour

$$= \frac{pv}{RT}$$

$$= \frac{0.039396 \times 100}{\frac{8.314}{18} \times (308)} = 0.277 \text{ kg/m}^3$$

$\therefore$ ρ = density of mixture

$= \rho_a + \rho_v$

$= 1.1297$ kg/m³ **Ans**

(vi) enthalpy of mixture

$= h_1 + wh_v$

$= C_{P_a} T_1 + w\, [h_{fg} + C_{P_w} \times T_{dp} + C_{P_v} (T - T_{dp})]$

$= 1.005 \times 35 + 0.02516\, [2418.6 + 4.187 \times 28.98 + 1.863\, (35 - 28.58]$

= 99.36 kJ/kg of dry air

[from steam table at 35°C, h_{fg} = 2418.6 kJ/kg]

Problem 15.19 : Atmospheric air at 40°C dry bulb temperature and 55% relative humidity is passed over a cooling coil at the rate of 600 m³/min. At exit from the coil the air is saturated and its temperature is 8°C. The condensate leaving the coil per minute and the refrigeration required in kW. The specific heat of superheated steam may be assumed to be 1.88 kJ/kg.K. Solve the problem without the use of Psychrometic chart. (IES – 1995)

Solution : Given :

$T_1 = 40°C$

$\phi_1 = 55\%$

$v = 600\ m^3/min$

$\phi_2 = 100\%$

$T_2 = 8°C$

Condensate at $= 8°C$

Let atmospheric pressure $= 101.32$ kPa

from steam table at 40°C,

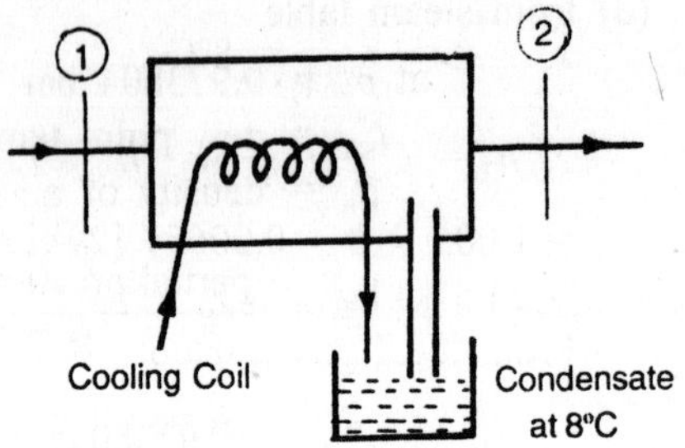

Fig. P. 15.19

$$p_{sat} = 0.07384 \text{ bar}$$

$$h_{fg} = 2406.7 \text{ kJ/kg}$$

$\therefore$ p_{v1} = partial pressure of water vapour at incoming air

$$= \phi \times p_{sat} = 0.040612 \text{ bar}$$

$\therefore$ w_1 = sp. humidity

$$= 0.622 \frac{p_{v1}}{p - p_{v1}}$$

$$= 0.622 \times \frac{0.040612}{1.0132 - 0.040612}$$

$$= 0.025973 \text{ kg/kg of d.a.}$$

from steam table

at $p_{v1} = 0.040612$ bar

$$t_{sat} = 28.98°C = t_{dp_1}$$

$$\therefore h_1 = C_{p_a} T_1 + w_1 [h_{fg} + C_{p_w} \times T_{dp_1} + 1.88 (T_1 - T_{dp_1})]$$

$$= 1.005 \times 40 + 0.025973 [2406.7 + 4.187 \times 28.98 + 1.88 (40 - 28.98)]$$

$$= 106.4 \text{ kJ/kg}$$

From steam table

at 8°C; $p_{sat} = 0.010724$ bar,

$$h_{fg} = 2482.5 \text{ kJ/kg}$$

$\therefore$ $\phi_2 = 100\%$

$$\therefore \quad pv_2 = p_{sat} = 0.01072 \text{ bar}$$

$$\therefore \quad w_2 = 0.622 \frac{pv_2}{p - pv_2}$$

$$= 0.622 \times \frac{0.010724}{1.0132 - 0.010724}$$

$$= 0.00654 \text{ kg/kg of d.a.}$$

$$\therefore \quad t_{dp_2} = 8°C$$

$$\therefore h_2 = C_{p_a} T_2 + w_2 \left[h_{fg} + t_{dp_1} \times C_{pw} + 1.88(T_2 - t_{dp_2})\right]$$

$$= 1.005 \times 8 + 0.00654 \; [2482.5 + 8 \times 4.187 + 1.88 \; (8 - 8)]$$

$$= 24.5 \text{ kJ/kg}$$

From ideal gas equation

$$\dot{m}_a = \text{mass of dry air}$$

$$= \frac{p_a V}{RT_1}$$

$$= \frac{(1.0132 - 0.07389)}{0.287 \times (273 + 40)} \times \frac{600}{60} \times 100$$

$$= 10.5 \text{ kg/sec}$$

$\therefore$ amount of condensate

$$= \dot{m}_a (w_1 - w_2)$$

$$= 10.5 \; (0.025973 - 0.00654)$$

$$= 12.19 \text{ kg/min}$$ **Ans**

Amount of refrigeration

$$= \dot{m}_a \; (h_1 - h_2)$$

$$= 10.5 \; (106.4 - 24.5)$$

$$= 859.95 \text{ kW}$$ **Ans**

Problem 15.20 : Outside air at 35°C, 60% *RH* is passed over a cooling coil having an apparatus dew point of 10°C and 0.06 bupass factor. The cooled air is then supplied to a room which is to be maintained at 22°C dbt and 60% *RH*. THe supply air rate 4500 kg/hr. Estimate the sensible and latent heat loads of the room and the heat removed in the cooling coil, in kW. (IES – 1993)

Solution : Given :

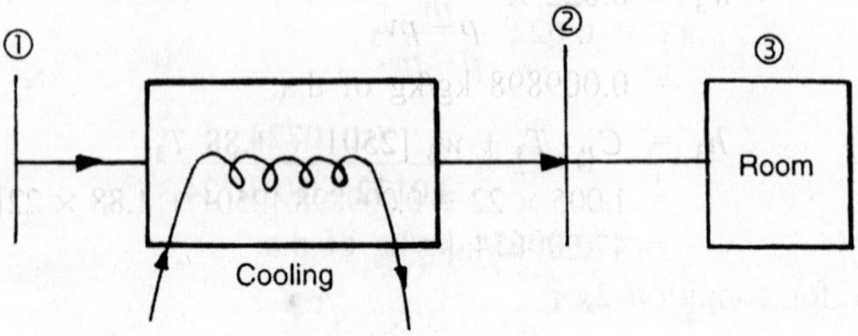

Fig. P. 15.20 (a)

$T_1 = 35°C$ dbt

$\phi_1 = 60\%$

A.D.P. = 10°C

B.P.F. = 0.06

$T_3 = 22°C$ dbt

$\phi_3 = 60\%$ *RH*

$\dot{m}$ = supply air rate = 4500 kg/hr

Let atmospheric pressure = p = 1.0132 bar

From steam table, at 35°C

$p_{sat} = 0.05628$ bar

$\therefore p v_1 = p_{sat} \times \phi_1 = .03377$ bar

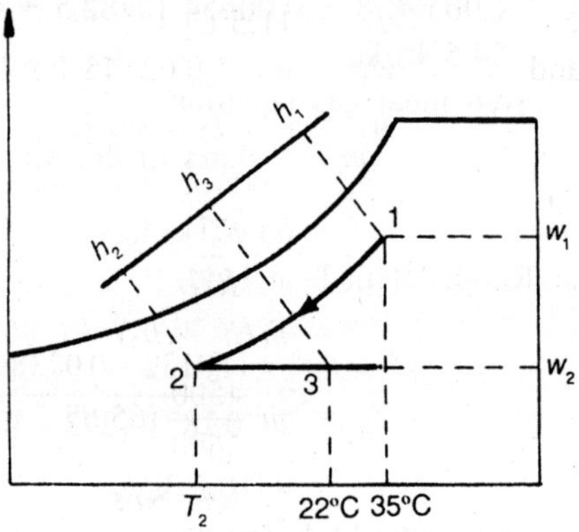

Fig. P. 15.20 (b)

$$\therefore \quad w_1 = 0.622 \times \frac{pv_1}{p - pv_1}$$

$$= 0.622 \times \frac{0.03377}{1.0132 - 0.03377}$$

$$= 0.02145 \text{ kg/kg of d.a.}$$

$$\therefore \quad h_1 = C_{p_a} T + w_1 [2501 + 1.88 \times T]$$

$$= 1.005 \times 35 + 0.02145 [2501 + 1.88 \times 35]$$

$$= 90.22 \text{ kJ/kg}$$

From steam table,

at $T_3 = 22°C$

$p_{sat} = 0.02645$ bar

$pv_3 = p_{sat} \times \phi_3 = 0.01587$ bar

$$\therefore \quad w_3 = 0.622 \times \frac{pv_3}{p - pv_3}$$

$$= 0.009898 \text{ kg/kg of d.a.}$$

$$\therefore \quad h_3 = C_{p_a} T_3 + w_3 [2501 + 1.88 T_3]$$

$$= 1.005 \times 22 + 0.009898 [2501 + 1.88 \times 22]$$

$$= 47.27 \text{ kJ/kg}$$

Now for condition 2,

$$\text{B.P.F.} = 0.06 = \frac{t_2 - t_{ADP}}{t_1 - t_{ADP}} = \frac{T_2 - 10}{35 - 10}$$

$$\therefore \quad T_2 = 11.5°C$$

and $w_1 = w_2 = 0.02145$ kg/kg of d.a.

$$\therefore \quad h_2 = C_{p_a} T_2 + w_2 [2501 + 1.88 T_2]$$

$$= 1.005 \times 11.5 + 0.02145 [2501 + 1.88 \times 11.5]$$

$$= 65.67 \text{ kJ/kg}$$

∴ Room latent heat load

$$= \dot{m} (h_2 - h_3)$$

$$= \dot{m} \frac{4500}{3600} [65.67 - 47.27]$$

$= 23$ kW **Ans**

Room sensible heat load

$$= \dot{m} (h_1 - h_2)$$

$= \frac{4500}{3600} (90.22 - 65.67) = 30.69$ kW **Ans**

∴ total heat removed

= S.L.H.L. + S.H.L.

$= 30.69 + 23$

$= 53.69$ kW **Ans**

Problem 15.21 : A small auditorium is required to be maintained at 22°C dry bulb temperature and 70% relative humidity. The ambient conditions are at 30°C dry bulb temperature and 75% relative humidity. The amount of free air circulated is 200 m^3/min. The required conditions are achieved by first cooling and dehumidifying through a cooling coil having apparatus dew point of 14°C and then by heating. With the help of psychrometric chart find :

(i) The cooling capacity of cooling coil in tons of refrigeration and its bypass factor.
(ii) The amount of water vapour removed by the cooling coil in kg/h.
(iii) Assuming the bypass factor as 0.2 determine the capacity of the heating coil in kW and its surface temperature. (IES – 1992)

Solution : Given :

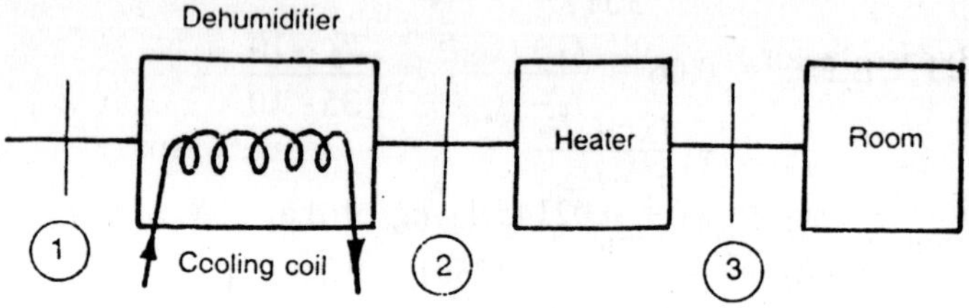

Fig. P. 15.21 (a)

$$t_3 = 22°C$$
$$\phi_3 = 70\%$$
$$t_1 = 30°C \text{ dbt}$$
$$\phi_1 = 75\%$$
$$v_a = \text{amount of free air circulated} = 200 \text{ m}^3\text{/min}$$

ADP temperature = t_{ADP} = 14°C

From psychrometric chart

at t_1 = 30°C dbt and ϕ_1 = 75%

h_1 = 85 kJ/kg

v_1 = specific volume

= 0.90154 m^3/kg

w_1 = 0.021 kg/kg of d.a.,

at t_3 = 22°C dbt and 70% *RH*

h_3 = 51 kJ/kg

w_3 = 0.011 kJ/kg of d.a.

$\therefore \dot{m}_a$ = mass rate of free air delivered

$$= \frac{v}{v_1}$$

$$= \frac{200}{60 \times 0.90154} \text{ kg/s}$$

h_2 = 44 kJ/kg

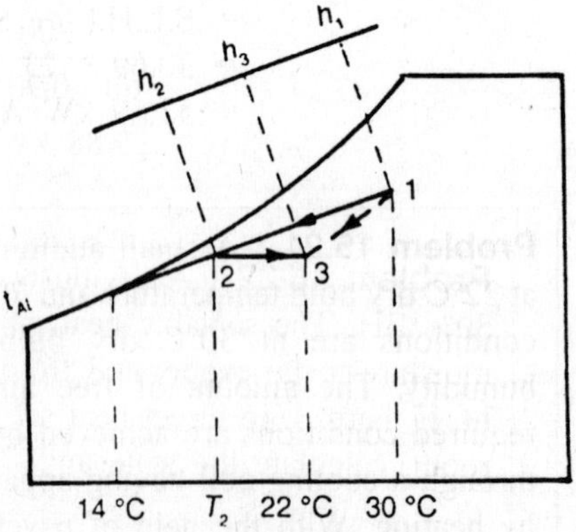

Fig. P. 15.21 (b)

$t_2 = 18°C$ dbt

(1) Cooling capacity of cooling coil

$$= \dot{m}_a \,(h_1 - h_2)$$

$$= \frac{200}{60 \times 0.90159} (85 - 44) = 151.5925 \text{ kW}$$

$$= \frac{151.5925}{3.517} = 43 \text{ Ton } [\because \text{ 1 Ton} = 3.517 \text{ kW}]$$

Bypass factor

$$= \frac{t_2 - t_{ADP}}{t_1 - t_{ADP}} = \frac{18 - 14}{30 - 14} = 0.25 \text{ Ans}$$

(2) amount of water removed by cooling coil

$$= \dot{m}_a \,(w_1 - w_2)$$

$$= \frac{200}{60 \times 0.90154} (0.021 - 0.011)$$

$$= 0.03697 \text{ kg/s} = 133.11 \text{ kg/hr}$$

(3) $$\text{B.P.F.} = 0.2 = \frac{T_{surface} - T_3}{T_{surface} - T_2}$$

$$= \frac{T_{surface} - 22}{T_{surface} - 18}$$

or $T_{surface} = 23°C$ **Ans**

Capacity of heating coil

$$= \dot{m} \,(h_3 - h_2)$$

$$= \frac{200}{60 \times 0.90154} (51 - 44)$$

$$= 25.88 \text{ kW } \textbf{Ans}$$

Problem 15.22 : A drying room is to be maintained at 32°C and 30% RH. The sensible heat gain to the roo mis 150000 kJ/hr. The moisture to be evaporated from the objects during drying is 18 kg/hr. If there is no direct heat source to provide for evaporation in the room, calculate the state and rate of supply air at 15°C dbt.

(IES – 1995)

Solution : Given :

$$t_2 = 32°C$$
$$\phi_2 = 30\%$$
$$Q = \text{sensible heat gain} = 150000 \text{ kJ/hr}$$

Moisture evaporated

$$= 18 \text{ kg/hr}$$
$$t_1 = 15°C$$

Now from steam table

$$\text{at } t_2 = 32°C,$$
$$p_{sat} = 0.04759 \text{ bar}$$

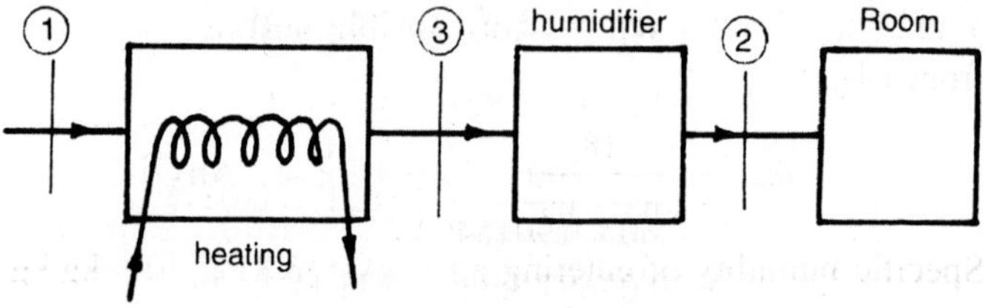

Fig. P. 15.22 (a)

$$\text{Let } p = \text{atmosphere pressure}$$
$$= 1.0132 \text{ bar}$$

$$\therefore \quad pv_2 = p_{sat} \times \phi = 0.014277 \text{ bar}$$

$$\therefore \quad w_2 = 0.622 \times \frac{pv_2}{p - pv_2}$$
$$= 8.89 \times 10^{-3} \text{ kg/kg of d.a.}$$

$$\therefore \quad h_2 = C_{p_a} t_2 + w_2 (2500 + 1.88\, t_2)$$
$$= 1.005 \times 32 + 8.89 \times 10^{-3} (2500 + 1.88 \times 32)$$
$$= 54.92 \text{ kJ/kg}$$

Sensible heat load $= (h_3 - h_1)\, \dot{m}_a$

or, $\dot{m}_a (h_3 - h_1) = 150000$ kg/hr ...(i)

In humidifier,

$\dot{m}_a (w_2 - w_3) = 18$ kg/hr ...(ii)

For sensible humidifying,

$$t_3 = t_2 = 32°C$$

$$\therefore \quad h_1 = C_{p_a} \times t_1 + w_3 (2500 + 1.88 \times t_1)$$

$\therefore\ 1.005 \times 15 + w_3\,(2500 + 1.88 \times 15)$

$= 15.08 + 2528.2\, w_3$

$$\therefore \quad h_3 = C_{p_a}\, t_3 + w_3\,(2500 + 1.88 \times 32)$$

$$= 1.005 \times 32 + w_3 \times 2560.16$$

$\therefore$ diving equation (i) and (ii), we get,

$$\frac{h_3 - h_1}{w_2 - w_3} = \frac{150000}{18}$$

$$\frac{[32.16 + w_3 \times 2560.16 - 15.08 - w_3 \times 2528.2]}{w_2 - w_3} = \frac{150000}{18}$$

or $\quad 17.08 + 31.96\, w_3 = 74.08 - 8333.33\, w_3$

or $\quad w_3 = 6.81 \times 10^{-3}$ kg/kg of d.a.

$w_1 =$ for sensible heating

from (ii),

$$\dot{m}_a = \frac{18}{w_2 - w_3} = 2.404 \text{ kg/sec } \textbf{Ans}$$

Specific humidity of entering air $= w_1 = 6.81 \times 10^{-3}$ kg/kg of d.a.

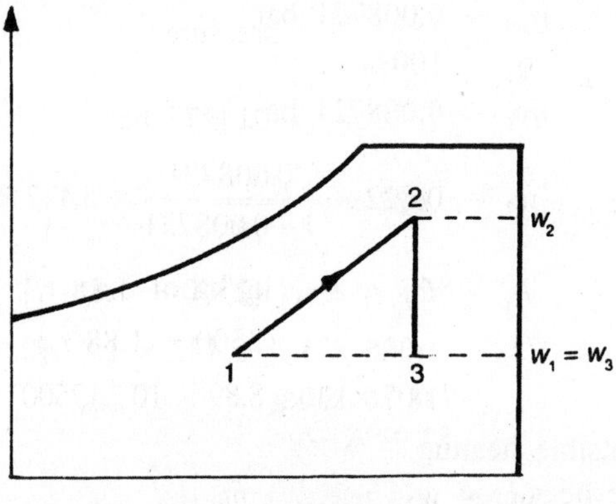

Fig. P. 15.22 (b)

Problem 15.23 : Moist air is heated by steam condensing inside the tubes of a heating coil as shown by Fig. P 15.23. Part of the air passes through the coil and part is by-passed around the coil. Barometric pressure is 1 bar.

Determine (i) the air per min (in 2) which by-pass the coil and (ii) the heat added by the coil. (IES – 1997)

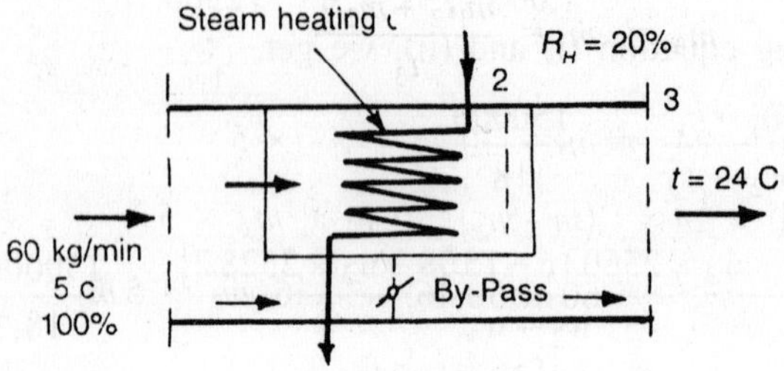

Fig. P. 15.23

Solution : Given :

$$\dot{m} = 60 \text{ kg/min},$$
$$p = \text{atmospheric pressure} = 1 \text{ bar}$$

Let $\dot{m}_1$ pass through the heating coil and $\dot{m}_2$ is bypassed.

For Section 1

at 5°C, from steam table

$$p_{sat} = 0.008721 \text{ bar}$$
$$\phi_1 = 100\%$$
$$\therefore \quad pv_1 = 0.008721 \text{ bar}$$
$$\therefore \quad w_1 = 0.622 \times \frac{0.008721}{1 - 0.008721} = 5.472 \times 10^{-3} \text{ kg/kg of d.a.}$$
$$\therefore \quad h_1 = C_{p_a} t_1 + w_1 [2500 + 1.88 \, t_1]$$
$$= 1.005 \times 5 + 5.472 \times 10^{-3} [2500 + 1.88 \times 5]$$
$$= 18.76 \text{ kJ/kg}$$

For sensible heating

$$w_1 = w_2$$
$$\therefore \quad pv_1 = pv_2 = 0.008721 \text{ bar}$$
$$\text{or} \quad p_{sat} = \frac{pv_1}{\phi_2} = \frac{0.008721}{0.2} = 0.04361 \text{ bar}$$

Now interpolatin from steam table,

$$t_2 = 30.46°\text{C}$$
$$\therefore \quad h_2 = C_{p_a} t_2 + w_2 [2500 + 1.88 \times t_2]$$

$= 1.005 \times 30.46 + 5.472 \times 10^{-3} [2500 + 1.88 \times 30.46]$
$= 44.61$ kJ/kg

Now after mixing of by pass air and heated air

$$m_3 = \dot{m} = \frac{\dot{m}_1 t_2 + \dot{m}_2 t_1}{t_3}$$

or $$60 \times 24 = \dot{m}_1 \times 30.46 + \dot{m}_2 \times 5$$

or $$60 \times 24 = (\dot{m} - \dot{m}_2)\ 30.46 + \dot{m}_2 \times 5$$
$$= 60 \times 30.46 - 30.46\ \dot{m}_2 + 5\dot{m}_2$$

or $$\dot{m}_2 = \frac{60[30.46 - 24]}{25.46} = 15.22 \text{ kg/min } \textbf{Ans}$$

$$\dot{m}_1 = \dot{m} - \dot{m}_2 = 60 - 15.22$$
$$= 44.78 \text{ kg/min}$$

(ii) heat added by the coil

$$= \dot{m}_1\ (h_2 - h_1)$$
$$= \frac{44.78}{60} \times (44.61 - 18.76)$$
$$= 19.29 \text{ kW } \textbf{Ans}$$

Chapter 16

Compressible Fluid Flow

Problem.16.1 : Air in a reservoir has a temperature of 27°C and a pressure of 0.8 MPa. The air is allowed to escape through a channel at a rate of 2.5 kg/s. Assuming that the air velocity in the reservoir is regligible and that the flow through the channel is isentropic, find the Mach number, the velocity, and the area at a section in the channel where the static pressure is 0.6 MPa.

Solution : Given data :

Reservoir temperature = T_0 = 27°C = 27 + 273 = 300 K

Reservoir pressure = p_0 = 0.8 MPa

Air velocity in reservoir = $V_0 = 0$

Mass flow rate in channel = $\dot{m}$ = 2.5 kg/s

Static pressure = p = 0.6 MPa

Flow through channel in isen tropic

Since inside the tank stagnation conditions are exit

so, p_0 and T_0 are stagnation pressure and temperature

$$\therefore \quad \frac{p_0}{p} = \left(1+\frac{\gamma-1}{2}M^2\right)^{\frac{\gamma}{\gamma-1}}$$

$$\text{or,} \quad \frac{0.8}{0.6} = \left(1+\frac{1.4-1}{2}M^2\right)^{\frac{1.4}{1.4-1}}$$

$$\therefore \quad M = 0.654 \textbf{ Ans}$$

Again $$\frac{T_0}{T} = \left(1+\frac{\gamma-1}{2}M^2\right)$$

or, $$T = \frac{T_0}{\left(1+\frac{\gamma-1}{2}M^2\right)} = \frac{300}{\left[1+\frac{0.4}{2}\times(0.654)^2\right]}$$

$$= 276.36 \text{ K}$$

$\therefore$ Sonic velocity $= C = \sqrt{\gamma RT}$

$$= \sqrt{1.4\times 287\times 276.36} = 333.3 \text{ m/s.}$$

$$M = \frac{V}{C}$$

$\therefore$ Velocity of air in the channel = M.C

$= 0.654 \times 333.3 = 217.98$ m/s **Ans**

$\dot{m} = \rho AV$, where ρ = density

and A = Area at a section in the channel

Applying ideal gas eqn.

$$V = \frac{RT}{p_1} = \frac{1}{\rho}$$

$$A = \frac{\dot{m}}{V}\frac{RT}{p_1} = \frac{25\times 287\times 276.36}{217.98\times 0.6\times 10^3}$$

$$= 0.1516 \times 10^{-4} \text{ m}^2 = 15.16 \text{ cm}^2$$

Problem.16.2 : A supersonic wind tunnel nozzle is to be designed for $M = 2$, with a throat section, 0.11 m^2 in area. The supply pressure and temperature and temperature at the nozzle inlet, where the velocity is negligible are 70 kPa and 37°C respectively. Compute the mass flow rate, the exit area, and the fluid properties at the throat and exit. Take γ = 1.4

Solution : Given data :

M_2 = Mach no at exit = 2

A_1 = Throat area = 0.11 m^2

p_0 = 70 kPa

$$T_0 = 37°C = 37 + 273 = 310 \text{ K}$$
$$V_0 = 0$$

Inlet conditions are taken as stagnation properties because velocity = 0

For a supersonic wind tunnel, throat Mach number is 1

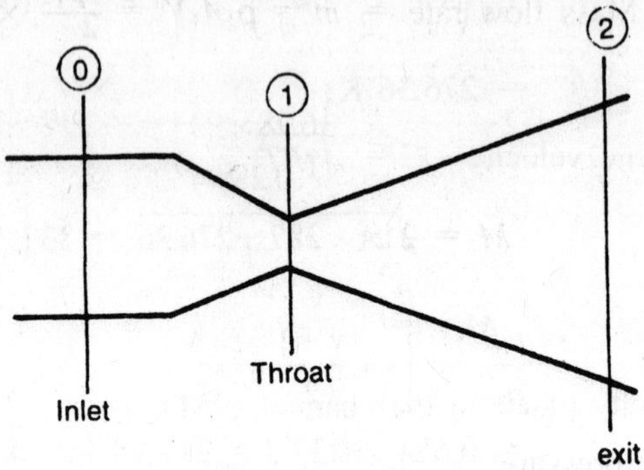

Fig. P. 16.2

$$\therefore \quad \frac{A_2}{A_1} = \left[\left(\frac{2}{\gamma+1}\right)\left(1+\frac{\gamma-1}{2}M_2^2\right)\right]^{\frac{\gamma+1}{2(\gamma-1)}} \times \frac{1}{M_2}$$

$$\therefore \text{exit area, } A_2 = \left[\left(\frac{2}{1+1.4}\right)\left(1+\frac{1.4-1}{2}\times 2^2\right)\right]^{\frac{1+1.4}{2(1.4-1)}} \times \frac{1}{2} \times 0.11$$

$$= 0.1856 \text{ m}^2$$

For $M_1 = 1$

$$\frac{p_1}{p_0} = \left(\frac{2}{\gamma+1}\right)^{\frac{\gamma}{\gamma-1}},$$

$$\frac{T_1}{T_0} = \left(\frac{2}{\gamma+1}\right)$$

$$\therefore \quad p_1 = \text{Pressure at throat} = 70\left(\frac{2}{1.4+1}\right)^{\frac{1.4}{1.4-1}}$$

$$= 36.98 \text{ kPa}$$

$$T_1 = \text{Temperature at throat} = 310\left(\frac{2}{1.4+1}\right)$$

$$= 258.33 \text{ K}$$

$$\text{Velocity at throat} = V_1 = MC_1 = 1 \times \sqrt{\gamma R T_1}$$

$$= \sqrt{1.4 \times 287 \times 258.33} = 322.18 \text{ m/s}$$

$$\text{Mass flow rate} = \dot{m} = \rho_1 A_1 V_1 = \frac{p_1}{RT_1} \times A_1 V_1$$

$$= \frac{36.98 \times 0.11 \times 322.18}{0.287 \times 258.33} = 17.677 \text{ kg/s}$$

For $M = 2$

$$\frac{p_0}{p_2} = \left[1 + \frac{\gamma - 1}{2} M_2^2\right]^{\frac{\gamma}{\gamma - 1}}$$

Exit pressure,

$$p_2 = \frac{p_0}{\left[1 + \frac{\gamma - 1}{2} M_2^2\right]^{\frac{\gamma}{\gamma - 1}}}$$

$$= \frac{70}{\left[1 + \frac{1.4 - 1}{2} \times 2^2\right]^{\frac{1.4}{0.4}}} = 8.95 \text{ kPa}$$

Similarly, exit temperature $= T_2$

$$= \frac{T_0}{1 + \frac{\gamma - 1}{2} M_2^2} = \frac{310}{1 + \frac{1.4 - 1}{2} \times 2^2}$$

$$= 172.22 \text{ K}$$

Sonic velocity,

$$C_2 = \sqrt{\gamma R T_2}$$

$$= \sqrt{1.4 \times 287 \times 172.22} = 263.06 \text{ m/s}$$

$$\therefore \text{ Exit velocity} = V_2 = M_2 C_2 = 2 \times 263.06 = 526.11 \text{ m/s}$$

Problem.16.3 : An ideal gas flows into a convergent nozzle at a pressure of 0.565 MPa, a temperature of 280°C, and negligible velocity. After reversible adiabatic expansion in the nozzle the gas flows directly into a large vessel. The gas in the vessel may be maintained at any specified state while the nozzle supply state is held constant. The exit area of the nozzle is 500 mm². For this gas γ = 1.3 and C_p = 1.172 kJ/kg.K. Determine (a) the pressure of the gas leaving the nozzle when its temperature is 225°C, and (b) the gas mass flow rate when the pressure in the vessel is 0.21 MPa.

Solution : Given data :

$$p_0 = 0.565 \text{ MPa}$$

$$T_0 = 280°C = 280 + 273 = 553 \text{ K}$$

$$V_0 = 0$$

$$A_1 = \text{exit area of nozzle} = 500 \text{ mm}^2$$

$$T_1 = 225°C = 225 + 273 = 498 \text{ K}$$

Inlet condition are taken as stagnation properties.

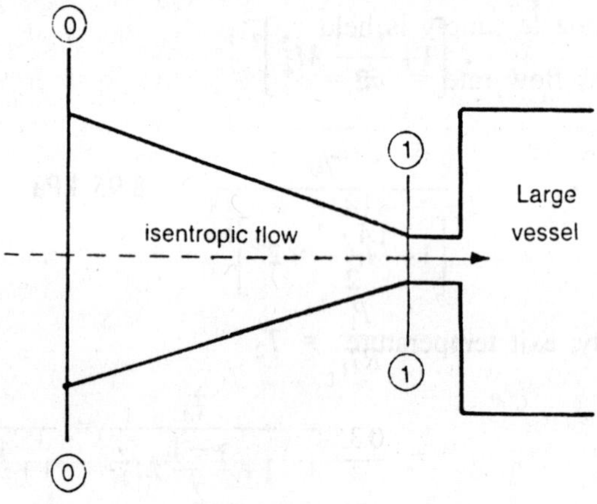

Fig. P. 16.3

$$\frac{C_p}{C_v} = \gamma = 1.3, \; C_P = 1.172 \text{ kJ/kg}$$

$$R = C_p - C_v = C_p - \frac{C_p}{\gamma} = \frac{C_p(\gamma - 1)}{\gamma}$$

$$= \frac{1.172 \times 0.3}{1.3} = 0.270$$

$$\frac{T_0}{T_1} = 1 + \frac{\gamma - 1}{2} M_1^2 = \left(\frac{P_0}{P_1}\right)^{\frac{\gamma-1}{\gamma}}$$

or, $$\frac{553}{498} = 1 + \frac{1.3 - 1}{2} M_1^2$$

$\therefore$ $$M_1 = 0.858$$

(a) Again, $$p_1 = \text{nozzle pressure} = \frac{P_0}{\left[1 + \frac{\gamma - 1}{2} M_1^2\right]^{\frac{\gamma}{\gamma-1}}}$$

$$= \frac{0.565}{\left[1 + \frac{0.3}{2} \times 0.858^2\right]^{\frac{1.3}{.3}}} = 0.358 \text{ MPa } \textbf{Ans}$$

(b) Since nozzle supply is held constant

gas mass flow rate $= \dot{m} = \rho_1 A_1 V_1$

sonic velocity, $$C_1 = \sqrt{\gamma R T_1} = \sqrt{1.3 \times 0.270 \times 10^3 \times 498}$$

$$= 418.09 \text{ m/s}$$

$\therefore$ $$V_1 = M_1 C_1 = 418.09 \times 0.858 = 358.72 \text{ m/s}$$

$\therefore$ $$\dot{m} = \frac{P_1}{RT_1} \times A_1 \times V_1$$

$$= \frac{0.358 \times 500 \times 10^{-6} \times 10^3 \times 358.72}{0.27 \times 498}$$

$$= 0.48 \text{ kg/s } \textbf{Ans}$$

Problem.16.4 : Air flows adiabatically through a pipe with a constant area. At point 1, the stagnation pressure is 0.35 MPa and the Mach number is 0.4. Further downstream the stagnation pressure is found to be 0.25 MPa. What is the mach number at the second point for subsonic flow ?

Solution : Given data :

At section

(1) – (1)

$p_{01} = 0.35$ MPa

$M_1 = 0.4$

At section

(2) – (2)

$p_{02} = 0.25$ MPa

$M_2 = ?$

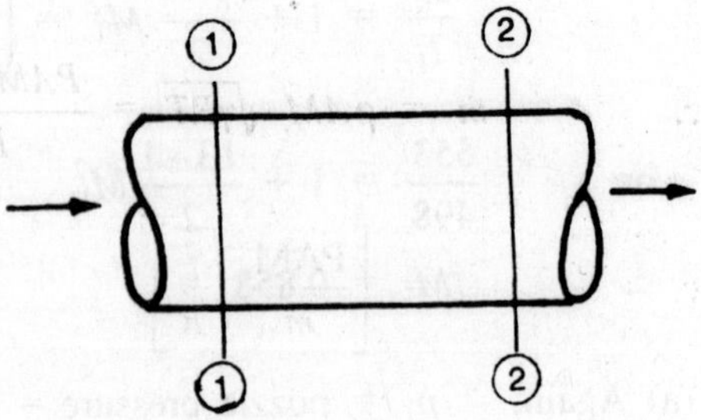

Fig. P. 16.4

we know that

$$M_2^2 = \frac{M_1^2 + \dfrac{2}{\gamma - 1}}{\dfrac{2\gamma}{\gamma - 1} M_1^2 - 1}$$

Above equation valid for adiabatic flow in constant duct with friction in one dimentional model

$$\therefore \quad M_2^2 = \frac{M_1^2(\gamma - 1) + 2}{2\gamma\left(M_1^2\right) - (\gamma - 1)} = \frac{0.4^2 \times 0.4 + 2}{2 \times 1.4 \times 0.4^2 - 0.4}$$

$$\therefore \quad M_2 = 6.56 \textbf{ Ans}$$

> **Problem.16.5 :** The intake duct to an axial flow air compressor has a diameter of 0.3 m and compresses air at 10 kg/s. The static pressure inside the duct is 67 kPa and the stagnation temperature is 40°C. Calculate the Mach number in the duct.

Solution : Given data :

duct dia $= d = 0.3$ m

$\dot{m}$ = mass flow rate = 10 kg/s

p = Static pressure in the duct = 67 kPa

T_0 = Stagnation temperature = 40°C

= 40 + 273 = 313 K

Flowing fluid in air

Let T be the static temperature inside the duct

Now $\quad \dot{m} = \rho AV$

$$\because \quad M = \frac{V}{C} = \frac{V}{\sqrt{\gamma RT}} \text{ and } \rho = \frac{p}{RT}$$

$$\therefore \quad \dot{m} = \rho AM\sqrt{\gamma RT} = \frac{PAM\sqrt{\gamma RT}}{RT} = PAM \times \sqrt{\frac{\gamma}{RT}}$$

$$\therefore \quad T = \left[\frac{PAM}{\dot{m}}\sqrt{\frac{\gamma}{R}}\right]^2$$

$$= \left(\frac{67}{10} \times \frac{\pi}{4} \times 0.3^2 \times M\sqrt{\frac{1.4}{287}} \times 10^3\right)^2$$

$$T = 1094.11\ M^2$$

$$\frac{T_0}{T} = 1 + \frac{\gamma - 1}{2} M^2$$

$$\Rightarrow \quad \frac{313}{1094.11\ M^2} = 1 + 0.2\ M^2$$

$$\Rightarrow \quad 0.2\ M^4 + M^2 = 0.2861$$

$$\Rightarrow \quad M^4 + 5\ M^2 - 1.4304 = 0$$

$$\Rightarrow \quad M^2 = \frac{-5 \pm \sqrt{(5)^2 + 4 \times 1 \times 1.4304}}{2}$$

$$= 0.2714 \text{ (taking } +v_e \text{ sign only)}$$

$$\therefore \quad M = 0.521$$

Problem.16.6 : Show that for an ideal gas the fraction change in pressure across a small pressure pulse is given by

$$\frac{dp}{p} = \gamma \frac{dV}{c}$$

and that the fractional change in absolute temperature is given by

$$\frac{dT}{T} = (\gamma - 1)\frac{dV}{c}$$

Solution : Given data :

(a) For an ideal gas, in an isentropic process

pv^{γ} = constant = k

$$\Rightarrow \frac{p}{\rho^{\gamma}} = k$$

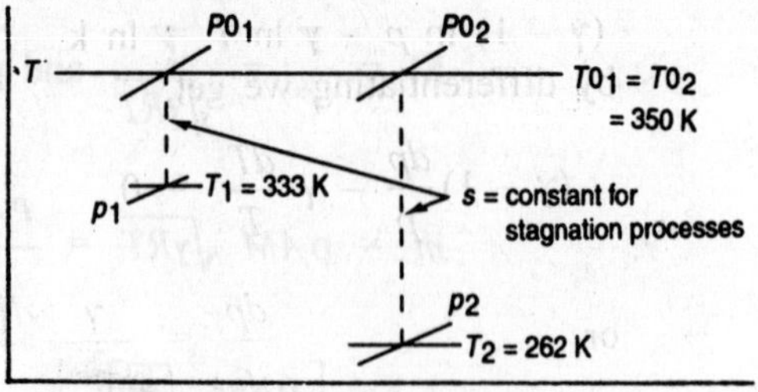

Fig. P. 16.6

taking logic then both the sides, we get,

$$\log p - \gamma \log \rho = \log k$$

by defferentiating we get,

$$\text{or,} \quad \frac{dp}{p} - \gamma \frac{d\rho}{\rho} = 0$$

$$\text{or,} \quad \frac{dp}{p} = \gamma \frac{d\rho}{\rho} \qquad ...(1)$$

Let us consider an infinitesimal pressure pulse initiated by a slight movement of a piston to right which change the property

pressure p to $p + dp$

density ρ to $\rho + d\rho$

velocity c to $c - dv$

Assuming cross sectional area of the duct constant. The continuity equation for the control volume is

$$\rho Ac = (\rho + d\rho) A (c - dv)$$

$$\Rightarrow \quad \rho c = \rho c + cd\rho - \rho dv - d\rho.dv$$

Neglecting $d\rho.dv$

$$\Rightarrow \quad cd\rho = \rho dv$$

$$\text{or} \quad \frac{d\rho}{\rho} = \frac{dV}{c}$$

∴ from (i)

$$\frac{dp}{p} = \gamma \frac{dV}{c} \quad \text{Proved}$$

(b) For an ideal gas, in an isentropic process

$$\frac{p^{\gamma - 1}}{T^{\gamma}} = \text{constasnt} = k$$

taking logarithem we get

$$(\gamma - 1) \ln p - \gamma \ln T = \ln k$$

by differentiating we get,

$$(\gamma - 1)\frac{dp}{p} - \gamma \frac{dT}{T} = 0$$

or $$\frac{dp}{p} = \frac{\gamma}{\gamma - 1} \frac{dT}{T} \quad \ldots(2)$$

From last proved relation

$$\frac{dp}{p} = \gamma \frac{dV}{c} \quad \ldots(3)$$

Equating (2) and (3), We get

$$\frac{dT}{T} = (\gamma - 1) \frac{dV}{c}$$ **Proved**

Problem.16.7 : An airplane flies at an altitude of 13,000 m (temperature −55°C, pressure 18.5 kPa) with a speed of 180 m/s. Neglecting frictional effects, calculate (a) the critical velocity of the air relative to the aircraft, and (b) the maximum possible velocity of the air relative to the aircraft.

Solution : Given :

V = speed of air plane

= 180 m/s

at 13,000 m attitude,

p = 18.5 kPa

T = −55°C = 218 K

T_0 = Stagnation temp.

$$= T + \frac{V^2}{2C_p}$$

$$= 218 + \frac{180 \times 180}{2 \times 1.005 \times 1000}$$

= 234.12 K

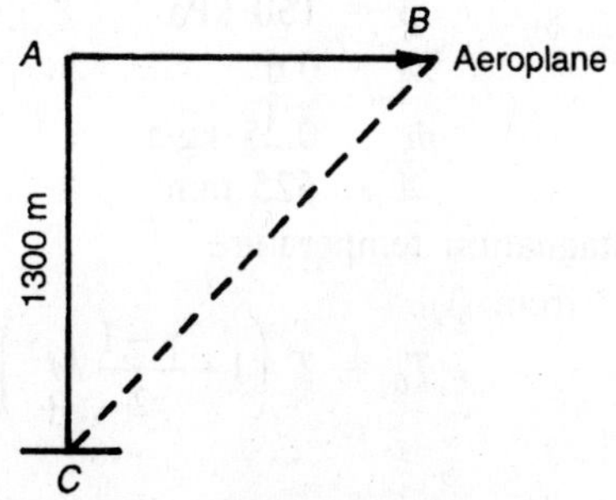

Fig. P. 16.7

(a) For critical velocity of air, $M = 1$

$$\frac{T^*}{T_0} = \left(\frac{Z}{\gamma+1}\right) = \frac{2}{2.4}$$

$$\Rightarrow \quad T^* = 195.1 \text{ K}$$

$\therefore$ Critical velocity $= C^* = \sqrt{\gamma R T^*} = 280$ m/s

(b) For maximum possible velocity of air, M

$$M^* = \sqrt{\frac{\gamma+1}{\gamma-1}} = \sqrt{\frac{1+1.4}{1.4-1}} = 2.4495$$

$$C^* = \sqrt{\gamma R T^*} = 280 \text{ m/s}$$

$$\therefore \quad V = M^* \, C^* = 2.4495 \times 280 = 685.80 \text{ m/s}$$

Problem.16.8 : A stream of air flowing in a duct is at a pressure of 150 kPa, has a Mach number of 0.6, and flows at a rate of 0.25 kg/s. The cross-sectional area of the duct is 625 mm². (a) Compute the stagnation temperature of the stream in °C. (b) What is the maximum percentage reduction in area which could be introduced without reducing the flow rate of the stream ? (c) For the maximum area reduction of part (b), find the velocity and pressure at the minimum area, assuming no friction and heat transfer.

Solution : Given data :

$$p = 150 \text{ kPa}$$
$$M = 0.6$$
$$\dot{m} = 0.25 \text{ kg/s}$$
$$A = 625 \text{ mm}^2$$

(a) Stagnation temperature

$$T_o = T\left(1 + \frac{\gamma-1}{2} M^2\right)$$

Now $$\dot{m} = \rho A V = \frac{p}{RT} \times AM\sqrt{\gamma R T}$$

$$= PAM\sqrt{\frac{\gamma}{RT}}$$

or, $$T = \left(\frac{PAM}{\dot{m}}\sqrt{\frac{\gamma}{R}}\right)^2$$

$$= \left[\frac{150 \times 1000 \times 625 \times 10^{-6} \times 0.6}{0.25}\sqrt{\frac{1.4}{287}}\right]^2$$

$$= 246.95 \text{ K}$$

$\therefore$ $$T_0 = 246.95\left[1 + \frac{1.4 - 1}{2}0.6^2\right]$$

$$= 264.73 \text{ K} = -8.27°\text{C } \textbf{Ans}$$

(b) The maximum area reduction possible is upto throat area from critical pressure ratio.

$\therefore$ $$\frac{A}{A^*} = \left[\left(\frac{2}{\gamma+1}\right)\left(1 + \frac{\gamma - 1}{2}M^2\right)\right]^{\frac{\gamma+1}{2(\gamma-1)}} \times \frac{1}{M}$$

$\Rightarrow$ $$\frac{625}{A^*} = \left[\left(\frac{2}{1.4+1}\right)\left(1 + \frac{0.4}{2} \times 0.6^2\right)\right]^{\frac{1+1.4}{2(1.4-1)}} \times \frac{1}{0.6}$$

$$A^* = 526.01 \text{ mm}^2$$

$\therefore$ % Reduction of area

$$= \frac{625 - 526.01}{625} = 15.84\% \textbf{ Ans}$$

(c) For critical condition at the throat

$$\frac{T^*}{T_0} = \frac{2}{\gamma + 1}$$

or, $$T^* = T_o\left(\frac{2}{\gamma + 1}\right)$$

$$= 264.73 \times \frac{2}{2.4} = 220.61 \text{ K}$$

$$V^* = M \times C = 1 \times \sqrt{\gamma R T^*}$$

$$= \sqrt{1.4 \times 287 \times 220.01} = 297.73 \text{ m/s}$$

$$\frac{p^*}{p_0} = \left(\frac{2}{\gamma+1}\right)^{\frac{\gamma}{\gamma-1}}$$

where p_0 = Stagnation pressure

$$= p\left[1+\frac{\gamma-1}{2}M^2\right]^{\frac{\gamma}{\gamma-1}}$$

$$= 150\left[1+\frac{0.4}{2}\times 0.6^2\right]^{\frac{1.4}{0.4}} = 191.33 \text{ kPa}$$

$$\therefore \quad p^* = p_0\left(\frac{2}{\gamma+1}\right)^{\frac{\gamma}{\gamma-1}}$$

$$= 191.33\left(\frac{2}{1.4+1}\right)^{\frac{1.4}{0.4}} = 101.07 \text{ kPa}$$

Problem.16.9 : A rocket combustion chamber is supplied with 12 kg/s of hydrogen and 38 kg/s of oxygen. Before entering the nozzle all the oxygen is consumed, the pressure is 2.3 MPa, and the temperature is 2800°C. Neglecting dissociation and friction, find the throat area of the nozzle required. Assume $\gamma = 1.25$

***Solution* :** Given data :

$$\dot{m}_{H_2} = 12 \text{ kg/s}$$

$$\dot{m}_{O_2} = 38 \text{ kg/s}$$

$$p = 2.3 \text{ MPa}$$

$$T = 2800°C = 3073 \text{ K}$$

For hydrogen,

$$\dot{m}_{H_2} = 12 \text{ kg/s}$$

$$R_{H_2} = \frac{\overline{R}}{M_{H_2}} = \frac{8.314}{32} = 4.157 \text{ kJ/kg.K}$$

For oxygen,

$$R_{O_2} = \frac{\bar{R}}{M_{O_2}} = \frac{8.314}{32} = 0.2598 \text{ kJ/kg.K}$$

$$\therefore \quad R_m = \frac{m_{H_2} \cdot R_{H_2} + m_{O_2} \cdot R_{O_2}}{(m_{H_2} + m_{O_2})}$$

$$= \frac{12 \times 4.157 + 38 \times 0.2598}{(12 + 38)}$$

$$= 1.195 \text{ kJ/kg.K}$$

$\therefore$ From ideal gas eqn

$$\rho = \frac{p}{R_m T} = \frac{2.3 \times 10^3}{1.195 \times 3073} = 0.6263 \text{ kg/m}^3$$

$$C^* = V^* = \sqrt{\gamma R_m T}$$

$$= \sqrt{1.25 \times 1.195 \times 1000 \times 3073}$$

$$= 2142.5 \text{ m/s}$$

Now $\quad \dot{m} = (\dot{m}_{H_2} + \dot{m}_{O_2}) = \rho A V^*$

$$A = \frac{\dot{m}}{\rho V^*} = \frac{50}{0.6263 \times 2142.5} = 0.03726 \text{ m}^2$$

Problem.16.10 : A gas with specific heat ratio $\gamma = 1.4$ passes through a plane normal shock immediately before the shock the Mach number is 2, and the stagnation pressure and temperature are 600 kPa and 400 K respectively. Find the static pressure and static temperature immediately downstream of the shock.

Solution : Given data :

$$M_1 = 2$$
$$p_{01} = 600 \text{ kPa}$$
$$T_{01} = 400 \text{ K}$$

Now static tempr. before the shock =

$$T_1 = \frac{To_1}{1 + \frac{\gamma - 1}{2} M_1^2}$$

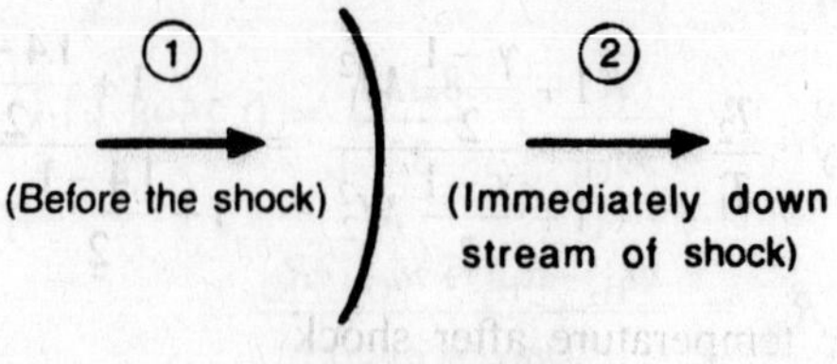

Fig. P. 16.10

$$= \frac{400}{\left(1 + \frac{0.4}{2} \times 4\right)^{\frac{1.4}{0.4}}} = 222.22 \text{ K}$$

Pressure before the shock

$$p_1 = \frac{Po_1}{\left(1 + \frac{\gamma - 1}{2} M_1^2\right)^{\frac{\gamma}{\gamma - 1}}} = \frac{600}{\left(1 + \frac{0.4}{2} \times 4\right)^{\frac{1.4}{0.4}}}$$

$$= 76.68 \text{ kPa}$$

M_2 = mach no. after the shock

$$= \sqrt{\frac{(\gamma - 1) M_1^2 + 2}{2\gamma M_1^2 - (\gamma - 1)}}$$

$$= \sqrt{\frac{(1.4 - 1) \times 4 + 2}{2 \times 1.4 \times 4 - (1.4 - 1)}} = 0.5774$$

Now from relation,

$$\frac{p_2}{p_1} = \frac{1 + \gamma M_1^2}{1 + \gamma M_2^2}$$

or static pressure,

$$p_2 = p_1 \left(\frac{1 + \gamma M_1^2}{1 + \gamma M_2^2}\right)$$

$$= 76.68 \left[\frac{1 + 1.4 \times 4}{1 + 1.4 \times (0.5774)^2}\right] = 345.04 \text{ kPa}$$

Similarly
$$\frac{T_2}{T_1} = \frac{1+\frac{\gamma-1}{2}M_1^2}{1+\frac{\gamma-1}{2}M_2^2} = \frac{1+\frac{1.4-1}{2}\times 4}{1+\frac{1.4-1}{2}\times(0.5774)^2}$$

∴ Static temperature after shock

$$T_2 = 374.99 \approx 375 \text{ K}$$

Problem.16.11 : Air flowing through a nozzle encounters a shock. The mach number upstream of the shock is 1.8, and the static temperature downstream of the shock is 450 K. How much has the velocity changed across the shock?

Solution : Given data :

For upstream of the shock → $M_1 = 1.8$

For downstream of the shock → T_2 = static temperature = 450 K

Down stream velocity,

$$V_2 = M_2 \times \sqrt{\gamma R T_2}$$

Now
$$M_2 = \sqrt{\frac{(\gamma-1)M_1^2+2}{2\gamma M_1^2-(\gamma-1)}}$$

$$= \sqrt{\frac{(1.4-1)\times 1.8^2+2}{2\times 1.4\times 1.8^2-(1.4-1)}} = 0.6165$$

∴
$$V_2 = 0.6165\times\sqrt{1.4\times 287\times 450} = 262.15 \text{ m/s}$$

Now
$$\frac{T_1}{T_2} = \left(\frac{1+\frac{\gamma-1}{2}M_2^2}{1+\frac{\gamma-1}{2}M_1^2}\right)$$

$$= \frac{1+\frac{1.4-1}{2}\times(0.6165)^2}{1+\frac{1.4-1}{2}\times(1.8)^2}$$

∴
$$T_1 = T_2\times 0.652921 = 293.81 \text{ K}$$

∴ Upstream velocity

$$V_2 = M_1 \times \sqrt{\gamma R T_1}$$

$$= 1.8\sqrt{1.4 \times 287 \times 293.81} = 618.46 \text{ m/s}$$

∴ Change of velocity

$$V_2 - V_1 = 356.31 \text{ m/s}$$

Problem.16.12 : A normal shock occurs in the diverging section of a nozzle under steady flow canditions at a point where the air is at 150 kPa, and 300 K travelling at 1000 m/s. What are the pressure and temperature on the subsonic side of the front 2. If the surrounding are at 20°C, what is the irreversibility caused by the shock process.

Solution : Given data :

p_x = 150 kPa

T_x = 300 K

V_x = 1000 m/s.

p_y = ?

T_y = ?

C_x = sonic velocity of air

$$= \sqrt{\gamma R T_x}$$

$$= \sqrt{1.4 \times 287 \times 300}$$

= 347.19 m/s

$\therefore M_x$ = mach no.

$$= \frac{V_x}{c_x}$$

$$= \frac{1000}{347.19} = 2.88$$

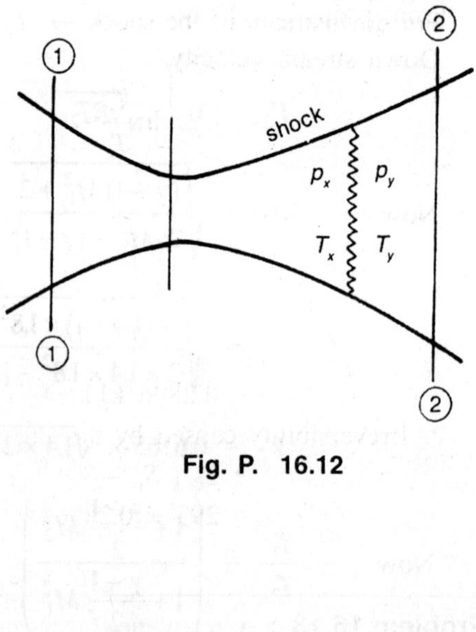

Fig. P. 16.12

∴ M_y = mach no of down stream of shock

$$\sqrt{\frac{(\gamma - 1)M_x^2 + 2}{2\gamma M_x^2 - (\gamma - 1)}} = 0.4827$$

or, T_y = temperature on subsonic side

$$= T_x \times \frac{1+\frac{0.4}{2}(2.88)^2}{1+\frac{0.4}{2}(0.4827)^2} = 762.1 \text{ K}$$

Now $$\frac{p_y}{p_x} = \frac{M_x}{M_y}\sqrt{\frac{1+\left(\frac{\gamma-1}{2}\right)M_x^2}{1+\left(\frac{\gamma-1}{2}\right)M_y^2}}$$

$$\therefore \quad p_y = p_x \frac{\left[1+0.4\times(2.88)^2\right]}{\left[1+0.4\times(0.4827)^2\right]} = 1426.5 \text{ kPa}$$

From property relation,

$$S_y - S_x = C_p\left[\ln\frac{T_y}{T_x} - \ln\left(\frac{p_y}{p_x}\right)^{\frac{\gamma-1}{\gamma}}\right]$$

$$= 1.005\left[\ln\frac{762.1}{300} - \ln\left(\frac{1426.5}{150}\right)^{\frac{0.4}{1.4}}\right]$$

$$= 0.2896 \text{ kJ/kg.K}$$

$\therefore$ Irreversibility caused by the shock

$$= T_0\,(S_y - S_x)$$

$$= 293 \times 0.2896 = 84.84 \text{ kJ/kg.}$$

Problem.16.13 : A jet plane travels through the air with a speed of 1000 km per hour at an altitude of 6000 m, where the pressure is 40 kPa and the temperature is –12°C. Consider the diffuser of the engine. The air leaves the diffuser with a velocity of 100 m/s. Determine the pressure and temperature leaving the diffuser, and the ratio of the inlet to the exit of the diffuser, assuming isentropic flow.

Solution : Given data :

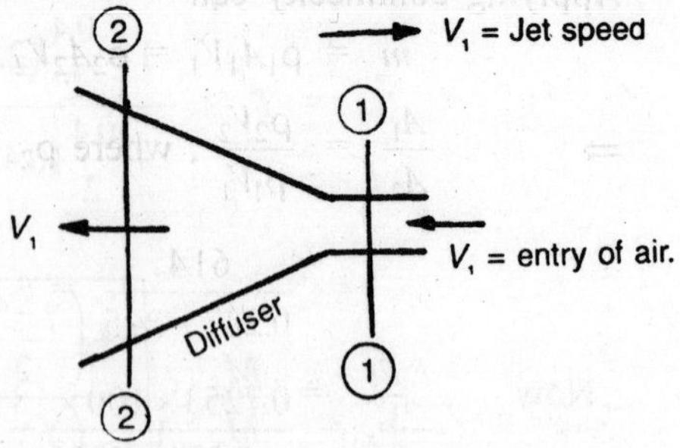

Fig. P. 16.13

V_1 = speed of jet plane

$= 1000 \times \frac{5}{18}$ m/s

A_0 = Altitude

= 6000 m

p_1 = 40 kPa

= static pressure of air

$T_1 = -12°C = 12 + 273$

= 261 K = static temperature of air

V_2 = Velocity of air leaving the diffusser

= 100 m/s

$$M_1 = \frac{V_1}{C_1} = \frac{5000/18}{\sqrt{\gamma RT}} = \frac{5000/18}{\sqrt{1.4 \times 287 \times 261}}$$

$$= 0.858 < 1$$

$$\rho_1 = \frac{P_1}{RT_1} = \frac{40}{0.287 \times 261} = 0.534 \text{ kg/m}^3$$

Stagnation temperature at entry

$$T_{01} = T_1 = \left[1 + \frac{\gamma - 1}{2} M_1^2\right] = 299.43 \text{ K}$$

Here $T_{01} = T_{02}$

Now $$\frac{T_{02}}{T_2} = \left(1 + \frac{\gamma - 1}{2} M_2^2\right) = 1 + \frac{V_2^2}{2C_p T_2}$$

$$\therefore \quad T_2 = T_{02} - \frac{V_2^2}{2C_p} = 299.43 - \frac{100^2}{2 \times 1.005 \times 1000}$$

$$= 294.45 \text{ K} \approx 295 \text{ K}$$

For an isentropic flow

$$\frac{p_2}{p_1} = \left(\frac{T_2}{T_1}\right)^{\frac{\gamma}{\gamma-1}}$$

$$\therefore \quad p_2 = p_1 \times \left(\frac{295}{261}\right)^{\frac{1.4}{0.4}} = 61.4 \text{ kPa}$$

Applying contineCity eqn

$$\dot{m} = \rho_1 A_1 V_1 = \rho_2 A_2 V_2$$

$$\Rightarrow \quad \frac{A_1}{A_2} = \frac{\rho_2 V_2}{\rho_1 V_1}, \text{ where } \rho_2 = \frac{p_2}{RT_2}$$

$$= \frac{61.4}{0.287 \times 295} = 0.7253 \text{ kg/m}^3$$

$$\therefore \quad \frac{A_1}{A_2} = \frac{0.7253 \times 100 \times 18}{0.534 \times 5000} = 0.489$$

Problem.16.14 : A nozzle is designed assuming reversible adiabatic flow with an exit Mach number of 2.6. Air flows through with a stagnation pressure and temperature of 2 MPa and 150° C respectively. The mass rate of flow is 5 kg/s. (a) Determine the exit pressure, temperature, area, and throat area. (b) If the back pressure at the nozzle exit is raised to 1.35 MPa, and the flow remains isentropic except for a normal shock wave, determine the exit Mach number and temperature, and the mass flow ratc through the nozzle.

Solution : Given data :

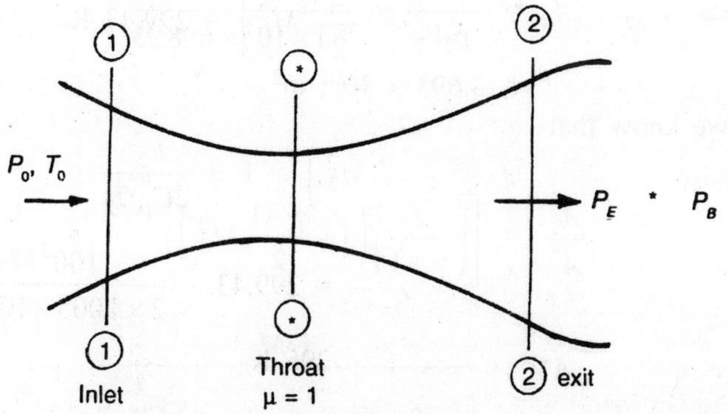

Fig. P. 16.14

$$M_2 = 2.6$$

$$p_0 = 2 \text{ MPa} = \text{stagnation pressure}$$

$$T_0 = 150°\text{C} = 423 \text{ K} = \text{stagnation temperature}$$

$$\dot{m} = 5 \text{ kg/s}$$

since T_0 = is constant through out the nozzle.

$$\therefore \quad \frac{T_0}{T_2} = 1 + \frac{\gamma - 1}{2} M_2^2 = 1 + \frac{0.4}{2} \times 2.6^2 = 2.352$$

$$\therefore \quad T_2 = \frac{T_0}{2.352} = \frac{423}{2.352} = 179.85 \text{ K}$$

since flow is isentropic

$$\therefore \quad \frac{p_2}{p_0} = \left(\frac{T_2}{T_0}\right)^{\frac{\gamma}{\gamma-1}}$$

$$\Rightarrow \quad p_2 = 2 \times \left(\frac{179.85}{423}\right)^{3.5} = 0.1 \text{ MPa}$$

$$V_2 = M_2 . C_2 = 2.6 \sqrt{\gamma R T_2}$$

$$= 2.6 \times \sqrt{1.4 \times 287 \times 179.85} = 698.93 \text{ m/s}$$

$$\text{mass flow rate } \dot{m} = \rho_2 A_2 V_2$$

$$\frac{p_2}{RT_2} \times A_2 \times V_2 = \frac{p_2 A_2 V_2}{RT_2}$$

$$\Rightarrow \quad A_2 = \frac{\dot{m} R T_2}{p_2 V_2} = \frac{5 \times 0.287 \times 179.85}{0.1 \times 10^3 \times 698.93}$$

$$= 3.693 \times 10^{-3} \text{ m}^2$$

we know that

$$\frac{A_2}{A^*} = \left[\left(\frac{2}{\gamma+1}\right)\left(1 + \frac{\gamma - 1}{2} M_2^2\right)\right]^{\frac{\gamma+1}{2(\gamma-1)}} \times \frac{1}{M_2}$$

$$\therefore \quad A^* = \frac{A_2 M_2}{\left[\left(\frac{2}{\gamma+1}\right)\left(1 + \frac{\gamma - 1}{2} M_2^2\right)\right]^{\frac{\gamma+1}{2(\gamma-1)}}}$$

$$= \frac{3.693 \times 10^{-3} \times 2.6}{\left[\frac{2}{2.4} \times \left(1 + 0.2 \times 2.6^2\right)\right]^{\frac{2.4}{0.8}}} = 1.2756 \times 10^{-3} \text{ m}^2$$

Problem.16.15 : The stagnation temperature and stagnation pressure of air in a reservoir supplying a convergent–divergent nozzle are 450 K and 400 kPa respectively. The nozzle throat area is 625 mm^2 and the nozzle exit area is 1875 mm^2. A shock is noted at a position in the divergent section where the area is 1250 mm^2. (a) What are the exit pressure, temperature, and velocity? (b) What value of the back pressure would cause the flow through the nozzle to be completely supersonic? (c) What value of back pressure would result in completely isentropic flow interior and exterior to the nozzle?

Solution : Given data :

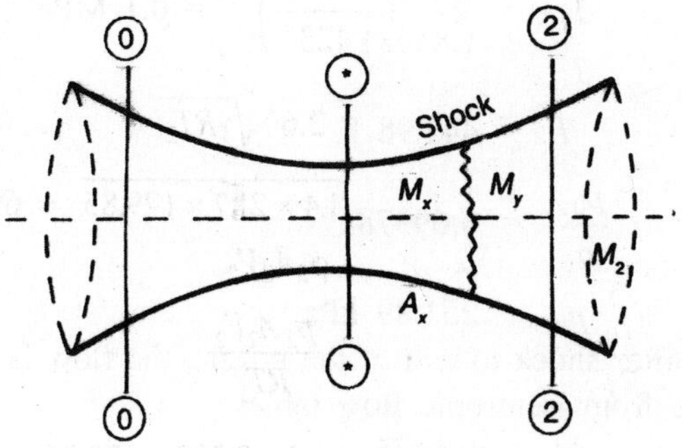

Fig. P. 16.15

T_0 = Stagnation temperature = 450 K

p_0 = Stagnation pressure = 400 kPa

A_* = Nozzle throat are = 625 mm^2

A_2 = Nozzle exit area = 1875 mm^2

A_x = 1250 mm^2

upto the shock, the flow is isentropic

$\therefore$ from isentropic flow tables, when

$$\frac{A_x}{A_*} = \frac{1250}{625} = 2,$$

M_x = 2.197 and 0.308

Considering the divergent section as nozzle

M_x = 2.197

$$\frac{p_x}{p_0} = 0.093991$$

$$\therefore \quad p_x = 37.5964 \text{ K}$$

$$\frac{T_x}{T_0} = 0.5321$$

$$T_x = 239.43 \text{ K}$$

From the gas tables on normal shocks

when

$$M_x = 2.197$$

$$M_y = 0.5475$$

$$\frac{p_y}{p_x} = 5.465952 \Rightarrow p_y = 205.5 \text{ kPa}$$

$$\frac{T_y}{T_x} = 1.85431$$

$$\Rightarrow \quad T_y = 443.98 \text{ K}$$

$$\frac{p_{0y}}{p_x} = 6.69978$$

$$\Rightarrow \quad p_{0y} = 251.89 \text{ kPa}$$

Now after shock to exit of the nozzle, the flow is again isentropic

Hence from isentropic flow tables

when $\quad M_y = 0.5475$

$$\frac{A_y}{A_*} = 1.26779$$

$$\therefore \quad \frac{A_2}{A^*} = \frac{A_2}{A_y} \times \frac{A_y}{A^*} = \frac{1875}{1250} \times 1.26779 = 1.901685$$

from isentropic flow tables,

when $\quad \dfrac{A_2}{A^*} = 1.901685$

$$M_2 = 2.1385$$

$$\frac{p_2}{p_{0y}} = 0.1032492$$

$$p_2 = 26.01 \text{ kPa}$$

$$\frac{T_2}{T_0} = 0.52240$$

$$\therefore \quad T_2 = 235.1 \text{ K}$$

At $\quad T_2 = 235.1$ K

$$C_2 = \text{sonic velocity} = \sqrt{\gamma R T_2}$$

$$= \sqrt{1.4 \times 287 \times 235.1} = 307.35 \text{ m/s}$$

$$\therefore \quad M_2 = \frac{V_2}{c_2}$$

$$\therefore \quad V_2 = 657.27 \text{ m/s}$$

Problem 16.16 : Air flows through a frictionless convergent–divergent nozzle. The area of the exit section is three times the area of the throat section, and the ratio of stagnation pressure at the entrance to the static pressure in the exit section is 2.5 Calculate (a) the Mach number M_x where the shock occurs, (b) the area A_x where the shock occurs, and (c) the entropy increase.

Solution : Given data :

$$\frac{A_2}{A_*} = 3$$

$$\frac{p_2}{p_0} = 0.4$$

or

$$\frac{p_2}{p_{0x}} = 0.4$$

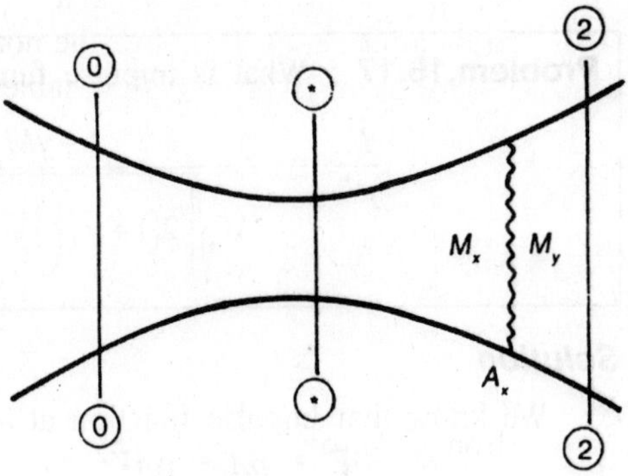

Fig. P. 16.16

Now after shock to exit of nozzle, flow is isentropic

$\therefore$ From isentropic flow tables

when $\quad \dfrac{A_2}{A_*} = 3$

$$M_2 = 0.1987$$

$$\frac{p_2}{p_{0y}} = 0.9728$$

$$\therefore \quad \frac{p_{0y}}{p_{0x}} = \frac{0.4}{0.9728} = 0.411$$

Now for normal shock

when $\frac{p_{0y}}{p_{0x}} = 0.411$

$M_x = 2.74$

and $M_y = 0.4929$

And from isentropic flow

when $M_y = 0.4929$

$$\frac{A_y}{A_*} = 1.441$$

(c) $\therefore$ Entropy increase $= S_y - S_x$

$$= -R \ln \frac{P_{0y}}{P_{0x}} = -0.287 \ln 0.411$$

$$= +0.2552 \text{ kJ/kg.K}$$

Problem.16.17 : What is impulse function. Show that

$$\frac{F}{F^*} = \frac{1+\gamma M^2}{M\sqrt{\left[2(1+\gamma)\left(1+\frac{\gamma-1}{2}M^2\right)\right]}} \qquad \text{IES - 1997}$$

Solution :

We know that impulse function at any section in given by

$$F = pA + \rho A V^2$$

at throat, impulse function

$$F^* = p^*A^* + \rho^*A^*V^{*2}$$

$$\therefore \quad \frac{F}{F^*} = \frac{A_2}{A^*}\left[\frac{p+\rho V^2}{p^* + \rho^* V^{*2}}\right]$$

$$= \frac{A}{A^*}\left[\frac{p + \frac{p}{RT} \times M^2\left(\sqrt{\gamma RT}\right)^2}{p^* + \frac{p^*}{RT^*} M^{*2}\left(\sqrt{\gamma RT_*}\right)^2}\right]$$

From ideal gas eqn

$$\left[\begin{array}{c} \because \rho = \dfrac{p}{RT} \\ \text{and } M = \dfrac{V}{C} \\ \text{where, } c = \sqrt{\gamma RT} \end{array}\right]$$

Assuming sonic velocity at throat

$$M^* = 1$$

$$\frac{F}{F^*} = \frac{A}{A^*} \times \frac{p}{p^*}\left[\frac{1+\gamma M^2}{1+\gamma}\right] \qquad \ldots(1)$$

Now we know that

$$\text{Area ratio} = \frac{A}{A^*}$$

$$= \left[\left(\frac{2}{\gamma+1}\right)\left(1+\frac{\gamma-1}{2}M^2\right)\right]^{\frac{\gamma+1}{2(\gamma-1)}} \times \frac{1}{M}$$

$$\frac{p}{p^*} = \frac{p}{p_0} \times \frac{p_0}{p^*} = \frac{1}{\left(\dfrac{p_0}{p}\right)} \times \frac{1}{\left(\dfrac{p^*}{p_0}\right)}$$

$$= \frac{1}{\left(1+\dfrac{\gamma-1}{2}M^2\right)^{\frac{\gamma}{\gamma-1}}} \times \frac{1}{\left(\dfrac{2}{\gamma+1}\right)^{\frac{\gamma}{\gamma-1}}}$$

$$= \frac{1}{\left[\left(\dfrac{2}{\gamma+1}\right)\left(1+\dfrac{\gamma-1}{2}M^2\right)\right]^{\frac{\gamma}{\gamma-1}}}$$

$\therefore$ From (1), putting the value of $\dfrac{A}{A^*}$ and $\dfrac{p}{p^*}$

$$\frac{F}{F^*} = \left[\left(\frac{2}{\gamma+1}\right)\left(1+\frac{\gamma-1}{2}M^2\right)\right]^{\frac{\gamma+1}{2(\gamma-1)}} \times \frac{1}{M} \times \frac{1\times\left[\dfrac{1+\gamma M^2}{1+\gamma}\right]}{\left[\left(\dfrac{2}{\gamma+1}\right)\left(1+\dfrac{\gamma-1}{2}M^2\right)\right]^{\frac{\gamma}{\gamma-1}}}$$

$$= \left[\left(\frac{2}{\gamma+1}\right)\left(1+\frac{\gamma-1}{2}M^2\right)\right]^{\frac{-1}{2}} \times \frac{1}{M} \times \frac{\left[1+\gamma M^2\right]}{\left[1+\gamma\right]}$$

$$\frac{F}{F^*} = \frac{\left(1+\gamma M^2\right)}{M \times \sqrt{(1+\gamma)\times 2 \times \left(1+\dfrac{\gamma-1}{2}M^2\right)}} \quad \textbf{Ans}$$

Problem 16.18 : Air flows through a long duct of constant area at 0.15 kg/s. A short section of the duct is cooled by liquid nitrogen that surrounds the duct. The rate of heat loss in this section is 15.0 kJ/s from the air. The absolute pressure, temperature, and velocity entering the cooled section are 188 kPa, 440 K, and 210 m/s, respectively.

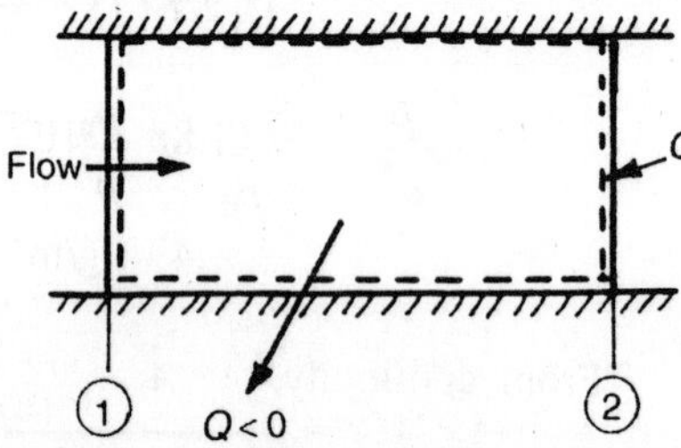

Fig. P. 16.18

At the outlet, the absolute pressure and temperature are 213 kPa and 351 K. Compute the duct cross-sectional area and the changes in enthalpy, internal energy, and entropy for this flow.

Find :

(a) duct areas (b) Δh.

(c) Δu. (d) Δs.

Solution : Given :

Air flows steadily through a short section for constant–area duct that is cooled by liquid nitrogen.

$$T_1 = 440 \text{ K}$$
$$T_2 = 351 \text{ K}$$
$$p_1 = 188 \text{ kPa (abs)}$$
$$p_2 = 213 \text{ kPa (abs)}$$
$$V_1 = 210 \text{ m/s}$$

The duct area may be found from the continuity equation.

Basic equation: $0 = \cancelto{=0(1)}{\dfrac{\partial}{\partial t}} \int_{CV} \rho d\forall + \int_{CS} \rho \vec{V} . d\vec{A}$

Assumption:

(1) Stead flow

(2) Uniform flow at each section

(3) Ideal gas

Then $\quad 0 = \{-|\rho_1 V_1 A_1|\} + \{-|\rho_2 V_2 A_2|\}$

or $\quad \dot{m} = \rho_1 V_1 A = \rho_2 V_2 A$

since $A = A_1 = A_2$ = constant. Using the ideal gas relation, $p = \rho RT$, we find

$$\rho_1 = \frac{p_1}{RT_1}$$

$$= 1.88 \times 10^5 \text{ N/m}^2 \times \frac{\text{kg.K}}{287 \text{ N.m}} \times \frac{1}{440 \text{ K}}$$

$$= 1.49 \text{ kg/m}^3$$

From continuity, $\quad A = \dfrac{\dot{m}}{\rho_1 V_1}$

$$= 0.15 \text{ kg/s} \times \frac{\text{m}^3}{1.49 \text{ kg}} \times \frac{\text{s}}{210 \text{ m}}$$

$$= 4.80 \times 10^{-4} \text{ m}^2$$

For an ideal gas, $\quad dh = C_p dT$, so

$$\Delta h = h_2 - h_1$$

$$= \int_{T_1}^{T_2} C_p \, dT = C_p (T_2 - T_1)$$

$$\Delta h = 1.00 \text{ kJ/kg.K} \times (351 - 440) \text{ K}$$

$$= -89.0 \text{ kJ/kg}$$

Also $du = c_v \, dT$, so

$$\Delta u = u_2 - u_1$$

$$= \int_{T_1}^{T_2} C_v \, dT = C_v (T_2 - T_1)$$

$$\Delta u = 0.717 \text{ kJ/kg.K} \times (351 - 440) \text{ K} = -63.8 \text{ kJ/kg}$$

The entropy change may be obtained from the *Tds* equation,

$$Tds = dh - v\,dp$$

$$ds = \frac{dh}{T} - \frac{v\,dp}{T}$$

$$= C_p \frac{dT}{T} - R \frac{dp}{p}$$

or

$$\Delta s = s_2 - s_1$$

$$= \int_{T_1}^{T_2} C_p \frac{dT}{T} - \int_{p_1}^{p_2} R \frac{dp}{p}$$

$$= C_p \ln \frac{T_2}{T_1} - R \ln \frac{p_2}{p_1}$$

$$= 1.00 \text{ kJ/kg.K} \ln \left(\frac{351}{440}\right) - 0.287 \text{ kJ/kg.K}$$

$$\ln \left(\frac{2.13 \times 10^5}{1.88 \times 10^5}\right)$$

$$\Delta s = -0.262 \text{ kJ/(kg.K)}$$

Problem 16.19 : For an ideal gas, determine the slope of (a) a constant volume line and (b) a constant pressure line in the *Ts* plane.

Find:

(a) Slope of constant volume line in *Ts* plane.

(b) Slope of constant pressure line in *Ts* plane.

Solution : Given :

An ideal gas.

The *Tds* equations may be applied.

$$Tds = dh + pdv$$

$$Tds = dh - vdp$$

Substitutin for an ideal gas,

$$du = C_v dT$$

and $$dh = C_p dT, \text{ we obtain}$$

$$Tds = C_v dT + pdv$$

$$Tds = C_p dT - vdp$$

For a constant volume process, $dv = 0$. From the first equation,

$$\left(\frac{dT}{ds}\right)_{\substack{\text{constant}\\ \text{volume}}} = \left(\frac{\partial T}{\partial s}\right)_v = \frac{T}{C_v} \qquad \text{constant volume slope}$$

For a constant pressure process, $dp = 0$. From the second equation,

$$\left(\frac{dT}{ds}\right)_{\substack{\text{constant}\\ \text{volume}}} = \left(\frac{\partial T}{\partial s}\right)_p = \frac{T}{C_p} \qquad \text{constant pressure slope}$$

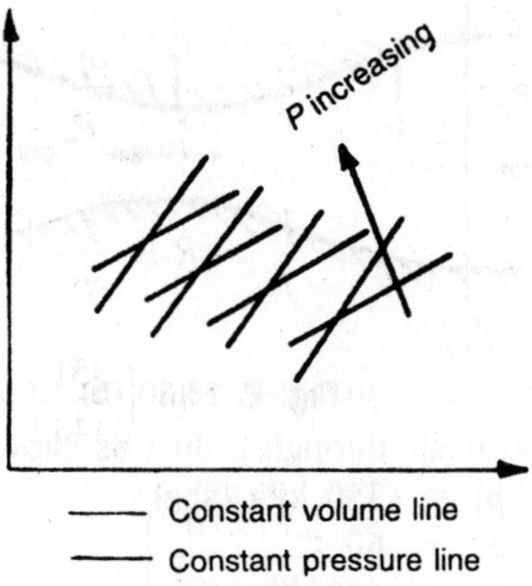

Fig. P. 16.19

Problem 16.20 : Air flows steadily through the duct shown from 350 kPa (abs), 60C, and 183 m/s at the initial state to $M = 1.3$ at the outlet, where local isentropic stagnation conditions are known to be 385 kPa (abs) and 350 K. Compute the local isentropic stagnation pressure and temperature at the inlet and the static pressure and temperature at the duct outlet. Locate the inlet and the static pressure and temperature at the duct outlet. Locate the inlet and outlet static state points on a *Ts* diagram, and indicate the stagnation process.

Find:

(a) p_{01} (b) T_{01}

(c) p_2 (d) T_2

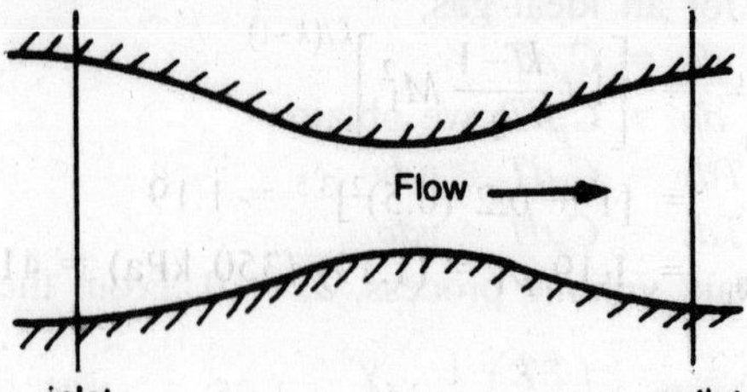

Fig. P. 16.20 (a)

(e) Locate state points (1) and (2) on a *Ts* diagram and indicate the stagnation process.

Solution : Given data :

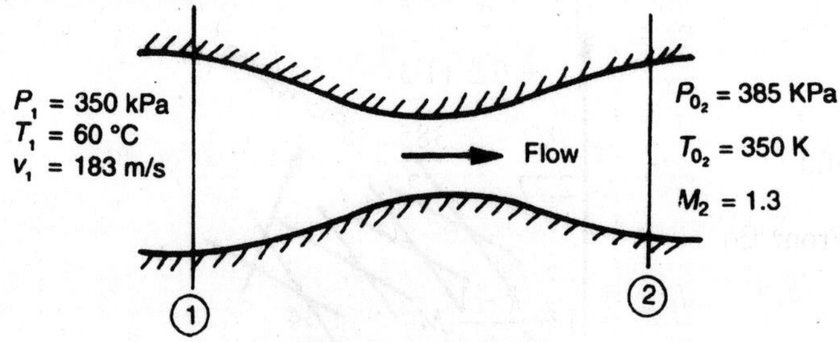

Fig. P. 16.20 (b)

Steady flow of air through a duct as shown in the sketch.

$$p_1 = 350 \text{ kPa (abs)}$$
$$T_1 = 60\,°\text{C}$$
$$V_1 = 183 \text{ m/s}$$
$$p_{02} = 385 \text{ kPa (abs)}$$
$$T_{02} = 350 \text{ K}$$
$$M_2 = 1.3$$

To evaluate local isentropic stagnation conditions at section (1), we must calculate the mach number, $M_1 = V_1/c_1$. For and ideal gas, $c = \sqrt{kRT}$. Then

$$c_1 = \sqrt{kRT_1}$$

$$= \left[1.4 \times 287 \frac{\text{N}\cdot\text{m}}{\text{kg}\cdot\text{K}} \times (273+60)\ \text{K} \times \frac{\text{kg}\cdot\text{m}}{\text{N}\cdot\text{s}^2}\right]^{1/2} = 366 \text{ m/s}$$

and
$$M_1 = \frac{V_1}{c_1} = \frac{183}{366} = 0.5$$

Local isentropic stagnation properties can be evaluated from Eqs. below

$$\frac{p_{0_1}}{p_1} = \left[1+\frac{k-1}{2}M_1^2\right]^{k/(k-1)}$$

$$= [1 + 0.2\,(0.5)^2]^{3.5} = 1.19$$

and $$p_{0_1} = 1.19\,p_1 = (1.19)\,(350\text{ kPa}) = 415\text{ kPa (abs)}$$

$$\frac{T_{0_1}}{T_1} = 1 + \frac{k-1}{2}M_1^2 = 1.05$$

and $$T_{0_1} = 1.05\,T_1 = (1.05)\,(333\text{ K}) = 350\text{ K}$$

At section (2) using equation

$$\frac{p_{0_2}}{p_2} = \left[1+\frac{k-1}{2}M_2^2\right]^{k/(k-1)}$$

$$= [1 + 0.2\,(1.3)^2]^{3.5} = 2.77$$

and $$p_2 = \frac{p_{0_2}}{2.77} = \frac{385\text{ kPa}}{2.77} = 139\text{ kPa (abs)}$$

From Eq.

$$\frac{T_{0_2}}{T_2} = 1+\frac{k-1}{2}M_2^2 = 1.34$$

and $$T_2 = \frac{T_{0_2}}{1.34} = \frac{350\text{ K}}{1.34} = 261\text{ K}$$

The entropy change must be evaluated to locate state (2) with respect to state (1). Using the *Tds* equation.

$$Tds = dh - vdp$$

or $$ds = \frac{dh}{T} - \frac{v\,dp}{T}$$

$$= C_p\frac{dT}{T} - R\frac{dp}{p}$$

Integrating gives

$$S_2 - S_1 = \int_{T_1}^{T_2} C_p\,\frac{dT}{T} - \int_{p_1}^{p_2} R\,\frac{dp}{p}$$

$$= C_p \ln\frac{T_2}{T_1} - R\ln\frac{p_2}{p_1}$$

$$S_2 - S_1 = 100 \text{ kJ/kg.K} \times \ln\left(\frac{262}{333}\right)$$

$$- 0.287 \text{ kJ/kg.K} \times \ln\left(\frac{139}{350}\right) = 0.0252 \text{ kJ/(kg.K)}$$

Therefore, state (2) lies to the right of state (1) on the *Ts* plane, as shown in the following sketch:

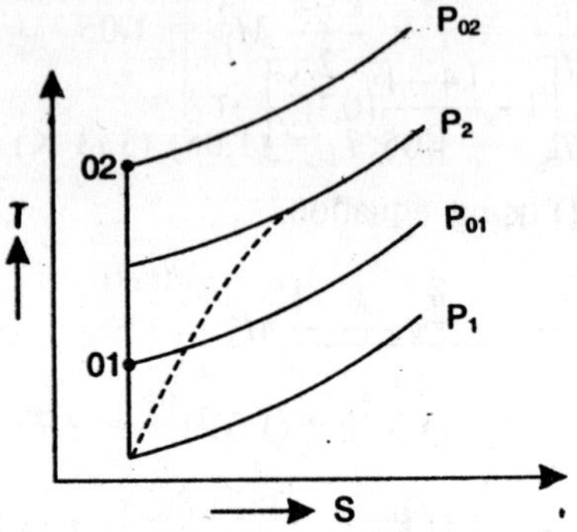

Fig. P. 16.20 (c)

The process the fluid follows between states (1) and (2) is not specified. However, it need not be known. A unique isentropic stagnation process is defined at each state point. Note that $S_{02} - S_{01} = S_2 - S_1$.

Problem 16.21 : Air flows isentropically in a channel. At sectin (1), the Mach number is 0.3, the area is 0.001 m^2, and the absolute pressure and the temperature are 650 kPa and 62°C, respectively. At section (2), the Mach number is 0.8. Sketch the channel shape, plot a *Ts* diagram for the process, and evaluate properties at section (2).

Find:

(a) Sketch the channel shape.
(b) Plot a *Ts* diagram for the process.
(c) Properties at section (2)

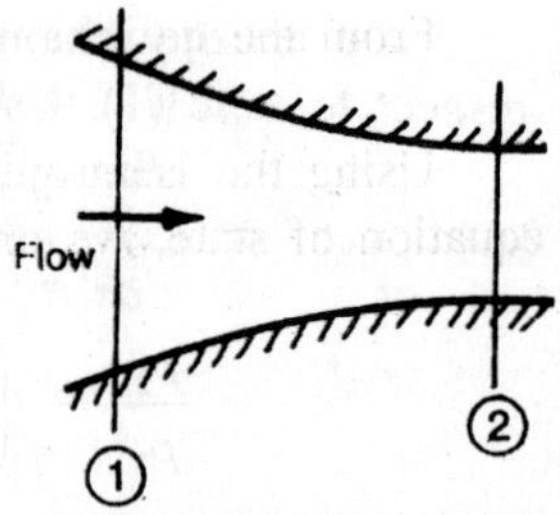

Fig. P. 16.21

Solution : *Given* : Isentropic flow of air in a channel. At sections (1) and (2), the following data are given: $M_1 = 0.3$, $T_1 = 62°C$, $p_1 = 650$ kPa (abs), $A_1 = 0.001$ m^2, and $M_2 = 0.8$.

To accelerate a subsonic flow requires a converging nozzle. The channel shape must be as shown.

On the *Ts* plane, the process follows a line, s = constant. Stagnation conditions remain fixed for isentropic flow.

Consequently, the stagnation temperature at section (2) can be calculated (for air, k = 1.4) from

$$T_{0_2} = T_{0_1} = T_1\left[1+\frac{k-1}{2}M_1^2\right]$$

$$= (62 + 273)\ \text{K}\left[1+\frac{1.4-1}{2}(0.3)^2\right]$$

$$T_{0_2} = T_{0_1} = 341\ \text{K}$$

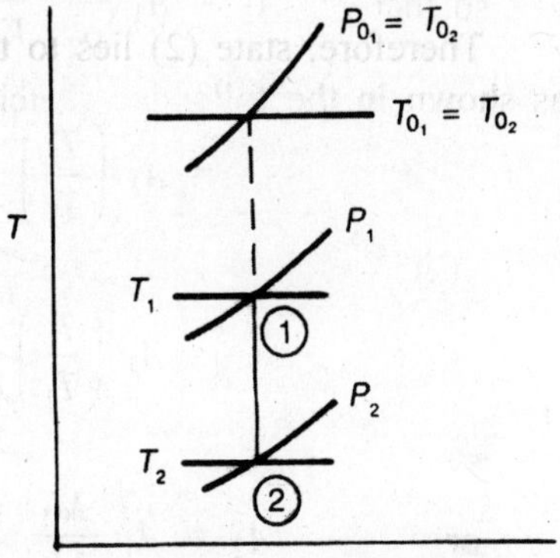

Fig. P. 16.21 (a)

And
$$T_2 = \frac{T_{0_2}}{\left[1+\frac{k-1}{2}M_2^2\right]}$$

$$= \frac{341\ \text{K}}{\left[1+0.2(0.8)^2\right]} = 302\ \text{K or } 29°\text{C}$$

Also, for an ideal gas,

$$c_2 = \sqrt{kRT_2} = \left[1.4\times 287\frac{\text{N.m}}{\text{kg.K}}\times 302\ \text{K}\times\frac{\text{kg.m}}{\text{N.s}^2}\right]^{1/2}$$

$$= 348\ \text{m/s}$$

From the definition of Mach number,

$$V_2 = M_2c_2 = (0.8)\ 348\ \text{m/s} = 278\ \text{m/s}$$

Using the isentropic relation, p/p^k = constant, and the ideal gas equation of state, we obtain

$$\frac{p_2}{p_1} = \left(\frac{T_2}{T_1}\right)^{k/(k-1)}$$

$$= \left(\frac{302\ \text{K}}{335\ \text{K}}\right)^{3.5} = 0.696$$

and
$$p_2 = 0.696\,p_1 = (0.696)\ 650\ \text{kPa (abs)} = 452\ \text{kPa (abs)}$$

Also
$$\rho_2 = \frac{p_2}{RT_2} = 4.52\times 10^5\ \text{N/m}^2\times\frac{\text{kg. K}}{287\ \text{N.m}}\times\frac{1}{302\ \text{K}}$$

$$= 5.21\ \text{kg/m}^3$$

From the continuity equation,

$$\dot{m} = \rho_1 V_1 A_1 = \rho_2 V_2 A_2 = \text{constant}$$

so that $A_2 = A_1 \dfrac{\rho_1}{\rho_2} \dfrac{V_1}{V_2}$

$$= A_1 \left(\frac{T_1}{T_2}\right)^{1/(k-1)} \times \frac{M_1 c_1}{M_2 c_2}$$

$$= A_1 \left(\frac{T_1}{T_2}\right)^{1/(k-1)} \frac{M_1}{M_2} \sqrt{\frac{kRT_1}{kRT_2}}$$

or

$$A_2 = A_1 \frac{M_1}{M_2} \left(\frac{T_1}{T_2}\right)^{(k+1)/2(k-1)}$$

$$= 0.001 \text{ m}^2 \times \frac{0.3}{0.8} \left(\frac{335}{302}\right)^3$$

$$= 5.12 \times 10^{-4} \text{ m}^2$$

Thus $A_2 < A_1$, as expected. Finally,

$$p_{0_2} = p_2 \left[1 + \frac{k-1}{2} M_2^2\right]^{k/(k-1)}$$

$$= 452 \text{ kPa } [1 + 0.2 (0.8)^2]^{3.5}$$

$$p_{0_2} = 689 \text{ kPa (abs)}$$

The stagnation pressure should be constant for isentropic flow. Checking gives

$$p_{0_1} = p_1 \left[1 + \frac{k-1}{2} M_1^2\right]^{k/(k-1)}$$

$$= 650 \text{ kPa } [1 + 0.2 (0.3)^2]^{3.5}$$

$$p_{0_1} = 692 \text{ kPa (abs)}$$

The discrepancy between the calculated p_{0_1} and p_{0_2} is caused by using temperatures rounded to three significant figures to calculate p_{0_2}. We note again for isentropic flow that $T_{0_1} = T_{0_2}$ and $p_{0_1} = p_{0_2}$.

Problem 16.22 : A converging nozzle, with a throat area of 0.001 m^2, is operated with air at a back pressure of 591 kPa (abs). The nozzle is fed from a large plenum chamber where the absolute stagnation pressure and temperature are 1.0 MPa and 60°C. The exit Mach number and mass flow rate are to be determined using isentropic flow relations.

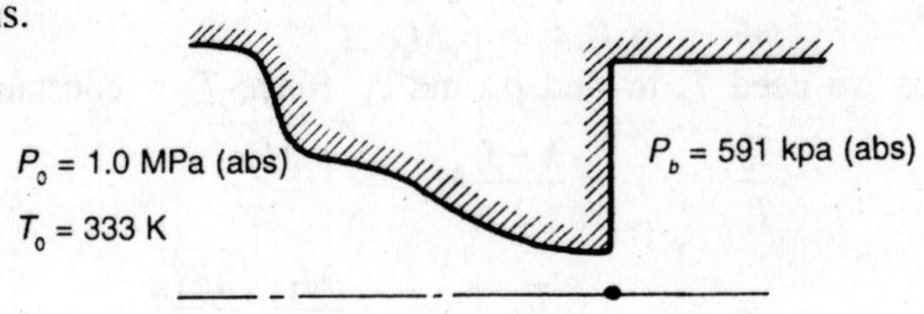

Fig. P. 16.22

Find: (a) M_e. (b) $\dot{m}$

Solution: *Given:* Air flow through a converging nozzle at the conditions shown:

Flow is isentropic.

The first step is to check for choking.

The pressure ratio is

$$\frac{p_b}{p_0} = \frac{5.91 \times 10^5}{1.0 \times 10^6} = 0.591 > 0.528$$

so the flow is not choked.

Thus $p_b = p_e$, and the flow is isentropic, as sketched on the *Ts* diagram.

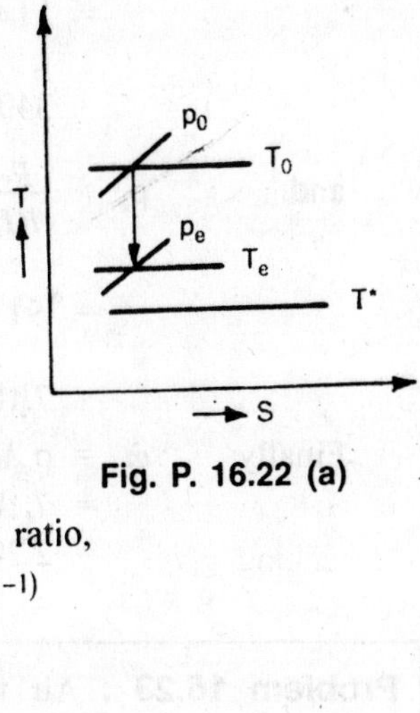

Fig. P. 16.22 (a)

(i) *Isentropic Flow Relations*

Since p_0 = constant,

M_e may be found from the pressure ratio,

$$\frac{p_0}{p_e} = \left[1 + \frac{k-1}{2} M_e^2\right]^{k/(k-1)}$$

Solving for M_e,

since $p_e = p_b$, we obtain

$$1 + \frac{k-1}{2} M_e^2 = \left(\frac{p_0}{p_b}\right)^{(k-1)/k}$$

and

$$M_e = \left\{\left[\left(\frac{p_0}{p_b}\right)^{(k-1)/k} - 1\right]\frac{2}{k-1}\right\}^{1/2}$$

$$= \left\{ \left[\left(\frac{1.0 \times 10^6}{5.91 \times 10^5} \right)^{0.286} - 1 \right] \frac{2}{1.4 - 1} \right\}^{1/2} = 0.9$$

The mass flow rate will be given by

$$\dot{m} = \rho_e V_e A_e = \rho_e M_c c_e A_e$$

Thus we need T_e to find ρ_e and c_e. Since T_0 = constant,

$$\frac{T_0}{T_e} = 1 + \frac{k-1}{2} M_e^2$$

or $$T_e = \frac{T_0}{1 + \frac{k-1}{2} M_e^2} = \frac{(273 + 60)\text{ K}}{1 + 0.2(0.9)^2} = 287 \text{ K}$$

$$c_e = \sqrt{kRT_e}$$

$$= \left[1.4 \times 287 \frac{\text{N} \cdot \text{m}}{\text{kg} \cdot \text{K}} \times 287 \text{ K} \times \frac{\text{kg} \cdot \text{m}}{\text{N} \cdot \text{s}^2} \right]^{1/2}$$

$$= 340 \text{ m/s}$$

and $$\rho_e = \frac{p_e}{RT_e}$$

$$= .91 \times 10^5 \text{ N/m}^2 \times \frac{\text{kg} \cdot \text{K}}{287 \text{ N} \cdot \text{m}} \times \frac{1}{287 \text{ K}}$$

$$= 7.18 \text{ kg/m}^3$$

Finally, $$\dot{m} = \rho_e M_e c_e A_e$$

$$= 7.18 \text{ kg/m}^3 \times 0.9 \times 340 \text{ m/s} \times 0.001 \text{ m}^2$$

$$= 2.20 \text{ kg/s}$$

Problem 16.23 : Air flows isentropically through a converging nozzle. At a section where the nozzle area is 1.2×10^{-3} m^2, the local pressure, temperature, and Mach number are 300 kPa (abs), 5°C, and 0.52, respectively. The back pressure is 200 kPa (abs). The Mach number at the throat, the mass flow rate, and the throat area are to be determined, using isentropic flow relations.

Find:

(a) M_t (b) $\dot{m}$ (c) A_t

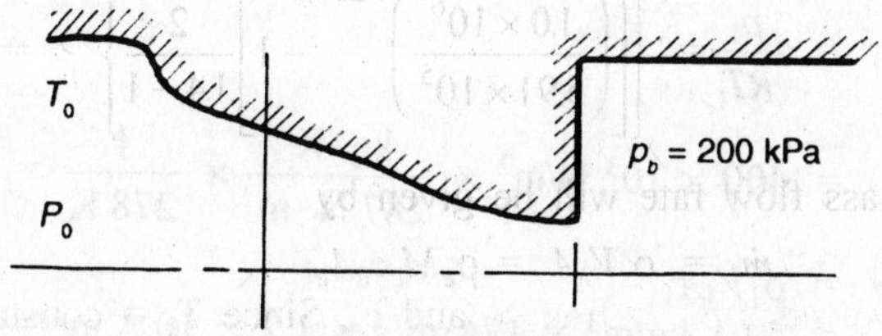

Fig. P. 16.23 (a)

Solution : *Given* : Air flow through a converging nozzle at the conditions shown:

$$M_1 = 0.52$$
$$T_1 = 5°\text{C}$$
$$p_1 = 400 \text{ kPa (abs)}$$
$$A_1 = 1.2 \times 10^{-3} \text{ m}^2$$

Isentropic Flow Relations

First we check for choking, to determine if flow is isentropic down to p_b. To check, we evaluate the stagnation conditions.

$$\frac{p_0}{p_1} = \left[1 + \frac{k-1}{2} M_1^2\right]^{k/(k-1)} = [1 + 0.2\,(0.52)^2]^{3.5} = 1.20$$

and $p_0 = 1.20\, p_1 = (1.2)\ 400 \text{ kPa (abs)} = 480 \text{ kPa (abs)}$

The back pressure ratio is

$$\frac{p_b}{p_0} = \frac{200}{480} = 0.417 < 0.528$$

so the flow is choked! For choked flow,

$$M_t = 1.0$$

The *Ts* diagram is given in Fig. P. 16.23(b).

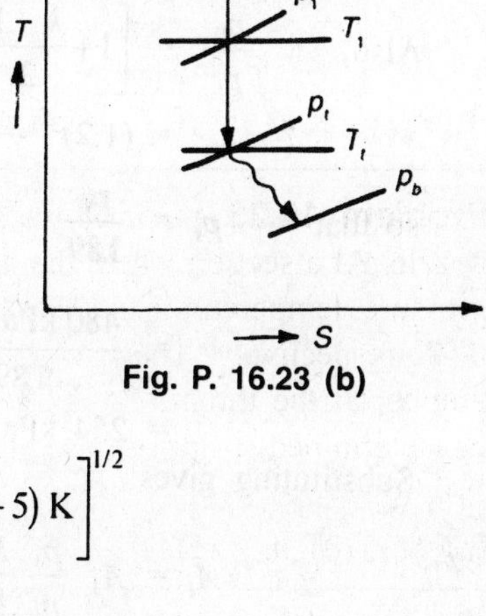

Fig. P. 16.23 (b)

The mass flow rate may be found from conditions at section (1), using $\dot{m} = \rho_1 V_1 A_1$.

$$V_1 = M_1 c_1 = M_1 \sqrt{kRT_1}$$

$$= 0.52 \left[1.4 \times 287 \frac{\text{N}\cdot\text{m}}{\text{kg}\cdot\text{K}} \times (273 + 5)\text{ K}\right]^{1/2}$$

$$V_1 = 174 \text{ m/s}$$

$$\rho_1 = \frac{p_1}{RT_1}$$

$$= 400 \times 10^3 \text{ N/m}^2 \times \frac{\text{kg}\cdot\text{K}}{287 \text{ N}\cdot\text{m}} \times \frac{1}{278 \text{ K}} = 5.01 \text{ kg/m}^3$$

$$\dot{m} = \rho_1 V_1 A_1$$

$$= 5.01 \text{ kg/m}^3 \times 174 \text{ m/s} \times 1.2 \times 10^{-3} \text{ m}^2 = 1.05 \text{ kg/s}$$

Thew throat area may be computed by applying continuity between section (1) and the throat, i.e., $\dot{m} = \rho_1 V_1 A_1 = \rho_t V_t A_t$, so

$$A_t = A_1 \frac{\rho_1}{\rho_t} \frac{V_1}{V_t}$$

$$= A_1 \frac{p_1}{RT} \frac{RT_t}{p_t} \frac{M_1\sqrt{kRT_1}}{M_t\sqrt{kRT_t}} = A_1 \frac{p_1}{p_t} \frac{M_1}{M_t} \sqrt{\frac{T_t}{T_1}}$$

For isentropic flow,

$$T_0 = \text{constant, so}$$

$$\frac{T_t}{T_1} = \frac{T_t}{T_0}\frac{T_0}{T_1}$$

$$= \frac{1+\frac{k-1}{2}M_1^2}{1+\frac{k-1}{2}M_t^2} = \frac{1+(0.2)(0.52)^2}{1.2}$$

$$= 0.878$$

Also, $$\frac{p_0}{p_t} = \left[1+\frac{k-1}{2}M_t^2\right]^{k/(k-1)}$$

$$= (1.2)^{3.5} = 1.89$$

so that $$p_t = \frac{p_0}{1.89}$$

$$= \frac{480 \text{ kPa (abs)}}{1.89}$$

$$= 254 \text{ kPa (abs)}$$

Substituting gives

$$A_t = A_1 \frac{p_1}{p_t} \frac{M_1}{M_t} \sqrt{\frac{T_t}{T_1}}$$

$$= 1.2 \times 10^{-3}\ \text{m}^2 \times \frac{400\ \text{kPa (abs)}}{254\ \text{kPa (abs)}} \times \frac{0.52}{1.0}\ \sqrt{0.878}$$

$$= 921 \times 10^{-6}\ \text{m}^2$$

Problem 16.24 : Air flows isentropically in a converging–diverging nozzle, with exit area of 0.001 m². The nozzle is fed from a large plenum where the stagnation conditions are 350 K and 1.0 MPa (abs). The exit pressure is 954 kPa (abs) and the Mach number at the throat is 0.68. Flow conditions at the throat and the exit Mach number are to be determined.

Find:

(a) Properties at nozzle throat. (b) M_e

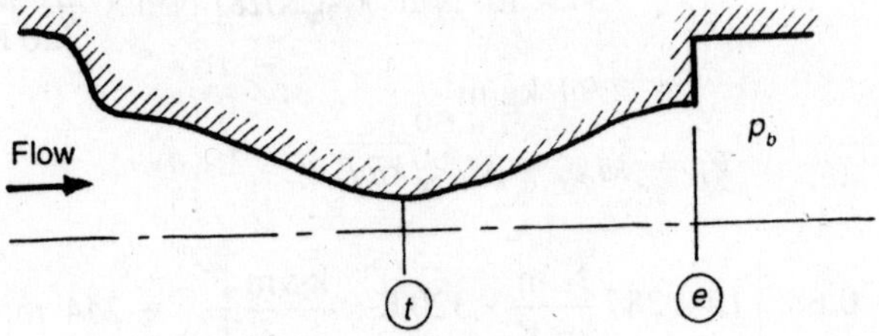

Fig. P. 16.24 (a)

Solution : *Given :*

$T_0 = 350$ K

$p_0 = 1.0$ MPa (abs)

$p_b = 954$ kPa (abs)

$M_t = 0.68$

$A_e = 0.001$ m²

Isentropic Flow Relations

Stagnation temperature is constant for isentropic flow. Thus, since

$$\frac{T_0}{T} = 1 + \frac{k-1}{2} M^2$$

then
$$T_t = \frac{T_0}{1 + \dfrac{k-1}{2} M_t^2} = \frac{350\ \text{K}}{1 + 0.2\,(0.68)^2} = 320\ \text{K}$$

Also, since p_0 is constant for isentropic flow,

$$p_t = p_0 \left(\frac{T_t}{T_0}\right)^{k/(k-1)} = p_0 \left[\frac{1}{1+\frac{k-1}{2}M_t^2}\right]^{k/(k-1)}$$

$$p_t = 1.0 \times 10^6 \ p_a \left[\frac{1}{1+0.2\,(0.68)^2}\right]^{3.5}$$

$$= 734 \text{ kPa (abs)}$$

so
$$\rho_t = \frac{p_t}{RT_t}$$

$$= 7.34 \times 10^5 \text{ N/m}^2 \times \text{kg.K/287 N.m} \times \frac{1}{320 \text{ K}}$$

$$= 7.99 \text{ kg/m}^3$$

and
$$V_t = M_t c_t = M_t \sqrt{kRT_t}$$

$$V_t = 0.68 \left[1.4 \times 287 \frac{\text{N.m}}{\text{kg.K}} \times 320 \text{ K} \times \frac{\text{kg.m}}{\text{n.s}^2}\right]^{1/2} = 244 \text{ m/s}$$

Since $M_t < 1$, flow at the exit must be subsonic. Therefore, $p_e = p_b$. Stagnation properties are constant, so

$$\frac{p_0}{p_e} = \left[1+\frac{k-1}{2}M_e^2\right]^{k/(k-1)}$$

Solving for M_e gives

$$M_e = \left\{\left[\left(\frac{p_o}{p_e}\right)^{(k-1)/k} - 1\right]\frac{2}{k-1}\right\}^{1/2}$$

$$M_e = \left\{\left[\left(\frac{1.0 \times 10^6}{9.54 \times 10^5}\right)^{0.286} - 1\right](5)\right\}^{1/2}$$

$$= 0.26$$

The Ts diagram for this flow is

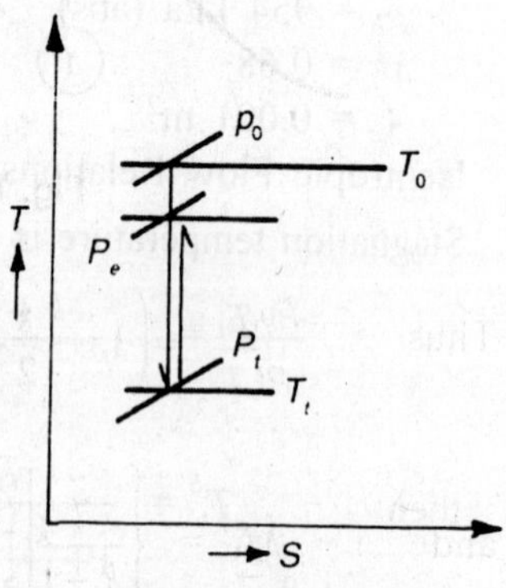

Fig. P. 16.24 (b)

Problem 16.25 : Air flow is induced in an insulated tube of 7.16 mm diameter by a vacuum pump. The air is drawn from a room, where p_0 = 101 kPa (abs) and T_0 = 23°C, through a smoothly contoured converging nozzle. At section (1), where the nozzle joins the constant–area tube, the static pressure is 98.5 kPa (abs). At section (2), located some distance downstream in the constant–area tube, the air temperature is 14°C. Determine the mass flow rate, the local isentropic stagnation pressure at seciton (2), and the friction force on the duct wall between sections (1) and (2).

Find:

(a) $\dot{m}$

(b) Stagnation pressure at section (2).

(c) Force on duct wall.

***Solution** : Given :*

Air flow in insulated tube.

The mass flow rate can be obtained from properties at section (1). For isentropic flow through the converging nozzle, local isentropic stagnation properties remain constant.

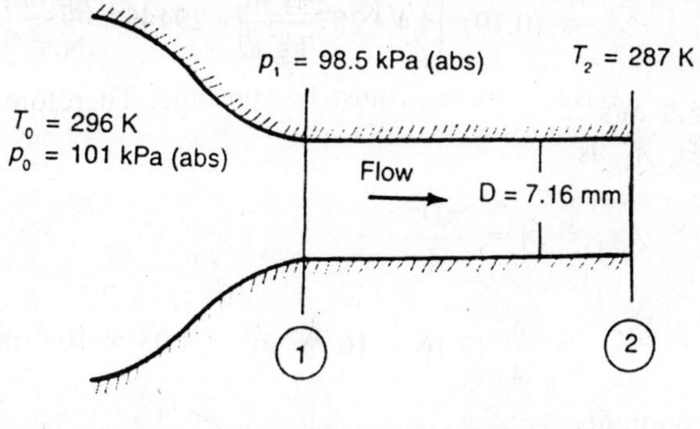

Fig. P. 16.25 (a)

Thus $$\frac{p_{0_1}}{p_1} = \left(1+\frac{k-1}{2}M_1^2\right)^{k/(k-1)}$$

and $$M_1 = \left\{\frac{2}{k-1}\left[\left(\frac{p_{0_1}}{p_1}\right)^{(k-1)/k} - 1\right]\right\}^{1/2}$$

$$= \left\{ \frac{2}{0.4} \left[\left(\frac{1.01 \times 10^5}{9.85 \times 10^4} \right)^{0.286} - 1 \right] \right\}^{1/2} = 0.190$$

$$T_1 = \frac{T_{0_1}}{1 + \frac{k-1}{2} M_1^2} = \frac{296 \text{ K}}{1 + 0.2\,(0.190)^2}$$

$$= 294 \text{ K}$$

For an ideal gas,

$$\rho_1 = \frac{p_1}{RT_1}$$

$$= 9.85 \times 10^4 \text{ N/m}^2 \times \frac{\text{kg.K}}{287 \text{ N.m}} \times \frac{1}{294 \text{ K}}$$

$$= 1.17 \text{ kg/m}^3$$

$$V_1 = M_1 c_1 = M_1 \sqrt{kRT_1}$$

$$= (0.19) \left[1.4 \times 287 \frac{\text{N.m}}{\text{kg.K}} \times 294 \text{ K} \times \frac{\text{kg.m}}{\text{N.s}^2} \right]^{1/2}$$

$V_1 = 65.3$ m/s

The area, A_1, is

$$A_1 = A = \frac{\pi D^2}{4}$$

$$= \frac{\pi}{4} (7.16 \times 10^{-3})^2 \text{ m}^2 = 4.03 \times 10^{-5} \text{ m}^2$$

From continuity,

$$\dot{m} = \rho_1 V_1 A_1$$

$$= 1.17 \text{ kg/m}^3 \times 65.3 \text{ m/s} \times 4.03 \times 10^{-5} \text{ m}^2$$

$$\dot{m} = 3.08 \times 10^{-3} \text{ kg/s}$$

Flow is adiabatic, so T_0 is constant, and

$$T_{0_2} = T_{0_1} = 296 \text{ K}$$

Then $$\frac{T_{0_2}}{T_2} = 1 + \frac{k-1}{2} M_2^2$$

Solving for M_2 gives

$$M_2 = \left[\frac{2}{k-1}\left(\frac{T_{0_2}}{T_2}-1\right)\right]^{1/2}$$

$$= \left[\frac{2}{0.4}\left(\frac{296}{287}-1\right)\right]^{1/2} = 0.396$$

$$V_2 = M_2 c_2 = M_2\sqrt{kRT_2}$$

$$= (0.396)\left[1.4 \times 287 \frac{\text{N.m}}{\text{kg.K}} \times 287\ \text{K} \times \frac{\text{kg.m}}{\text{N.s}^2}\right]^{1/2}$$

$V_2 = 134$ m/s

From continuity, $\rho_1 V_1 = \rho_2 V_2$, so

$$\rho_2 = \rho_1 \frac{V_1}{V_2}$$

$$= 1.17\ \text{kg/m}^3 \times \frac{65.3}{134} = 0.570\ \text{kg/m}^3$$

and
$$p_2 = \rho_2 R T_2$$
$$= 0.570\ \text{kg/m}^3 \times 287\ \text{Nm/kg.K} \times 287\ \text{K}$$
$$= 47.0\ \text{kPa (abs)}$$

The local isentropic stagnation pressure is

$$p_{0_2} = p_2\left(1+\frac{k-1}{2}M_2^2\right)^{k/(k-1)}$$

$$= 4.70 \times 10^4\ P_a\,[1 + 0.2\,(0.396)^2]^{3.5}$$

$$p_{0_2} = 52.4\ \text{kPa (abs)}$$

The friction force may be found by applying the momentum equation to the control volume shown :

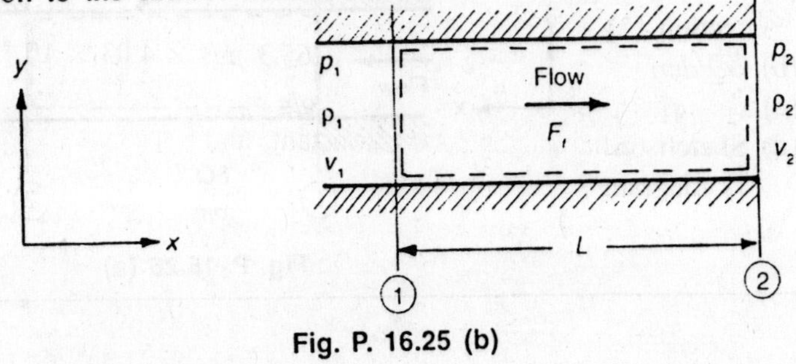

Fig. P. 16.25 (b)

Basic equation :

$$F_{S_x} + \cancel{F_{B_x}}^{=0\,(1)} = \cancel{\frac{\partial}{\partial t}}^{=0\,(2)} \int_{CV} V_x \rho\, dV + \int_{CS} V_x \rho \vec{V} \cdot d\vec{A}$$

Assumptions:

(1) $F_{B_x} = 0$

(2) Steady flow

(3) Uniform flow at each section

Then

$$-F_f + p_1 A - p_2 A = V_1 \{-|\rho_1 V_1 A|\} + V_2 \{|\rho_2 V_2 A|\} = \dot{m}\ (V_2 - V_1)$$

and

$$-F_f = (p_2 - p_1)\ A + \dot{m}\ (V_2 - V_1)$$

$$-F_f = (4.70 - 9.85)\ 10^4\ \text{N/m}^2 \times 4.03 \times 10^{-5}\ \text{m}^2$$
$$+\ 3.08 \times 10^{-3}\ \text{kg/s} \times (134 - 65.3)\ \text{m/s} \times \text{N.s}^2/\text{kg.m}$$

or

$F_f = 1.86$ N (to the left, as shown)

This is the force exerted on the control volume by the duct wall. The force of the *fluid* on the *duct* is

$$K_x = F_f = 1.86 \text{ N (to the right)}$$

Problem 16.26 : Air flows with negligible friction through a duct of area $A = 0.025$ m². At section (1), flow properties are $T_1 = 330$ K, $p_1 = 140$ kPa (abs), and $V_1 = 110$ m/s. At section (2), $p_2 = 70$ kPa (abs). The flow is heated between sections (1) and (2). Determine the properties at section (2), the energy added, and the entropy change. Finally, plot the process on a *Ts* diagram.

Find :

(a) Properties at section (2).

(b) $\delta Q/dm$.

(c) $s_2 - s_1$.

(d) Sketch on *Ts* diagram

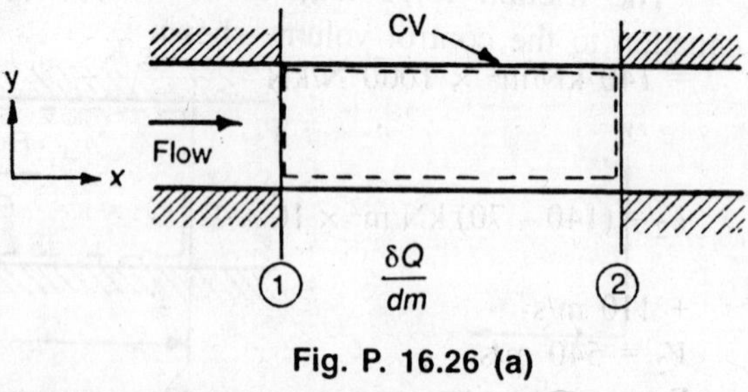

Fig. P. 16.26 (a)

Solution : *Given* :

$T_1 = 330$ K
$p_1 = 140$ kPa (abs)
$p_2 = 70$ kPa (abs)
$V_1 = 110$ m/s
$A_1 = A_2 = A = 0.025$ m^2

Apply the x component of the momentum equation, using the coordinates and control volume shown.

Basic equation:

$$F_{S_x} + \cancel{F_{B_x}}^{=0(1)} = \cancel{\frac{\partial}{\partial t}\int_{CV} V_x \rho \, d\forall}^{=0(2)} + \int_{CS} V_x \rho \vec{V} \cdot d\vec{A}$$

Assumptions:

(1) $F_{B_x} = 0$

(2) Steady flow

(3) Uniform flow at each section

Then

$$p_1A - p_2A = V_1 \{-|\rho_1 V_1 A|\} + V_2 \{|\rho_2 V_2 A|\} = \dot{m} \ (V_2 - V_1)$$

or $$p_1 - p_2 = \frac{\dot{m}}{A} (V_2 - V_1) = \rho_1 V_1 \ (V_2 - V_1)$$

Solving for V_2 gives

$$V_2 = \frac{p_1 - p_2}{\rho_1 V_1} + V_1$$

For an ideal gas,

$$\rho_1 = \frac{p_1}{RT_1}$$

$$= 140 \text{ kN/m}^2 \times 1000 \text{ N/kN} \times \frac{\text{kg.K}}{287 \text{ N.m}} \times \frac{1}{330 \text{ K}} = 1.48 \text{ kg/m}^3$$

$$V_2 = (140 - 70) \text{ kN/m}^2 \times 1000 \text{ N/kN} \times \frac{\text{m}^3}{1.48 \text{ kg}} \times \frac{s}{110 \text{ m}} \times \frac{\text{kg.m}}{\text{N.s}^2}$$

$+ 110$ m/s

$V_2 = 540$ m/s

From continuity,

$$G = \rho_1 V_1 = \rho_2 V_2, \text{ so}$$

$$\rho_2 = \rho_1 \frac{V_1}{V_2} = 1.48 \text{ kg/m}^3 \left(\frac{110}{540}\right) = 0.301 \text{ kg/m}^3$$

Solving for T_2, we obtain

$$T_2 = \frac{p_2}{\rho_2 R}$$

$$= 70 \text{ kN/m}^2 \times 1000 \text{ N/kN} \times \frac{\text{m}^3}{0.301 \text{ kg}} \times \frac{\text{kg.K}}{287 \text{ N.m}} = 810 \text{ K}$$

The local isentropic stagnation temperature is given by

$$T_{0_2} = T_2 \left(1 + \frac{k-1}{2} M_2^2\right)$$

$$c_2 = \sqrt{kRT_2} = 570 \text{ m/s;}$$

$$M_2 = \frac{V_2}{c_2} = \frac{540}{570} = 0.947$$

$$T_{0_2} = 810 \text{ k } [1 + 0.2\ (0.947)^2] = 955 \text{ K}$$

and $$p_{0_2} = p_2 \left(\frac{T_{0_2}}{T_2}\right)^{k/(k-1)} = 70 \text{ kPa (abs)} \left(\frac{955}{810}\right)^{3.5}$$

$$= 125 \text{ kPa (abs)}$$

The heat addition is determined from the energy equation.
Basic equation:

$$\dot{Q} - \underbrace{\dot{W}_s}_{=0(4)} - \underbrace{\dot{W}_{shear}}_{=0(5)} - \underbrace{\dot{W}_{other}}_{=0(5)} = \underbrace{\frac{\partial}{\partial t}\int_{CV} e\rho\, d\forall}_{=0(2)} + \int_{CS} (e + pv)\rho \vec{V} \cdot d\vec{A}$$

where

$$e = u + \frac{V^2}{2} \times \underbrace{gz}_{\approx 0(6)}$$

Assumptions:

(4) $\dot{W}_s = 0$

(5) $\dot{W}_{sheer} = \dot{W}_{other} = 0$

(6) Neglect changes in z

Then

$$\dot{Q} = \left(u_1 + p_1 v_1 + \frac{V_1^2}{2}\right)\{-|\rho_1 V_1 A|\} + \left(u_2 + p_2 v_2 + \frac{V_2^2}{2}\right)\{|\rho_2 V_2 A|\}$$

$$\dot{Q} = \dot{m}\left(h_2 + \frac{V_2^2}{2} - h_1 - \frac{V_1^2}{2}\right)$$

$$= \dot{m}\left(h_{0_2} - h_{0_1}\right)$$

$$= \dot{m} C_P \left(T_{0_2} - T_{0_1}\right)$$

and

$$\frac{\delta Q}{dm} = \frac{1}{\dot{m}}\dot{Q} = C_P\left(T_{0_2} - T_{0_1}\right)$$

$$T_{0_1} = T_1\left(1 + \frac{k-1}{2} M_1^2\right)$$

$$c_1 = \sqrt{kRT_1} = 364 \text{ m/s};$$

$$M_1 = \frac{V_1}{c_1} = \frac{110}{364} = 0.302$$

$$T_{0_1} = 330 \text{ K } [1 + 0.2\,(0.302)^2] = 366 \text{ K}$$

so

$$\frac{\delta Q}{dm} = 1.00 \text{ kJ/kg.K } (955 - 336) \text{ K}$$

$$= 619 \text{ kJ/kg}$$

Using the Tds equation, $Tds = dh - vdp$, we obtain, for an ideal gas with constant specific heats,

$$s_2 - s_1 = C_P \ln \frac{T_2}{T_1} - R \ln \frac{p_2}{p_1}$$

$$= C_P \ln \frac{T_2}{T_1} - (C_P - C_V) \ln \frac{p_2}{p_1}$$

Then

$$s_2 - s_1 = 1.00 \text{ kJ/kg.K} \times \ln\left(\frac{810}{330}\right)$$

$$- (1.004 - 0.717) \text{ kJ/kg.K} \times \ln\left(\frac{70}{140}\right)$$

$$s_2 - s_1 = 1.10 \text{ kJ/(kg.K)}$$

The process follows a Rayleigh line :

To complete our analysis, let us examine the change in p_0 by comparing p_{0_2} with p_{0_1}.

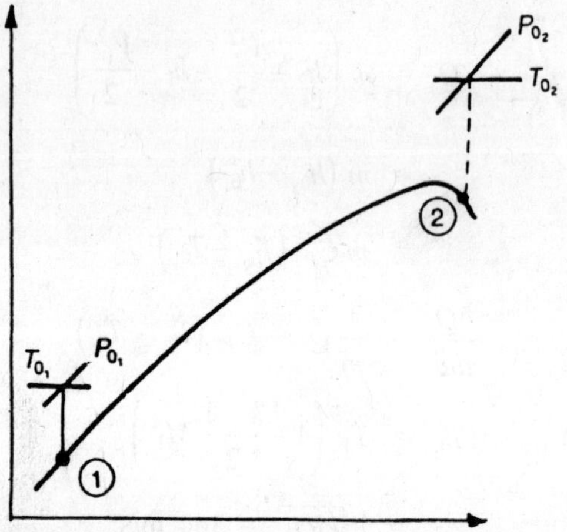

Fig. P. 16.26 (b)

$$p_{0_1} = p_1 \left(\frac{T_{0_1}}{T_1}\right)^{k/(k-1)}$$

$$= 140 \text{ kPa (abs)} \left(\frac{336}{330}\right)^{3.5}$$

$$= 149 \text{ kPa (abs)}$$

Comparing, we see that p_{0_2} is less than p_{0_1}.

In general, stagnation pressure is decreased by heating and increased by cooling in Rayleigh line flow.